W9-BWH-251

An Introduction to the Theory of Numbers

An Introduction to the Theory of Numbers

FIFTH EDITION

Ivan Niven
University of Oregon

Herbert S. Zuckerman
University of Washington

Hugh L. Montgomery
University of Michigan

John Wiley & Sons, Inc.

New York • Chichester • Brisbane • Toronto • Singapore

Acquisitions Editor: Bob Macek
Designer: Laura Nicholls
Copyediting Supervisor: Gilda Stahl
Production Manager: Katherine Rubin
Production Supervisor: Micheline Frederick
Manufacturing Manager: Denis Clarke
Marketing Manager: Susan Elbe

Library of Congress Cataloging in Publication Data:

Niven, Ivan Morton, 1915–
An introduction to the theory of numbers / Ivan Niven, Herbert S. Zuckerman, Hugh L. Montgomery.—5th ed.
p. cm.
Includes bibliographical references (p.).
Includes index.
ISBN 0-471-62546-9
1. Number theory. I. Zuckerman, Herbert S. II. Montgomery, Hugh L. III. Title.
QA241.N56 1991
512′.7—dc20 90-13013
CIP

Printed in the United States of America

10 9

Preface

This text is intended for use in a first course in number theory, at the upper undergraduate or beginning graduate level. To make the book appropriate for a wide audience, we have included large collections of problems of varying difficulty. Some effort has been devoted to make the first chapters less demanding. In general, the chapters become gradually more challenging. Similarly, sections within a given chapter are progressively more difficult, and the material within a given section likewise. At each juncture the instructor must decide how deeply to pursue a particular topic before moving ahead to a new subject. It is assumed that the reader has a command of material covered in standard courses on linear algebra and on advanced calculus, although in the early chapters these prerequisites are only slightly used. A modest course requiring only freshman mathematics could be constructed by covering Sections 1.1, 1.2, 1.3 (Theorem 1.19 is optional), 1.4 through Theorem 1.21, 2.1, 2.2, 2.3, 2.4 through Example 9, 2.5, 2.6 through Example 12, 2.7 (the material following Corollary 2.30 is optional), 2.8 through Corollary 2.38, 4.1, 4.2, 4.3, 5.1, 5.3, 5.4, 6.1, 6.2. The instructor may wish to obtain from the publisher the *Solution Manual*, which provides solutions to the starred problems. The publisher will also provide the instructor with a complimentary copy of *Computational Laboratories in Number Theory*, and also with a set of 70 instructional computer programs for IBM PC-compatible machines. The *Solution Manual* should be kept confidential, but the lab manual and software may be freely copied and distributed to students using the text. The lab manual and software can also be obtained by anonymous ftp over the internet at the address ftp.math.lsa.umich.edu in the subdirectory /pub/clint.

New in this edition are accounts of the binomial theorem (Section 1.4), public-key cryptography (Section 2.5), the singular situation in Hensel's lemma (Section 2.6), simultaneous systems of linear Diophantine equations (Section 5.2), rational points on curves (Section 5.6), elliptic curves (Section 5.7), description of Faltings' theorem (Section 5.9), the geometry of numbers (Section 6.4), Mertens' estimates of prime number sums (in Section 8.1), Dirichlet series (Section 8.2), and asymptotic estimates of arithmetic functions (Section 8.3). Many other parts of the books have also been extensively revised, and many new starred problems have been introduced. We address a number of calculational issues, most notably in Section 1.2 (Euclidean algorithm), Section 2.3 (the Chinese remainder theorem), Section 2.4 (pseudoprime tests and Pollard rho factorization), Section 2.9 (Shanks' RESSOL algorithm), Section 3.6 (sums of two squares), Section 4.4 (linear recurrences and Lucas pseudoprimes), Section 5.8 (Lenstra's elliptic curve method of factorization), and Section 7.9 (the continued fraction of a quadratic irrational). In the Appendixes we have provided some important material that all too often is lost in the cracks of the undergraduate curriculum.

Number theory is a broad subject with many strong connections with other branches of mathematics. Our desire is to present a balanced view of the area. Each subspecialty possesses a personality uniquely its own, which we have sought to portray accurately. Although much may be learned by exploring the extent to which advanced theorems may be proved using only elementary techniques, we believe that many such arguments fail to convey the spirit of current research, and thus are of less value to the beginner who wants to develop a feel for the subject. In an effort to optimize the instructional value of the text, we sometimes avoid the shortest known proof of a result in favor of a longer proof that offers greater insights.

While revising the book we sought advice from many friends and colleagues, and we would most especially like to thank G. E. Andrews, A. O. L. Atkin, P. T. Bateman, E. Berkove, P. Blass, A. Bremner, J. D. Brillhart, J. W. S. Cassels, T. Cochrane, R. K. Guy, H. W. Lenstra Jr., D. J. Lewis, D. G. Malm, D. W. Masser, J. E. McLaughlin, A. M. Odlyzko, C. Pomerance, J. Rickert, K. A. Ross, L. Schoenfeld, J. L. Selfridge, R. C. Vaughan, S. S. Wagstaff Jr., C. Williams, K. S. Williams, and M. C. Wunderlich for their valuable suggestions. We hope that readers will contact us with further comments and suggestions.

Ivan Niven
Hugh L. Montgomery

Contents

Notation

Items are listed in order of appearance.

$\mathbb{Z}$	The set of integers, **4**
$\mathbb{Q}$	The set of rational numbers, **4**
$\mathbb{R}$	The set of real numbers, **4**
$a \mid b$	a divides b, **4**
$a \nmid b$	a does not divide b, **4**
$a^k \Vert b$	$a^k \mid b$ but $a^{k+1} \nmid b$, **4**
$[x]$	Integer part, **6, 134, 180**
(b, c)	The greatest common divisor of b and c, **7** (Alternatively, depending on the context, a point in the plane, or an open interval)
$gcd(b, c)$	The greatest common divisor of b and c, **7**
$(b_1, b_2, \ldots, b_n)$	The greatest common divisor of the b_i, **7**
$[a_1, a_2, \ldots, a_n]$	The least common multiple of the a_i, **16**
$\mathrm{N}(\alpha)$	Norm of α, **22, 427**
$\pi(x)$	Number of primes $p \leqslant x$, **26**
$f(x) \sim g(x)$	Asymptotic equivalence, **28** (Alternatively, depending on the context, an equivalence relation, **128, 157**)
F_n	Fermat number, **33** (Alternatively, Fibonacci number, **199**)
M_p	Mersenne prime, **33**
$\mathscr{A} \setminus \mathscr{B}$	Subtraction of sets, **34**
$n \in \mathscr{N}$	n is an element of the set $\mathscr{N}$, **34**
$\binom{\alpha}{k}$	Binomial coefficient, **35**
$\mathscr{A} \cup \mathscr{B}$	Union of sets, **41**
$\mathscr{A} \cap \mathscr{B}$	Intersection of sets, **41**
$\Delta^k f(x)$	kth forward difference, **42**
$a \equiv b \pmod{m}$	a congruent to b modulo m, **48**
$a \not\equiv b \pmod{m}$	a not congruent to b modulo m, **48**

$\phi(n)$	Euler's totient function, **50**
$\bar{a}$	Congruential inverse of a, **52**
$\det(A)$	Determinant of square matrix A, **59**
$\operatorname{card}(\mathscr{A})$	Number of elements in the set $\mathscr{A}$, **64**
$\mathscr{S}_1 \times \mathscr{S}_2 \times \cdots \times \mathscr{S}_n$	Cartesian product of sets, **68**
$\mathbb{R}^2$	Set of pairs (x, y) of real numbers, **68**
$\{x\}$	Fractional part, $\{x\} = x - [x]$, **75**
$\operatorname{spsp}(a)$	Strong pseudoprime base a, **78**
$\approx$	Approximately equal, **85**
$\cong$	Isomorphic, as with groups, **118**
$\left(\frac{a}{p}\right)$	Legendre symbol, **132**
$\left(\frac{P}{Q}\right)$	Jacobi symbol, **142**
d	Discriminant of a quadratic form, **150**
Γ	The modular group, **157**
$f \sim g$	Equivalence of quadratic forms, **157**
$H(d)$	The class number, **161**
$h(d)$	The number of primitive classes, **163**
$R(n)$	Number of representations of n as a sum of two squares, **163**
$r(n)$	Number of representations of n as a sum of two relatively prime squares, **163**
$P(n)$	Restricted representations of n as a sum of two squares, **163**
$N(n)$	Number of solutions of $x^2 \equiv -1 \pmod{n}$, **163**
$w(f)$	Number of automorphs of a quadratic form, **173**
I	The identity matrix, **173**
$R_f(n)$	Number of representations by a quadratic form, **174**
$r_f(n)$	Number of proper representations of n by a quadratic form, **174**
$H_f(n)$	Generalization of $N(n)$, **174**
$GL(n, F)$	General linear group, **177**
$SL(n, R)$	Special linear group, **177**
$d(n)$	Number of positive divisors, **188**
$\sigma(n)$	Sum of the positive divisors, **188**
$\sigma_k(n)$	Sum of the kth powers of the positive divisors, **188**
$\omega(n)$	Number of distinct prime factors, **188**
$\Omega(n)$	Total number of primes dividing n, counting multiplicity, **188**
$\lambda(n)$	Liouville's lambda function, **192**

$\mu(n)$	Möbius mu function, **193**
F_n	Fibonacci number, **199**
	(Alternatively, depending on the context, a Fermat number, **33**)
L_n	Lucas number, **199**
$U_n(a, b), V_n(a, b)$	Lucas functions, **201**
C1, C2, C3	Column operations on matrices, **217**
R1, R2, R3	Row operations on matrices, **221**
M^t	Transpose of a matrix, **229**
$\mathscr{C}_f(\mathbb{R})$	Curve, **249**
AB	Point on a curve, determined by the points **A** and **B**, **256**
$\mathbb{P}_2(\mathbb{R})$	Projective plane, **259**
$E_f(\mathbb{R})$	Group of points on an elliptic curve, **270**
ECM	Elliptic curve method of factoring, **281**
$\langle a_0, a_1, \cdots, a_j \rangle$	Continued fraction, **326**
$\langle b_0, b_1, \cdots, b_j, \overline{a_0, a_1, \cdots, a_{n-1}} \rangle$	Periodic continued fraction, **345**
$\Lambda(n)$	von Mangoldt lambda function, **361**
$\psi(x)$	Sum of $\Lambda(n)$ over $n \leqslant x$, **361**
$\vartheta(x)$	Sum of $\log p$ over primes $p \leqslant x$, **361**
$f(x) = O(g(x))$	f is of the order of g, **365**
$\zeta(s)$	Riemann zeta function, **375**
γ	Euler's constant, **392**
$\mathbb{Q}[x]$	Set of polynomials in x with rational coefficients, **410**
$\mathbb{Z}[x]$	Set of polynomials in x with integral coefficients, **410**
$\mathbb{Q}(\xi)$	Algebraic number field, **419**
modulo $G(x)$	Polynomial congruences, **420**
$p(n)$	Partition function, **446**
$p_m(n), p^o(n), p^d(n), q^e(n), q^o(n)$	Restricted partition functions, **446**
$\mathscr{A} \supset \mathscr{B}$	Same as $\mathscr{B} \subset \mathscr{A}$, $\mathscr{B}$ is a subset of $\mathscr{A}$, **472**
$\overline{\mathscr{A}}$	Complement of $\mathscr{A}$, **472**
$\delta_1(\mathscr{A})$	Asymptotic density of $\mathscr{A}$, **473**
$\delta(\mathscr{A})$	Natural density of $\mathscr{A}$, **473**
$d(\mathscr{A})$	Schnirelmann density of $\mathscr{A}$, **476**
$\mathscr{A} + \mathscr{B}$	Sum of two sets, **477**
$\sigma_1, \sigma_2, \cdots, \sigma_n$	Elementary symmetric polynomials, **485**
$D(f)$	Discriminant of a polynomial f, **487**

CHAPTER 1

Divisibility

1.1 INTRODUCTION

The theory of numbers is concerned with properties of the *natural numbers* $1, 2, 3, 4, \cdots$, also called the *positive integers*. These numbers, together with the negative integers and zero, form the set of integers. Properties of these numbers have been studied from earliest times. For example, an integer is divisible by 3 if and only if the sum of its digits is divisible by 3, as in the number 852 with sum of digits $8 + 5 + 2 = 15$. The equation $x^2 + y^2 = z^2$ has infinitely many solutions in positive integers, such as $3^2 + 4^2 = 5^2$, whereas $x^3 + y^3 = z^3$ and $x^4 + y^4 = z^4$ have none. There are infinitely many prime numbers, where a prime is a natural number such as 31 that cannot be factored into two smaller natural numbers. Thus, 33 is not a prime, because $33 = 3 \cdot 11$.

The fact that the sequence of primes, $2, 3, 5, 7, 11, 13, 17, \cdots$, is endless was known to Euclid, who lived about 350 B.C. Also known to Euclid was the result that $\sqrt{2}$ is an *irrational number*, that is, a number that cannot be expressed as the quotient a/b of two integers. The numbers $2/7$, $13/5$, $-14/9$, and $99/100$ are examples of *rational numbers*. The integers are themselves rational numbers because, for example, 7 can be written in the form $7/1$. Another example of an irrational number is π, the ratio of the circumference to the diameter of any circle. The rational number $22/7$ is a good approximation to π, close but not precise. The fact that π is irrational means that there is no fraction a/b that is exactly equal to π, with a and b integers.

In addition to known results, number theory abounds with unsolved problems. Some background is needed just to state these problems in many cases. But there are a few unsolved problems that can be understood with essentially no prior knowledge. Perhaps the most famous of these is the conjecture known as *Fermat's last theorem*, which is not really a theorem at all because it has not yet been proved. Pierre de Fermat (1601–1665) stated that he had a truly wondrous proof that the equation $x^n + y^n = z^n$ has no solutions in positive integers x, y, z for any exponent $n > 2$. Fermat added that the margin of the book was too small to hold the proof. Whether Fermat really had a proof is not known, but it now seems unlikely, as the question has eluded mathematicians since his time.

Results in number theory often have their sources in empirical observations. We might notice, for example, that every natural number up to 1000 can be expressed as a sum of four squares of natural numbers, as illustrated by

$$1000 = 30^2 + 10^2 + 0^2 + 0^2, \qquad 999 = 30^2 + 9^2 + 3^2 + 3^2.$$

We might then feel confident enough to make the conjecture that every natural number is expressible as a sum of four squares. This turns out to be correct; it is presented as Theorem 6.25 in Chapter 6. The first proof of this result was given by J. L. Lagrange (1736–1813). We say that the four square theorem is *best possible*, because not every positive integer is expressible as a sum of three squares of integers, 7 for example.

Of course, a conjecture made on the basis of a few examples may turn out to be incorrect. For example, the expression $n^2 - n + 41$ is a prime number for $n = 1, 2, 3, \cdots, 40$ because it is easy to verify that $41, 43, 47, 53, \cdots, 1601$ are indeed prime numbers. But it would be hasty to conjecture that $n^2 - n + 41$ is a prime for every natural number n, because for $n = 41$ the value is 41^2. We say that the case $n = 41$ is a *counterexample* to the conjecture.

Leonhard Euler (1707–1783) conjectured that no nth power is a sum of fewer than n nth powers (the Swiss name Euler is pronounced "Oiler"). For $n = 3$, this would assert that no cube is the sum of two smaller cubes. This is true; it is proved in Theorem 9.35. However, a counterexample to Euler's conjecture was provided in 1968 by L. J. Lander and Thomas Parkin. As the result of a detailed computer search, they found that

$$144^5 = 27^5 + 84^5 + 110^5 + 133^5.$$

In 1987, N. J. Elkies used the arithmetic of elliptic curves to discover that

$$20615673^4 = 2682440^4 + 15365639^4 + 18796760^4,$$

and a subsequent computer search located the least counterexample to Euler's conjecture for fourth powers.

The *Goldbach conjecture* asserts that every even integer greater than 2 is the sum of two primes, as in the examples

$$4 = 2 + 2, \qquad 6 = 3 + 3, \qquad 20 = 7 + 13,$$

$$50 = 3 + 47, \qquad 100 = 29 + 71.$$

Stated by Christian Goldbach in 1742, verified up to 100,000 at least, this conjecture has evaded all attempts at proof.

Because it is relatively easy to make conjectures in number theory, the person whose name gets attached to a problem has often made a lesser contribution than the one who later solves it. For example, John Wilson (1741–1793) stated that every prime p is a divisor of $(p-1)!+1$, and this result has henceforth been known as Wilson's theorem, although the first proof was given by Lagrange.

However, empirical observations *are* important in the discovery of general results and in testing conjectures. They are also useful in understanding theorems. In studying a book on number theory, you are well advised to construct numerical examples of your own devising, especially if a concept or a theorem is not well understood at first.

Although our interest centers on integers and rational numbers, not all proofs are given within this framework. For example, the proof that π is irrational makes use of the system of real numbers. The proof that $x^3+y^3=z^3$ has no solution in positive integers is carried out in the setting of complex numbers.

Number theory is not only a systematic mathematical study but also a popular diversion, especially in its elementary form. It is part of what is called *recreational mathematics*, including numerical curiosities and the solving of puzzles. This aspect of number theory is not emphasized in this book, unless the questions are related to general propositions. Nevertheless, a systematic study of the theory is certainly helpful to anyone looking at problems in recreational mathematics.

The theory of numbers is closely tied to the other areas of mathematics, most especially to abstract algebra, but also to linear algebra, combinatorics, analysis, geometry, and even topology. Consequently, proofs in the theory of numbers rely on many different ideas and methods. Of these, there are two basic principles to which we draw especial attention. The first is that any set of positive integers has a smallest element if it contains any members at all. In other words, if a set $\mathscr{S}$ of positive integers is not empty, then it contains an integer s such that for any member a of $\mathscr{S}$, the relation $s \leqslant a$ holds. The second principle, *mathematical induction*, is a logical consequence of the first.[1] It can be stated as follows: If a set $\mathscr{S}$ of positive integers contains the integer 1, and contains $n+1$ whenever it contains n, then $\mathscr{S}$ consists of all the positive integers.

It also may be well to point out that a simple statement which asserts that there is an integer with some particular property may be easy to prove, by simply citing an example. For example, it is easy to demonstrate the proposition, "There is a positive number that is not the sum of three squares," by noting that 7 is such a number. On the other hand, a

[1]Compare G. Birkhoff and S. MacLane, *A Survey of Modern Algebra*, 4th ed., Macmillan (New York), 1977, 10–13.

statement which asserts that all numbers possess a certain property cannot be proved in this manner. The assertion, "Every prime number of the form $4n + 1$ is a sum of two squares," is substantially more difficult to establish (see Lemma 2.13 in Section 2.1).

Finally, it is presumed that you are familiar with the usual formulation of mathematical propositions. In particular, if A and B are two assertions, the following statements are logically equivalent—they are just different ways of saying the same thing.

A implies B.

If A is true, then B is true.

In order that A be true it is necessary that B be true.

B is a necessary condition for A.

A is a sufficient condition of B.

If A implies B and B implies A, then one can say that B is a necessary and sufficient condition for A to hold.

In general, we shall use letters of the roman alphabet, $a, b, c, \cdots$, $m, n, \cdots, x, y, z$ to designate integers unless otherwise specified. We let $\mathbb{Z}$ denote the set $\{\cdots, -2, -1, 0, 1, 2, \cdots\}$ of all integers, $\mathbb{Q}$ the set of all rational numbers, $\mathbb{R}$ the set of all real numbers, and $\mathbb{C}$ the set of all complex numbers.

1.2 DIVISIBILITY

Divisors, multiples, and prime and composite numbers are concepts that have been known and studied at least since the time of Euclid, about 350 B.C. The fundamental ideas are developed in this and the next section.

Definition 1.1 *An integer b is divisible by an integer a, not zero, if there is an integer x such that $b = ax$, and we write $a \mid b$. In case b is not divisible by a, we write $a \nmid b$.*

Other language for the divisibility property $a \mid b$ is that a divides b, that a is a divisor of b, and that b is a multiple of a. If $a \mid b$ and $0 < a < b$, then a is called a *proper divisor* of b. It is understood that we never use 0 as the left member of the pair of integers in $a \mid b$. On the other hand, not only may 0 occur as the right member of the pair, but also in such instances we always have divisibility. Thus $a \mid 0$ for every integer a not zero. The notation $a^K \| b$ is sometimes used to indicate that $a^K \mid b$ but $a^{K+1} \nmid b$.

Theorem 1.1

(1) $a \mid b$ *implies* $a \mid bc$ *for any integer* c;
(2) $a \mid b$ *and* $b \mid c$ *imply* $a \mid c$;
(3) $a \mid b$ *and* $a \mid c$ *imply* $a \mid (bx + cy)$ *for any integers* x *and* y;
(4) $a \mid b$ *and* $b \mid a$ *imply* $a = \pm b$;
(5) $a \mid b$, $a > 0$, $b > 0$, *imply* $a \leqslant b$;
(6) *if* $m \neq 0$, $a \mid b$ *implies and is implied by* $ma \mid mb$.

Proof The proofs of these results follow at once from the definition of divisibility. Property 3 admits an obvious extension to any finite set, thus:

$$a \mid b_1, a \mid b_2, \cdots, a \mid b_n \text{ imply } a \,\Big|\, \sum_{j=1}^{n} b_j x_j \text{ for any integers } x_j.$$

Property 2 can be extended similarly.

To give a sample proof, consider item 3. Since $a \mid b$ and $a \mid c$ are given, this implies that there are integers r and s such that $b = ar$ and $c = as$. Hence, $bx + cy$ can be written as $a(rx + sy)$, and this proves that a is a divisor of $bx + cy$.

The next result is a formal statement of the outcome when any integer b is divided by any positive integer. For example, if 25 is divided by 7, the quotient is 3 and the remainder is 4. These numbers are related by the equality $25 = 7 \cdot 3 + 4$. Now we formulate this in the general case.

Theorem 1.2 *The division algorithm. Given any integers* a *and* b, *with* $a > 0$, *there exist unique integers* q *and* r *such that* $b = qa + r$, $0 \leqslant r < a$. *If* $a \nmid b$, *then* r *satisfies the stronger inequalities* $0 < r < a$.

Proof Consider the arithmetic progression

$$\cdots, b - 3a, b - 2a, b - a, b, b + a, b + 2a, b + 3a, \cdots$$

extending indefinitely in both directions. In this sequence, select the smallest non-negative member and denote it by r. Thus by definition r satisfies the inequalities of the theorem. But also r, being in the sequence, is of the form $b - qa$, and thus q is defined in terms of r.

To prove the uniqueness of q and r, suppose there is another pair q_1 and r_1 satisfying the same conditions. First we prove that $r_1 = r$. For if not, we may presume that $r < r_1$ so that $0 < r_1 - r < a$, and then we see that $r_1 - r = a(q - q_1)$ and so $a|(r_1 - r)$, a contradiction to Theorem 1.1, part 5. Hence $r = r_1$, and also $q = q_1$.

We have stated the theorem with the assumption $a > 0$. However, this hypothesis is not necessary, and we may formulate the theorem without it: given any integers a and b, with $a \neq 0$, there exist integers q and r such that $b = qa + r$, $0 \leqslant r < |a|$.

Theorem 1.2 is called the *division algorithm*. An *algorithm* is a mathematical procedure or method to obtain a result. We have stated Theorem 1.2 in the form "there exist integers q and r," and this wording suggests that we have a so-called existence theorem rather than an algorithm. However, it may be observed that the proof does give a method for obtaining the integers q and r, because the infinite arithmetic progression $\cdots, b - a, b, b + a, \cdots$ need be examined only in part to yield the smallest positive member r.

In actual practice the quotient q and the remainder r are obtained by the arithmetic division of a into b.

Remark on Calculation Given integers a and b, the values of q and r can be obtained in two steps by use of a hand-held calculator. As a simple example, if $b = 963$ and $a = 428$, the calculator gives the answer 2.25 if 428 is divided into 963. From this we know that the quotient $q = 2$. To get the remainder, we multiply 428 by 2, and subtract the result from 963 to obtain $r = 107$. In case $b = 964$ and $a = 428$ the calculator gives 2.2523364 as the answer when 428 is divided into 964. This answer is approximate, not exact; the exact answer is an infinite decimal. Nevertheless, the value of q is apparent, because q is the largest integer not exceeding $964/428$; in this case $q = 2$. In symbols we write $q = [964/428]$. (In general, if x is a real number then $[x]$ denotes the largest integer not exceeding x. That is, $[x]$ is the unique integer such that $[x] \leqslant x < [x] + 1$. Further properties of the function $[x]$ are discussed in Section 4.1.) The value of r can then also be determined, as $r = b - qa = 964 - 2 \cdot 428 = 108$. Because the value of q was obtained by rounding down a decimal that the calculator may not have determined to sufficient precision, there may be a question as to whether the calculated value of q is correct. Assuming that the calculator performs integer arithmetic accurately, the proposed value of q is confirmed by checking that the proposed remainder $b - qa = 108$ lies in the interval $0 \leqslant r < a = 428$. In case r alone is of interest, it would be tempting to note that 428 times 0.2523364 is 107.99997, and then round to the nearest integer. The method we have described, though longer, is more reliable, as it depends only on integer arithmetic.

Definition 1.2 *The integer a is a* common divisor *of b and c in case $a|b$ and $a|c$. Since there is only a finite number of divisors of any nonzero integer, there is only a finite number of common divisors of b and c, except in the case $b = c = 0$. If at least one of b and c is not 0, the greatest among their common divisors is called the* greatest common divisor *of b and c and is denoted by (b, c). Similarly, we denote the greatest common divisor g of the integers $b_1, b_2, \cdots, b_n$, not all zero, by $(b_1, b_2, \cdots, b_n)$.*

Thus the greatest common divisor (b, c) is defined for every pair of integers b, c except $b = 0$, $c = 0$, and we note that $(b, c) \geqslant 1$.

Theorem 1.3 *If g is the greatest common divisor of b and c, then there exist integers x_0 and y_0 such that $g = (b, c) = bx_0 + cy_0$.*

Another way to state this very fundamental result is that the greatest common divisor (abbreviated g.c.d.) of two integers b and c is expressible as a linear combination of b and c with integral multipliers x_0 and y_0. This assertion holds not just for two integers but for any finite collection, as we shall see in Theorem 1.5.

Proof Consider the linear combinations $bx + cy$, where x and y range over all integers. This set of integers $\{bx + cy\}$ includes positive and negative values, and also 0 by the choice $x = y = 0$. Choose x_0 and y_0 so that $bx_0 + cy_0$ is the least positive integer l in the set; thus $l = bx_0 + cy_0$.

Next we prove that $l|b$ and $l|c$. We establish the first of these, and the second follows by analogy. We give an indirect proof that $l|b$, that is, we assume $l \nmid b$ and obtain a contradiction. From $l \nmid b$ it follows that there exist integers q and r, by Theorem 1.2, such that $b = lq + r$ with $0 < r < l$. Hence we have $r = b - lq = b - q(bx_0 + cy_0) = b(1 - qx_0) + c(-qy_0)$, and thus r is in the set $\{bx + cy\}$. This contradicts the fact that l is the least positive integer in the set $\{bx + cy\}$.

Now since g is the greatest common divisor of b and c, we may write $b = gB$, $c = gC$, and $l = bx_0 + cy_0 = g(Bx_0 + Cy_0)$. Thus $g|l$, and so by part 5 of Theorem 1.1, we conclude that $g \leqslant l$. Now $g < l$ is impossible, since g is the *greatest* common divisor, so $g = l = bx_0 + cy_0$.

Theorem 1.4 *The greatest common divisor g of b and c can be characterized in the following two ways:* (1) *It is the least positive value of $bx + cy$ where x and y range over all integers;* (2) *it is the positive common divisor of b and c that is divisible by every common divisor.*

Proof Part 1 follows from the proof of Theorem 1.3. To prove part 2, we observe that if d is any common divisor of b and c, then $d|g$ by part 3 of Theorem 1.1. Moreover, there cannot be two distinct integers with property 2, because of Theorem 1.1, part 4.

If an integer d is expressible in the form $d = bx + cy$, then d is not necessarily the g.c.d. (b, c). However, it does follow from such an equation that (b, c) is a divisor of d. In particular, if $bx + cy = 1$ for some integers x and y, then $(b, c) = 1$.

Theorem 1.5 *Given any integers $b_1, b_2, \cdots, b_n$ not all zero, with greatest common divisor g, there exist integers $x_1, x_2, \cdots, x_n$ such that*

$$g = (b_1, b_2, \cdots, b_n) = \sum_{j=1}^{n} b_j x_j.$$

Furthermore, g is the least positive value of the linear form $\sum_{j=1}^{n} b_j y_j$ where the y_j range over all integers; also g is the positive common divisor of $b_1, b_2, \cdots, b_n$ that is divisible by every common divisor.

Proof This result is a straightforward generalization of the preceding two theorems, and the proof is analogous without any complications arising in the passage from two integers to n integers.

Theorem 1.6 *For any positive integer m,*

$$(ma, mb) = m(a, b).$$

Proof By Theorem 1.4 we have

$$\begin{aligned}(ma, mb) &= \text{least positive value of } max + mby \\ &= m \cdot \{\text{least positive value of } ax + by\} \\ &= m(a, b).\end{aligned}$$

Theorem 1.7 *If $d|a$ and $d|b$ and $d > 0$, then*

$$\left(\frac{a}{d}, \frac{b}{d}\right) = \frac{1}{d}(a, b).$$

If $(a, b) = g$, then

$$\left(\frac{a}{g}, \frac{b}{g}\right) = 1.$$

Proof The second assertion is the special case of the first obtained by using the greatest common divisor g of a and b in the role of d. The first assertion in turn is a direct consequence of Theorem 1.6 obtained by replacing m, a, b in that theorem by d, $a/d, b/d$ respectively.

Theorem 1.8 *If* $(a, m) = (b, m) = 1$, *then* $(ab, m) = 1$.

Proof By Theorem 1.3 there exist integers x_0, y_0, x_1, y_1 such that $1 = ax_0 + my_0 = bx_1 + my_1$. Thus we may write $(ax_0)(bx_1) = (1 - my_0)(1 - my_1) = 1 - my_2$ where y_2 is defined by the equation $y_2 = y_0 + y_1 - my_0y_1$. From the equation $abx_0x_1 + my_2 = 1$ we note, by part 3 of Theorem 1.1, that any common divisor of ab and m is a divisor of 1, and hence $(ab, m) = 1$.

Definition 1.3 *We say that a and b are* relatively prime *in case* $(a, b) = 1$, *and that* $a_1, a_2, \cdots, a_n$ *are* relatively prime *in case* $(a_1, a_2, \cdots, a_n) = 1$. *We say that* $a_1, a_2, \cdots, a_n$ *are* relatively prime in pairs *in case* $(a_i, a_j) = 1$ *for all* $i = 1, 2, \cdots, n$ *and* $j = 1, 2, \cdots, n$ *with* $i \neq j$.

The fact that $(a, b) = 1$ is sometimes expressed by saying that a and b are coprime, or by saying that a is prime to b.

Theorem 1.9 *For any integer* x, $(a, b) = (b, a) = (a, -b) = (a, b + ax)$.

Proof Denote (a, b) by d and $(a, b + ax)$ by g. It is clear that $(b, a) = (a, -b) = d$.

By Theorem 1.3, we know that there exist integers x_0 and y_0 such that $d = ax_0 + by_0$. Then we can write

$$d = a(x_0 - xy_0) + (b + ax)y_0.$$

It follows that the greatest common divisor of a and $b + ax$ is a divisor of d, that is, $g|d$. Now we can also prove that $d|g$ by the following argument. Since $d|a$ and $d|b$, we see that $d|(b + ax)$ by Theorem 1.1, part 3. And from Theorem 1.4, part 2, we know that every common divisor of a and $b + ax$ is a divisor of their g.c.d., that is, a divisor of g. Hence, $d|g$. From $d|g$ and $g|d$, we conclude that $d = \pm g$ by Theorem 1.1, part 4. However, d and g are both positive by definition, so $d = g$.

Theorem 1.10 *If $c|ab$ and $(b, c) = 1$, then $c|a$.*

Proof By Theorem 1.6, $(ab, ac) = a(b, c) = a$. By hypothesis $c|ab$ and clearly $c|ac$, so $c|a$ by Theorem 1.4, part 2.

Given two integers b and c, how can the greatest common divisor g be found? Definition 1.2 gives no answer to this question. The investigation of the set of integers $\{bx + cy\}$ to find a smallest positive element is not practical for large values of b and c. If b and c are small, values of g, x_0, and y_0 such that $g = bx_0 + cy_0$, can be found by inspection. For example, if $b = 10$ and $c = 6$, it is obvious that $g = 2$, and one pair of values for x_0, y_0 is $2, -3$. But if b and c are large, inspection is not adequate except in rather obvious cases such as $(963, 963) = 963$ and $(1000, 600) = 200$. However, Theorem 1.9 can be used to calculate g effectively and also to get values of x_0 and y_0. (The reason we want values of x_0 and y_0 is to find integral solutions of linear equations. These turn up in many simple problems in number theory.) We now discuss an example to show how Theorem 1.9 can be used to calculate the greatest common divisor.

Consider the case $b = 963$, $c = 657$. If we divide c into b, we get a quotient $q = 1$, and remainder $r = 306$. Thus $b = cq + r$, or $r = b - cq$, in particular $306 = 963 - 1 \cdot 657$. Now $(b, c) = (b - cq, c)$ by replacing a and x by c and $-q$ in Theorem 1.9, so we see that

$$(963, 657) = (963 - 1 \cdot 657, 657) = (306, 657).$$

The integer 963 has been replaced by the smaller integer 306, and this suggests that the procedure be repeated. So we divide 306 into 657 to get a quotient 2 and a remainder 45, and

$$(306, 657) = (306, 657 - 2 \cdot 306) = (306, 45).$$

Next 45 is divided into 306 with quotient 6 and remainder 36, then 36 is divided into 45 with quotient 1 and remainder 9. We conclude that

$$(963, 657) = (306, 657) = (306, 45) = (36, 45) = (36, 9).$$

Thus $(963, 657) = 9$, and we can express 9 as a linear combination of 963 and 657 by sequentially writing each remainder as a linear combination of the two original numbers:

$$306 = 963 - 657;$$
$$45 = 657 - 2 \cdot 306 = 657 - 2 \cdot (963 - 657)$$
$$= 3 \cdot 657 - 2 \cdot 963;$$
$$36 = 306 - 6 \cdot 45 = (963 - 657) - 6 \cdot (3 \cdot 657 - 2 \cdot 963)$$
$$= 13 \cdot 963 - 19 \cdot 657;$$
$$9 = 45 - 36 = 3 \cdot 657 - 2 \cdot 963 - (13 \cdot 963 - 19 \cdot 657)$$
$$= 22 \cdot 657 - 15 \cdot 963.$$

In terms of Theorem 1.3, where $g = (b, c) = bx_0 + cy_0$, beginning with $b = 963$ and $c = 657$ we have used a procedure called the *Euclidean algorithm* to find $g = 9$, $x_0 = -15$, $y_0 = 22$. Of course, these values for x_0 and y_0 are not unique: $-15 + 657k$ and $22 - 963k$ will do where k is any integer.

To find the greatest common divisor (b, c) of *any* two integers b and c, we now generalize what is done in the special case above. The process will also give integers x_0 and y_0 satisfying the equation $bx_0 + cy_0 = (b, c)$. The case $c = 0$ is special: $(b, 0) = |b|$. For $c \neq 0$, we observe that $(b, c) = (b, -c)$ by Theorem 1.9, and hence, we may presume that c is positive.

Theorem 1.11 *The Euclidean algorithm. Given integers b and $c > 0$, we make a repeated application of the division algorithm, Theorem 1.2, to obtain a series of equations*

$$\begin{aligned} b &= cq_1 + r_1, & 0 < r_1 < c,\\ c &= r_1 q_2 + r_2, & 0 < r_2 < r_1,\\ r_1 &= r_2 q_3 + r_3, & 0 < r_3 < r_2,\\ &\cdots & \cdots\\ r_{j-2} &= r_{j-1} q_j + r_j, & 0 < r_j < r_{j-1},\\ r_{j-1} &= r_j q_{j+1}. \end{aligned}$$

The greatest common divisor (b, c) of b and c is r_j, the last nonzero remainder in the division process. Values of x_0 and y_0 in $(b, c) = bx_0 + cy_0$ can be obtained by writing each r_i as a linear combination of b and c.

Proof The chain of equations is obtained by dividing c into b, r_1 into c, r_2 into $r_1, \cdots, r_j$ into r_{j-1}. The process stops when the division is exact, that is, when the remainder is zero. Thus in our application of Theorem 1.2 we have written the inequalities for the remainder without an equality sign. Thus, for example, $0 < r_1 < c$ in place of $0 \leqslant r_1 < c$, because if r_1

were equal to zero, the chain would stop at the first equation $b = cq_1$, in which case the greatest common divisor of b and c would be c.

We now prove that r_j is the greatest common divisor g of b and c. By Theorem 1.9, we observe that

$$\begin{aligned}(b,c) &= (b - cq_1, c) = (r_1, c) = (r_1, c - r_1q_2)\\ &= (r_1, r_2) = (r_1 - r_2q_3, r_2) = (r_3, r_2).\end{aligned}$$

Continuing by mathematical induction, we get $(b,c) = (r_{j-1}, r_j) = (r_j, 0) = r_j$.

To see that r_j is a linear combination of b and c, we argue by induction that each r_i is a linear combination of b and c. Clearly, r_1 is such a linear combination, and likewise r_2. In general, r_i is a linear combination of r_{i-1} and r_{i-2}. By the inductive hypothesis we may suppose that these latter two numbers are linear combinations of b and c, and it follows that r_i is also a linear combination of b and c.

Example 1 Find the greatest common divisor of 42823 and 6409.

Solution We apply the Euclidean algorithm, using a calculator. We divide c into b, where $b = 42823$ and $c = 6409$, following the notation of Theorem 1.11. The quotient q_1 and remainder r_1 are $q_1 = 6$ and $r_1 = 4369$, with the details of this division as follows. Assuming the use of the simplest kind of hand-held calculator with only the four basic operations $+, -, \times, \div$, when 6409 is divided into 42823 the calculator gives 6.6816976, or some version of this with perhaps fewer decimal places. So we know that the quotient is 6. To get the remainder, we multiply 6 by 6409 to get 38454, and we subtract this from 42823 to get the remainder 4369.

Continuing, if we divide 4369 into 6409 we get a quotient $q_2 = 1$ and remainder $r_2 = 2040$. Dividing 2040 into 4369 gives $q_3 = 2$ and $r_3 = 289$. Dividing 289 into 2040 gives $q_4 = 7$ and $r_4 = 17$. Since 17 is an exact divisor of 289, the solution is that the g.c.d. is 17.

This can be put in tabular form as follows:

$$\begin{aligned}
42823 &= 6 \cdot 6409 + 4369 & &(42823, 6409)\\
6409 &= 1 \cdot 4369 + 2040 & &= (6409, 4369)\\
4369 &= 2 \cdot 2040 + 289 & &= (4369, 2040)\\
2040 &= 7 \cdot 289 + 17 & &= (2040, 289)\\
289 &= 17 \cdot 17 & &= (289, 17) = 17
\end{aligned}$$

Example 2 Find integers x and y to satisfy

$$42823x + 6409y = 17.$$

Solution We find integers x_i and y_i such that

$$42823x_i + 6409y_i = r_i.$$

Here it is natural to consider $i = 1, 2, \cdots$, but to initiate the process we also consider $i = 0$ and $i = -1$. We put $r_{-1} = 42823$, and write

$$42823 \cdot 1 + 6409 \cdot 0 = 42823.$$

Similarly, we put $r_0 = 6409$, and write

$$42823 \cdot 0 + 6409 \cdot 1 = 6409.$$

We multiply the second of these equations by $q_1 = 6$, and subtract the result from the first equation, to obtain

$$42823 \cdot 1 + 6409 \cdot (-6) = 4369.$$

We multiply this equation by $q_2 = 1$, and subtract it from the preceding equation to find that

$$42823 \cdot (-1) + 6409 \cdot 7 = 2040.$$

We multiply this by $q_3 = 2$, and subtract the result from the preceding equation to find that

$$42823 \cdot 3 + 6409 \cdot (-20) = 289.$$

Next we multiply this by $q_4 = 7$, and subtract the result from the preceding equation to find that

$$42823 \cdot (-22) + 6409 \cdot 147 = 17.$$

On dividing 17 into 289, we find that $q_5 = 17$ and that $289 = 17 \cdot 1$. Thus r_4 is the last positive remainder, so that $g = 17$, and we may take $x = -22$, $y = 147$. These values of x and y are not the only ones possible. In Section 5.1, an analysis of *all* solutions of a linear equation is given.

Remark on Calculation. We note that x_i is determined from x_{i-1} and x_{i-2} by the same formula that r_i is determined from r_{i-1} and r_{i-2}. That is,

$$r_i = r_{i-2} - q_i r_{i-1},$$

$$x_i = x_{i-2} - q_i x_{i-1},$$

and similarly

$$y_i = y_{i-2} - q_i y_{i-1}.$$

The only distinction between the three sequences r_i, x_i, and y_i is that they start from different initial conditions:

$$r_{-1} = b, \qquad r_0 = c,$$

$$x_{-1} = 1, \qquad x_0 = 0,$$

and

$$y_{-1} = 0, \qquad y_0 = 1.$$

Just as polynomial division may be effected symbolically, omitting the powers of the variable, we may generate the q_i, r_i, x_i, y_i in a compact table. In the numerical example just considered, this would take the following form:

i	q_{i+1}	r_i	x_i	y_i
-1		42823	1	0
0	6	6409	0	1
1	1	4369	1	-6
2	2	2040	-1	7
3	7	289	3	-20
4	17	17	-22	147
5		0		

When implemented on a computer, it is unnecessary to record the entire table. Each row is generated solely from the two preceding rows, so it suffices to keep only the two latest rows. In the numerical cases we have considered it has been the case that $b > c$. Although it is natural to start in this way, it is by no means necessary. If $b < c$, then $q_1 = 0$ and $r_1 = b$, which has the effect of interchanging b and c.

Example 3 Find $g = (b, c)$ where $b = 5033464705$ and $c = 3137640337$, and determine x and y such that $bx + cy = g$.

Solution We calculate:

	5033464705	1	0
1	3137640337	0	1
1	1895824368	1	−1
1	1241815969	−1	2
1	654008399	2	−3
1	587807570	−3	5
8	66200829	5	−8
1	58200938	−43	69
7	7999891	48	−77
3	2201701	−379	608
1	1394788	1185	−1901
1	806913	−1564	2509
1	587875	2749	−4410
2	219038	−4313	6919
1	149799	11375	−18248
2	69239	−15688	25167
6	11321	42751	−68582
8	1313	−272194	436659
1	817	2220303	−3561854
1	496	−2492497	3998513
1	321	4712800	−7560367
1	175	−7205297	11558880
1	146	11918097	−19119247
5	29	−19123394	30678127
29	1	107535067	−172509882

Thus $g = 1$, and we may take $x = 107535067$, $y = -172509882$.

The exact number of iterations j of the Euclidean algorithm required to calculate (b, c) depends in an intricate manner on b and c, but it is easy to establish a rough bound for j as follows: If r_i is small compared with r_{i-1}, say $r_i \leqslant r_{i-1}/2$, then substantial progress has been made at this step. Otherwise $r_{i-1}/2 < r_i < r_{i-1}$, in which case $q_{i+1} = 1$, and $r_{i+1} = r_{i-1} - r_i < r_{i-1}/2$. Thus we see that $r_{i+1} < r_{i-1}/2$ in either case. From this it can be deduced that $j < 3 \log c$. (Here, and throughout this book, we employ the natural logarithm, to the base e. Some writers denote this function $\ln x$.) With more care we could improve on the constant 3 (see Problem 10 in Section 4.4), but it is nevertheless the case that j is comparable to $\log c$

for most pairs b, c. Since the logarithm increases very slowly, the practical consequence is that one can calculate the g.c.d. quickly, even when b and c are very large.

Definition 1.4 *The integers $a_1, a_2, \cdots, a_n$, all different from zero, have a common multiple b if $a_i | b$ for $i = 1, 2, \cdots, n$. (Note that common multiples do exist; for example the product $a_1 a_2 \cdots a_n$ is one.) The least of the positive common multiples is called the* least common multiple, *and it is denoted by $[a_1, a_2, \cdots, a_n]$.*

Theorem 1.12 *If b is any common multiple of $a_1, a_2, \cdots, a_n$, then $[a_1, a_2, \cdots, a_n] | b$. This is the same as saying that if h denotes $[a_1, a_2, \cdots, a_n]$, then $0, \pm h, \pm 2h, \pm 3h, \cdots$ comprise all the common multiples of $a_1, a_2, \cdots, a_n$.*

Proof Let m be any common multiple and divide m by h. By Theorem 1.2 there is a quotient q and a remainder r such that $m = qh + r$, $0 \leqslant r < h$. We must prove that $r = 0$. If $r \neq 0$ we argue as follows. For each $i = 1, 2, \cdots, n$ we know that $a_i | h$ and $a_i | m$, so that $a_i | r$. Thus r is a positive common multiple of $a_1, a_2, \cdots, a_n$ contrary to the fact that h is the least of all the positive common multiples.

Theorem 1.13 *If $m > 0$, $[ma, mb] = m[a, b]$. Also $[a, b] \cdot (a, b) = |ab|$.*

Proof Let $H = [ma, mb]$, and $h = [a, b]$. Then mh is a multiple of ma and mb, so that $mh \geqslant H$. Also, H is a multiple of both ma and mb, so H/m is a multiple of a and b. Thus, $H/m \geqslant h$, from which it follows that $mh = H$, and this establishes the first part of the theorem.

It will suffice to prove the second part for positive integers a and b, since $[a, -b] = [a, b]$. We begin with the special case where $(a, b) = 1$. Now $[a, b]$ is a multiple of a, say ma. Then $b | ma$ and $(a, b) = 1$, so by Theorem 1.10 we conclude that $b | m$. Hence $b \leqslant m$, $ba \leqslant ma$. But ba, being a positive common multiple of b and a, cannot be less than the least common multiple, so $ba = ma = [a, b]$.

Turning to the general case where $(a, b) = g > 1$, we have $(a/g, b/g) = 1$ by Theorem 1.7. Applying the result of the preceding paragraph, we obtain

$$\left[\frac{a}{g}, \frac{b}{g}\right]\left(\frac{a}{g}, \frac{b}{g}\right) = \frac{a}{g}\frac{b}{g}.$$

Multiplying by g^2 and using Theorem 1.6 as well as the first part of the present theorem, we get $[a, b](a, b) = ab$.

PROBLEMS

1. By using the Euclidean algorithm, find the greatest common divisor (g.c.d.) of
   (*a*) 7469 and 2464; (*b*) 2689 and 4001;
   (*c*) 2947 and 3997; (*d*) 1109 and 4999.
2. Find the greatest common divisor g of the numbers 1819 and 3587, and then find integers x and y to satisfy
$$1819x + 3587y = g.$$
3. Find values of x and y to satisfy
   (*a*) $423x + 198y = 9$;
   (*b*) $71x - 50y = 1$;
   (*c*) $43x + 64y = 1$;
   (*d*) $93x - 81y = 3$;
   (*e*) $6x + 10y + 15z = 1$.
4. Find the least common multiple (l.c.m.) of (*a*) 482 and 1687, (*b*) 60 and 61.
5. How many integers between 100 and 1000 are divisible by 7?
6. Prove that the product of three consecutive integers is divisible by 6; of four consecutive integers by 24.
7. Exhibit three integers that are relatively prime but not relatively prime in pairs.
8. Two integers are said to be of the same *parity* if they are both even or both odd; if one is even and the other odd, they are said to be of opposite parity, or of different parity. Given any two integers, prove that their sum and their difference are of the same parity.
9. Show that if $ac \mid bc$ then $a \mid b$.
10. Given $a \mid b$ and $c \mid d$, prove that $ac \mid bd$.
11. Prove that $4 \nmid (n^2 + 2)$ for any integer n.
12. Given that $(a, 4) = 2$ and $(b, 4) = 2$, prove that $(a + b, 4) = 4$.
13. Prove that $n^2 - n$ is divisible by 2 for every integer n; that $n^3 - n$ is divisible by 6; that $n^5 - n$ is divisible by 30.
14. Prove that if n is odd, $n^2 - 1$ is divisible by 8.
15. Prove that if x and y are odd, then $x^2 + y^2$ is even but not divisible by 4.
16. Prove that if a and b are positive integers satisfying $(a, b) = [a, b]$ then $a = b$.
17. Evaluate $(n, n + 1)$ and $[n, n + 1]$ where n is a positive integer.

18. Find the values of (a, b) and $[a, b]$ if a and b are positive integers such that $a|b$.

19. Prove that any set of integers that are relatively prime in pairs are relatively prime.

20. Given integers a and b, a number n is said to be of the form $ak + b$ if there is an integer k such that $ak + b = n$. Thus the numbers of the form $3k + 1$ are $\cdots -8, -5, -2, 1, 4, 7, 10, \cdots$. Prove that every integer is of the form $3k$ or of the form $3k + 1$ or of the form $3k + 2$.

21. Prove that if an integer is of the form $6k + 5$, then it is necessarily of the form $3k - 1$, but not conversely.

22. Prove that the square of any integer of the form $5k + 1$ is of the same form.

23. Prove that the square of any integer is of the form $3k$ or $3k + 1$ but not of the form $3k + 2$.

24. Prove that no integers x, y exist satisfying $x + y = 100$ and $(x, y) = 3$.

25. Prove that there are infinitely many pairs of integers x, y satisfying $x + y = 100$ and $(x, y) = 5$.

26. Let s and $g > 0$ be given integers. Prove that integers x and y exist satisfying $x + y = s$ and $(x, y) = g$ if and only if $g|s$.

27. Find positive integers a and b satisfying the equations $(a, b) = 10$ and $[a, b] = 100$ simultaneously. Find all solutions.

28. Find all triples of positive integers a, b, c satisfying $(a, b, c) = 10$ and $[a, b, c] = 100$ simultaneously.

29. Let g and l be given positive integers. Prove that integers x and y exist satisfying $(x, y) = g$ and $[x, y] = l$ if and only if $g|l$.

30. Let b and $g > 0$ be given integers. Prove that the equations $(x, y) = g$ and $xy = b$ can be solved simultaneously if and only if $g^2|b$.

31. Let $n \geqslant 2$ and k be any positive integers. Prove that $(n - 1)|(n^k - 1)$.

32. Let $n \geqslant 2$ and k be any positive integers. Prove that $(n - 1)^2|(n^k - 1)$ if and only if $(n - 1)|k$. (H)[†]

33. Prove that $(a, b) = (a, b, a + b)$, and more generally that $(a, b) = (a, b, ax + by)$ for all integers x, y.

34. Prove that $(a, a + k)|k$ for all integers a, k not both zero.

35. Prove that $(a, a + 2) = 1$ or 2 for every integer a.

[†] The designation (H) indicates that a Hint is provided at the end of the book.

36. Prove that $(a, b, c) = ((a, b), c)$.

37. Prove that $(a_1, a_2, \cdots, a_n) = ((a_1, a_2, \cdots, a_{n-1}), a_n)$.

38. Extend Theorems 1.6, 1.7, and 1.8 to sets of more than two integers.

39. Suppose that the method used in the proof of Theorem 1.11 is employed to find x and y so that $bx + cy = g$. Thus $bx_i + cy_i = r_i$. Show that $(-1)^i x_i \leqslant 0$ and $(-1)^i y_i \geqslant 0$ for $i = -1, 0, 1, 2, \cdots, j+1$. Deduce that $|x_{i+1}| = |x_{i-1}| + q_{i+1}|x_i|$ and $|y_{i+1}| = |y_{i-1}| + q_{i+1}|y_i|$ for $i = 0, 1, \cdots, j$.

40. With the x_i and y_i determined as in Problem 39, show that $x_{i-1}y_i - x_i y_{i-1} = (-1)^i$ for $i = 0, 1, 2, \cdots, j+1$. Deduce that $(x_i, y_i) = 1$ for $i = -1, 0, 1, \cdots, j+1$. (H)

41. In the foregoing notation, if $g = (b, c)$, show that $|x_{j+1}| = c/g$ and $|y_{j+1}| = b/g$. (H)

42. In the foregoing notation, show that $|x_j| \leqslant c/(2g)$, with equality if and only if $q_{j+1} = 2$ and $x_{j-1} = 0$. Show similarly that $|y_j| \leqslant b/(2g)$.

43. Prove that $a|bc$ if and only if $\dfrac{a}{(a, b)} \bigg| c$.

44. Prove that every positive integer is uniquely expressible in the form

$$2^{j_0} + 2^{j_1} + 2^{j_2} + \cdots + 2^{j_m}$$

where $m \geqslant 0$ and $0 \leqslant j_0 < j_1 < j_2 < \cdots < j_m$.

45. Prove that any positive integer a can be uniquely expressed in the form

$$a = 3^m + b_{m-1}3^{m-1} + b_{m-2}3^{m-2} + \cdots + b_0$$

where each $b_j = 0$, 1, or -1.

***46.** Prove that there are no positive integers $a, b, n > 1$ such that $(a^n - b^n)|(a^n + b^n)$.

***47.** If a and $b > 2$ are any positive integers, prove that $2^a + 1$ is not divisible by $2^b - 1$.

***48.** The integers $1, 3, 6, 10, \cdots, n(n+1)/2, \cdots$ are called the *triangular numbers* because they are the numbers of dots needed to make successive triangular arrays of dots. For example, the number 10 can be perceived as the number of acrobats in a human triangle, 4 in a row at the bottom, 3 at the next level, then 2, then 1 at the top. The *square numbers* are $1, 4, 9, \cdots, n^2, \cdots$. The *pentagonal numbers*, $1, 5, 12, 22, \cdots, (3n^2 - n)/2, \cdots$, can be seen in a geometric array in the following way. Start with n equally spaced dots $P_1, P_2, \cdots, P_n$ on a straight line in a plane, with distance 1 between consecutive dots. Using P_1P_2 as a base side, draw a regular pentagon in the plane. Similarly, draw $n - 2$ additional regular pentagons on the base sides $P_1P_3, P_1P_4, \cdots, P_1P_n$, all pentagons lying on the same side of the line P_1P_n. Mark dots at each vertex and at unit intervals along the sides of these pentagons. Prove that the total number of dots in the array is $(3n^2 - n)/2$. In general, if regular k-gons are constructed on the sides $P_1P_2, P_1P_3, \cdots, P_1P_n$, with dots marked again at unit intervals, prove that the total number of dots is $1 + kn(n-1)/2 - (n-1)^2$. This is the nth k-gonal number.

***49.** Prove that if $m > n$ then $a^{2^n} + 1$ is a divisor of $a^{2^m} - 1$. Show that if a, m, n are positive with $m \neq n$, then

$$(a^{2^m} + 1, a^{2^n} + 1) = \begin{cases} 1 \text{ if } a \text{ is even} \\ 2 \text{ if } a \text{ is odd.} \end{cases}$$

***50.** Show that if $(a, b) = 1$ then $(a + b, a^2 - ab + b^2) = 1$ or 3.

***51.** Show that if $(a, b) = 1$ and p is an odd prime, then

$$\left(a + b, \frac{a^p + b^p}{a + b}\right) = 1 \text{ or } p.$$

***52.** Suppose that $2^n + 1 = xy$, where x and y are integers > 1 and $n > 0$. Show that $2^a | (x - 1)$ if and only if $2^a | (y - 1)$.

***53.** Show that $(n! + 1, (n + 1)! + 1) = 1$.

****54.** Let a and b be positive integers such that $(1 + ab) | (a^2 + b^2)$. Show that the integer $(a^2 + b^2)/(1 + ab)$ must be a perfect square.

1.3 PRIMES

Definition 1.5 *An integer $p > 1$ is called a* prime number, *or a* prime, *in case there is no divisor d of p satisfying $1 < d < p$. If an integer $a > 1$ is not a prime, it is called a* composite number.

Thus, for example, 2, 3, 5, and 7 are primes, whereas 4, 6, 8, and 9 are composite.

Theorem 1.14 *Every integer n greater than 1 can be expressed as a product of primes (with perhaps only one factor).*

Proof If the integer n is a prime, then the integer itself stands as a "product" with a single factor. Otherwise n can be factored into, say,

**Problems marked with a double asterisk are much more difficult.

n_1n_2, where $1 < n_1 < n$ and $1 < n_2 < n$. If n_1 is a prime, let it stand; otherwise it will factor into, say, n_3n_4 where $1 < n_3 < n_1$ and $1 < n_4 < n_1$; similarly for n_2. This process of writing each composite number that arises as a product of factors must terminate because the factors are smaller than the composite number itself, and yet each factor is an integer greater than 1. Thus we can write n as a product of primes, and since the prime factors are not necessarily distinct, the result can be written in the form

$$n = p_1^{\alpha_1} p_2^{\alpha_2} \cdots p_r^{\alpha_r}$$

where $p_1, p_2, \cdots, p_r$ are distinct primes and $\alpha_1, \alpha_2, \cdots, \alpha_r$ are positive.

This representation of n as a product of primes is called the *canonical factoring of n into prime powers*. It turns out that the representation is unique in the sense that, for fixed n, any other representation is merely a reordering or permutation of the factors. Although it may appear obvious that the factoring of an integer into a product of primes is unique, nevertheless, it requires proof. Historically, mathematicians took the unique factorization theorem for granted, but the great mathematician Gauss stated the result and proved it in a systematic way. It is proved later in the chapter as Theorem 1.16. The importance of this result is suggested by one of the names given to it, *the fundamental theorem of arithmetic*. This unique factorization property is needed to establish much of what comes later in the book. There are mathematical systems, notably in algebraic number theory, which is discussed in Chapter 9, where unique factorization fails to hold, and the absence of this property causes considerable difficulty in a systematic analysis of the subject. To demonstrate that unique factorization need not hold in a mathematical system, we digress from the main theme for a moment to present two examples in which factorization is not unique. The first example is easy; the second is much harder to follow, so it might well be omitted on a first reading of this book.

First consider the class $\mathscr{E}$ of positive even integers, so that the elements of $\mathscr{E}$ are $2, 4, 6, 8, 10, \cdots$. Note that $\mathscr{E}$ is a multiplicative system, the product of any two elements in $\mathscr{E}$ being again in $\mathscr{E}$. Now let us confine our attention to $\mathscr{E}$ in the sense that the only "numbers" we know are members of $\mathscr{E}$. Then $8 = 2 \cdot 4$ is "composite," whereas 10 is a "prime" since 10 is not the product of two or more "numbers." The "primes" are $2, 6, 10, 14, \cdots$, the "composite numbers" are $4, 8, 12, \cdots$. Now the "number" 60 has two factorings into "primes," namely $60 = 2 \cdot 30 = 6 \cdot 10$, and so factorization is not unique.

A somewhat less artificial, but also rather more complicated, example is obtained by considering the class $\mathscr{C}$ of numbers $a + b\sqrt{-6}$ where a and b range over all integers. We say that this system $\mathscr{C}$ is *closed* under addition and multiplication, meaning that the sum and product of two elements in $\mathscr{C}$ are elements of $\mathscr{C}$. By taking $b = 0$ we note that the integers form a subset of the class $\mathscr{C}$.

First we establish that there are primes in $\mathscr{C}$, and that every number in $\mathscr{C}$ can be factored into primes. For any number $a + b\sqrt{-6}$ in $\mathscr{C}$ it will be convenient to have a norm, $N(a + b\sqrt{-6})$, defined as

$$N(a + b\sqrt{-6}) = (a + b\sqrt{-6})(a - b\sqrt{-6}) = a^2 + 6b^2.$$

Thus the norm of a number in $\mathscr{C}$ is the product of the complex number $a + b\sqrt{-6}$ and its conjugate $a - b\sqrt{-6}$. Another way of saying this, perhaps in more familiar language, is that the norm is the square of the absolute value. Now the norm of every number in $\mathscr{C}$ is a positive integer greater than 1, except for the numbers $0, 1, -1$ for which we have $N(0) = 0$, $N(1) = 1$, $N(-1) = 1$. We say that we have a factoring of $a + b\sqrt{-6}$ if we can write

$$a + b\sqrt{-6} = (x_1 + y_1\sqrt{-6})(x_2 + y_2\sqrt{-6}) \tag{1.1}$$

where $N(x_1 + y_1\sqrt{-6}) > 1$ and $N(x_2 + y_2\sqrt{-6}) > 1$. This restriction on the norms of the factors is needed to rule out such trivial factorings as $a + b\sqrt{-6} = (1)(a + b\sqrt{-6}) = (-1)(-a - b\sqrt{-6})$. The norm of a product can be readily calculated to be the product of the norms of the factors, so that in the factoring (1.1) we have $N(a + b\sqrt{-6}) = N(x_1 + y_1\sqrt{-6})N(x_2 + y_2\sqrt{-6})$. It follows that

$$1 < N(x_1 + y_1\sqrt{-6}) < N(a + b\sqrt{-6}),$$

$$1 < N(x_2 + y_2\sqrt{-6}) < N(a + b\sqrt{-6})$$

so any number $a + b\sqrt{-6}$ will break up into only a finite number of factors since the norm of each factor is an integer.

We remarked above that the norm of any number in $\mathscr{C}$, apart from 0 and ± 1, is greater than 1. More can be said. Since $N(a + b\sqrt{-6})$ has the value $a^2 + 6b^2$, we observe that

$$N(a + b\sqrt{-6}) \geqslant 6 \qquad \text{if } b \neq 0, \tag{1.2}$$

that is, the norm of any nonreal number in $\mathscr{C}$ is not less than 6.

A number of $\mathscr{C}$ having norm > 1, but that cannot be factored in the sense of (1.1), is called a *prime in* $\mathscr{C}$. For example, 5 is a prime in $\mathscr{C}$, for in the first place, 5 cannot be factored into real numbers in $\mathscr{C}$. In the second place, if we had a factoring $5 = (x_1 + y_1\sqrt{-6})(x_2 + y_2\sqrt{-6})$ into complex numbers, we could take norms to get

$$25 = N(x_1 + y_1\sqrt{-6})N(x_2 + y_2\sqrt{-6}),$$

which contradicts (1.2). Thus, 5 is a prime in $\mathscr{C}$, and a similar argument establishes that 2 is a prime.

We are now in a position to show that not all numbers of $\mathscr{C}$ factor uniquely into primes. Consider the number 10 and its two factorings:

$$10 = 2 \cdot 5 = (2 + \sqrt{-6})(2 - \sqrt{-6}).$$

The first product $2 \cdot 5$ has factors that are prime in $\mathscr{C}$, as we have seen. Thus we can conclude that there is not unique factorization of the number 10 in $\mathscr{C}$. Note that this conclusion does not depend on our knowing that $2 + \sqrt{-6}$ and $2 - \sqrt{-6}$ are primes; they actually are, but it is unimportant in our discussion.

This example may also seem artificial, but it is, in fact, taken from an important topic, algebraic number theory, discussed in Chapter 9.

We now return to the discussion of unique factorization in the ordinary integers $0, \pm 1, \pm 2, \cdots$. It will be convenient to have the following result.

Theorem 1.15 *If $p \mid ab$, p being a prime, then $p \mid a$ or $p \mid b$. More generally, if $p \mid a_1 a_2 \cdots a_n$, then p divides at least one factor a_i of the product.*

Proof If $p \nmid a$, then $(a, p) = 1$ and so by Theorem 1.10, $p \mid b$. We may regard this as the first step of a proof of the general statement by mathematical induction. So we assume that the proposition holds whenever p divides a product with fewer than n factors. Now if $p \mid a_1 a_2 \cdots a_n$, that is, $p \mid a_1 c$ where $c = a_2 a_3 \cdots a_n$, then $p \mid a_1$ or $p \mid c$. If $p \mid c$ we apply the induction hypothesis to conclude that $p \mid a_i$ for some subscript i from 2 to n.

Theorem 1.16 *The fundamental theorem of arithmetic, or the unique factorization theorem. The factoring of any integer $n > 1$ into primes is unique apart from the order of the prime factors.*

First Proof Suppose that there is an integer n with two different factorings. Dividing out any primes common to the two representations, we would have an equality of the form

$$p_1 p_2 \cdots p_r = q_1 q_2 \cdots q_s \tag{1.3}$$

where the factors p_i and q_j are primes, not necessarily all distinct, but where no prime on the left side occurs on the right side. But this is impossible because $p_1|q_1q_2 \cdots q_s$, so by Theorem 1.15, p_1 is a divisor of at least one of the q_j. That is, p_1 must be identical with at least one of the q_j.

Second Proof Suppose that the theorem is false and let n be the smallest positive integer having more than one representation as the product of primes, say

$$n = p_1p_2 \cdots p_r = q_1q_2 \cdots q_s. \tag{1.4}$$

It is clear that r and s are greater than 1. Now the primes $p_1, p_2, \cdots, p_r$ have no members in common with $q_1, q_2, \cdots, q_s$ because if, for example, p_1 were a common prime, then we could divide it out of both sides of (1.4) to get two distinct factorings of n/p_1. But this would contradict our assumption that all integers smaller than n are uniquely factorable.

Next, there is no loss of generality in presuming that $p_1 < q_1$, and we define the positive integer N as

$$N = (q_1 - p_1)q_2q_3 \cdots q_s = p_1(p_2p_3 \cdots p_r - q_2q_3 \cdots q_s). \tag{1.5}$$

It is clear that $N < n$, so that N is uniquely factorable into primes. But $p_1 \nmid (q_1 - p_1)$, so (1.5) gives us two factorings of N, one involving p_1 and the other not, and thus we have a contradiction.

In the application of the fundamental theorem we frequently write any integer $a \geqslant 1$ in the form

$$a = \prod_p p^{\alpha(p)}$$

where $\alpha(p)$ is a non-negative integer, and it is understood that $\alpha(p) = 0$ for all sufficiently large primes p. If $a = 1$ then $\alpha(p) = 0$ for all primes p, and the product may be considered to be empty. For brevity we sometimes write $a = \prod p^\alpha$, with the tacit understanding that the exponents α depend on p and, of course on a. If

$$a = \prod_p p^{\alpha(p)}, \qquad b = \prod_p p^{\beta(p)}, \qquad c = \prod_p p^{\gamma(p)}, \tag{1.6}$$

and $ab = c$, then $\alpha(p) + \beta(p) = \gamma(p)$ for all p, by the fundamental theorem. Here $a|c$, and we note that $\alpha(p) \leqslant \gamma(p)$ for all p. If, conversely, $\alpha(p) \leqslant \gamma(p)$ for all p, then we may define an integer $b = \prod p^{\beta(p)}$ with

$\beta(p) = \gamma(p) - \alpha(p)$. Then $ab = c$, which is to say that $a|c$. Thus we see that the divisibility relation $a|c$ is equivalent to the family of inequalities $\alpha(p) \leqslant \gamma(p)$. As a consequence, the greatest common divisor and the least common multiple can be written as

$$(a, b) = \prod_p p^{\min(\alpha(p), \beta(p))}, \qquad [a, b] = \prod_p p^{\max(\alpha(p), \beta(p))}. \quad (1.7)$$

For example, if $a = 108$ and $b = 225$, then

$$a = 2^2 3^3 5^0, \qquad b = 2^0 3^2 5^2,$$

$$(a, b) = 2^0 3^2 5^0 = 9, \qquad [a, b] = 2^2 3^3 5^2 = 2700.$$

The first part of Theorem 1.13, like many similar identities, follows easily from the fundamental theorem in conjunction with (1.7). Since $\min(\alpha, \beta) + \max(\alpha, \beta) = \alpha + \beta$ for any real numbers α, β, the relations (1.7) also provide a means of establishing the second part of Theorem 1.13. On the other hand, for calculational purposes the identifies (1.7) should only be used when the factorizations of a and b are already known, as in general the task of factoring a and b will involve much more computation than is required if one determines (a, b) by the Euclidean algorithm.

We call a a *square* (or alternatively a *perfect square*) if it can be written in the form n^2. By the fundamental theorem we see that a is a square if and only if all the exponents $\alpha(p)$ in (1.6) are even. We say that a is *square-free* if 1 is the largest square dividing a. Thus a is square-free if and only if the exponents $\alpha(p)$ take only the values 0 and 1. Finally, we observe that if p is prime, then the assertion $p^k \| a$ is equivalent to $k = \alpha(p)$.

Theorem 1.17 *Euclid. The number of primes is infinite. That is, there is no end to the sequence of primes*

$$2, 3, 5, 7, 11, 13, \cdots.$$

Proof Suppose that $p_1, p_2, \cdots, p_r$ are the first r primes. Then form the number

$$n = 1 + p_1 p_2 \cdots p_r.$$

Note that n is not divisible by p_1 or p_2 or $\cdots$ or p_r. Hence any prime divisor p of n is a prime distinct from $p_1, p_2, \cdots, p_r$. Since n is either a prime or has a prime factor p, this implies that there is a prime distinct from $p_1, p_2, \cdots, p_r$. Thus we see that for any finite r, the number of primes is not exactly r. Hence the number of primes is infinite.

Students often note that the first few of the numbers n here are primes. However, $1 + 2 \cdot 3 \cdot 5 \cdot 7 \cdot 11 \cdot 13 = 59 \cdot 509$.

Theorem 1.18 *There are arbitrarily large gaps in the series of primes. Stated otherwise, given any positive integer k, there exist k consecutive composite integers.*

Proof Consider the integers

$$(k+1)!+2, (k+1)!+3, \cdots, (k+1)!+k, (k+1)!+k+1.$$

Every one of these is composite because j divides $(k+1)!+j$ if $2 \leqslant j \leqslant k+1$.

The primes are spaced rather irregularly, as the last theorem suggests. If we denote the number of primes that do not exceed x by $\pi(x)$, we may ask about the nature of this function. Because of the irregular occurrence of the primes, we cannot expect a simple formula for $\pi(x)$, but we may seek to estimate its rate of growth. The proof of Theorem 1.17 can be used to derive a lower bound for $\pi(x)$, but the estimate obtained, $\pi(x) > c \log\log x$, is very weak. We now derive an inequality that is more suggestive of the true state of affairs.

Theorem 1.19 *For every real number $y \geqslant 2$,*

$$\sum_{p \leqslant y} \frac{1}{p} > \log\log y - 1.$$

Here it is understood that the sum is over all primes $p \leqslant y$. From this it follows that the infinite series $\Sigma 1/p$ diverges, which provides a second proof of Theorem 1.17.

Proof Let y be given, $y \geqslant 2$, and let $\mathcal{N}$ denote the set of all those positive integers n that are composed entirely of primes p not exceeding y. Since there are only finitely many primes $p \leqslant y$, and since the terms of an absolutely convergent infinite series may be arbitrarily rearranged, we see that

$$\prod_{p \leqslant y} \left(1 + \frac{1}{p} + \frac{1}{p^2} + \frac{1}{p^3} + \cdots\right) = \sum_{n \in \mathcal{N}} \frac{1}{n}. \tag{1.8}$$

If n is a positive integer $\leqslant y$ then $n \in \mathscr{N}$, and thus the sum above includes the sum $\Sigma_{n \leqslant y} 1/n$. Let N denote the largest integer not exceeding y. By the integral test,

$$\sum_{n=1}^{N} \frac{1}{n} \geqslant \int_1^{N+1} \frac{dx}{x} = \log(N+1) > \log y.$$

Thus the right side of (1.8) is $> \log y$. On the other hand, the sum on the left side of (1.8) is a geometric series, whose value is $(1 - 1/p)^{-1}$, so we see that

$$\prod_{p \leqslant y} \left(1 - \frac{1}{p}\right)^{-1} > \log y.$$

We assume for the moment that the inequality

$$e^{v+v^2} \geqslant (1-v)^{-1} \tag{1.9}$$

holds for all real numbers v in the interval $0 \leqslant v \leqslant 1/2$. Taking $v = 1/p$, we deduce that

$$\prod_{p \leqslant y} \exp\left(\frac{1}{p} + \frac{1}{p^2}\right) > \log y.$$

Since $\Pi \exp(a_i) = \exp(\Sigma a_i)$, and since the logarithm function is monotonically increasing, we may take logarithms of both sides and deduce that

$$\sum_{p \leqslant y} \frac{1}{p} + \sum_{p \leqslant y} \frac{1}{p^2} > \log\log y.$$

By the comparison test we see that the second sum is

$$< \sum_{n=2}^{\infty} \frac{1}{n^2},$$

and by the integral test this is

$$< \int_1^{\infty} \frac{dx}{x^2} = 1.$$

This gives the stated inequality, but it remains to prove (1.9). We need to show that $f(v) \geqslant 1$ for $0 \leqslant v \leqslant 1/2$, where $f(v) = (1 - v)\exp(v + v^2)$. Since $f(0) = 1$, it suffices to show that $f(v)$ is increasing for $0 \leqslant v \leqslant 1/2$. To this end it is enough to observe that

$$f'(v) = v(1 - 2v)\exp(v + v^2) \geqslant 0.$$

Thus we have (1.9), and the proof is complete.

With more work it can be shown that the difference

$$\sum_{p \leqslant y} \frac{1}{p} - \log\log y$$

is a bounded function of y, for $y \geqslant 2$. Deeper still lies the *Prime Number Theorem*, which asserts that

$$\lim_{x \to \infty} \frac{\pi(x)}{x/\log x} = 1.$$

We say that $f(x)$ is asymptotic to $g(x)$, or write $f(x) \sim g(x)$, if $\lim_{x \to \infty} f(x)/g(x) = 1$. Thus the prime number theorem may be expressed by writing $\pi(x) \sim x/\log x$. This is one of the most important results of analytic number theory. We do not prove it in this book, but in Section 8.1 we establish a weaker estimate in this direction.

PROBLEMS

1. With a and b as in (1.6) what conditions on the exponents must be satisfied if $(a, b) = 1$?
2. What is the largest number of consecutive square-free positive integers? What is the largest number of consecutive cube-free positive integers, where a is cube-free if it is divisible by the cube of no integer greater than 1?
3. In any positive integer, such as 8347, the last digit is called the *units* digit, the next the *tens* digit, the next the *hundreds* digit, and so forth. In the example 8347, the units digit is 7, the tens digit is 4, the hundreds digit is 3, and the thousands digit is 8. Prove that a number is divisible by 2 if and only if its units digit is divisible by 2; that a number is divisible by 4 if and only if the integer formed by its tens digit and its units digit is divisible by 4; that a number is divisible by 8 if and only if the integer formed by its last three digits is divisible by 8.

4. Prove that an integer is divisible by 3 if and only if the sum of its digits is divisible by 3. Prove that an integer is divisible by 9 if and only if the sum of its digits is divisible by 9.

5. Prove that an integer is divisible by 11 if and only if the difference between the sum of the digits in the odd places and the sum of the digits in the even places is divisible by 11.

6. Show that every positive integer n has a unique expression of the form $n = 2^r m$, $r \geqslant 0$, m a positive odd integer.

7. Show that every positive integer n can be written uniquely in the form $n = ab$, where a is square-free and b is a square. Show that b is then the largest square dividing n.

8. A test for divisibility by 7. Starting with any positive integer n, subtract double the units digit from the integer obtained from n by removing the units digit, giving a smaller integer r. For example, if $n = 41283$ with units digit 3, we subtract 6 from 4128 to get $r = 4122$. The problem is to prove that if either n or r is divisible by 7, so is the other. This gives a test for divisibility by 7 by repeating the process. From 41283 we pass to 4122, then to 408 by subtracting 4 from 412, and then to 24 by subtracting 16 from 40. Since 24 is not divisible by 7, neither is 41283. (H)

9. Prove that any prime of the form $3k + 1$ is of the form $6k + 1$.

10. Prove that any positive integer of the form $3k + 2$ has a prime factor of the same form; similarly for each of the forms $4k + 3$ and $6k + 5$.

11. If x and y are odd, prove that $x^2 + y^2$ cannot be a perfect square.

12. If x and y are prime to 3, prove that $x^2 + y^2$ cannot be a perfect square.

13. If $(a, b) = p$, a prime, what are the possible values of (a^2, b)? Of (a^3, b)? Of (a^2, b^3)?

14. Evaluate (ab, p^4) and $(a + b, p^4)$ given that $(a, p^2) = p$ and $(b, p^3) = p^2$ where p is a prime.

15. If a and b are represented by (1.6), what conditions must be satisfied by the exponents if a is to be a cube? For $a^2 | b^2$?

16. Find a positive integer n such that $n/2$ is a square, $n/3$ is a cube, and $n/5$ is a fifth power.

17. Twin primes are those differing by 2. Show that 5 is the only prime belonging to two such pairs. Show also that there is a one-to-one correspondence between twin primes and numbers n such that $n^2 - 1$ has just four positive divisors.

18. Prove that $(a^2, b^2) = c^2$ if $(a, b) = c$.

19. Let a and b be positive integers such that $(a, b) = 1$ and ab is a perfect square. Prove that a and b are perfect squares. Prove that the result generalizes to kth powers.

20. Given $(a, b, c)[a, b, c] = abc$, prove that $(a, b) = (b, c) = (a, c) = 1$.

21. Prove that $[a, b, c](ab, bc, ca) = |abc|$.

22. Determine whether the following assertions are true or false. If true, prove the result, and if false, give a counterexample.

(1) If $(a, b) = (a, c)$ then $[a, b] = [a, c]$.
(2) If $(a, b) = (a, c)$ then $(a^2, b^2) = (a^2, c^2)$.
(3) If $(a, b) = (a, c)$ then $(a, b) = (a, b, c)$.
(4) If p is a prime and $p|a$ and $p|(a^2 + b^2)$ then $p|b$.
(5) If p is a prime and $p|a^7$ then $p|a$.
(6) If $a^3|c^3$ then $a|c$.
(7) If $a^3|c^2$ then $a|c$.
(8) If $a^2|c^3$ then $a|c$.
(9) If p is a prime and $p|(a^2 + b^2)$ and $p|(b^2 + c^2)$ then $p|(a^2 - c^2)$.
(10) If p is a prime and $p|(a^2 + b^2)$ and $p|(b^2 + c^2)$ then $p|(a^2 + c^2)$.
(11) If $(a, b) = 1$ then $(a^2, ab, b^2) = 1$.
(12) $[a^2, ab, b^2] = [a^2, b^2]$.
(13) If $b|(a^2 + 1)$ then $b|(a^4 + 1)$.
(14) If $b|(a^2 - 1)$ then $b|(a^4 - 1)$.
(15) $(a, b, c) = ((a, b), (a, c))$.

23. Given integers a, b, c, d, m, n, u, v satisfying $ad - bc = \pm 1$, $u = am + bn$, $v = cm + dn$, prove that $(m, n) = (u, v)$.

24. Prove that if n is composite, it must have a prime factor $p \leqslant \sqrt{n}$. (Note that a straightforward implication of this problem is that if we want to test whether an integer n is a prime, it suffices to check whether it is divisible by any of the primes $\leqslant \sqrt{n}$. For example, if $n = 1999$, we check divisibility by the primes $2, 3, 5, \cdots, 43$. This is easy to do with a hand calculator. It turns out that 1999 is divisible by none of these primes, so it is itself a prime.)

25. Obtain a complete list of the primes between 1 and n, with $n = 200$ for convenience, by the following method, known as the *sieve of Eratosthenes*. By the *proper* multiples of k we mean all positive multiples of k except k itself. Write all numbers from 2 to 200. Cross out all proper multiples of 2, then of 3, then of 5. At each stage the next larger remaining number is a prime. Thus 7 is now the next remaining larger than 5. Cross out the proper multiples of 7. The next remaining number larger than 7 is 11. Continuing, we cross out the proper multiples of 11 and then of 13. Now we observe that the next remaining number greater than 13 exceeds $\sqrt{200}$, and hence by the previous problem all the numbers remaining in our list are prime.

26. Prove that there are infinitely many primes of the form $4n + 3$; of the form $6n + 5$. (H)

Remark The last problem can be stated thus: each of the arithmetic progressions $3, 7, 11, 15, 19, \cdots$, and $5, 11, 17, 23, 29, \cdots$ contains an infinitude of primes. One of the famous theorems of number theory (the proof of which lies deeper than the methods of this book), due to Dirichlet, is that the arithmetic progression $a, a + b, a + 2b, a + 3b, \cdots$ contains infinitely many primes if the integers a and $b > 0$ are relatively prime, that is if $(a, b) = 1$.

27. Show that $n|(n - 1)!$ for all composite $n > 4$.

28. Suppose that $n > 1$. Show that the sum of the positive integers not exceeding n divides the product of the positive integers not exceeding n if and only if $n + 1$ is composite.

29. Suppose that m and n are integers > 1, and that $(\log m)/(\log n)$ is rational, say equal to a/b with $(a, b) = 1$. Show that there must be an integer c such that $m = c^a$, $n = c^b$.

30. Prove that $n^2 - 81n + 1681$ is a prime for $n = 1, 2, 3, \cdots, 80$, but not for $n = 81$. (Note that this problem shows that a sequence of propositions can be valid for many beginning cases, and then fail.)

31. Prove that no polynomial $f(x)$ of degree > 1 with integral coefficients can represent a prime for every positive integer x. (H)

Remark Let $f(x)$ be a nonconstant polynomial with integral coefficients. If there is an integer $d > 1$ such that $d|f(n)$ for all integers n, then there exist at most finitely many integers n such that $f(n)$ is prime. (For example, if $f(x) = x^2 + x + 2$, then $2|f(n)$ for all n, and $f(n)$ is prime only for $n = -1, 0$.) Similarly, if there exist nonconstant polynomials $g(x)$ and $h(x)$ with integral coefficients such that $f(x) = g(x)h(x)$ for all x, then $f(n)$ is prime for at most finitely many integers n, since $g(n)$ will be a proper divisor of $f(n)$ when $|n|$ is large. (For example, if $f(x) = x^2 + 8x + 15$, then $n + 3$ is a proper divisor of $f(n)$ except when $n = -2$, -4, or -6.) It is conjectured that if neither of these two situations applies to $f(x)$, then there exist infinitely many integers n such that $f(n)$ is prime. If f is of degree 1, then this is precisely the theorem of Dirichlet concerning primes in arithmetic progressions, alluded to earlier, but the conjecture has not been proved for any polynomial of degree greater than 1. In particular, it has not been proved that there exist infinitely many integers n such that $n^2 + 1$ is prime.

32. Show that $n^4 + 4$ is composite for all $n > 1$.

33. Show that $n^4 + n^2 + 1$ is composite if $n > 1$.

***34.** Show that if $m^4 + 4^n$ is prime, then m is odd and n is even, except when $m = n = 1$.

***35.** Show that there exist non-negative integers x and y such that $x^2 - y^2 = n$ if and only if n is odd or is a multiple of 4. Show that there is exactly one such representation of n if and only if $n = 1, 4$, an odd prime, or four times a prime.

***36.** Consider the set $\mathscr{S}$ of integers $1, 2, \cdots, n$. Let 2^k be the integer in $\mathscr{S}$ that is the highest power of 2. Prove that 2^k is not a divisor of any other integer in $\mathscr{S}$. Hence, prove that $\Sigma_{j=1}^{n} 1/j$ is not an integer if $n > 1$.

***37.** Prove that in any block of consecutive positive integers there is a unique integer divisible by a higher power of 2 than any of the others. Then use this, or any other method, to prove that there is no integer among the 2^{n+1} numbers

$$\pm\frac{1}{k} \pm \frac{1}{k+1} \pm \frac{1}{k+2} \pm \cdots \pm \frac{1}{k+n}$$

where all possible combinations of plus and minus signs are allowed, and where n and k are any positive integers. (Note that this result is a sweeping generalization of the preceding problem.)

***38.** Consider the set $\mathscr{T}$ of integers $1, 3, 5, \cdots, 2n - 1$. Let 3^r be the integer in $\mathscr{T}$ that is the highest power of 3. Prove that 3^r is not a divisor of any other integer in $\mathscr{T}$. Hence, prove that $\Sigma_{j=1}^{n} 1/(2j - 1)$ is not an integer if $n > 1$.

***39.** Prove that

$$1 - \frac{1}{2} + \frac{1}{3} - \frac{1}{4} + \cdots + \frac{1}{1999} - \frac{1}{2000} = \frac{1}{1001} + \frac{1}{1002} + \cdots + \frac{1}{2000}$$

where the signs are alternating on the left side of the equation but are all alike on the right side. (This is an example of a problem where it is easier to prove a general result than a special case.)

***40.** Say that a positive integer n is a sum of consecutive integers if there exist positive integers m and k such that $n = m + (m + 1) + \cdots$

$+ (m + k)$. Prove that n is so expressible if and only if it is not a power of 2.

***41.** Prove that an odd integer $n > 1$ is a prime if and only if it is not expressible as a sum of three or more consecutive positive integers.

42. If $2^n + 1$ is an odd prime for some integer n, prove that n is a power of 2. (H)

43. The numbers $F_n = 2^{2^n} + 1$ in the preceding problem are called the *Fermat numbers*, after Pierre Fermat who thought they might all be primes. Show that F_5 is composite by verifying that

$$(2^9 + 2^7 + 1)(2^{23} - 2^{21} + 2^{19} - 2^{17} + 2^{14} - 2^9 - 2^7 + 1) = 2^{32} + 1.$$

(It is not hard to show that F_n is prime for $n = 0, 1, \cdots, 4$; these are the only n for which F_n is known to be prime. It is now known that F_n is composite for $n = 5, 6, \cdots, 21$. It is conjectured that only finitely many Fermat numbers are prime.)

44. If $2^n - 1$ is a prime for some integer n, prove that n is itself a prime. (Numbers of the form $2^p - 1$, where p is a prime, are called the *Mersenne numbers* M_p because the Frenchman Father Marin Mersenne (1588–1648) stated the M_p is a prime for $p = 2, 3, 5, 7, 13, 17, 19, 31, 67, 127, 257$, but is composite for all other primes $p < 257$. It took some 300 years before the details of this assertion could be checked completely, with the following outcome: M_p *is not* a prime for $p = 67$ and $p = 257$, and M_p *is* a prime for $p = 61$, $p = 89$, and $p = 107$. Thus, there are 12 primes $p < 257$ such that M_p is a prime. It is now known that M_p is a prime in the following additional cases, $p = 521, 607, 1279, 2203, 2281, 3217, 4253, 4423, 9689, 9941, 11213, 19937, 21701, 23209, 44497, 86243, 110503, 132049, 216091$. The Mersenne prime M_{756839} is the largest specific number that is known to be prime. It is conjectured that infinitely many of the Mersenne numbers are prime.)

***45.** Let positive integers g and l be given with $g|l$. Prove that the number of pairs of positive integers x, y satisfying $(x, y) = g$ and $[x, y] = l$ is 2^k, where k is the number of distinct prime factors of l/g. (Count x_1, y_1 and x_2, y_2 as different pairs if $x_1 \neq x_2$ or $y_1 \neq y_2$.)

***46.** Let $k \geqslant 3$ be a fixed integer. Find all sets $a_1, a_2, \cdots, a_k$ of positive integers such that the sum of any triplet is divisible by each member of the triplet.

***47.** Prove that $2 + \sqrt{-6}$ and $2 - \sqrt{-6}$ are primes in the class $\mathscr{C}$ of numbers $a + b\sqrt{-6}$.

***48.** Prove that there are infinitely many primes by considering the sequence $2^{2^1}+1, 2^{2^2}+1, 2^{2^3}+1, 2^{2^4}+1, \cdots$. (H)

***49.** If g is a divisor of each of ab, cd, and $ac+bd$, prove that it is also a divisor of ac and bd, where a, b, c, d are integers.

***50.** Show that

$$(ab, cd) = (a,c)(b,d)\left(\frac{a}{(a,c)}, \frac{d}{(b,d)}\right)\left(\frac{c}{(a,c)}, \frac{b}{(b,d)}\right).$$

***51.** Show that 24 is the largest integer divisible by all integers less than its square root. (H)

***52.** (For readers familiar with the rudiments of point-set topology.) We topologize the integers as follows: a set $\mathscr{N}$ of integers is open if for every $n \in \mathscr{N}$ there is an arithmetic progression $\mathscr{A}$ such that $n \in \mathscr{A} \subseteq \mathscr{N}$. (An arithmetic progression is a set of the form $\{dk + r: k \in \mathbb{Z}\}$ with $d \neq 0$.) Prove that arbitrary unions of open sets are open, and that finite intersections of open sets are open, so that these open sets define a topology in the usual sense. (From a more advanced perspective, this is known as a *profinite* topology.) As is usual in topology, we call a set $\mathscr{N}$ *closed* if its complement $\mathbb{Z} \setminus \mathscr{N}$ is open. Let $\mathscr{A}$ be an arithmetic progression. Prove that the complement of $\mathscr{A}$ is a union of arithmetic progressions. Deduce that $\mathscr{A}$ is both open and closed. Let $\mathscr{U}$ denote the union over all prime numbers p of the arithmetic progressions $\{np: n \in \mathbb{Z}\}$, and let $\mathscr{V}$ denote the complement of $\mathscr{U}$. In symbols, $\mathscr{U} = \bigcup_p p\mathbb{Z}$ and $\mathscr{V} = \mathbb{Z} \setminus \mathscr{U}$. Show that $\mathscr{V} = \{-1, 1\}$. Show that if there were only finitely many prime numbers then the set $\mathscr{U}$ would be closed. From the observation that $\mathscr{V}$ is not an open set, conclude that there exist infinitely many prime numbers.

***53.** Let $\pi(x)$ denote the number of primes not exceeding x. Show that

$$\sum_{p \leqslant x} 1/p = \frac{\pi(x)}{x} + \int_2^x \pi(u)/u^2\, du.$$

Using Theorem 1.19, deduce that

$$\limsup_{x \to \infty} \frac{\pi(x)}{x/\log x} \geqslant 1.$$

1.4 THE BINOMIAL THEOREM

We first define the *binomial coefficients* and describe them combinatorially.

Definition 1.6 *Let α be any real number, and let k be a non-negative integer. Then the* binomial coefficient $\binom{\alpha}{k}$ *is given by the formula*

$$\binom{\alpha}{k} = \frac{\alpha(\alpha - 1) \cdots (\alpha - k + 1)}{k!}.$$

Suppose that n and k are both integers. From the formula we see that if $0 \leqslant k \leqslant n$ then $\binom{n}{k} = \dfrac{n!}{k!(n-k)!}$, whereas if $0 \leqslant n < k$, then $\binom{n}{k} = 0$. Here we employ the convention $0! = 1$.

Theorem 1.20 *Let $\mathscr{S}$ be a set containing exactly n elements. For any non-negative integer k, the number of subsets of $\mathscr{S}$ containing precisely k elements is $\binom{n}{k}$.*

By the definition, $\binom{4}{2} = \dfrac{4 \cdot 3}{2!} = 6$, whereas if $\mathscr{S} = \{1, 2, 3, 4\}$ then the subsets containing two elements are $\{1, 2\}, \{1, 3\}, \{1, 4\}, \{2, 3\}, \{2, 4\}, \{3, 4\}$. Because of this combinatorial interpretation, the binomial coefficient $\binom{n}{k}$ is read "n choose k."

Proof Suppose that $\mathscr{S} = \{1, 2, \cdots, n\}$. These numbers may be listed in various orders, called *permutations*, here denoted by π. There are $n!$ of these permutations π, because the first term may be any one of the n numbers, the second term any one of the $n - 1$ remaining numbers, and the third term any one of the still remaining $n - 2$ numbers, and so on. We count the permutations in a way that involves the number X of subsets containing precisely k elements. Let $\mathscr{A}$ be a specific subset of $\mathscr{S}$ with k elements. There are $k!$ permutations of the elements of $\mathscr{A}$, each permutation having k terms. Similarly there are $(n - k)!$ permutations of the $n - k$ elements not in $\mathscr{A}$. If we attach any one of these $(n - k)!$ permutations to the right end of any one of the $k!$ previous permutations, the ordered sequence of n elements thus obtained is one of the permutations π of $\mathscr{S}$. Thus we can generate $k!(n - k)!$ of the permutations π in this way. To get all the permutations π of $\mathscr{S}$, we repeat this procedure with $\mathscr{A}$ replaced by each of the subsets in question. Let X denote the number of these subsets. Then there are $k!(n-k)!X$ permutations π, and equating this to $n!$ we find that $X = \binom{n}{k}$.

We now see that the quotient $\dfrac{n!}{k!(n-k)!}$ is an integer, because it represents the number of ways of doing something. In this way, combinatorial interpretations can be useful in number theory. We now use Theorem 1.20 to derive the following result, which we shall need in Section 2.6.

Theorem 1.21 *The product of any k consecutive integers is divisible by $k!$.*

Proof Write the product as $n(n-1)\cdots(n-k+1)$. If $n \geqslant k$, then we write this in the form $\binom{n}{k}k!$, and note that $\binom{n}{k}$ is an integer, by Theorem 1.20. If $0 \leqslant n < k$, then one of the factors of our product is 0, so the product vanishes, and is therefore a multiple of $k!$ in this case also. Finally, if $n < 0$, we note that the product may be written

$$(-1)^k(-n)(-n+1)\cdots(-n+k-1) = (-1)^k\binom{-n+k-1}{k}k!.$$

Note that in this case the upper member $-n+k-1$ is at least k, so that by Theorem 1.20 the binomial coefficient is an integer.

In the formula for the binomial coefficients we note a symmetry:

$$\binom{n}{k} = \binom{n}{n-k}. \tag{1.10}$$

This is also evident from the combinatorial interpretation, since the subsets of $\mathscr{A}$ containing k elements are in one-to-one correspondence with the complementary subsets $\mathscr{S} \setminus \mathscr{A} = \{i \in \mathscr{S} : i \notin \mathscr{A}\}$ containing $n-k$ elements.

Theorem 1.22 *The binomial theorem. For any integer $n \geqslant 1$ and any real numbers x and y,*

$$(x+y)^n = \sum_{k=0}^{n}\binom{n}{k}x^k y^{n-k}. \tag{1.11}$$

Proof We consider first the product

$$\prod_{i=1}^{n}(x_i + y_i).$$

On multiplying this out, we obtain 2^n monomial terms of the form

$$\prod_{i \in \mathscr{A}} x_i \prod_{i \notin \mathscr{A}} y_i$$

where $\mathscr{A}$ is any subset of $\{1, 2, \cdots, n\}$. For each fixed k, $0 \leqslant k \leqslant n$, we consider the monomial terms obtained from those subsets $\mathscr{A}$ of $\{1, 2, \cdots, n\}$ having exactly k elements. We set $x_i = x$ and $y_i = y$ for all i and note that such a monomial has value $x^k y^{n-k}$ for the subsets in question. Since there are $\binom{n}{k}$ such subsets, we see that the contribution of such subsets is $\binom{n}{k} x^k y^{n-k}$, which gives (1.11).

The binomial theorem can also be proved analytically by appealing to the following simple result.

Lemma 1.23 *Let* $P(z) = \sum_{k=0}^{n} a_k z^k$ *be a polynomial with real coefficients. Then* $a_r = P^{(r)}(0)/r!$ *for* $0 \leqslant r \leqslant n$, *where* $P^{(r)}(0)$ *is the rth derivative of* $P(z)$ *at* $z = 0$.

Proof By differentiating repeatedly, we see that

$$P^{(r)}(z) = \sum_{k=r}^{n} k(k-1) \cdots (k-r+1) a_k z^{k-r}.$$

On setting $z = 0$ we see that $P^{(r)}(0) = r! a_r$, as desired.

If we take $P(z) = (1 + z)^n$, then

$$P^{(r)}(z) = n(n-1) \cdots (n-r+1)(1+z)^{n-r},$$

so that $P^{(r)}(0) = n(n-1) \cdots (n-r+1)$, and hence by the Lemma, $a_r = n(n-1) \cdots (n-r+1)/r! = \binom{n}{r}$. That is,

$$(1 + z)^n = \sum_{k=0}^{n} \binom{n}{k} z^k. \tag{1.12}$$

This is a form of the binomial theorem. We can recover (1.11) by taking $z = x/y$, and then multiplying both sides by y^n. This gives the identity when $y \neq 0$. The case $y = 0$ of (1.11) is obvious. In our first (combinatorial) proof of this theorem, the binomial coefficients arose in the context of Theorem 1.20, but in our second (analytic) proof, they occurred in the form described in Definition 1.6. Thus the two proofs of Theorem 1.22 may be combined to provide a second proof of Theorem 1.20.

As a matter of logic, we require only one proof of each theorem, but additional proofs often provide new insights, and the various proofs may generalize in different directions. In the present case, the first proof can be used whenever x and y are members of a commutative ring, whereas the second proof can be used to derive a more general form of the binomial theorem, which asserts that

$$(1+z)^\alpha = \sum_{k=0}^{\infty} \binom{\alpha}{k} z^k \tag{1.13}$$

for $|z| < 1$. Here α is an arbitrary real or complex number. This is consistent with (1.12) if α is a non-negative integer. As a function of α, the quantity $\binom{\alpha}{k}$ is a polynomial of degree k with rational coefficients. By Theorem 1.21 we see that this polynomial takes integral values whenever α is an integer. A polynomial with this property is called *integer-valued*.

The series (1.13) is the Taylor series of the function on the left. To demonstrate that it converges to the desired value, one may use the integral form of the remainder, which states that if $f(z)$ is a function for which $f^{(K+1)}(z)$ is continuous, then

$$f(z) = \sum_{k=0}^{K} \frac{f^{(k)}(0)}{k!} z^k + R_K(z)$$

where

$$R_K(z) = \frac{z^{K+1}}{K!} \int_0^1 (1-t)^K f^{(K+1)}(tz)\, dt.$$

We take $f(z) = (1+z)^\alpha$, so that

$$f^{(k)}(z) = \alpha(\alpha-1)\cdots(\alpha-k+1)(1+z)^{\alpha-k}.$$

Hence

$$R_K(z) = \alpha\binom{\alpha-1}{K} z^{K+1} \int_0^1 (1-t)^K (1+tz)^{\alpha-K-1}\, dt.$$

From the hypothesis $|z| < 1$ it follows that $|1+tz| \geqslant 1 - |tz| \geqslant 1 - t$.

Hence $|1 + tz|^{-K} \leqslant (1 - t)^{-K}$, and we see that

$$|R_K(z)| \leqslant \left|\alpha\binom{\alpha - 1}{K} z^{K+1}\right| \int_0^1 \left|(1 + tz)^{\alpha - 1}\right| dt = T_K,$$

say. Here the integral is independent of K, and

$$\frac{T_{K+1}}{T_K} = \left|\frac{(\alpha - K - 1)z}{K + 1}\right| \to |z|$$

as $K \to \infty$. Taking r so that $|z| < r < 1$, we deduce that $T_{K+1} \leqslant rT_K$ for all large K, say $K \geqslant L$. By induction it follows that $T_K \leqslant Cr^K$ for $K \geqslant L$, where $C = T_L / r^L$. Thus $T_K \to 0$ as $K \to \infty$, and we conclude that $R_K(z) \to 0$ as $K \to \infty$. Thus (1.13) holds when $|z| < 1$.

The binomial coefficients arise in many identities, both in analysis and in combinatorics. One of the simplest of these is the recursion

$$\binom{n}{k} + \binom{n}{k + 1} = \binom{n + 1}{k + 1}, \tag{1.14}$$

used in many ways, for example, to construct *Pascal's triangle*. We define this triangle below, but first we give three short proofs of identity (1.14). Since all members vanish if $k > n$, and since the identity is clear when $k = -1$, we may assume that $0 \leqslant k \leqslant n$. First, we may simply use the formula of Definition 1.6, and then simplify the expressions. Second, we can interpret the identity combinatorially. To this end, observe that if $\mathscr{A}$ contains $k + 1$ elements of $\mathscr{S} = \{1, 2, \cdots, n + 1\}$, then one can consider two cases: either $n + 1 \in \mathscr{A}$, or $n + 1 \notin \mathscr{A}$. In the first case, $\mathscr{A}$ is determined by choosing k of the numbers $1, 2, \cdots, n$; there are $\binom{n}{k}$ ways of doing this. In the second case, $\mathscr{A}$ is determined by choosing $k + 1$ numbers from among $1, 2, \cdots, n$, which gives $\binom{n}{k + 1}$ subsets of this type. This again gives the identity, by Theorem 1.20. Third, we note that the right side is the coefficient of z^{k+1} in $(1 + z)^{n+1}$. But this polynomial may be written

$$(1 + z)(1 + z)^n = (1 + z)^n + z(1 + z)^n = \sum_{k=0}^{n} \binom{n}{k} z^k + \sum_{k=0}^{n} \binom{n}{k} z^{k+1}.$$

In this last expression, the coefficient of z^{k+1} is $\binom{n}{k + 1} + \binom{n}{k}$. From Lemma 1.23 we see that the coefficient of z^{k+1} is uniquely defined. Thus we again have (1.14).

Pascal's triangle is the infinite array of numbers

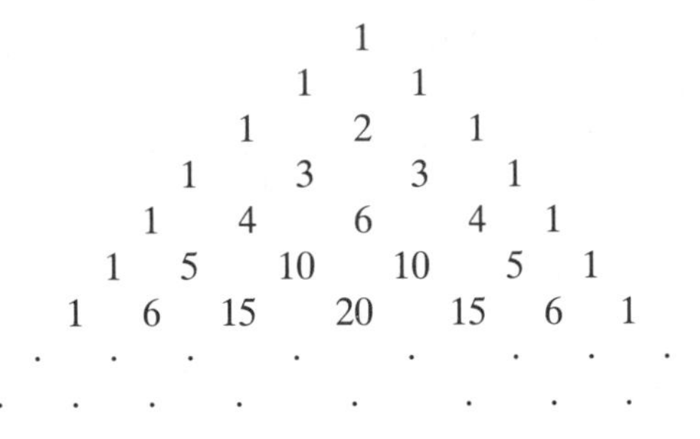

where, for example, the last row exhibited gives the binomial coefficients in the expansion of $(x + y)^6$. The identity (1.14) can be used to generate as many further rows as we please. Apart from the 1's at the ends of each row, the numbers can be obtained by adding the two integers on the preceding row, one just to the left and one just to the right. For example the next row is $1, 1 + 6, 6 + 15, 15 + 20$, and so on, or $1, 7, 21, 35, 35, 21, 7, 1$. The nth row has n entries, namely the coefficients in the binomial expansion of $(x + y)^{n-1}$.

PROBLEMS

1. Use the binomial theorem to show that

$$\sum_{k=0}^{n} \binom{n}{k} = 2^n.$$

Can you give a combinatorial proof of this?

2. Show that if $n \geqslant 1$ then $\sum_{k=0}^{n} (-1)^k \binom{n}{k} = 0$.

3. (*a*) By comparing the coefficient of z^k in the polynomial identity

$$\sum_{k=0}^{m+n} \binom{m+n}{k} z^k = (1+z)^{m+n} = (1+z)^m (1+z)^n$$

$$= \left(\sum_{k=0}^{m} \binom{m}{k} z^k\right)\left(\sum_{k=0}^{n} \binom{n}{k} z^k\right)$$

show that

$$\sum_{i=0}^{k} \binom{m}{i}\binom{n}{k-i} = \binom{m+n}{k}.$$

(b) Let $\mathscr{U}$ and $\mathscr{V}$ be disjoint sets containing m and n elements, respectively, and put $\mathscr{S} = \mathscr{U} \cup \mathscr{V}$. Show that the number of subsets $\mathscr{A}$ of $\mathscr{S}$ that contain k elements and that also have the property that $\mathscr{A} \cap \mathscr{U}$ contains i elements is $\binom{m}{i}\binom{n}{k-i}$. Interpret this identity combinatorially.

(c) Show that for $n \geqslant 0$,

$$\sum_{k=0}^{n} \binom{n}{k}^2 = \binom{2n}{n}.$$

4. (a) Suppose that $\mathscr{S}$ contains $2n$ elements, and that $\mathscr{S}$ is partitioned into n disjoint subsets each one containing exactly two elements of $\mathscr{S}$. Show that this can be done in precisely

$$(2n-1)(2n-3)\cdots 5\cdot 3\cdot 1 = \frac{(2n)!}{2^n n!}$$

ways.

(b) Show that $(n+1)(n+2)\cdots(2n)$ is divisible by 2^n, but not by 2^{n+1}.

5. Show that if a and b are positive integers, then $a!^b b! \mid (ab)!$. (H)

6. Let $f(x)$ and $g(x)$ be n-times differentiable functions. Show that the nth derivative of $f(x)g(x)$ is

$$\sum_{k=0}^{n} \binom{n}{k} f^{(k)}(x) g^{(n-k)}(x).$$

7. Show that $\binom{-\alpha-1}{k} = (-1)^k \binom{\alpha+k}{k}$ for $k \geqslant 0$. Deduce that if $|z| < 1$ then

$$\frac{1}{(1-z)^{\alpha+1}} = \sum_{k=0}^{\infty} \binom{\alpha+k}{k} z^k. \tag{1.15}$$

8. Give three proofs that

$$\sum_{m=0}^{M} \binom{m+k}{k} = \binom{k+M+1}{k+1}.$$

(a) With k fixed, induct on M, using Theorem 1.20.

(b) Let $\mathscr{S} = \{1, 2, \cdots, k+M+1\}$. Count the number of subsets $\mathscr{A}$ of $\mathscr{S}$ containing $k+1$ elements, with the maximum one being $k+m+1$.

(c) Compute the coefficient of z^M in the identity

$$(1 + z + z^2 + \cdots)\cdot\frac{1}{(1-z)^{k+1}} = \frac{1}{(1-z)^{k+2}}.$$

9. Let $f(x)$ be a function of a real variable, and let Δf be the function $\Delta f(x) = f(x+1) - f(x)$. For $k > 1$, put $\Delta^k f = \Delta(\Delta^{k-1} f)$. The function $\Delta^k f(x)$ is called the kth *forward difference* of f. Show that

$$\Delta^k f(x) = \sum_{j=0}^{k} (-1)^j \binom{k}{j} f(x+k-j).$$

***10.** Let $\mathscr{S}$ be a set of n elements. Count the number of ordered pairs $(\mathscr{A}, \mathscr{B})$ of subsets such that $\varnothing \subseteq \mathscr{A} \subseteq \mathscr{B} \subseteq \mathscr{S}$. Let $c(j,k)$ denote the number of such ordered pairs for which $\mathscr{A}$ contains j elements and $\mathscr{B}$ contains k elements. Show that

$$(1 + y + xy)^n = \sum_{0 \leqslant j \leqslant k \leqslant n} c(j,k) x^j y^k.$$

What does this give if $x = y = 1$?

***11.** Show that $\binom{x}{k}$ is a polynomial in x of degree k and leading coefficient $1/k!$. Let $P(x)$ be an arbitrary polynomial with real coefficients and degree at most n. Show that there exist real numbers c_k such that

$$P(x) = \sum_{k=0}^{n} c_k \binom{x}{k} \tag{1.16}$$

for all x, and that such c_k are uniquely determined.

***12.** Show that $\binom{x+1}{k} - \binom{x}{k} = \binom{x}{k-1}$ when k is a positive integer and x is a real number. Show that if $P(x)$ is given by (1.16), then

$$\Delta P(x) = \sum_{k=1}^{n} c_k \binom{x}{k-1}.$$

Note the similarity to the formula for the derivative of a polynomial. Show that if $P(x)$ is a polynomial with real coefficients and of degree n, then ΔP is a polynomial of degree $n-1$.

***13.** Show that if x is a real number and k is a non-negative integer, then

$$\sum_{m=0}^{M} \binom{x+m}{k} = \binom{x+M+1}{k+1} - \binom{x}{k+1}.$$

Show that if $P(x)$ is a polynomial written in the form (1.16), then

$$\sum_{m=0}^{M} P(x+m) = Q(x+M+1) - Q(x),$$

where

$$Q(x) = \sum_{k=0}^{n} c_k \binom{x}{k+1}.$$

Note the similarity to the formula for the integral of a polynomial and that $Q(x)$ is a polynomial of degree $n + 1$.

***14.** Suppose that $P(x)$ is a polynomial written in the form (1.16). Show that if the c_k are integers, then $P(x)$ is an integer-valued polynomial.

***15.** Suppose that $P(x)$ is a polynomial written in the form (1.16). Show that if $P(0), P(1), \cdots, P(n)$ are integers then the c_k are integers and $P(x)$ is integer-valued.

***16.** Show that if $f(x)$ is a polynomial of degree n with real coefficients, which takes integral values on a certain set of $n + 1$ consecutive integers, then $f(x)$ is integer-valued.

***17.** Show that if $f(x)$ is an integer-valued polynomial of degree n, then $n!f(x)$ is a polynomial with integral coefficients.

***18.** Suppose that $f(x)$ is an integer-valued polynomial of degree n and that $g = \text{g.c.d.}\ (f(0), f(1), \cdots, f(n))$. Show that $g | f(k)$ for all integers k.

***19.** Show that if m and n are non-negative integers then

$$\sum_{k=0}^{n} (-1)^k \binom{m+1}{k}\binom{m+n-k}{m} = \begin{cases} 1 \text{ if } n = 0, \\ 0 \text{ if } n > 0. \end{cases}$$

***20.** Show that if m and n are integers with $0 \leqslant m < n$, then

$$\sum_{k=m+1}^{n} (-1)^k \binom{n}{k}\binom{k-1}{m} = (-1)^{m+1}.$$

***21.** Show that if n is a positive integer then

$$\sum_{k=1}^{n} \frac{(-1)^{k+1}}{k}\binom{n}{k} = \sum_{k=1}^{n} \frac{1}{k}.$$

***22.** Show that if m and n are integers, $0 \leqslant m < n$, then

$$\sum_{k=0}^{m} (-1)^k \binom{n}{k} = (-1)^m \binom{n-1}{m}.$$

***23.** (*a*) Show that

$$\sum_{k=0}^{\infty} \binom{k+n}{k} 2^{-k} = 2^{n+1}.$$

(*b*) Show that

$$\sum_{k=0}^{n} \binom{k+n}{k} 2^{-k} = 2^{n}.$$

***24.** Show that

$$\sum_{k=0}^{2n} (-1)^k \binom{2n}{k}^2 = (-1)^n \binom{2n}{n}.$$

***25.** Show that

$$\sum_{k=0}^{n+1} \left(\binom{n}{k} - \binom{n}{k-1} \right)^2 = \frac{2}{n+1} \binom{2n}{n}.$$

***26.** Show that

$$\sum_{k=1}^{n} k \binom{n}{2k+1} = (n-2)2^{n-3}.$$

NOTES ON CHAPTER 1

For "most" pairs b, c, the Euclidean algorithm requires approximately $\frac{12 \log 2}{\pi^2} \log c$ steps. A precise formulation of this is given by J. Dixon, "The number of steps in the Euclidean algorithm," *J. Number Theory*, 2 (1970), 414–422.

When seeking to write (b, c) as a linear combination of b and c, an alternative method is obtained by solving up from the bottom. In Example 2 this would be done by writing

$$\begin{aligned} 17 &= 2040 - 7 \cdot 289 \\ &= 2040 - 7 \cdot (4369 - 2 \cdot 2040) = (-7) \cdot 4369 + 15 \cdot 2040 \\ &= (-7) \cdot 4369 + 15 \cdot (6409 - 1 \cdot 4369) = 15 \cdot 6409 - 22 \cdot 4369 \\ &= 15 \cdot 6409 - 22 \cdot (42823 - 6 \cdot 6409) = (-22) \cdot 42823 + 147 \cdot 6409. \end{aligned}$$

In general, we set $s_{j+1} = 0$, $s_j = 1$ and determine the numbers s_{j-1}, $s_{j-2}, \cdots, s_0$ successively by the relation $s_{i-1} = -q_i s_i + s_{i+1}$. Put $t_i = s_{i+1} r_{i-1} + s_i r_i$. Since

$$t_{i-1} = (-q_i s_i + s_{i+1}) r_{i-1} + s_i(r_i + q_i r_{i-1}) = t_i$$

it follows that the value of t_i is independent of i. As $t_j = r_j = \text{g.c.d.}(b, c)$, we conclude that $t_0 = bs_1 + cs_0 = \text{g.c.d.}(b, c)$. The advantage of this method is we need construct only the one sequence $\{s_i\}$, whereas in our former method we constructed two sequences, $\{u_i\}$ and $\{v_i\}$. The disadvantage of this new method is that all the q_i must be saved, as the s_i are computed in reverse order. Thus if memory is limited (as on a programmable pocket calculator), the former method is preferable, whereas on larger machines it is faster to follow the method above. However, this new method is advantageous only in situations in which both the coefficients of b and of c are desired. In most of the applications that arise later (e.g., Theorems 2.9, 2.17, 2.18), only the coefficient of b is needed.

It can be noted that the second proof of Theorem 1.16 does not depend on Theorem 1.15 or indeed on any previous theorem. Thus the logical arrangement of this chapter could be altered considerably by putting Theorems 1.14 and 1.16 in an early position, and then using the formulas for (b, c) and $[b, c]$ in (1.7) to prove such results as Theorems 1.6, 1.7, 1.8, 1.10, and 1.15.

Many special cases of the Dirichlet theorem, that is, that there are infinitely many primes in the arithmetic progression $a, a + b, a + 2b, \cdots$ if a and b are relatively prime integers, are given throughout the book. The cases $a = 3$, $b = 4$ and $a = 5$, $b = 6$ (or, what is the same thing, $a = 2$, $b = 3$) are given in Problem 26 of Section 1.3; $a = 1$, $b = 4$ in Problem 38 of Section 2.1; $a = 1$ in Problem 36 in Section 2.8; $a = 1, 3, 5, 7$, $b = 8$ in Problem 20 of Section 3.1; $a = 1, 2$, $b = 3$ in Problem 13 of Section 3.2. In Section 8.4 we develop a different method that can be used to prove the theorem in general. The full details are found in Chapter 7 of Apostol or Section 4 of Davenport (1980). (Books referred to briefly by the author's surnames are listed in the General References on page 500.)

The prime number theorem, stated at the end of Section 1.3, was first proved in 1896, independently by Jacques Hadamard and C. J. de la Vallée Poussin. They used the theory of functions of a complex variable to derive the theorem from properties of the Riemann zeta function $\zeta(s)$. The account in Sections 8 through 18 of Davenport (1980) follows the original method quite closely. A shorter proof, which still uses the theory of a complex variable but which requires less information concerning the zeta function, is given in Chapter 13 of Apostol. In 1949, Atle Selberg gave an elementary proof of an identity involving prime numbers, which led him and Pál Erdős to give elementary (though complicated) proofs of the prime number theorem. A readable account of the elementary proof of the prime number theorem has been given by N. Levinson, "A motivated account of an elementary proof of the prime number theorem," *Amer. Math. Monthly*, 76 (1969), 225–245.

Because it dates back to antiquity, the most famous result in this chapter is Euclid's proof in Theorem 1.17 that there are infinitely many primes. The argument given is essentially the same as that by Euclid in the third century B.C. Many variations on this argument can be given, such as the simple observation that for any positive integer n, the number $n! + 1$

must have a prime factor exceeding n. Other proofs of Euclid's theorem are outlined in Problems 48 and 52 of Section 1.3. Euler argued that $\Sigma 1/p = \infty$ because $\Pi(1 - 1/p)^{-1} = \Pi(1 + 1/p + 1/p^2 + \cdots) = \Sigma 1/n = \infty$. Our proof of Theorem 1.19 presents Euler's reasoning in a more precise (and rigorous) form.

Except when α is a non-negative integer, the series in (1.13) diverges when $|z| > 1$. We do not address the more subtle question of whether the identity (1.13) holds when $|z| = 1$. Further material concerning binomial coefficients is found in Chapter 1 §2 of Pólya and Szegö. The "q-binomial theorem" of Gauss is introduced in Chapter 1 §5 of the same book.

CHAPTER 2

Congruences

2.1 CONGRUENCES

It is apparent from Chapter 1 that divisibility is a fundamental concept of number theory, one that sets it apart from many other branches of mathematics. In this chapter we continue the study of divisibility, but from a slightly different point of view. A *congruence* is nothing more than a statement about divisibility. However, it is more than just a convenient notation. It often makes it easier to discover proofs, and we shall see that congruences can suggest new problems that will lead us to new and interesting topics.

The theory of congruences was introduced by Carl Friedrich Gauss (1777–1855), one of the greatest mathematicians of all time. Gauss contributed to the theory of numbers in many outstanding ways, including the basic ideas of this chapter and the next. Although Pierre de Fermat (1601–1665) had earlier studied number theory in a somewhat systematic way, Gauss was the first to develop the subject as a branch of mathematics rather than just a scattered collection of interesting problems. In his book *Disquisitiones Arithmeticae*, written at age 24, Gauss introduced the theory of congruences, which gained ready acceptance as a fundamental tool for the study of number theory.

Some fundamental ideas of congruences are included in this first section. The theorems of Fermat and Euler are especially noteworthy, providing powerful techniques for analyzing the multiplicative aspects of congruences. These two pioneers in number theory worked in widely contrasting ways. Mathematics was an avocation for Fermat, who was a lawyer by profession. He communicated his mathematical ideas by correspondence with other mathematicians, giving very few details of the proofs of his assertions. (One of his claims is known as Fermat's "last theorem," although it is not a theorem at all as yet, having never been proved. This situation is discussed in Section 5.4.) Leonard Euler (1707–1783), on the other hand, wrote prolifically in almost all the known branches of mathematics of his time. For example, although Fermat undoubtedly was able to prove the result attributed to him as Theorem 2.7 below, Euler in 1736 was the first to publish a proof. Years later, in 1760, Euler stated and proved his generalization of Fermat's result, which is given as Theorem 2.8 here.

Definition 2.1 *If an integer m, not zero, divides the difference $a - b$, we say that a is* congruent *to b modulo m and write $a \equiv b \pmod{m}$. If $a - b$ is not divisible by m, we say that a is* not congruent *to b modulo m, and in this case we write $a \not\equiv b \pmod{m}$.*

Since $a - b$ is divisible by m if and only if $a - b$ is divisible by $-m$, we can generally confine our attention to a positive modulus. Indeed, we shall assume throughout the present chapter that the modulus m is a positive integer.

Congruences have many properties in common with equalities. Some properties that follow easily from the definition are listed in the following theorem.

Theorem 2.1 *Let a, b, c, d denote integers. Then:*

(1) $a \equiv b \pmod{m}$, $b \equiv a \pmod{m}$, and $a - b \equiv 0 \pmod{m}$ are equivalent statements.

(2) If $a \equiv b \pmod{m}$ and $b \equiv c \pmod{m}$, then $a \equiv c \pmod{m}$.

(3) If $a \equiv b \pmod{m}$ and $c \equiv d \pmod{m}$, then $a + c \equiv b + d \pmod{m}$.

(4) If $a \equiv b \pmod{m}$ and $c \equiv d \pmod{m}$, then $ac \equiv bd \pmod{m}$.

(5) If $a \equiv b \pmod{m}$ and $d \mid m$, $d > 0$, then $a \equiv b \pmod{d}$.

(6) If $a \equiv b \pmod{m}$ then $ac \equiv bc \pmod{mc}$ for $c > 0$.

Theorem 2.2 *Let f denote a polynomial with integral coefficients. If $a \equiv b \pmod{m}$ then $f(a) \equiv f(b) \pmod{m}$.*

Proof We can suppose $f(x) = c_n x^n + c_{n-1}x^{n-1} + \cdots + c_0$ where the c_i are integers. Since $a \equiv b \pmod{m}$ we can apply Theorem 2.1, part 4, repeatedly to find $a^2 \equiv b^2$, $a^3 \equiv b^3, \cdots, a^n \equiv b^n \pmod{m}$, and then $c_j a^j \equiv c_j b^j \pmod{m}$, and finally $c_n a^n + c_{n-1}a^{n-1} + \cdots + c_0 \equiv c_n b^n + c_{n-1}b^{n-1} + \cdots + c_0 \pmod{m}$, by Theorem 2.1 part 3.

You are, of course, well aware of the property of real numbers that if $ax = ay$ and $a \neq 0$ then $x = y$. More care must be used in dividing a congruence through by a.

Theorem 2.3

(1) $ax \equiv ay \pmod{m}$ *if and only if* $x \equiv y \left(\bmod \dfrac{m}{(a,m)}\right)$.

(2) If $ax \equiv ay \pmod{m}$ *and* $(a, m) = 1$, *then* $x \equiv y \pmod{m}$.

(3) $x \equiv y \pmod{m_i}$ *for* $i = 1, 2, \cdots, r$ *if and only if* $x \equiv y \pmod{[m_1, m_2, \cdots, m_r]}$.

Proof (1) If $ax \equiv ay \pmod{m}$ then $ay - ax = mz$ for some integer z. Hence we have

$$\frac{a}{(a,m)}(y - x) = \frac{m}{(a,m)}z,$$

and thus

$$\frac{m}{(a,m)} \Bigg| \frac{a}{(a,m)}(y - x).$$

But $(a/(a, m), m/(a, m)) = 1$ by Theorem 1.7 and therefore $\{m/(a, m)\}|(y - x)$ by Theorem 1.10. That is,

$$x \equiv y \left(\bmod \frac{m}{(a,m)}\right).$$

Conversely, if $x \equiv y \pmod{m/(a, m)}$, we multiply by a to get $ax \equiv ay \pmod{am/(a, m)}$ by use of Theorem 2.1, part 6. But (a, m) is a divisor of a, so we can write $ax \equiv ay \pmod{m}$ by Theorem 2.1, part 5.

For example, $15x \equiv 15y \pmod{10}$ is equivalent to $x \equiv y \pmod{2}$, which amounts to saying that x and y have the same parity.

(2) This is a special case of part 1. It is listed separately because we shall use it very often.

(3) If $x \equiv y \pmod{m_i}$ for $i = 1, 2, \cdots, r$, then $m_i|(y - x)$ for $i = 1, 2, \cdots, r$. That is, $y - x$ is a common multiple of $m_1, m_2, \cdots, m_r$, and therefore (see Theorem 1.12) $[m_1, m_2, \cdots, m_r]|(y - x)$. This implies $x \equiv y \pmod{[m_1, m_2, \cdots, m_r]}$.

If $x \equiv y \pmod{[m_1, m_2, \cdots, m_r]}$ then $x \equiv y \pmod{m_i}$ by Theorem 2.1 part 5, since $m_i|[m_1, m_2, \cdots, m_r]$.

In dealing with integers modulo m, we are essentially performing the ordinary operations of arithmetic but are disregarding multiples of m. In a sense we are not distinguishing between a and $a + mx$, where x is any integer. Given any integer a, let q and r be the quotient and remainder on division by m; thus $a = qm + r$ by Theorem 1.2. Now $a \equiv r \pmod{m}$ and, since r satisfies the inequalities $0 \leqslant r < m$, we see that every integer is congruent modulo m to one of the values $0, 1, 2, \cdots, m - 1$. Also it is clear that no two of these m integers are congruent modulo m. These m values constitute a complete residue system modulo m, and we now give a general definition of this term.

Definition 2.2 *If $x \equiv y \pmod{m}$ then y is called a* residue *of x modulo m. A set $x_1, x_2, \cdots, x_m$ is called a* complete residue system *modulo m if for every integer y there is one and only one x_j such that $y \equiv x_j \pmod{m}$.*

It is obvious that there are infinitely many complete residue systems modulo m, the set $1, 2, \cdots, m - 1, m$ being another example.

A set of m integers forms a complete residue system modulo m if and only if no two integers in the set are congruent modulo m.

For fixed integers a and $m > 0$, the set of all integers x satisfying $x \equiv a \pmod{m}$ is the arithmetic progression

$$\cdots, a - 3m, a - 2m, a - m, a, a + m, a + 2m, a + 3m, \cdots.$$

This set is called a *residue class*, or *congruence class*, modulo m. There are m distinct residue classes modulo m, obtained for example by taking successively $a = 1, 2, 3, \cdots, m$.

Theorem 2.4 *If $b \equiv c \pmod{m}$, then $(b, m) = (c, m)$.*

Proof We have $c = b + mx$ for some integer x. To see that $(b, m) = (b + mx, m)$, take $a = m$ in Theorem 1.9.

Definition 2.3 *A* reduced residue system *modulo m is a set of integers r_i such that $(r_i, m) = 1$, $r_i \not\equiv r_j \pmod{m}$ if $i \neq j$, and such that every x prime to m is congruent modulo m to some member r_i of the set.*

In view of Theorem 2.4 it is clear that a reduced residue system modulo m can be obtained by deleting from a complete residue system modulo m those members that are not relatively prime to m. Furthermore, all reduced residue systems modulo m will contain the same number of members, a number that is denoted by $\phi(m)$. This function is called *Euler's ϕ-function*, sometimes the *totient*. By applying this definition of $\phi(m)$ to the complete residue system $1, 2, \cdots, m$ mentioned in the paragraph following Definition 2.2, we can get what amounts to an alternative definition of $\phi(m)$, as given in the following theorem.

Theorem 2.5 *The number $\phi(m)$ is the number of positive integers less than or equal to m that are relatively prime to m.*

Euler's function $\phi(m)$ is of considerable interest. We shall consider it further in Sections 2.3, 4.2, 8.2, and 8.3.

Theorem 2.6 *Let $(a, m) = 1$. Let $r_1, r_2, \cdots, r_n$ be a complete, or a reduced, residue system modulo m. Then $ar_1, ar_2, \cdots, ar_n$ is a complete, or a reduced, residue system, respectively, modulo m.*

For example, since $1, 2, 3, 4$ is a reduced residue system modulo 5, so also is $2, 4, 6, 8$. Since $1, 3, 7, 9$ is a reduced residue system modulo 10, so is $3, 9, 21, 27$.

Proof If $(r_i, m) = 1$, then $(ar_i, m) = 1$ by Theorem 1.8.

There are the same number of $ar_1, ar_2, \cdots, ar_n$ as of $r_1, r_2, \cdots, r_n$. Therefore we need only show that $ar_i \not\equiv ar_j \pmod{m}$ if $i \neq j$. But Theorem 2.3, part 2, shows that $ar_i \equiv ar_j \pmod{m}$ implies $r_i \equiv r_j \pmod{m}$ and hence $i = j$.

Theorem 2.7 *Fermat's theorem. Let p denote a prime. If $p \nmid a$ then $a^{p-1} \equiv 1 \pmod{p}$. For every integer a, $a^p \equiv a \pmod{p}$.*

We shall postpone the proof of this theorem and shall obtain it as a corollary to Theorem 2.8.

Theorem 2.8 *Euler's generalization of Fermat's theorem. If $(a, m) = 1$, then*

$$a^{\phi(m)} \equiv 1 \pmod{m}.$$

Proof Let $r_1, r_2, \cdots, r_{\phi(m)}$ be a reduced residue system modulo m. Then by Theorem 2.6, $ar_1, ar_2, \cdots, ar_{\phi(m)}$ is also a reduced residue system modulo m. Hence, corresponding to each r_i there is one and only one ar_j such that $r_i \equiv ar_j \pmod{m}$. Furthermore, different r_i will have different corresponding ar_j. This means that the numbers $ar_1, ar_2, \cdots, ar_{\phi(m)}$ are just the residues modulo m of $r_1, r_2, \cdots, r_{\phi(m)}$, but not necessarily in the same order. Multiplying and using Theorem 2.1, part 4, we obtain

$$\prod_{j=1}^{\phi(m)} (ar_j) \equiv \prod_{i=1}^{\phi(m)} r_i \pmod{m},$$

and hence

$$a^{\phi(m)} \prod_{j=1}^{\phi(m)} r_j \equiv \prod_{j=1}^{\phi(m)} r_j \pmod{m}.$$

Now $(r_j, m) = 1$, so we can use Theorem 2.3, part 2, to cancel the r_j and we obtain $a^{\phi(m)} \equiv 1 \pmod{m}$.

Proof of Theorem 2.7 If $p \nmid a$, then $(a, p) = 1$ and $a^{\phi(p)} \equiv 1 \pmod{p}$. To find $\phi(p)$, we refer to Theorem 2.5. All the integers $1, 2, \cdots, p-1, p$ with the exception of p are relatively prime to p. Thus we have $\phi(p) = p - 1$, and the first part of Fermat's theorem follows. The second part is now obvious.

Theorem 2.9 *If $(a, m) = 1$ then there is an x such that $ax \equiv 1 \pmod{m}$. Any two such x are congruent* (mod m). *If $(a, m) > 1$ then there is no such x.*

Proof If $(a, m) = 1$, then there exist x and y such that $ax + my = 1$. That is, $ax \equiv 1 \pmod{m}$. Conversely, if $ax \equiv 1 \pmod{m}$, then there is a y such that $ax + my = 1$, so that $(a, m) = 1$. Thus if $ax_1 \equiv ax_2 \equiv 1 \pmod{m}$, then $(a, m) = 1$, and it follows from part 2 of Theorem 2.3 that $x_1 \equiv x_2 \pmod{m}$.

The relation $ax \equiv 1 \pmod{m}$ is equivalent to the assertion that the residue class $x \pmod{m}$ is the multiplicative inverse of the residue class $a \pmod{m}$. To avoid confusion with the rational number $a^{-1} = 1/a$, we denote this residue class by $\bar{a} \pmod{m}$. The value of $\bar{a}$ is quickly found by employing the Euclidean algorithm, as described in Section 1.2. The existence of $\bar{a}$ is also evident from Theorem 2.6, for if $(a, m) = 1$, then the numbers $a, 2a, \ldots, ma$ form a complete system of residues, which is to say that one of them is $\equiv 1 \pmod{m}$. In addition, the existence of $\bar{a}$ can also be inferred from Theorem 2.8, by taking $\bar{a} = a^{\phi(m)-1}$.

Lemma 2.10 *Let p be a prime number. Then $x^2 \equiv 1 \pmod{p}$ if and only if $x \equiv \pm 1 \pmod{p}$.*

In Section 2.7 we establish a more general result (Theorem 2.26) from which the foregoing is easily derived, but we give a direct proof now, since this observation has many useful applications.

Proof This quadratic congruence may be expressed as $x^2 - 1 \equiv 0 \pmod{p}$. That is, $(x-1)(x+1) \equiv 0 \pmod{p}$, which is to say that

$p|(x-1)(x+1)$. By Theorem 1.15 it follows that $p|(x-1)$ or $p|(x+1)$. Equivalently, $x \equiv 1 \pmod p$ or $x \equiv -1 \pmod p$. Conversely, if either one of these latter congruences holds, then $x^2 \equiv 1 \pmod p$.

Theorem 2.11 *Wilson's theorem. If p is a prime, then $(p-1)! \equiv -1 \pmod p$.*

Proof If $p = 2$ or $p = 3$, the congruence is easily verified. Thus we may assume that $p \geqslant 5$. Suppose that $1 \leqslant a \leqslant p-1$. Then $(a, p) = 1$, so that by Theorem 2.9 there is a unique integer $\bar{a}$ such that $1 \leqslant \bar{a} \leqslant p-1$ and $a\bar{a} \equiv 1 \pmod p$. By a second application of Theorem 2.9 we find that if $\bar{a}$ is given then there is exactly one a, $1 \leqslant a \leqslant p-1$, such that $a\bar{a} \equiv 1 \pmod p$. Thus a and $\bar{a}$ form a pair whose combined contribution to $(p-1)!$ is $\equiv 1 \pmod p$. However, a little care is called for because it may happen that $a = \bar{a}$. This is equivalent to the assertion that $a^2 \equiv 1 \pmod p$, and by Lemma 2.10 we see that this is in turn equivalent to $a = 1$ or $a = p-1$. That is, $\bar{1} = 1$ and $\overline{p-1} = p-1$, but if $2 \leqslant a \leqslant p-2$ then $\bar{a} \neq a$. By pairing these latter residues in this manner we find that $\prod_{a=2}^{p-2} a \equiv 1 \pmod p$, so that $(p-1)! = 1 \cdot (\prod_{a=2}^{p-2} a) \cdot (p-1) \equiv -1 \pmod p$.

We give a second proof of Wilson's theorem in our remarks following Corollary 2.30 in Section 2.7, and a third proof is outlined in Problem 22 of Section 2.8.

Theorem 2.12 *Let p denote a prime. Then $x^2 \equiv -1 \pmod p$ has solutions if and only if $p = 2$ or $p \equiv 1 \pmod 4$.*

Proof If $p = 2$ we have the solution $x = 1$.

For any odd prime p, we can write Wilson's theorem in the form

$$\left(1 \cdot 2 \cdots j \cdots \frac{p-1}{2}\right)\left(\frac{p+1}{2} \cdots (p-j) \cdots (p-2)(p-1)\right)$$
$$\equiv -1 \pmod p.$$

The product on the left has been divided into two parts, each with the same number of factors. Pairing off j in the first half with $p-j$ in the second half, we can rewrite the congruence in the form

$$\prod_{j=1}^{(p-1)/2} j(p-j) \equiv -1 \pmod p.$$

But $j(p-j) \equiv -j^2 \pmod p$, and so the above is

$$\equiv \prod_{j=1}^{(p-1)/2} (-j^2) \equiv (-1)^{(p-1)/2} \left(\prod_{j=1}^{(p-1)/2} j \right)^2 \pmod p.$$

If $p \equiv 1 \pmod 4$ then the first factor on the right is 1, and we see that $x = \left(\frac{p-1}{2}\right)!$ is a solution of $x^2 \equiv -1 \pmod p$.

Suppose, conversely, that there is an x such that $x^2 \equiv -1 \pmod p$. We note that for such an x, $p \nmid x$. We suppose that $p > 2$, and raise both sides of the congruence to the power $(p-1)/2$ to see that

$$(-1)^{(p-1)/2} \equiv (x^2)^{(p-1)/2} = x^{p-1} \pmod p.$$

By Fermat's congruence, the right side here is $\equiv 1 \pmod p$. The left side is ± 1, and since $-1 \not\equiv 1 \pmod p$, we deduce that

$$(-1)^{(p-1)/2} = 1.$$

Thus $(p-1)/2$ is even; that is, $p \equiv 1 \pmod 4$.

In case $p \equiv 1 \pmod 4$, we have explicitly constructed a solution of the congruence $x^2 \equiv -1 \pmod p$. However, the amount of calculation required to evaluate $\left(\frac{p-1}{2}\right)! \pmod p$ is no smaller than would be required by exhaustively testing $x = 2, x = 3, \cdots, x = (p-1)/2$. In Section 2.9 we develop a method by which the desired x can be quickly determined.

Theorem 2.12 provides the key piece of information needed to determine which integers can be written as the sum of the squares of two integers. We begin by showing that a certain class of prime numbers can be represented in this manner.

Lemma 2.13 *If p is a prime number and $p \equiv 1 \pmod 4$, then there exist positive integers a and b such that $a^2 + b^2 = p$.*

This was first stated in 1632 by Albert Girard, on the basis of numerical evidence. The first proof was given by Fermat in 1654.

Proof Let p be a prime number, $p \equiv 1 \pmod 4$. By Theorem 2.12 we know that there exists an integer x such that $x^2 \equiv -1 \pmod p$. Define $f(u, v) = u + xv$, and $K = [\sqrt{p}]$. Since $\sqrt{p}$ is not an integer, it follows that

$K < \sqrt{p} < K + 1$. We consider pairs (u, v) of integers for which $0 \leqslant u \leqslant K$ and $0 \leqslant v \leqslant K$. Since u and v each take on $K + 1$ values, we have $(K + 1)^2$ pairs. Since $K + 1 > \sqrt{p}$, the number of pairs is $> p$. If we consider $f(u, v) \pmod{p}$, we have more numbers under consideration than we have residue classes to put them in, so there must be some residue class that contains the number $f(u, v)$ for two different pairs (u, v). (This is known as the *pigeonhole principle*, which we discuss in greater detail in Section 4.5.) Suppose, for example, that (u_1, v_1) and (u_2, v_2) are distinct pairs with coordinates in the interval $[0, K]$, for which $f(u_1, v_1) \equiv f(u_2, v_2) \pmod{p}$. That is, $u_1 + xv_1 \equiv u_2 + xv_2 \pmod{p}$, which gives $(u_1 - u_2) \equiv -x(v_1 - v_2) \pmod{p}$. Take $a = u_1 - u_2$ and $b = v_1 - v_2$. Then $a \equiv -xb \pmod{p}$, and on squaring both sides we see that $a^2 \equiv (-xb)^2 \equiv x^2b^2 \equiv -b^2 \pmod{p}$ since $x^2 \equiv -1 \pmod{p}$. That is, $a^2 + b^2 \equiv 0 \pmod{p}$, which is to say that $p \mid (a^2 + b^2)$. Since the ordered pair (u_1, v_1) is distinct from the pair (u_2, v_2), it follows that not both a and b vanish, so that $a^2 + b^2 > 0$. On the other hand, $u_1 \leqslant K$ and $u_2 \geqslant 0$, so that $a = u_1 - u_2 \leqslant K$. Similarly, we may show that $a \geqslant -K$, and in the same manner that $-K \leqslant b \leqslant K$. But $K < \sqrt{p}$, so this gives $|a| < \sqrt{p}$ and $|b| < \sqrt{p}$. On squaring these inequalities we find that $a^2 < p$ and $b^2 < p$, which gives $a^2 + b^2 < 2p$. Thus altogether we have shown that $0 < a^2 + b^2 < 2p$ and that $p \mid (a^2 + b^2)$. But the only multiple of p in the interval $(0, 2p)$ is p, so we conclude that $a^2 + b^2 = p$.

We now establish a similar result in the converse direction.

Lemma 2.14 *Let q be a prime factor of $a^2 + b^2$. If $q \equiv 3 \pmod{4}$ then $q \mid a$ and $q \mid b$.*

Proof We prove the contrapositive, that is, that if q does not divide both a and b then $q \not\equiv 3 \pmod{4}$. By interchanging a and b, if necessary, we may suppose that $(a, q) = 1$. Let $\bar{a}$ be chosen so that $a\bar{a} \equiv 1 \pmod{q}$. We multiply both sides of the congruence $a^2 \equiv -b^2 \pmod{q}$ by $\bar{a}^2$ to see that $1 \equiv (a\bar{a})^2 \equiv -(b\bar{a})^2 \pmod{q}$. Thus if $x = b\bar{a}$ then x is a solution of the congruence $x^2 \equiv -1 \pmod{q}$, and by Theorem 2.12 it follows that $q = 2$ or $q \equiv 1 \pmod{4}$.

Theorem 2.15 *Fermat. Write the canonical factorization of n in the form*

$$n = 2^\alpha \prod_{p \equiv 1(4)} p^\beta \prod_{q \equiv 3(4)} q^\gamma.$$

Then n can be expressed as a sum of two squares of integers if and only if all the exponents γ are even.

Proof We note that the identity

$$(a^2 + b^2)(c^2 + d^2) = (ac - bd)^2 + (ad - bc)^2$$

holds for any real numbers. In particular, it follows that if m and n are both sums of two squares then mn is also a sum of two squares. The prime number $2 = 1^2 + 1^2$ is a sum of two squares, and every prime number $p \equiv 1 \pmod 4$ is a sum of two squares. If q is a prime number, $q \equiv 3 \pmod 4$, then $q^2 = q^2 + 0^2$ is a sum of two squares. Hence any number that may be expressed as a product of 2's, p's, and q^2's is a sum of two squares. Conversely, suppose that n is a sum of two squares, say $n = a^2 + b^2$. If q is a prime number, $q \equiv 3 \pmod 4$, for which $\gamma > 0$, then $q|n$, and by Lemma 2.14 it follows that $q|a$ and $q|b$, which implies that $q^2|n$. That is, $\gamma \geqslant 2$, and we may write $n/q^2 = (a/q)^2 + (b/q)^2$. By applying this same argument to n/q^2 we discover that if $\gamma > 2$ then $\gamma \geqslant 4$ and that $q^2|a$ and $q^2|b$. Since this process must terminate, we conclude that γ must be even, and additionally that $q^{\gamma/2}|a$ and $q^{\gamma/2}|b$.

This theorem of Fermat is the first of many similar such theorems. The object of constructing a coherent theory of quadratic forms was the primary influence on research in number theory for several centuries. The first step in the theory is to generalize Theorem 2.12. This is accomplished in the law of quadratic reciprocity, which we study in the initial sections of Chapter 3. With this tool in hand, we develop some of the fundamentals concerning quadratic forms in the latter part of Chapter 3. In Section 3.6 we apply the general theory to sums of two squares, to give not only a second proof of Theorem 2.15, but also some further results.

PROBLEMS

1. List all integers x in the range $1 \leqslant x \leqslant 100$ that satisfy $x \equiv 7 \pmod{17}$.
2. Exhibit a complete residue system modulo 17 composed entirely of multiples of 3.
3. Exhibit a reduced residue system for the modulus 12; for 30.
4. If an integer x is even, observe that it must satisfy the congruence $x \equiv 0 \pmod 2$. If an integer y is odd, what congruence does it satisfy? What congruence does an integer z of the form $6k + 1$ satisfy?
5. Write a single congruence that is equivalent to the pair of congruences $x \equiv 1 \pmod 4$, $x \equiv 2 \pmod 3$.
6. Prove that if p is a prime and $a^2 \equiv b^2 \pmod p$, then $p|(a + b)$ or $p|(a - b)$.

7. Show that if $f(x)$ is a polynomial with integral coefficients and if $f(a) \equiv k \pmod{m}$, then $f(a + tm) \equiv k \pmod{m}$ for every integer t.
8. Prove that any number that is a square must have one of the following for its units digit: 0, 1, 4, 5, 6, 9.
9. Prove that any fourth power must have one of 0, 1, 5, 6 for its units digit.
10. Evaluate $\phi(m)$ for $m = 1, 2, 3, \cdots, 12$.
11. Find the least positive integer x such that $13 | (x^2 + 1)$.
12. Prove that 19 is not a divisor of $4n^2 + 4$ for any integer n.
13. Exhibit a reduced residue system modulo 7 composed entirely of powers of 3.
14. Show that $7 | (3^{2n+1} + 2^{n+2})$ for all n.
15. Find integers $a_1, \cdots, a_5$ such that every integer x satisfies at least one of the congruences $x \equiv a_1 \pmod{2}$, $x \equiv a_2 \pmod{3}$, $x \equiv a_3 \pmod{4}$, $x \equiv a_4 \pmod{6}$, $x \equiv a_5 \pmod{12}$.
16. Illustrate the proof of Theorem 2.11 for $p = 11$ and $p = 13$ by actually determining the pairs of associated integers.
17. Show that $61! + 1 \equiv 63! + 1 \equiv 0 \pmod{71}$.
18. Show that if $p \equiv 3 \pmod{4}$, then $\left(\frac{p-1}{2}\right)! \equiv \pm 1 \pmod{p}$.
19. Prove that $n^6 - 1$ is divisible by 7 if $(n, 7) = 1$.
20. Prove that $n^7 - n$ is divisible by 42, for any integer n.
21. Prove that $n^{12} - 1$ is divisible by 7 if $(n, 7) = 1$.
22. Prove that $n^{6k} - 1$ is divisible by 7 if $(n, 7) = 1$, k being any positive integer.
23. Prove that $n^{13} - n$ is divisible by 2, 3, 5, 7 and 13 for any integer n.
24. Prove that $n^{12} - a^{12}$ is divisible by 13 if n and a are prime to 13.
25. Prove that $n^{12} - a^{12}$ is divisible by 91 if n and a are prime to 91.
26. Show that the product of three consecutive integers is divisible by 504 if the middle one is a cube.
27. Prove that $\frac{1}{5}n^5 + \frac{1}{3}n^3 + \frac{7}{15}n$ is an integer for every integer n.
28. What is the last digit in the ordinary decimal representation of 3^{400}? (H)
29. What is the last digit in the ordinary decimal representation of 2^{400}?
30. What are the last two digits in the ordinary decimal representation of 3^{400}? (H)
31. Show that $-(m-1)/2, -(m-3)/2, \cdots, (m-3)/2, (m-1)/2$ is a complete residue system modulo m if m is odd, and that $-(m-2)/2, -(m-4)/2, \cdots, (m-2)/2, m/2$ is a complete residue system modulo m if m is even.

32. Show that $2, 4, 6, \cdots, 2m$ is a complete residue system modulo m if m is odd.

33. Show that $1^2, 2^2, \cdots, m^2$ is not a complete residue system modulo m if $m > 2$.

34. Show that an integer $m > 1$ is a prime if and only if m divides $(m - 1)! + 1$.

35. If n is composite, prove that $(n - 1)! + 1$ is not a power of n.

36. If p is a prime, prove that $(p - 1)! + 1$ is a power of p if and only if $p = 2$, 3, or 5. (H)

37. Show that there exist infinitely many n such that $n! + 1$ is divisible by at least two distinct primes.

38. Prove that there are infinitely many primes of the form $4n + 1$. (H)

39. If a and b are real numbers such that $a^2 = b^2$, it is well known that $a = b$ or $a = -b$. Give an example to show that if $a^2 \equiv b^2 \pmod{m^2}$ for integers a, b and $m \geqslant 2$, it does not necessarily follow that $a \equiv b \pmod{m}$ or $a \equiv -b \pmod{m}$.

40. For m odd, prove that the sum of the elements of any complete residue system modulo m is congruent to zero modulo m; prove the analogous result for any reduced residue system for $m > 2$.

41. Find all sets of positive integers a, b, c satisfying all three congruences $a \equiv b \pmod{c}$, $b \equiv c \pmod{a}$, $c \equiv a \pmod{b}$. (H)

42. Find all triples a, b, c of nonzero integers such that $a \equiv b \pmod{|c|}$, $b \equiv c \pmod{|a|}$, $c \equiv a \pmod{|b|}$.

43. If p is an odd prime, prove that:

$$1^2 \cdot 3^2 \cdot 5^2 \cdots (p-2)^2 \equiv (-1)^{(p+1)/2} \pmod{p}$$

and

$$2^2 \cdot 4^2 \cdot 6^2 \cdots (p-1)^2 \equiv (-1)^{(p+1)/2} \pmod{p}.$$

***44.** Show that if p is prime then $\binom{p-1}{k} \equiv (-1)^k \pmod{p}$ for $0 \leqslant k \leqslant p - 1$.

***45.** Show that if p is prime then $\binom{p}{k} \equiv 0 \pmod{p}$ for $1 \leqslant k \leqslant p - 1$.

***46.** For any prime p, if $a^p \equiv b^p \pmod{p}$, prove that $a^p \equiv b^p \pmod{p^2}$.

47. If $r_1, r_2, \cdots, r_{p-1}$ is any reduced residue system modulo a prime p, prove that

$$\prod_{j=1}^{p-1} r_j \equiv -1 \pmod{p}.$$

***48.** If $r_1, r_2, \cdots, r_p$ and $r'_1, r'_2, \cdots, r'_p$ are any two complete residue systems modulo a prime $p > 2$, prove that the set $r_1r'_1, r_2r'_2, \cdots, r_pr'_p$ cannot be a complete residue system modulo p.

***49.** If p is any prime other than 2 or 5, prove that p divides infinitely many of the integers $9, 99, 999, 9999, \cdots$. If p is any prime other than 2 or 5, prove that p divides infinitely many of the integers $1, 11, 111, 1111, \cdots$.

***50.** Given a positive integer n, prove that there is a positive integer m that to base ten contains only the digits 0 and 1 such that $n|m$. Prove that the same holds for digits 0 and 2, or 0 and $3, \cdots,$ or 0 and 9, but for no other pair of digits.

51. Prove that $(p-1)! \equiv p - 1 \pmod{1 + 2 + \cdots + (p-1)}$ if p is a prime.

***52.** Show that if p is prime then $p|((p-2)! - 1)$, but that if $p > 5$ then $(p-2)! - 1$ is not a power of p. (H)

53. Show that there are infinitely many n such that $n! - 1$ is divisible by at least two distinct primes.

54. (*a*) Noting the factoring $341 = 11 \cdot 31$, verify that $2^5 \equiv 1 \pmod{31}$ and hence that $2^{341} \equiv 2 \pmod{341}$, but that $3^{341} \not\equiv 3 \pmod{341}$. (*b*) Using the factoring $561 = 3 \cdot 11 \cdot 17$, prove that $a^{561} \equiv a \pmod{561}$ holds for every integer a.

Remarks A composite integer m such that $a^{m-1} \equiv 1 \pmod{m}$ is called a *pseudoprime to the base a*. There are infinitely many pseudoprimes to the base 2 (see Problem 19 in Section 2.4), 341 and 561 being the smallest two. A composite integer m that is a pseudoprime to base a whenever $(a, m) = 1$ is called a *Carmichael number*, the smallest being 561. All Carmichael numbers $< 10^{13}$ are known. It is not known that there are infinitely many, but it is conjectured that for any $\varepsilon > 0$ there is an $x_0(\varepsilon)$ such that if $x > x_0(\varepsilon)$, then the number of Carmichael numbers not exceeding x is $> x^{1-\varepsilon}$.

***55.** Let $A = [a_{ij}]$ and $B = [b_{ij}]$ be two $n \times n$ matrices with integral entries. Show that if $a_{ij} \equiv b_{ij} \pmod{m}$ for all i, j, then $\det(A) \equiv \det(B) \pmod{m}$. Show that

$$\det \begin{bmatrix} 4771 & 1452 & 8404 & 3275 & 9163 \\ 6573 & 8056 & 7312 & 2265 & 3639 \\ 9712 & 2574 & 4612 & 4321 & 7196 \\ 8154 & 2701 & 6007 & 2147 & 7465 \\ 2158 & 7602 & 5995 & 2327 & 8882 \end{bmatrix} \neq 0. \qquad \text{(H)}$$

***56.** Let p be a prime number, and suppose that x is an integer such that $x^2 \equiv -2 \pmod{p}$. By considering the numbers $u + xv$ for various pairs (u, v), show that at least one of the equations $a^2 + 2b^2 = p$, $a^2 + 2b^2 = 2p$ has a solution.

57. Show that $(a + b\sqrt{-2})(c + d\sqrt{-2}) = (ac - 2bd) + (bc + ad)\sqrt{-2}$. Thus or otherwise show that $(a^2 + 2b^2)(c^2 + 2d^2) = (ac - 2bd)^2 + 2(bc + ad)^2$.

58. Show that if p is an odd prime and $a^2 + 2b^2 = 2p$, then a is even and b is odd. Deduce that $(2b)^2 + 2a^2 = 4p$, and hence that $b^2 + 2(a/2)^2 = p$.

59. Let p be a prime factor of $a^2 + 2b^2$. Show that if p does not divide both a and b then the congruence $x^2 \equiv -2 \pmod{p}$ has a solution.

60. Combine the results of the foregoing problems to show that a prime number p can be expressed in the form $a^2 + 2b^2$ if and only if the congruence $x^2 \equiv -2 \pmod{p}$ is solvable. (In Chapter 3 we show that this congruence is solvable if and only if $p = 2$ or $p \equiv 1$ or $3 \pmod{8}$.)

2.2 SOLUTIONS OF CONGRUENCES

In analogy with the solution of algebraic equations it is natural to consider the problem of solving a congruence. In the rest of this chapter we shall let $f(x)$ denote a polynomial with integral coefficients, and we shall write $f(x) = a_n x^n + a_{n-1} x^{n-1} + \cdots + a_0$. If u is an integer such that $f(u) \equiv 0 \pmod{m}$, then we say that u is a solution of the congruence $f(x) \equiv 0 \pmod{m}$. Whether or not an integer is a solution of a congruence depends on the modulus m as well as on the polynomial $f(x)$. If the integer u is a solution of $f(x) \equiv 0 \pmod{m}$, and if $v \equiv u \pmod{m}$, Theorem 2.2 shows that v is also a solution. Because of this we shall say that $x \equiv u \pmod{m}$ is a solution of $f(x) \equiv 0 \pmod{m}$, meaning that every integer congruent to u modulo m satisfies $f(x) \equiv 0 \pmod{m}$.

For example, the congruence $x^2 - x + 4 \equiv 0 \pmod{10}$ has the solution $x = 3$ and also the solution $x = 8$. It also has the solutions $x = 13$, $x = 18$, and all other numbers obtained from 3 and 8 by adding and subtracting 10 as often as we wish. In counting the number of solutions of a congruence, we restrict attention to a complete residue system belonging to the modulus. In the example $x^2 - x + 4 \equiv 0 \pmod{10}$, we say that there are two solutions because $x = 3$ and $x = 8$ are the only numbers among $0, 1, 2, \cdots, 9$ that are solutions. The two solutions can be written in equation form, $x = 3$ and $x = 8$, or in congruence form, $x \equiv 3 \pmod{10}$ and $x \equiv 8 \pmod{10}$. As a second example, the congruence $x^2 - 7x + 2 \equiv 0 \pmod{10}$ has exactly four solutions $x = 3, 4, 8, 9$. The reason for counting the number of solutions in this way is that if $f(x) \equiv 0 \pmod{m}$ has a solution $x = a$, then it follows that all integers x satisfying $x \equiv a \pmod{m}$

are automatically solutions, so this entire congruence class is counted as a single solution.

Definition 2.4 *Let $r_1, r_2, \cdots, r_m$ denote a complete residue system modulo m. The number of solutions of $f(x) \equiv 0 \pmod{m}$ is the number of the r_i such that $f(r_i) \equiv 0 \pmod{m}$.*

It is clear from Theorem 2.2 that the number of solutions is independent of the choice of the complete residue system. Furthermore, the number of solutions cannot exceed the modulus m. If m is small it is a simple matter to just compute $f(r_i)$ for each of the r_i and thus to determine the number of solutions. In the foregoing example the congruence has just two solutions. Some other examples are

$$x^2 + 1 \equiv 0 \pmod{7} \text{ has no solution,}$$

$$x^2 + 1 \equiv 0 \pmod{5} \text{ has two solutions,}$$

$$x^2 - 1 \equiv 0 \pmod{8} \text{ has four solutions.}$$

Definition 2.5 *Let $f(x) = a_n x^n + a_{n-1} x^{n-1} + \cdots + a_0$. If $a_n \not\equiv 0 \pmod{m}$ the degree of the congruence $f(x) \equiv 0 \pmod{m}$ is n. If $a_n \equiv 0 \pmod{m}$, let j be the largest integer such that $a_j \not\equiv 0 \pmod{m}$; then the degree of the congruence is j. If there is no such integer j, that is, if all the coefficients of $f(x)$ are multiples of m, no degree is assigned to the congruence.*

It should be noted that the degree of the congruence $f(x) \equiv 0 \pmod{m}$ is not the same thing as the degree of the polynomial $f(x)$. The degree of the congruence depends on the modulus; the degree of the polynomial does not. Thus if $g(x) = 6x^3 + 3x^2 + 1$, then $g(x) \equiv 0 \pmod{5}$ is of degree 3, and $g(x) \equiv 0 \pmod{2}$ is of degree 2, whereas $g(x)$ is of degree 3.

Theorem 2.16 *If $d \mid m$, $d > 0$, and if u is a solution of $f(x) \equiv 0 \pmod{m}$, then u is a solution of $f(x) \equiv 0 \pmod{d}$.*

Proof This follows directly from Theorem 2.1, part 5.

There is a distinction made in the theory of algebraic equations that has an analogue for congruences. A conditional equation, such as $x^2 - 5x + 6 = 0$, is true for only certain values of x, namely $x = 2$ and $x = 3$. An identity or identical equation, such as $(x - 2)^2 = x^2 - 4x + 4$, holds for all real numbers x, or for all complex numbers for that matter.

Similarly, we say that $f(x) \equiv 0 \pmod{m}$ is an *identical congruence* if it holds for all integers x. If $f(x)$ is a polynomial all of whose coefficients are divisible by m, then $f(x) \equiv 0 \pmod{m}$ is an identical congruence. A different type of identical congruence is illustrated by $x^p \equiv x \pmod{p}$, true for all integers x by Fermat's theorem.

Before considering congruences of higher degree, we first describe the solutions in the linear case.

Theorem 2.17 *Let a, b, and $m > 0$ be given integers, and put $g = (a, m)$. The congruence $ax \equiv b \pmod{m}$ has a solution if and only if $g|b$. If this condition is met, then the solutions form an arithmetic progression with common difference m/g, giving g solutions* $\pmod{m}$.

Proof The question is whether there exist integers x and y such that $ax + my = b$. Since g divides the left side, for such integers to exist we must have $g|b$. Suppose that this condition is met, and write $a = g\alpha$, $b = g\beta$, $m = g\mu$. Then by the first part of Theorem 2.3, the desired congruence holds if and only if $\alpha x \equiv \beta \pmod{\mu}$. Here $(\alpha, \mu) = 1$ by Theorem 1.7, so by Theorem 2.9 there is a unique number $\bar{\alpha} \pmod{\mu}$ such that $\alpha\bar{\alpha} \equiv 1 \pmod{\mu}$. On multiplying through by $\bar{\alpha}$, we find that $x \equiv \bar{\alpha}\alpha x \equiv \bar{\alpha}\beta \pmod{\mu}$. Thus the set of integers x for which $ax \equiv b \pmod{m}$ is precisely the arithmetic progression of numbers of the form $\bar{\alpha}\beta + k\mu$. If we allow k to take on the values $0, 1, \ldots, g - 1$, we obtain g values of x that are distinct $\pmod{m}$. All other values of x are congruent $\pmod{m}$ to one of these, so we have precisely g solutions.

Since $\bar{\alpha}$ can be located by an application of the Euclidean algorithm, the solutions are easily found.

PROBLEMS

1. If $f(x) \equiv 0 \pmod{p}$ has exactly j solutions with p a prime, and $g(x) \equiv 0 \pmod{p}$ has no solution, prove that $f(x)g(x) \equiv 0 \pmod{p}$ has exactly j solutions.
2. Denoting the number of solutions of $f(x) \equiv k \pmod{m}$ by $N(k)$, prove that $\sum_{k=1}^{m} N(k) = m$.
3. If a polynomial congruence $f(x) \equiv 0 \pmod{m}$ has m solutions, prove that any integer whatsoever is a solution.
4. The fact that the product of any three consecutive integers is divisible by 3 leads to the identical congruence $x(x + 1)(x + 2) \equiv 0 \pmod{3}$. Generalize this, and write an identical congruence modulo m.

5. Find all solutions of the congruences

(a) $20x \equiv 4 \pmod{30}$;
(b) $20x \equiv 30 \pmod{4}$;
(c) $353x \equiv 254 \pmod{400}$;
(d) $57x \equiv 87 \pmod{105}$;
(e) $64x \equiv 83 \pmod{105}$:
(f) $589x \equiv 209 \pmod{817}$;
(g) $49x \equiv 5000 \pmod{999}$.

6. How many solutions are there to each of the following congruences:

(a) $15x \equiv 25 \pmod{35}$;
(b) $15x \equiv 24 \pmod{35}$;
(c) $15x \equiv 0 \pmod{35}$?

7. If a is selected at random from $1, 2, 3, \cdots, 14$, and b is selected at random from $1, 2, 3, \cdots, 15$, what is the probability that $ax \equiv b \pmod{15}$ has at least one solution? Exactly one solution?

8. Show that if p is an odd prime then the congruence $x^2 \equiv 1 \pmod{p^\alpha}$ has only the two solutions $x \equiv 1$, $x \equiv -1 \pmod{p^\alpha}$.

9. Show that the congruence $x^2 \equiv 1 \pmod{2^\alpha}$ has one solution when $\alpha = 1$, two solutions when $\alpha = 2$, and precisely the four solutions $1, 2^{\alpha-1} - 1, 2^{\alpha-1} + 1, -1$ when $\alpha \geqslant 3$.

10. Show that if p is an odd prime then the number of solutions (ordered pairs) of the congruence $x^2 - y^2 \equiv a \pmod{p}$ is $p - 1$ unless $a \equiv 0 \pmod{p}$, in which case the number of solutions is $2p - 1$. (H)

11. Suppose $(a, m) = 1$, and let x_1 denote a solution of $ax \equiv 1 \pmod{m}$. For $s = 1, 2, \cdots$, let $x_s = 1/a - (1/a)(1 - ax_1)^s$. Prove that x_s is an integer and that it is a solution of $ax \equiv 1 \pmod{m^s}$.

***12.** Suppose that $(a, m) = 1$. If $a = \pm 1$, the solution of $ax \equiv 1 \pmod{m^s}$ is obviously $x \equiv a \pmod{m^s}$. If $a = \pm 2$, then m is odd and $x \equiv \frac{1}{2}(1 - m^s)\frac{1}{2}a \pmod{m^s}$ is the solution of $ax \equiv 1 \pmod{m^s}$. For all other a use Problem 11 to show that the solution of $ax \equiv 1 \pmod{m^s}$ is $x \equiv k \pmod{m^s}$ where k is the nearest integer to $-(1/a)(1 - ax_1)^s$.

13. Solve $3x \equiv 1 \pmod{125}$ by Problem 12, taking $x_1 = 2$.

***14.** Show that $\binom{p^\alpha}{k} \equiv 0 \pmod{p}$ for $0 < k < p^\alpha$. (H)

***15.** Show that $\binom{p^\alpha - 1}{k} \equiv (-1)^k \pmod{p}$ for $0 \leqslant k \leqslant p^\alpha - 1$. (H)

***16.** Show that if r is a non-negative integer then all coefficients of the polynomial $(1 + x)^{2^r} - (1 + x^{2^r})$ are even. Write a positive integer n in binary, $n = \sum_{r \in \mathscr{R}} 2^r$. Show that all coefficients of the polynomial $(1 + x)^n - \prod_{r \in \mathscr{R}} (1 + x^{2^r})$ are even. Write $k = \sum_{s \in \mathscr{S}} 2^s$ in binary. Show that $\binom{n}{k}$ is odd if and only if $\mathscr{S} \subseteq \mathscr{R}$. Conclude that if n is given, then $\binom{n}{k}$ is odd for precisely $2^{w(n)}$ values of k, where $w(n)$, called the *binary weight* of n, is the number of 1's in the binary expansion of n. In symbols, $w(n) = \operatorname{card}(\mathscr{R})$.

Note This is a special case of a result of E. Lucas, proved in 1891. See N. J. Fine, "Binomial coefficients modulo a prime," *Amer. Math. Monthly*, 54 (1947), 589–592.

***17.** Let the numbers c_i be defined by the power series identity

$$(1 + x + \cdots + x^{p-1})/(1 - x)^{p-1} = 1 + c_1 x + c_2 x^2 + \cdots .$$

Show that $c_i \equiv 0 \pmod p$ for all $i \geqslant 1$.

2.3 THE CHINESE REMAINDER THEOREM

We now consider the important problem of solving simultaneous congruences. The simplest case of this is to find those x (if there are any) that satisfy the simultaneous congruences

$$\begin{aligned} x &\equiv a_1 \pmod{m_1}, \\ x &\equiv a_2 \pmod{m_2}, \\ &\vdots \\ x &\equiv a_r \pmod{m_r}. \end{aligned} \tag{2.1}$$

This is the subject of the next result, called the *Chinese Remainder Theorem* because the method was known in China in the first century A.D.

Theorem 2.18 *The Chinese Remainder Theorem. Let $m_1, m_2, \cdots, m_r$ denote r positive integers that are relatively prime in pairs, and let $a_1, a_1, \cdots, a_r$ denote any r integers. Then the congruences* (2.1) *have common solutions. If x_0 is one such solution, then an integer x satisfies the congruences* (2.1) *if and only if x is of the form $x = x_0 + km$ for some integer k. Here $m = m_1 m_2 \cdots m_r$.*

Using the terminology introduced in the previous section, the last assertion of the Theorem would be expressed by saying that the solution x is unique modulo $m_1 m_2 \cdots m_r$.

Proof Writing $m = m_1 m_2 \cdots m_r$, we see that m/m_j is an integer and that $(m/m_j, m_j) = 1$. Hence by Theorem 2.9 for each j there is an integer

b_j such that $(m/m_j)b_j \equiv 1 \pmod{m_j}$. Clearly $(m/m_j)b_j \equiv 0 \pmod{m_i}$ if $i \neq j$. Put

$$x_0 = \sum_{j=1}^{r} \frac{m}{m_j} b_j a_j. \tag{2.2}$$

We consider this number modulo m_i, and find that

$$x_0 \equiv \frac{m}{m_i} b_i a_i \equiv a_i \pmod{m_i}.$$

Thus x_0 is a solution of the system (2.1).

If x_0 and x_1 are two solutions of the system (2.1), then $x_0 \equiv x_1 \pmod{m_i}$ for $i = 1, 2, \cdots, r$, and hence $x_0 \equiv x_1 \pmod{m}$ by part 3 of Theorem 2.3. This completes the proof.

Example 1 Find the least positive integer x such that $x \equiv 5 \pmod 7$, $x \equiv 7 \pmod{11}$, and $x \equiv 3 \pmod{13}$.

Solution We follow the proof of the theorem, taking $a_1 = 5$, $a_2 = 7$, $a_3 = 3$, $m_1 = 7$, $m_2 = 11$, $m_3 = 13$, and $m = 7 \cdot 11 \cdot 13 = 1001$. Now $(m_2 m_3, m_1) = 1$, and indeed by the Euclidean algorithm we find that $(-2) \cdot m_2 m_3 + 21 \cdot m_1 = 1$, so we may take $b_1 = -2$. Similarly, we find that $4 \cdot m_1 m_3 + (-33) \cdot m_2 = 1$, so we take $b_2 = 4$. By the Euclidean algorithm a third time we find that $(-1) \cdot m_1 m_2 + 6 \cdot m_3 = 1$, so we may take $b_3 = -1$. Then by (2.2) we see that $11 \cdot 13 \cdot (-2) \cdot 5 + 7 \cdot 13 \cdot 4 \cdot 7 + 7 \cdot 11 \cdot (-1) \cdot 3 = 887$ is a solution. Since this solution is unique modulo m, this is the only solution among the numbers $1, 2, \cdots, 1001$. Thus 887 is the least positive solution.

In the Chinese Remainder Theorem, the hypothesis that the moduli m_j should be pairwise relatively prime is absolutely essential. When this hypothesis fails, the existence of a solution x of the simultaneous system (2.1) is no longer guaranteed, and when such an x does exist, we see from Part 3 of Theorem 2.3 that it is unique modulo $[m_1, m_2, \cdots, m_r]$, not modulo m. In case there is no solution of (2.1), we call the system *inconsistent*. In the following two examples we explore some of the possibilities that arise when the m_j are allowed to have common factors. An extension of the Chinese Remainder Theorem to the case of unrestricted m_j is laid out in Problems 19–23.

Example 2 Show that there is no x for which both $x \equiv 29 \pmod{52}$ and $x \equiv 19 \pmod{72}$.

Solution Since $52 = 4 \cdot 13$, we see by Part 3 of Theorem 2.3 that the first congruence is equivalent to the simultaneous congruences $x \equiv 29 \pmod 4$ and $x \equiv 29 \pmod{13}$, which reduces to $x \equiv 1 \pmod 4$ and $x \equiv 3 \pmod{13}$. Similarly, $72 = 8 \cdot 9$, and the second congruence given is equivalent to the simultaneous congruences $x \equiv 19 \pmod 8$ and $x \equiv 19 \pmod 9$. These reduce to $x \equiv 3 \pmod 8$ and $x \equiv 1 \pmod 9$. By the Chinese Remainder Theorem we know that the constraints (mod 13) and (mod 9) are independent of those (mod 8). The given congruences are inconsistent because there is no x for which both $x \equiv 1 \pmod 4$ and $x \equiv 3 \pmod 8$.

Once an inconsistency has been identified, a brief proof can be constructed: The first congruence implies that $x \equiv 1 \pmod 4$ while the second congruence implies that $x \equiv 3 \pmod 4$.

Example 3 Determine whether the system $x \equiv 3 \pmod{10}$, $x \equiv 8 \pmod{15}$, $x \equiv 5 \pmod{84}$ has a solution, and find them all, if any exist.

First Solution We factor each modulus into prime powers. By Part 3 of Theorem 2.3, we see that the first congruence of the system is equivalent to the two simultaneous congruences $x \equiv 3 \pmod 2$, $x \equiv 3 \pmod 5$. Similarly, the second congruence of the system is equivalent to the two conditions $x \equiv 8 \pmod 3$, $x \equiv 8 \pmod 5$, while the third congruence is equivalent to the three congruences $x \equiv 5 \pmod 4$, $x \equiv 5 \pmod 3$, $x \equiv 5 \pmod 7$. The new system of seven simultaneous congruences is equivalent to the ones given, but now all moduli are prime powers. We consider the powers of 2 first. The two conditions are $x \equiv 3 \pmod 2$ and $x \equiv 1 \pmod 4$. These two are consistent, but the second one implies the first, so that the first one may be dropped. The conditions modulo 3 are $x \equiv 8 \pmod 3$ and $x \equiv 5 \pmod 3$. These are equivalent, and may be expressed as $x \equiv 2 \pmod 3$. Third, the conditions modulo 5 are $x \equiv 3 \pmod 5$, $x \equiv 8 \pmod 5$. These are equivalent, so we drop the second of them. Finally, we have the condition $x \equiv 5 \pmod 7$. Hence our system of seven congruences is equivalent to the four conditions $x \equiv 1 \pmod 4$, $x \equiv 2 \pmod 3$, $x \equiv 3 \pmod 5$, and $x \equiv 5 \pmod 7$. Here the moduli are relatively prime in pairs, so we may apply the formula (2.2) used in the proof of the Chinese Remainder Theorem. Proceeding as in the solution of Example 1, we find that x satisfies the given congruences if and only if $x \equiv 173 \pmod{420}$.

The procedure we employed here provides useful insights concerning the way that conditions modulo powers of the same prime must mesh, but when the numbers involved are large, it requires a large amount of computation (because the moduli must be factored). A superior method is provided by the iterative use of Theorem 2.17. This avoids the need to factor the moduli, and requires only $r - 1$ applications of the Euclidean algorithm.

Second Solution The x that satisfy the third of the given congruences are precisely those x of the form $5 + 84u$ where u is an integer. On substituting this into the second congruence, we see that the requirement is that $5 + 84u \equiv 8 \pmod{15}$. That is, $84u \equiv 3 \pmod{15}$. By the Euclidean algorithm we find that $(84, 15) = 3$, and indeed we find that $2 \cdot 84 + (-11) \cdot 15 = 3$. By Theorem 2.17 we deduce that u is a solution of the congruence if and only if $u \equiv 2 \pmod 5$. That is, u is of the form $u = 2 + 5v$, and hence x satisfies both the second and the third of the given congruences if and only if x is of the form $5 + 84(2 + 5v) = 173 + 420v$. The first congruence now requires that $173 + 420v \equiv 3 \pmod{10}$. That is, $420v \equiv -170 \pmod{10}$. By the Euclidean algorithm we find that $(420, 10) = 10$. Since $10 \mid 170$, we deduce that this congruence holds for all v. That is, in this example, any x that satisfies the second and third of the given congruences also satisfies the first. The set of solutions consists of those x of the form $173 + 420v$. That is, $x \equiv 173 \pmod{420}$.

This procedure can be applied to general systems of the sort (2.1). In case the system is inconsistent, the inconsistency is revealed by a failure of the condition $g \mid b$ in Theorem 2.17. Alternatively, if it happens that the moduli are pairwise relatively prime, then $g = 1$ in each application of Theorem 2.17, and we obtain a second (less symmetric) proof of the Chinese Remainder Theorem.

Returning to Theorem 2.18, we take a fixed set of positive integers $m_1, m_2, \cdots, m_r$, relatively prime in pairs, with product m. But instead of considering just one set of equations (2.1), we consider all possible systems of this type. Thus a_1 may be any integer in a complete residue system modulo m_1, a_2 any integer in a complete residue system modulo m_2, and so on. To be specific, let us consider a_1 to be any integer among $1, 2, \cdots, m_1$, and a_2 any integer among $1, 2, \cdots, m_2, \cdots$, and a_r any integer among $1, 2, \cdots, m_r$. The number of such r-tuples $(a_1, a_2, \cdots, a_r)$ is $m_1 m_2 \cdots m_r = m$. By the Chinese Remainder Theorem, each r-tuple determines precisely one residue class x modulo m. Moreover, distinct r-tuples determine different residue classes. To see this, suppose that $(a_1, a_2, \cdots, a_r) \neq (a'_1, a'_2, \cdots, a'_r)$. Then $a_i \neq a'_i$ for some i, and we see that no integer x satisfies both the congruences $x \equiv a_i \pmod{m_i}$ and $x \equiv a'_i \pmod{m_i}$.

Thus we have a one-to-one correspondence between the r-tuples $(a_1, a_2, \cdots, a_r)$ and a complete residue system modulo m, such as the integers $1, 2, \cdots, m$. It is perhaps not surprising that two sets, each having

m elements, can be put into one-to-one correspondence. However, this correspondence is particularly natural, and we shall draw some important consequences from it.

For any positive integer n let $\mathscr{C}(n)$ denote the complete residue system $\mathscr{C}(n) = \{1, 2, \cdots, n\}$. The r-tuples we have considered are precisely the members of the *Cartesian product* (or *direct product*) of the sets $\mathscr{C}(m_1)$, $\mathscr{C}(m_2), \cdots, \mathscr{C}(m_r)$. In symbols, this Cartesian product is denoted $\mathscr{C}(m_1) \times \mathscr{C}(m_2) \times \cdots \times \mathscr{C}(m_r)$. For example, if $\mathbb{R}$ denotes the set of real numbers, then $\mathbb{R} \times \mathbb{R}$, abbreviated $\mathbb{R}^2$, describes the ordinary Euclidean plane with the usual rectangular coordinates belonging to any point (x, y). In this notation, we may express the one-to-one correspondence in question by writing

$$\mathscr{C}(m_1) \times \mathscr{C}(m_2) \times \cdots \times \mathscr{C}(m_r) \leftrightarrow \mathscr{C}(m).$$

Example 4 Exhibit the foregoing one-to-one correspondence explicitly, when $m_1 = 7$, $m_2 = 9$, $m = 63$.

Solution Consider the following matrix with 7 rows and 9 columns. At the intersection of the ith row and jth column we place the element c_{ij}, where $c_{ij} \equiv i \pmod 7$ and $c_{ij} \equiv j \pmod 9$. According to Theorem 2.18 we can select the element c_{ij} from the complete residue system $\mathscr{C}(63) = \{1, 2, \cdots, 63\}$. Thus the element 40, for example, is at the intersection of the fifth row and the fourth column, because $40 \equiv 5 \pmod 7$ and $40 \equiv 4 \pmod 9$. Note that the element 41 is at the intersection of the sixth row and fifth column, since $41 \equiv 6 \pmod 7$ and $41 \equiv 5 \pmod 9$. Thus the element $c + 1$ in the matrix is just southeast from the element c, allowing for periodicity when c is in the last row or column. For example, 42 is in the last row, so 43 turns up in the first row, one column later. Similarly, 45 is in the last column, so 46 turns up in the first column, one row lower. This gives us an easy way to construct the matrix: just write 1 in the c_{11} position and proceed downward and to the right with 2, 3, and so on.

1	**29**	57	**22**	**50**	15	**43**	**8**	36
37	**2**	30	**58**	**23**	51	**16**	**44**	9
10	**38**	3	**31**	**59**	24	**52**	**17**	45
46	**11**	39	**4**	**32**	60	**25**	**53**	18
19	**47**	12	**40**	**5**	33	**61**	**26**	54
55	**20**	48	**13**	**41**	6	**34**	**62**	27
28	56	21	49	14	42	7	35	63

Here the correspondence between the pair (i, j) and the entry c_{ij} provides a solution to the problem.

In the matrix, the entry c_{ij} is entered in boldface if $(c_{ij}, 63) = 1$. We note that these entries are precisely those for which i is one of the numbers $\{1, 2, \cdots, 6\}$, and j is one of the numbers $\{1, 2, 4, 5, 7, 8\}$. That is, $(c_{ij}, 63) = 1$ if and only if $(i, 7) = 1$ and $(j, 9) = 1$. Since there are exactly 6 such i, and for each such i there are precisely 6 such j, we deduce that $\phi(63) = 36 = \phi(7)\phi(9)$. We now show that this holds in general, and we derive a formula for $\phi(m)$ in terms of the prime factorization of m.

Theorem 2.19 *If m_1 and m_2 denote two positive, relatively prime integers, then $\phi(m_1 m_2) = \phi(m_1)\phi(m_2)$. Moreover, if m has the canonical factorization $m = \prod p^\alpha$, then $\phi(m) = \prod_{p|m} (p^\alpha - p^{\alpha-1}) = m \prod_{p|m} (1 - 1/p)$.*

If $m = 1$, then the products are empty, and by convention an empty product has value 1. Thus the formula gives $\phi(1) = 1$ in this case, which is correct.

Proof Put $m = m_1 m_2$, and suppose that $(x, m) = 1$. By reducing x modulo m_1 we see that there is a unique $a_1 \in \mathscr{C}(m_1)$ for which $x \equiv a_1 \pmod{m_1}$. Here, as before, $\mathscr{C}(m_1)$ is the complete system of residues $\mathscr{C}(m_1) = \{1, 2, \cdots, m_1\}$. Similarly, there is a unique $a_2 \in \mathscr{C}(m_2)$ for which $x \equiv a_2 \pmod{m_2}$. Since $(x, m_1) = 1$, it follows by Theorem 2.4 that $(a_1, m_1) = 1$. Similarly $(a_2, m_2) = 1$. For any positive integer n, let $\mathscr{R}(n)$ be the system of reduced residues formed of those numbers $a \in \mathscr{C}(n)$ for which $(a, n) = 1$. That is, $\mathscr{R}(n) = \{a \in \mathscr{C}(n) \colon (a, n) = 1\}$. Thus we see that any $x \in \mathscr{R}(m)$ gives rise to a pair (a_1, a_2) with $a_i \in \mathscr{R}(m_i)$ for $i = 1, 2$. Suppose, conversely, that we start with such a pair. By the Chinese Remainder Theorem (Theorem 2.18) there exists a unique $x \in \mathscr{C}(m)$ such that $x \equiv a_i \pmod{m_i}$ for $i = 1, 2$. Since $(a_1, m_1) = 1$ and $x \equiv a_1 \pmod{m_1}$, it follows by Theorem 2.4 that $(x, m_1) = 1$. Similarly we find that $(x, m_2) = 1$, and hence $(x, m) = 1$. That is, $x \in \mathscr{R}(m)$. In this way we see that the Chinese Remainder Theorem enables us to establish a one-to-one correspondence between the reduced residue classes modulo m and pairs of reduced residue classes modulo m_1 and m_2, provided that $(m_1, m_2) = 1$. Since $a_1 \in \mathscr{R}(m_1)$ can take any one of $\phi(m_1)$ values, and $a_2 \in \mathscr{R}(m_2)$ can take any one of $\phi(m_2)$ values, there are $\phi(m_1)\phi(m_2)$ pairs, so that $\phi(m) = \phi(m_1)\phi(m_2)$.

We have now established the first identity of the theorem. If $m = \prod p^\alpha$ is the canonical factorization of m, then by repeated use of this identity we see that $\phi(m) = \prod \phi(p^\alpha)$. To complete the proof it remains to determine the value of $\phi(p^\alpha)$. If a is one of the p^α numbers $1, 2, \cdots, p^\alpha$, then $(a, p^\alpha) = 1$ unless a is one of the $p^{\alpha-1}$ numbers $p, 2p, \cdots, p^{\alpha-1} \cdot p$. On subtracting, we deduce that the number of reduced residue classes modulo p^α is $p^\alpha - p^{\alpha-1} = p^\alpha(1 - 1/p)$. This gives the stated formulae.

We shall derive further properties of Euler's ϕ-function in Sections 4.2, 4.3, and an additional proof of the formula for $\phi(n)$ will be given in Section 4.5, by means of the inclusion–exclusion principle of combinatorial mathematics.

Let $f(x)$ denote a polynomial with integral coefficients, and let $N(m)$ denote the number of solutions of the congruence $f(x) \equiv 0 \pmod{m}$ as counted in Definition 2.4. We suppose that $m = m_1 m_2$, where $(m_1, m_2) = 1$. By employing the same line of reasoning as in the foregoing proof, we show that the roots of the congruence $f(x) \equiv 0 \pmod{m}$ are in one-to-one correspondence with pairs (a_1, a_2) in which a_1 runs over all roots of the congruence $f(x) \equiv 0 \pmod{m_1}$ and a_2 runs over all roots of the congruence $f(x) \equiv 0 \pmod{m_2}$. In this way we are able to relate $N(m)$ to $N(m_1)$ and $N(m_2)$.

Theorem 2.20 *Let $f(x)$ be a fixed polynomial with integral coefficients, and for any positive integer m let $N(m)$ denote the number of solutions of the congruence $f(x) \equiv 0 \pmod{m}$. If $m = m_1 m_2$ where $(m_1, m_2) = 1$, then $N(m) = N(m_1)N(m_2)$. If $m = \prod p^\alpha$ is the canonical factorization of m, then $N(m) = \prod N(p^\alpha)$.*

The possibility that one or more of the $N(p^\alpha)$ may be 0 is not excluded in this formula. Indeed, from Theorem 2.16 we see that if $d \mid m$ and $N(d) = 0$, then $N(m) = 0$. One immediate consequence of this is that the congruence $f(x) \equiv 0 \pmod{m}$ has solutions if and only if it has solutions (mod p^α) for each prime-power p^α exactly dividing m.

Proof Suppose that $x \in \mathscr{C}(m)$, where $\mathscr{C}(m)$ is the complete residue system $\mathscr{C}(m) = \{1, 2, \cdots, m\}$. If $f(x) \equiv 0 \pmod{m}$ and $m = m_1 m_2$, then by Theorem 2.16 it follows that $f(x) \equiv 0 \pmod{m_1}$. Let a_1 be the unique member of $\mathscr{C}(m_1) = \{1, 2, \ldots, m_1\}$ for which $x \equiv a_1 \pmod{m_1}$. By Theorem 2.2 it follows that $f(a_1) \equiv 0 \pmod{m_1}$. Similarly, there is a unique $a_2 \in \mathscr{C}(m_2)$ such that $x \equiv a_2 \pmod{m_2}$, and $f(a_2) \equiv 0 \pmod{m_2}$. Thus for each solution of the congruence modulo m we construct a pair (a_1, a_2) in which a_i is a solution of the congruence modulo m_i, for $i = 1, 2$. Thus far we have not used the hypothesis that m_1 and m_2 are relatively prime. It is in the converse direction that this latter hypothesis becomes vital.

Suppose now that $m = m_1 m_2$, where $(m_1, m_2) = 1$, and that for $i = 1$ and 2, numbers $a_i \in \mathscr{C}(m_i)$ are chosen so that $f(a_i) \equiv 0 \pmod{m_i}$. By the Chinese Remainder Theorem (Theorem 2.18), there is a unique $x \in \mathscr{C}(m)$ such that $x \equiv a_i \pmod{m_i}$ for $i = 1, 2$. By Theorem 2.2 we see that this x is a solution of the congruence $f(x) \equiv 0 \pmod{m_i}$, for $i = 1, 2$. Then by Part 3 of Theorem 2.3 we conclude that $f(x) \equiv 0 \pmod{m}$. We have now established a one-to-one correspondence between the solutions x of the congruence modulo m and pairs (a_1, a_2) of solutions modulo m_1 and m_2, respectively. Since a_1 runs over $N(m_1)$ values, and a_2 runs over $N(m_2)$ values, there are $N(m_1)N(m_2)$ such pairs, and we have the first assertion of the theorem. The second assertion follows by repeated application of the first part.

Example 5 Let $f(x) = x^2 + x + 7$. Find all roots of the congruence $f(x) \equiv 0 \pmod{15}$.

Solution Trying the values $x = 0, \pm 1, \pm 2$, we find that $f(x) \equiv 0 \pmod 5$ has no solution. Since $5 \mid 15$, it follows that there is no solution (mod 15).

Example 6 Let $f(x)$ be as in Example 5. Find all roots of $f(x) \equiv 0 \pmod{189}$, given that $189 = 3^3 \cdot 7$, that the roots (mod 27) are 4, 13, and 22, and that the roots (mod 7) are 0 and 6.

Solution In a situation of this kind it is more efficient to proceed as we did in the solution of Example 1, rather than employ the method adopted in the second solution of Example 3. By the Euclidean algorithm and (2.2), we find that $x \equiv a_1 \pmod{27}$ and $x \equiv a_2 \pmod 7$ if and only if $x \equiv 28a_1 - 27a_2 \pmod{189}$. We let a_1 take on the three values 4, 13, and 22, while a_2 takes on the values 0 and 6. Thus we obtain the six solutions $x \equiv 13, 49, 76, 112, 139, 175 \pmod{189}$.

We have now reduced the problem of locating the roots of a polynomial congruence modulo m to the case in which the modulus is a prime power. In Section 2.6 we reduce this further, to the case of a prime modulus, and finally in Section 2.7 we consider some of the special properties of congruences modulo a prime number p.

PROBLEMS

1. Find the smallest positive integer (except $x = 1$) that satisfies the following congruences simultaneously: $x \equiv 1 \pmod 3$, $x \equiv 1 \pmod 5$, $x \equiv 1 \pmod 7$.
2. Find all integers that satisfy simultaneously: $x \equiv 2 \pmod 3$, $x \equiv 3 \pmod 5$, $x \equiv 5 \pmod 2$.
3. Solve the set of congruences: $x \equiv 1 \pmod 4$, $x \equiv 0 \pmod 3$, $x \equiv 5 \pmod 7$.

4. Find all integers that give the remainders 1, 2, 3 when divided by 3, 4, 5, respectively.

5. Solve Example 2 using the technique that was applied to Example 4.

6. Solve Example 1 by the method used in the second solution of Example 3.

7. Determine whether the congruences $5x \equiv 1 \pmod 6$, $4x \equiv 13 \pmod{15}$ have a common solution, and find them if they exist.

8. Find the smallest positive integer giving remainders 1, 2, 3, 4, and 5 when divided by 3, 5, 7, 9, and 11, respectively.

9. For what values of n is $\phi(n)$ odd?

10. Find the number of positive integers $\leqslant 3600$ that are prime to 3600.

11. Find the number of positive integers $\leqslant 3600$ that have a factor greater than 1 in common with 3600.

12. Find the number of positive integers $\leqslant 7200$ that are prime to 3600.

13. Find the number of positive integers $\leqslant 25200$ that are prime to 3600. (Observe that $25200 = 7 \times 3600$.)

14. Solve the congruences:

$$x^3 + 2x - 3 \equiv 0 \pmod 9;$$

$$x^3 + 2x - 3 \equiv 0 \pmod 5;$$

$$x^3 + 2x - 3 \equiv 0 \pmod{45}.$$

15. Solve the congruence $x^3 + 4x + 8 \equiv 0 \pmod{15}$.

16. Solve the congruence $x^3 - 9x^2 + 23x - 15 \equiv 0 \pmod{503}$ by observing that 503 is a prime and that the polynomial factors into $(x - 1)(x - 3)(x - 5)$.

17. Solve the congruence $x^3 - 9x^2 + 23x - 15 \equiv 0 \pmod{143}$.

18. Given any positive integer k, prove that there are k consecutive integers each divisible by a square > 1.

19. Let $m_1, m_2, \cdots, m_r$ be relatively prime in pairs. Assuming that each of the congruences $b_i x \equiv a_i \pmod{m_i}$, $i = 1, 2, \cdots, r$, is solvable, prove that the congruences have a simultaneous solution.

20. Let m_1 and m_2 be arbitrary positive integers, and let a_1 and a_2 be arbitrary integers. Show that there is a simultaneous solution of the congruences $x \equiv a_1 \pmod{m_1}$, $x \equiv a_2 \pmod{m_2}$, if and only if $a_1 \equiv a_2 \pmod g$, where $g = (m_1, m_2)$. Show that if this condition is met, then the solution is unique modulo $[m_1, m_2]$.

***21.** Let p be a prime number, and suppose that $m_j = p^{\alpha_j}$ in (2.1), where $1 \leqslant \alpha_1 \leqslant \alpha_2 \leqslant \cdots \leqslant \alpha_r$. Show that the system has a simultaneous solution if and only if $a_i \equiv a_r \pmod{p^{\alpha_i}}$ for $i = 1, 2, \cdots, r$.

***22.** Let the m_j be as in the preceding problem. Show that the system (2.1) has a simultaneous solution if and only if $a_i \equiv a_j \pmod{p^{\alpha_i}}$ for all pairs of indices i, j for which $1 \leqslant i < j \leqslant r$.

***23.** Let the m_j be arbitrary positive integers in (2.1). Show that there is a simultaneous solution of this system if and only if $a_i \equiv a_j \pmod{(m_i, m_j)}$ for all pairs of the indices i, j for which $1 \leqslant i < j \leqslant r$.

***24.** Suppose that $m_1, m_2, \cdots, m_r$ are pairwise relatively prime positive integers. For each j, let $\mathscr{C}(m_j)$ denote a complete system of residues modulo m_j. Show that the numbers $c_1 + c_2m_1 + c_3m_1m_2 + \cdots + c_rm_1m_2 \cdots m_{r-1}$, $c_j \in \mathscr{C}(m_j)$, form a complete system of residues modulo $m = m_1m_2 \cdots m_r$.

25. If m and k are positive integers, prove that the number of positive integers $\leqslant mk$ that are prime to m is $k\phi(m)$.

26. Show that $\phi(nm) = n\phi(m)$ if every prime that divides n also divides m.

27. If P denotes the product of the primes common to m and n, prove that $\phi(mn) = P\phi(m)\phi(n)/\phi(P)$. Hence if $(m, n) > 1$, prove $\phi(mn) > \phi(m)\phi(n)$.

28. If $\phi(m) = \phi(mn)$ and $n > 1$, prove that $n = 2$ and m is odd.

29. Characterize the set of positive integers n satisfying $\phi(2n) = \phi(n)$.

30. Characterize the set of positive integers satisfying $\phi(2n) > \phi(n)$.

31. Prove that there are infinitely many integers n so that $3 \nmid \phi(n)$.

32. Find all solutions x of $\phi(x) = 24$.

33. Find the smallest positive integer n so that $\phi(x) = n$ has no solution; exactly two solutions; exactly three solutions; exactly four solutions. (It has been conjectured that there is no integer n such that $\phi(x) = n$ has exactly one solution, but this is an unsolved problem.)

34. Prove that there is no solution of the equation $\phi(x) = 14$ and that 14 is the least positive even integer with this property. Apart from 14, what is the next smallest positive even integer n such that $\phi(x) = n$ has no solution?

35. If n has k distinct odd prime factors, prove that $2^k | \phi(n)$.

36. What are the last two digits, that is, the tens and units digits, of 2^{1000}? of 3^{1000}? (H)

***37.** Let $a_1 = 3$, $a_{i+1} = 3^{a_i}$. Describe this sequence (mod 100).

***38.** Let $(a, b) = 1$ and $c > 0$. Prove that there is an integer x such that $(a + bx, c) = 1$.

39. Prove that for a fixed integer n the equation $\phi(x) = n$ has only a finite number of solutions.

40. Prove that for $n \geqslant 2$ the sum of all positive integers less than n and prime to n is $n\phi(n)/2$.

***41.** Define $f(n)$ as the sum of the positive integers less than n and prime to n. Prove that $f(m) = f(n)$ implies that $m = n$.

***42.** Find all positive integers n such that $\phi(n)|n$.

***43.** If $d|n$ and $0 < d < n$, prove that $n - \phi(n) > d - \phi(d)$.

***44.** Prove the following generalization of Euler's theorem:

$$a^m \equiv a^{m-\phi(m)} \pmod{m}$$

for any integer a.

***45.** Find the number of solutions of $x^2 \equiv x \pmod{m}$ for any positive integer m.

***46.** Let $\psi(n)$ denote the number of integers a, $1 \leqslant a \leqslant n$, for which both $(a, n) = 1$ and $(a + 1, n) = 1$. Show that $\psi(n) = n\prod_{p|n}(1 - 2/p)$. For what values of n is $\psi(n) = 0$?

***47.** Let $f(x)$ be a polynomial with integral coefficients, let $N(m)$ denote the number of solutions of the congruence $f(x) \equiv 0 \pmod{m}$, and let $\phi_f(m)$ denote the number of integers a, $1 \leqslant a \leqslant m$, such that $(f(a), m) = 1$. Show that if $(m, n) = 1$ then $\phi_f(mn) = \phi_f(m)\phi_f(n)$. Show that if $\alpha > 1$ then $\phi_f(p^\alpha) = p^{\alpha-1}\phi_f(p)$. Show that $\phi_f(p) = p - N(p)$. Conclude that for any positive integer n, $\phi_f(n) = n\prod_{p|n}(1 - N(p)/p)$. Show that for an appropriate choice of $f(x)$, this reduces to Theorem 2.19.

2.4 TECHNIQUES OF NUMERICAL CALCULATION

When investigating properties of integers, it is often instructive to examine a few examples. The underlying patterns may be more evident if one extends the numerical data by the use of a programmable calculator or electronic computer. For example, after considering a long list of those odd primes p for which the congruence $x^2 \equiv 2 \pmod{p}$ has a solution, one might arrive at the conjecture that it is precisely those primes that are congruent to ± 1 modulo 8. (This is true, and forms an important part of quadratic reciprocity, proved in Section 3.2.) By extending the range of the calculation, one may provide further evidence in favor of a conjecture. Computers are also useful in constructing proofs. For example, one might formulate an argument to show that there is a particular number n_0 such that if $n > n_0$, then n is not divisible by all numbers less than $\sqrt{n}$ (recall Problem 50 in Section 1.3). Then by direct calculation one might show that this is also true if n lies in the interval $24 < n \leqslant n_0$, in order to conclude that 24 is the largest number divisible by all numbers less than its square root. In this example, it is not hard to show that one may take $n_0 = 210$, and hence one might check the intermediate range by hand, but in other cases of this kind the n_0 may be very large, making a computer essential.

We assume that our calculators and computers perform integer arithmetic accurately, as long as the integers involved have at most d digits. We refer to d as the *word length*. This assumption applies not only to addition, subtraction, and multiplication, but also to division, provided that the resulting quotient is also an integer. That is, if $a|b$, the computer will accurately find b/a, with no round-off error. We also assume that our computer has a facility for determining the integral part $[x]$ of a real number. Thus in the division algorithm, $b = qa + r$, the computer will accurately find $q = [b/a]$. Use of the fractional part $\{x\} = x - [x]$ should be avoided, since in general the decimal (or binary) expansion of $\{x\}$ will not terminate, with the result that the computer will provide only an approximation to this function. In particular, as we indicated earlier, the remainder in the division algorithm should be calculated as $r = b - a[b/a]$, not as $r = a\{b/a\}$.

We have noted that the Euclidean algorithm does not require many steps. Indeed, when it is applied to very large numbers, the main constraint is the time involved in performing accurate multiple-precision arithmetic. The Euclidean algorithm provides a very efficient means of locating the solutions of linear congruences, and also of finding the root in the Chinese remainder theorem. Since the Euclidean algorithm has so many applications, it is worth spending some effort to optimize it. One way of improving the Euclidean algorithm is to form q_{i+1} by rounding to the nearest integer, rather than rounding down. The resulting r_i is generally smaller, although it may be negative. This modified form of the Euclidean algorithm requires fewer iterations to determine (b, c), but the order of magnitude is still usually $\log c$ when $b > c$. Example 3 of Section 1.2 required 24 iterations, but with the modified algorithm only 15 would be needed. (*Warning*: The integral part function conveniently provided on most machines rounds toward 0. That is, when asked for the integer part of a decimal (or binary) number $\pm a_k a_{k-1} \cdots a_0 . b_1 b_2 \cdots b_r$, the machine will return $\pm a_k a_{k-1} \cdots a_0$. This is $[x]$ when x is non-negative, but it is $-[-x]$ when x is negative. For example, $[-3.14159] = -4$, but the machine will round toward 0, giving an answer -3. To avoid this trap, ensure that a number is non-negative before asking a machine to give you the integer part. Alternatively, one could employ a conditional instruction, "Put $y = \text{int}(x)$. If $y > x$, then replace y by $y - 1$." This has the effect of setting $y = [x]$.)

In performing congruence arithmetic, we observe that if $0 \leqslant a < m$ and $0 \leqslant b < m$ then either $a + b$ is already reduced or else $m \leqslant a + b <$

$2m$, in which case $a + b - m$ is reduced. To calculate $ab \pmod m$, we may set $c = ab$, and then reduce $c \pmod m$. However, c may be as large as $(m-1)^2$, which means that if we are limited to integers $< 10^d$ then we can calculate $ab \pmod m$ in this way only for $m < 10^{d/2}$, that is, half the word length. The sensible solution to this problem is to employ multiple-precision arithmetic, but in the short term one may instead use an algorithm such as that described in Problem 21 at the end of this section.

Another situation in which we may introduce a modest saving is in the evaluation of a polynomial $f(x) = a_n x^n + a_{n-1}x^{n-1} + \cdots + a_0$. The naive approach would involve constructing the sequence of powers x^k, and as one does so, forming the partial sums $a_0, a_0 + a_1 x, \cdots$, until one arrives at $f(x)$. This requires n additions and $2n - 1$ multiplications. A more efficient process is suggested by observing that

$$f(x) = (\cdots((a_n x + a_{n-1})x + a_{n-2})x + \cdots)x + a_0.$$

Here we still have n additions, but now only n multiplications. This procedure is known as *Horner's method.*

A much greater saving can be introduced when computing a power a^k, when k is large. The naive approach would involve $k - 1$ multiplications. This is fine if k is small, but for large k one should repeatedly square to form the sequence of numbers $d_j = a^{2^j}$. Writing the binary expansion of k in the form $k = \sum_{j \in \mathscr{J}} 2^j$, we see that $a^k = \prod_{j \in \mathscr{J}} d_j$. Here the number of multiplications required is of the order of magnitude $\log k$, a great savings if k is large. This procedure can be made still more efficient if the machine in use automatically converts numbers to binary, for then the binary digits of k can be accessed, rather than computed. It might seem at first that this device is of limited utility. After all, if a^k is encountered in the context of real arithmetic, one would simply compute $\exp(k \log a)$. Even if a and k are integers, one is unlikely to examine a^k when k is large, unless one is willing to perform multiple-precision arithmetic. However, this device is extremely useful when computing $a^k \pmod m$.

Example 7 Determine the value of $999^{179} \pmod{1763}$.

Solution We find that $179 = 1 + 2 + 2^4 + 2^5 + 2^7$, that $999^2 \equiv 143 \pmod{1763}$, $999^4 \equiv 143^2 \equiv 1056 \pmod{1763}$, $999^8 \equiv 1056^2 \equiv 920 \pmod{1763}$, $999^{16} \equiv 920^2 \equiv 160 \pmod{1763}$, $999^{32} \equiv 160^2 \equiv 918 \pmod{1763}$, $999^{64} \equiv 918^2 \equiv 10 \pmod{1763}$, so that $999^{128} \equiv 10^2 \equiv 100 \pmod{1763}$. Hence $999^{179} \equiv 999 \cdot 143 \cdot 160 \cdot 918 \cdot 100 \equiv 54 \cdot 160 \cdot 918 \cdot 100 \equiv 1588 \cdot 918 \cdot 100 \equiv 1546 \cdot 100 \equiv 1219 \pmod{1763}$.

When implemented, it would be a mistake to first list the binary digits of k, then form a list of the numbers d_j, and finally multiply the appropriate d_j together, as we have done above. Instead, one should perform these three tasks concurrently, as follows:

1. Set $x = 1$. (Here x is the product being formed.)

2. While $k > 0$, repeat the following steps:

(*a*) Set $e = k - 2[k/2]$. (Thus $e = 0$ or 1, according as k is even or odd.)

(*b*) If $e = 1$ then replace x by ax, and reduce this (mod m). (If $e = 0$ then x is not altered.)

(*c*) Replace a by a^2, and reduce this (mod m).

(*d*) Replace k by $(k - e)/2$. (i.e., drop the unit digit in the binary expansion, and shift the remaining digits one place to the right.)

When this is completed, we see that $x \equiv a^k \pmod{m}$.

Our ability to evaluate $a^k \pmod{m}$ quickly can be applied to provide an easy means of establishing that a given number is composite.

Example 8 Show that 1763 is composite.

Solution By Fermat's congruence, if p is an odd prime number then $2^{p-1} \equiv 1 \pmod{p}$. In other words, if n is an odd number for which $2^{n-1} \not\equiv 1 \pmod{n}$, then n is composite. We calculate that $2^{1762} \equiv 742 \pmod{1763}$, and deduce that 1763 is composite. Alternatively, we might search for a divisor of 1763, but the use we have made here of Fermat's congruence provides a quicker means of establishing compositeness when n is large, provided, of course that the test succeeds. Since the empirical evidence is that the test detects most composite numbers, if $2^{n-1} \equiv 1 \pmod{n}$ then we call n a *probable prime to the base* 2. A composite probable prime is called a *pseudoprime*. That such numbers exist is seen in the following example.

Example 9 Show that 1387 is composite.

Solution We may calculate that $2^{1386} \equiv 1 \pmod{1387}$. Thus 1387 is a probable prime to the base 2. To demonstrate that it is composite, we may try a different base, but a more efficient procedure is provided by applying Lemma 2.10. We have a number $x = 2^{693}$ with the property that $x^2 \equiv 1 \pmod{1387}$. Since $2^{693} \equiv 512 \not\equiv \pm 1 \pmod{1387}$, we conclude that 1387 is composite.

When used systematically, this technique yields the *strong pseudoprime test*. If we wish to show that an odd number m is composite, we divide $m - 1$ by 2 repeatedly, in order to write $m - 1 = 2^j d$, with d odd. We form $a^d \pmod{m}$, and by repeatedly squaring and reducing, we construct the numbers

$$a^d, a^{2d}, a^{4d}, \cdots, a^{2^j d} \pmod{m}.$$

If the last number here is $\not\equiv 1 \pmod{m}$, then m is composite. If this last member is $\equiv 1 \pmod{m}$, then m is a probable prime to the base a, but if the entry immediately preceding the first 1 is $\not\equiv -1 \pmod{m}$, then we may still conclude (by Lemma 2.10) that m is composite. When this test is inconclusive, we call m a *strong probable prime*. An odd, composite, strong probable prime is called a *strong pseudoprime to the base* a, abbreviated spsp(a). Such numbers exist, but numerical evidence suggests that they are much rarer than pseudoprimes. In our remarks following Problem 54 in Section 2.1, we noted the existence of numbers m, called Carmichael numbers, which are pseudoprime to every base a that is relatively prime to m. Such a phenomenon does not persist with strong pseudoprimes, as it can be shown that if m is odd and composite then m is a spsp(a) for at most $m/4$ values of $a \pmod{m}$. For most m, the number of such a is much smaller. Expressed as an algorithm, the strong pseudoprime test for m takes the following shape:

1. Find j and d with d odd, so that $m - 1 = 2^j d$.
2. Compute $a^d \pmod{m}$. If $a^d \equiv \pm 1 \pmod{m}$, then m is a strong probable prime; stop.
3. Square a^d to compute $a^{2d} \pmod{m}$. If $a^{2d} \equiv 1 \pmod{m}$, then m is composite; stop. If $a^{2d} \equiv -1$, then m is a strong probable prime; stop.
4. Repeat step 3 with a^{2d} replaced by $a^{4d}, a^{8d}, \cdots, a^{2^{j-1}d}$.
5. If the procedure has not already terminated, then m is composite.

Let $X = 25 \cdot 10^9$. Integers in the interval $[1, X]$ have been examined in detail, and it has been found that the number of prime numbers in this interval is $\pi(X) = 1{,}091{,}987{,}405$, that the number of odd pseudoprimes in this interval is 21,853, and that the number of Carmichael numbers in this interval is 2163. On the other hand, in this interval there are 4842 numbers of the class spsp(2), 184 that are both spsp(2) and spsp(3), 13 that are spsp(a) for $a = 2, 3, 5$, only 1 that is spsp(a) for $a = 2, 3, 5, 7$, and none that is spsp(a) for $a = 2, 3, 5, 7, 11$.

The strong pseudoprime test provides a very efficient means for proving that an odd integer m is composite. With further information one can sometimes use it to demonstrate that a number is prime. If m is a strong probable prime base 2, and if $m < 2047$, then m is prime. Here 2047 is the least spsp (2). If m is larger, apply the test to the base 3. If m is again found to be a strong probable prime, then m is prime provided that $m < 1{,}373{,}653$. This latter number is the least integer that is both spsp (2) and spsp (3). If m is larger, then apply the test to the base 5. If m is yet again found to be a strong probable prime, then m is prime provided that $m < 25{,}326{,}001$. This is the least number that is simultaneously spsp(a) for $a = 2, 3$, and 5. If m is still larger, then apply the test to the base 7. If m is once more found to be a probable prime, then m is prime provided that $m < X = 25 \cdot 10^9$ and that $m \neq 3{,}215{,}031{,}751$. This last number is the only number $< X$ that is spsp(a) for $a = 2, 3, 5$, and 7. It is not known in general how many applications of the strong test suffice to ensure that a number m is prime. but it is conjectured that if m is a strong probable prime for all bases a in the range $1 < a \leqslant 2(\log m)^2$ then m is prime.

Suppose that m is a large composite number. By the strong pseudoprime test we may establish that m is composite without exhibiting a proper divisor of m. In general, finding the factorization of m involves much more calculation. If p denotes the least prime factor of m, then we locate the proper divisor p after p trial divisions. Since p may be nearly as large as $\sqrt{m}$, this may require up to $\sqrt{m}$ operations. We now describe a method which usually locates the smallest prime factor p in just a little more than $\sqrt{p}$ steps. As in many such factoring algorithms, our estimate for the running time is not proved, but is instead based on heuristics, probabilistic models, and experience. For our present purposes, the relevant probabilistic result is expressed in the following lemma.

Lemma 2.21 *Suppose that $1 \leqslant k \leqslant n$, and that the numbers $u_1, u_2, \cdots, u_k$ are independently chosen from the set $\{1, 2, \cdots, n\}$. Then the probability that the numbers u_k are distinct is*

$$\left(1 - \frac{1}{n}\right)\left(1 - \frac{2}{n}\right) \cdots \left(1 - \frac{k-1}{n}\right).$$

Proof Consider a sequence $u_1, \cdots, u_k$ in which each u_i is one of the numbers $1, 2, \cdots, n$. Since each u_i is one of n numbers, there are n^k such sequences. From among these, we count those for which the u_i are distinct. We see that u_1 can be any one of n numbers. If u_2 is to be distinct from u_1, then u_2 is one of $n - 1$ numbers. If u_3 is to be distinct from both u_1 and u_2, then u_3 is one of $n - 2$ numbers, and so on. Hence the total number of such sequences is $n(n-1)\cdots(n-k+1)$. We divide this by n^k to obtain the stated probability.

As an application, we note that if $n = 365$ and $k = 23$, then the probability in question is less than $1/2$. That is, if 23 people are chosen at random, then the probability of two of them having the same birthday is greater than $1/2$. It may seem counterintuitive that such a small number of people suffices, but it can be shown that the product is approximately $\exp(-k^2/(2n))$. (A derivation of a precise estimate of this sort is outlined in Problem 22 at the end of this section.) Hence the u_i are likely to be distinct if k is small compared with $\sqrt{n}$, but unlikely to be distinct if k is large compared with $\sqrt{n}$.

Suppose that m is a large composite number whose smallest prime divisor is p. If we choose k integers $u_1, u_2, \cdots, u_k$ "at random," with k large compared to $\sqrt{p}$ but small compared to $\sqrt{m}$, then it is likely that the u_i will be distinct $(\bmod\ m)$, but not distinct $(\bmod\ p)$. That is, there probably are integers i, j, with $1 \leqslant i < j \leqslant k$ such that $1 < (u_i - u_j, m) < m$. Each pair (i, j) is easily tested by the Euclidean algorithm, but the task of inspecting all $\binom{k}{2}$ pairs is painfully long. To shorten our work, we adopt the following scheme: We generate the u_i by a recursion of the form $u_{i+1} = f(u_i)$ where $f(u)$ is a polynomial with integral coefficients. The precise choice of $f(u)$ is unimportant, except that it should be easy to compute, and it should give rise to a sequence of numbers that "looks random." Here some experimentation is called for, but it has been found that $f(u) = u^2 + 1$ works well. (In general, polynomials of first degree do not.)

The advantage of generating the u_i in this way is that if $u_i \equiv u_j \pmod d$, then $u_{i+1} = f(u_i) \equiv f(u_j) = u_{j+1} \pmod d$, so the sequence u_i becomes periodic $(\bmod\ d)$ with period $j - i$. In other words, if we put $r = j - i$, then $u_s \equiv u_t \pmod d$ whenever $s \equiv t \pmod r$, $s \geqslant i$, and $t \geqslant i$. In particular, if we let s be the least multiple of r that is $\geqslant i$, and we take $t = 2s$, then $u_s \equiv u_{2s} \pmod d$. That is, among the numbers $u_{2s} - u_s$ we expect to find one for which $1 < (u_{2s} - u_s, m) < m$, with s of size roughly comparable to $\sqrt{p}$.

Example 10 Use this method to locate a proper divisor of the number $m = 36{,}287$.

Solution We take $u_0 = 1$, $u_{i+1} \equiv u_i^2 + 1 \pmod m$, $0 \leqslant u_{i+1} < m$. Then the numbers u_i, $i = 1, 2, \ldots, 14$ are 2, 5, 26, 677, 22886, 2439, 33941, 24380, 3341, 22173, 25652, 26685, 29425, 22806. We find that $(u_{2s} - u_s, m) = 1$ for $s = 1, 2, \cdots, 6$, but that $(u_{14} - u_7, m) = 131$. That is, 131 is a divisor of m. In this example, it turns out that 131 is the smallest prime divisor of m, because the division of 36,287 by 131 gives the other prime factor, 277.

If we reduce the $u_i \pmod{131}$, we obtain the numbers $2, 5, 26, 22, 92, 81, 12, 14, 66, 34, 109, 92, 81, 12$. Hence $u_{12} \equiv u_5 \pmod{131}$, and the sequence has period 7 from u_5 on. We might diagram this as follows:

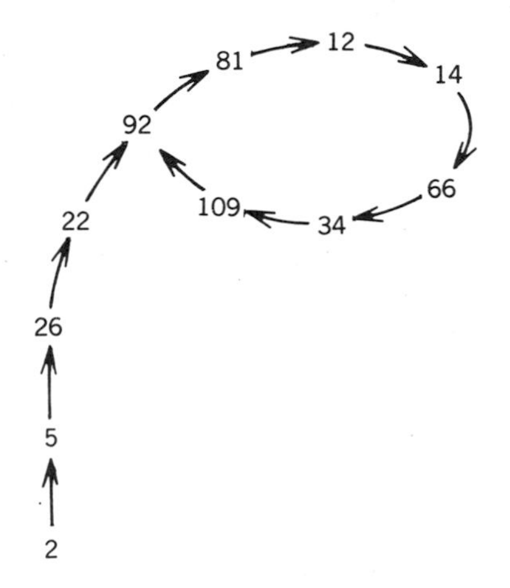

This method was proposed by J. M. Pollard in 1975. Since the pattern above resembles the Greek letter ρ ("rho"), this approach is known as the *Pollard rho method*. It should be applied only to numbers m that are already known to be composite (e.g., by the strong pseudoprime test), for if m is prime then the method will run for roughly $\sqrt{m}$ cycles, without proving anything. Since the method may be expected to disclose the smallest prime factor p of m in roughly $\sqrt{p}$ cycles, this method is faster than trial division for large composite m. Note that there is no guarantee that the divisor found will be the smallest prime factor of m. The divisor located may be some other prime factor, it may be composite, and it may even be m itself. In the latter eventuality, one may start over with a new value of u_0, or with a new function $f(u)$, say $f(u) = u^2 + c$ with some new value for c. (The two values $c = 0$, $c = -2$ should be avoided.)

As of this writing, the most efficient factoring strategies are expected to locate a proper divisor of a composite number m in no more than $\exp(c(\log m)^{1/2}(\log\log m)^{1/2})$ bit operations. (Here c is some positive constant.) In Section 5.8 we use elliptic curves to find proper divisors this quickly. If ε is a given positive number, then the function of m above is $< m^{\varepsilon}$ for all sufficiently large m. Nevertheless, it remains the case that we can perform congruence arithmetic, compositeness tests, and so forth for much larger m than we can factor.

PROBLEMS

1. Verify that $bx + cy = 1$ where b, c, x, y are the numbers given in Example 3 in Section 1.2. Use no number of more than 10 digits. (H).

2. Show that $2^{45} \equiv 57 \pmod{91}$. Deduce that 91 is composite.

3. (*a*) Let $m = 11111$. Show that $2^{m-1} \equiv 10536 \pmod{m}$. Deduce that m is composite.

(*b*) Let $m = 1111111$. Show that $2^{m-1} \equiv 553891 \pmod{m}$. Deduce that m is composite.

(*c*) Let $m = 11111111111$. Show that $2^{m-1} \equiv 1496324899 \pmod{m}$. Deduce that m is composite.

(*d*) Let $m = 1111111111111$. Show that $2^{m-1} \equiv 1015669396877 \pmod{m}$. Deduce that m is composite.

4. Show that the Carmichael number 561 is composite by showing that it is not a spsp(2).

5. Show that 2047 is a strong probable prime to the base 2.

6. Show that 2047 is composite by applying the strong pseudoprime test to the base 3.

7. Some earlier authors called a composite number m a pseudoprime to the base a if $a^m \equiv a \pmod{m}$. To distinguish this definition from the one we adopted (at the end of Section 2.1), call such a number m an *old pseudoprime to base a*. Explain why the set of pseudoprimes to base a lies in the set of old pseudoprimes to base a. Demonstrate that the two definitions do not coincide by showing that $m = 161{,}038$ is an old pseudoprime to base 2, but not a pseudoprime to base 2.

8. Note that if the algorithmic form of the strong pseudoprime test does not terminate prematurely, then the last number examined is $a^{2^{j-1}d} = a^{(m-1)/2}$. Explain why it is not necessary to consider a^{m-1}.

9. Show that if $x^2 \equiv 1 \pmod{m}$ but $x \not\equiv \pm 1 \pmod{m}$, then $1 < (x - 1, m) < m$, and that $1 < (x + 1, m) < m$.

10. Note that $85 = (341 - 1)/4$. Show that $2^{85} \equiv 32 \not\equiv \pm 1 \pmod{341}$, and that $2^{170} \equiv 1 \pmod{341}$. Deduce that 341 is a pseudoprime base 2, but not a spsp(2). Apply the Euclidean algorithm to calculate $(32 \pm 1, 341)$, and thus find numbers $d, e, 1 < d < 341$, such that $de = 341$.

11. Show that if m is a pseudoprime to the base a, but not a spsp(a), then the strong pseudoprime test in conjunction with the Euclidean algorithm provides an efficient means of locating a proper divisor d of m.

12. Let $m = 3215031751$. Observe that $d = (m - 1)/2$ is odd. Show that $11^d \equiv 2129160099 \not\equiv \pm 1 \pmod{m}$. Deduce that m is composite.

13. Let $f(u)$ be a given function. Suppose that a sequence u_i of real numbers is generated iteratively by putting $u_{i+1} = f(u_i)$. Suppose also that $u_1, u_2, \cdots, u_{17}$ are distinct, but that $u_{18} = u_{11}$. What is the least value of s such that $u_{2s} = u_s$?

14. Use the Pollard rho method to locate proper divisors of the following numbers:

(*a*) 8,131; (*d*) 16,019;
(*b*) 7,913; (*e*) 10,277;
(*c*) 7,807; (*f*) 199,934,971.

15. Show that if $(a, m) = 1$ and m has a prime factor p such that $(p-1)|Q$, then $(a^Q - 1, m) > 1$.

16. *The Pollard* $p - 1$ *method*. Suppose that m is an odd integer > 1, and let $d_n = (2^{n!} - 1, m)$. Explain why $d_n \mid d_{n+1}$ for $n = 1, 2, \cdots$. Show that $d_n > 1$ if m has a prime factor p such that $(p-1) \mid n!$. Apply this approach to find a proper divisor of 403. What is the least n that yields a factor? What is the least n for which $d_n = 403$?

***17.** Find a proper divisor of $m = 387058387$ by evaluating d_{100}, in the notation of the preceding problem.

***18.** Apply the Pollard $p - 1$ method to the number 1891. Explain what difficulties are encountered and how they might be overcome.

***19.** Let k be a positive integer such that $2^{k-1} \equiv 1 \pmod{k}$, and put $m = 2^k - 1$. Observe that $d = (m-1)/2$ is odd. Show that $2^d \equiv 1 \pmod{m}$. Show also that if k is composite then m is composite. Deduce that if k is a pseudoprime base 2 then m is a spsp (2). Conclude that there exist infinitely many numbers of the class spsp (2).

***20.** Let k be a positive integer such that $6k + 1 = p_1$, $12k + 1 = p_2$, and $18k + 1 = p_3$ are all prime numbers, and put $m = p_1 p_2 p_3$. Show that $(p_i - 1)|(m - 1)$ for $i = 1, 2, 3$. Deduce that if $(a, p_i) = 1$ then $a^{m-1} \equiv 1 \pmod{p_i}$, $i = 1, 2, 3$. Conclude that if $(a, m) = 1$ then $a^{m-1} \equiv 1 \pmod{m}$, that is, that m is a Carmichael number. (It is conjectured that there are infinitely many k for which the numbers p_i are all prime; the first three are $k = 1, 6, 35$.)

***21.** Let X be a large positive integer. Suppose that $m \leqslant X/2$, and that $0 \leqslant a < m$, $0 \leqslant b < m$. Explain why the number c determined by the following algorithm satisfies $0 \leqslant c < m$, and $c \equiv ab \pmod{m}$. Verify that in executing the algorithm, all numbers encountered lie in the interval $[0, X)$.

1. Set $k = b$, $c = 0$, $g = [X/m]$.
2. As long as $a > 0$, perform the following operations:
   (*a*) Set $r = a - g[a/g]$.
   (*b*) Choose s so that $s \equiv kr \pmod{m}$ and $0 \leqslant s < m$.

(*c*) Replace c by $c + s$.
(*d*) If $c \geqslant m$, replace c by $c - m$.
(*e*) Replace k by $gk - m[gk/m]$.
(*f*) Replace a by $(a - r)/g$.

***22.** Show that the product in Lemma 2.21 is smaller than $\exp\left(-\frac{k^2}{2n} + \frac{k}{2n}\right)$, but larger than $\exp\left(-\frac{k^2}{2n} - \frac{k^3}{3n^2}\right)$. (H)

2.5 PUBLIC-KEY CRYPTOGRAPHY

We now apply our knowledge of congruence arithmetic to construct a method of encrypting messages. The mathematical principle we use is formulated in the following lemma.

Lemma 2.22 *Suppose that m is a positive integer and that $(a, m) = 1$. If k and $\bar{k}$ are positive integers such that $k\bar{k} \equiv 1 \pmod{\phi(m)}$, then $a^{k\bar{k}} \equiv a \pmod{m}$.*

Proof Write $k\bar{k} = 1 + r\phi(m)$, where r is a non-negative integer. Then by Euler's congruence

$$a^{k\bar{k}} = a \cdot a^{r\phi(m)} = a\left(a^{\phi(m)}\right)^r \equiv a \cdot 1^r = a \pmod{m}.$$

If $(a, m) = 1$ and k is a positive integer, then $(a^k, m) = 1$. Thus if $n = \phi(m)$ and $r_1, r_2, \cdots, r_n$ is a system of reduced residues (mod m), then the numbers $r_1^k, r_2^k, \cdots, r_n^k$ are also relatively prime to m. These kth powers may not all be distinct (mod m), as we see by considering the special case $k = \phi(m)$. On the other hand, from Lemma 2.22 we can deduce that these kth powers are distinct (mod m) provided that $(k, \phi(m)) = 1$. For, suppose that $r_i^k \equiv r_j^k \pmod{m}$ and $(k, \phi(m)) = 1$. By Theorem 2.9 we may determine a positive integer $\bar{k}$ such that $k\bar{k} \equiv 1 \pmod{\phi(m)}$, and then it follow from the lemma that

$$r_i \equiv r_i^{k\bar{k}} = \left(r_i^k\right)^{\bar{k}} \equiv \left(r_j^k\right)^{\bar{k}} = r_j^{k\bar{k}} \equiv r_j \pmod{m}.$$

This implies that $i = j$. (From our further analysis in Section 2.8 it will become apparent that the converse also holds: the numbers $r_1^k, r_2^k, \cdots, r_n^k$ are distinct (mod m) only if $(k, \phi(m)) = 1$.) Suppose that $(k, \phi(m)) = 1$. Since the numbers $r_1^k, r_2^k, \cdots, r_n^k$ are distinct (mod m), they form a system of reduced residues (mod m). That is, the map $a \mapsto a^k$ permutes the reduced residues (mod m) if $(k, \phi(m)) = 1$. The significance of the lemma is that the further map $b \mapsto b^{\bar{k}}$ is the inverse permutation.

To apply these observations to cryptography, we take two distinct large primes, p_1, p_2, say each one with about 100 digits, and multiply them to form a composite modulus $m = p_1 p_2$ of about 200 digits. Since we know the prime factorization of m, from Theorem 2.19 we see that $\phi(m) = (p_1 - 1)(p_2 - 1)$. Here $\phi(m)$ is somewhat smaller than m. We choose a big number, k, from the interval $0 < k < \phi(m)$, and check by the Euclidean algorithm that $(k, \phi(m)) = 1$. If a proposed k does not have this property, we try another, until we obtain one for which this holds. We make the numbers m and k publicly available, but keep p_1, p_2, and $\phi(m)$ secret. Suppose now that some associate of ours wants to send us a message, say "Gauss was a genius!" The associate first converts the characters of the message to numbers in some standard way, say by employing the three digit American Standard Code for Information Interchange (ASCII) used on many computers. Then "G" becomes 071, "a" becomes 097, $\cdots$, and "!" becomes 033. Concatenate these codes to form a number

$$a = 07109711711511512611909711512609712610310111010511711503 3.$$

Since a has only 56 digits, we see that $0 < a < m$. If the message were longer, it could be divided into a number of blocks. Our associate could send us the number a, and then we could reconstruct the original characters, but suppose that the message contains some sensitive material that would make it desirable to ensure the privacy of the transmission. In that case, our associate would use the numbers k and m that we have provided. Being acquainted with the ideas discussed in the preceding section, our associate quickly finds the unique number b, $0 \leqslant b < m$, such that $b \equiv a^k \pmod{m}$, and sends this b to us. We use the Euclidean algorithm to find a positive number $\bar{k}$ such that $k\bar{k} \equiv 1 \pmod{\phi(m)}$, and then we find the unique number c such that $0 \leqslant c < m$, $c \equiv b^{\bar{k}} \pmod{m}$. From Lemma 2.22 we deduce that $a = c$. In theory it might happen that $(a, m) > 1$, in which case the lemma does not apply, but the chances of this are remote ($\approx 1/p_i \approx 10^{-100}$). (In this unlikely event, one could still appeal to Problem 4 at the end of this section.) Suppose that some inquisitive third party gains access to the numbers m, k, and b, and seeks to recover the number a. In principle, all that need be done is to factor m, which yields $\phi(m)$, and hence $\bar{k}$, just as we have done. In practice, however, the task of locating the factors of m is prohibitively long. Using the best algorithms known and fastest computers, it would take centuries to factor our 200 digit modulus m. Of course, we hope that faster factoring algorithms may yet be discovered, but here one can only speculate.

PROBLEMS

1. Suppose that $b \equiv a^{67} \pmod{91}$, and that $(a, 91) = 1$. Find a positive number $\bar{k}$ such that $b^{\bar{k}} \equiv a \pmod{91}$. If $b = 53$, what is $a \pmod{91}$?
2. Suppose that $m = pq$, and $\phi = (p-1)(q-1)$ where p and q are real numbers. Find a formula for p and q, in terms of m and ϕ. Supposing that $m = 39{,}247{,}771$ is the product of two distinct primes, deduce the factors of m from the information that $\phi(m) = 39{,}233{,}944$.
3. Show that if $d|m$, then $\phi(d)|\phi(m)$.
4. Suppose that m is square-free, and that k and $\bar{k}$ are positive integers such that $k\bar{k} \equiv 1 \pmod{\phi(m)}$. Show that $a^{k\bar{k}} \equiv a \pmod{m}$ for all integers a. (H)
5. Suppose that m is a positive integer that is not square-free. Show that there exist integers a_1 and a_2 such that $a_1 \not\equiv a_2 \pmod{m}$, but $a_1^k \equiv a_2^k \pmod{m}$ for all integers $k > 1$.

2.6 PRIME POWER MODULI

The problem of solving a congruence was reduced in Section 2.3 to the case of a prime-power modulus. To solve a polynomial congruence $f(x) \equiv 0 \pmod{p^k}$, we start with a solution modulo p, then move on to modulo p^2, then to p^3, and by iteration to p^k. Suppose that $x = a$ is a solution of $f(x) \equiv 0 \pmod{p^j}$ and we want to use it to get a solution modulo p^{j+1}. The idea is to try to get a solution $x = a + tp^j$, where t is to be determined, by use of Taylor's expansion

$$f(a + tp^j) = f(a) + tp^j f'(a) + t^2 p^{2j} f''(a)/2! + \cdots + t^n p^{nj} f^{(n)}(a)/n! \qquad (2.3)$$

where n is the presumed degree of the polynomial $f(x)$. All derivatives beyond the nth are identically zero.

Now with respect to the modulus p^{j+1}, equation (2.3) gives

$$f(a + tp^j) \equiv f(a) + tp^j f'(a) \pmod{p^{j+1}} \qquad (2.4)$$

as the following argument shows. What we want to establish is that the coefficients of $t^2, t^3, \cdots t^n$ in equation (2.3) are divisible by p^{j+1} and so can be omitted in (2.4). This is almost obvious because the powers of p in those terms are $p^{2j}, p^{3j}, \cdots, p^{nj}$. But this is not quite immediate because of the denominators $2!, 3!, \cdots, n!$ in these terms. The explanation is that

$f^{(k)}(a)/k!$ is an integer for each value of k, $2 \leqslant k \leqslant n$. To see this, let cx^r be a representative term from $f(x)$. The corresponding term in $f^{(k)}(a)$ is

$$cr(r-1)(r-2)\cdots(r-k+1)a^{r-k}.$$

According to Theorem 1.21, the product of k consecutive integers is divisible by $k!$, and the argument is complete. Thus, we have proved that the coefficients of $t^2, t^3, \cdots$ in (2.3) are divisible by p^{j+1}.

The congruence (2.4) reveals how t should be chosen if $x = a + tp^j$ is to be a solution of $f(x) \equiv 0 \pmod{p^{j+1}}$. We want t to be a solution of

$$f(a) + tp^j f'(a) \equiv 0 \pmod{p^{j+1}}.$$

Since $f(x) \equiv 0 \pmod{p^j}$ is presumed to have the solution $x = a$, we see that p^j can be removed as a factor to give

$$tf'(a) \equiv -\frac{f(a)}{p^j} \pmod{p} \tag{2.5}$$

which is a linear congruence in t. This congruence may have no solution, one solution, or p solutions. If $f'(a) \not\equiv 0 \pmod{p}$, then this congruence has exactly one solution, and we obtain

Theorem 2.23 *Hensel's lemma. Suppose that $f(x)$ is a polynomial with integral coefficients. If $f(a) \equiv 0 \pmod{p^j}$ and $f'(a) \not\equiv 0 \pmod{p}$, then there is a unique $t \pmod{p}$ such that $f(a + tp^j) \equiv 0 \pmod{p^{j+1}}$.*

If $f(a) \equiv 0 \pmod{p^j}$, $f(b) \equiv 0 \pmod{p^k}$, $j < k$, and $a \equiv b \pmod{p^j}$, then we say that *b lies above a*, or *a lifts to b*. If $f(a) \equiv 0 \pmod{p^j}$, then the root a is called *nonsingular* if $f'(a) \not\equiv 0 \pmod{p}$; otherwise it is *singular*. By Hensel's lemma we see that a nonsingular root $a \pmod{p}$ lifts to a unique root $a_2 \pmod{p^2}$. Since $a_2 \equiv a \pmod{p}$, it follows (by Theorem 2.2) that $f'(a_2) \equiv f'(a) \not\equiv 0 \pmod{p}$. By a second application of Hensel's lemma we may lift a_2 to form a root a_3 of $f(x)$ modulo p^3, and so on. In general we find that a nonsingular root a modulo p lifts to a unique root a_j modulo p^j for $j = 2, 3, \cdots$. By (2.5) we see that this sequence is generated by means of the recursion

$$a_{j+1} = a_j - f(a_j)\overline{f'(a)} \tag{2.6}$$

where $\overline{f'(a)}$ is an integer chosen so that $f'(a)\overline{f'(a)} \equiv 1 \pmod{p}$. This is entirely analogous to Newton's method for locating the root of a differentiable function.

Example 11 Solve $x^2 + x + 47 \equiv 0 \pmod{7^3}$.

Solution First we note that $x \equiv 1 \pmod 7$ and $x \equiv 5 \pmod 7$ are the only solutions of $x^2 + x + 47 \equiv 0 \pmod 7$. Since $f'(x) = 2x + 1$, we see that $f'(1) = 3 \not\equiv 0 \pmod 7$ and $f'(5) = 1 \not\equiv 0 \pmod 7$, so these roots are nonsingular. Taking $\overline{f'(1)} = 5$, we see by (2.6) that the root $a \equiv 1 \pmod 7$ lifts to $a_2 = 1 - 49 \cdot 5$. Since a_2 is considered $\pmod{7^2}$, we may take instead $a_2 = 1$. Then $a_3 = 1 - 49 \cdot 5 \equiv 99 \pmod{7^3}$. Similarly, we take $\overline{f'(5)} = 2$, and see by (2.6) that the root $5 \pmod 7$ lifts to $5 - 77 \cdot 2 = -149 \equiv 47 \pmod{7^2}$, and that $47 \pmod{7^2}$ lifts to $47 - f(47) \cdot 2 = 47 - 2303 \cdot 2 = -4559 \equiv 243 \pmod{7^3}$. Thus we conclude that 99 and 243 are the desired roots and that there are no others.

We now turn to the more difficult problem of lifting singular roots. Suppose that $f(a) \equiv 0 \pmod{p^j}$ and that $f'(a) \equiv 0 \pmod p$. From the Taylor expansion (2.3) we see that $f(a + tp^j) \equiv f(a) \pmod{p^{j+1}}$ for all integers t. Thus if $f(a) \equiv 0 \pmod{p^{j+1}}$ then $f(a + tp^j) \equiv 0 \pmod{p^{j+1}}$, so that the single root $a \pmod{p^j}$ lifts to p roots $\pmod{p^{j+1}}$. But if $f(a) \not\equiv 0 \pmod{p^{j+1}}$, then none of the p residue classes $a + tp^j$ is a solution $\pmod{p^{j+1}}$, and then there are no roots $\pmod{p^{j+1}}$ lying above $a \pmod{p^j}$.

Example 12 Solve $x^2 + x + 7 \equiv 0 \pmod{81}$.

Solution Starting with $x^2 + x + 7 \equiv 0 \pmod 3$, we note that $x = 1$ is the only solution. Here $f'(1) = 3 \equiv 0 \pmod 3$, and $f(1) \equiv 0 \pmod 9$, so that we have the roots $x = 1$, $x = 4$, and $x = 7 \pmod 9$. Now $f(1) \not\equiv 0 \pmod{27}$, and hence there is no root $x \pmod{27}$ for which $x \equiv 1 \pmod 9$. As $f(4) \equiv 0 \pmod{27}$, we obtain three roots, 4, 13, and $22 \pmod{27}$, which are $\equiv 4 \pmod 9$. On the other hand, $f(7) \not\equiv 0 \pmod{27}$, so there is no root $\pmod{27}$ that is $\equiv 7 \pmod 9$. We are now in a position to determine which, if any, of the roots 4, 13, 22 (mod 27) can be lifted to roots (mod 81). We find that $f(4) = 27 \not\equiv 0 \pmod{81}$, $f(13) = 189 \equiv 27 \not\equiv 0 \pmod{81}$, and that $f(22) = 513 \equiv 27 \not\equiv 0 \pmod{81}$, from which we deduce that the congruence has no solution (mod 81).

In this example, we see that a singular solution $a \pmod p$ may lift to some higher powers of p, but not necessarily to arbitrarily high powers of p. We now show that if the power of p dividing $f(a)$ is sufficiently large compared with the power of p in $f'(a)$, then the solution can be lifted without limit.

Theorem 2.24 *Let $f(x)$ be a polynomial with integral coefficients. Suppose that $f(a) \equiv 0 \pmod{p^j}$, that $p^\tau \| f'(a)$, and that $j \geqslant 2\tau + 1$. If $b \equiv a \pmod{p^{j-\tau}}$ then $f(b) \equiv f(a) \pmod{p^j}$ and $p^\tau \| f'(b)$. Moreover, there is a unique $t \pmod{p}$ such that $f(a + tp^{j-\tau}) \equiv 0 \pmod{p^{j+1}}$.*

In this situation, a collection of p^τ solutions $\pmod{p^j}$ give rise to p^τ solutions $\pmod{p^{j+1}}$, while the power of p dividing f' remains constant. Since the hypotheses of the theorem apply with a replaced by $a + tp^{j-\tau}$ and $\pmod{p^j}$ replaced by $\pmod{p^{j+1}}$ but with τ unchanged, the lifting may be repeated and continues indefinitely.

Proof By Taylor's expansion (2.3), we see that

$$f(b) = f(a + tp^{j-\tau}) \equiv f(a) + tp^{j-\tau}f'(a) \pmod{p^{2j-2\tau}}.$$

Here the modulus is divisible by p^{j+1}, since $2j - 2\tau = j + (j - 2\tau) \geqslant j + 1$. Hence

$$f(a + tp^{j-\tau}) \equiv f(a) + tp^{j-\tau}f'(a) \pmod{p^{j+1}}.$$

Since both terms on the right side are divisible by p^j, the left side is also. Moreover, on dividing through by p^j we find that

$$\frac{f(a + tp^{j-\tau})}{p^j} \equiv \frac{f(a)}{p^j} + t\frac{f'(a)}{p^\tau} \pmod{p},$$

and the coefficient of t is relatively prime to p, so that there is a unique $t \pmod{p}$ for which the right side is divisible by p. This establishes the final assertion of the theorem. To complete the proof, we note that $f'(x)$ is a polynomial with integral coefficients, so that

$$f'(a + tp^{j-\tau}) \equiv f'(a) \pmod{p^{j-\tau}}$$

for any integer t. But $j - \tau \geqslant \tau + 1$, so this congruence holds $\pmod{p^{\tau+1}}$. Since p^τ exactly divides $f'(a)$ (in symbols, $p^\tau \| f'(a)$), we conclude that $p^\tau \| f'(a + tp^{j-\tau})$.

Example 13 Discuss the solutions of $x^2 + x + 223 \equiv 0 \pmod{3^j}$.

Solution Since $223 \equiv 7 \pmod{27}$, the solutions $\pmod{27}$ are the same as in Example 12. For this new polynomial, we find that $f(4) \equiv 0 \pmod{81}$, and thus we have three solutions $4, 31, 58 \pmod{81}$. Similarly $f(13) \equiv 0 \pmod{81}$, giving three solutions $13, 40, 67 \pmod{81}$. Moreover, $f(22) \equiv$

Table 1 Solutions of $x^2 + x + 223 \equiv 0 \pmod{3^j}$.

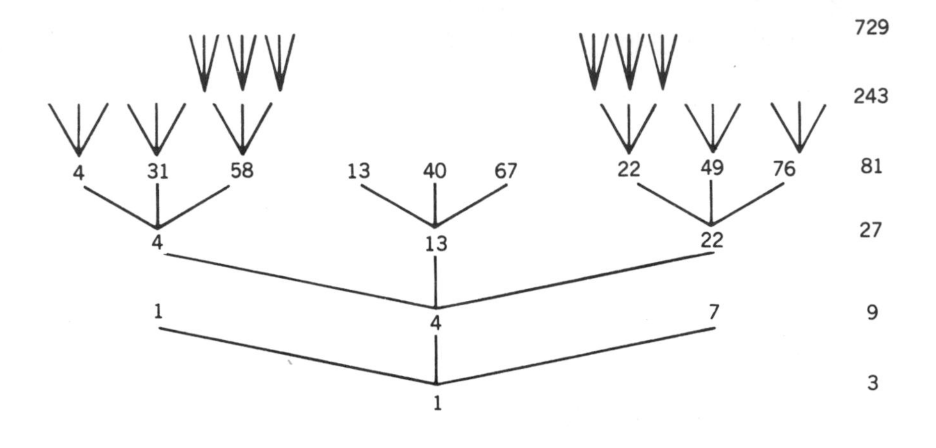

$0 \pmod{81}$, yielding the solutions $22, 49, 76 \pmod{81}$. Thus we find that the congruence has exactly nine solutions (mod 81). In fact we note that $f(4) \equiv 0 \pmod{3^5}$, $3^2 \| f'(4)$, so by Theorem 2.20 the solution $4 \pmod{243}$ is one of nine solutions of the form $4 + 27t \pmod{243}$. We may further verify that there is precisely one value of $t \pmod 3$, namely $t = 2$, for which $f(4 + 27t) \equiv 0 \pmod{3^6}$. This gives nine solutions of the form $58 + 81t \pmod{3^6}$. Similarly, $f(22) \equiv 0 \pmod{3^5}$, $3^2 \| f'(22)$, so that $22 \pmod{243}$ is one of nine solutions of the form $22 + 27t \pmod{243}$. Moreover, we can verify that there is precisely one value of $t \pmod 3$, namely $t = 0$, for which $22 + 27t$ is a solution $\pmod{3^6}$. That is, we have nine solutions $\pmod{3^6}$ of the form $22 + 81t$. On the other hand, $f'(13) \equiv 0 \pmod{27}$, so that $f(13 + 27t) \equiv f(13) \pmod{3^6}$. As $3^4 \| f(13)$, we find that none of the three solutions $13 + 27t \pmod{81}$ lifts to a solution (mod 243). In conclusion, we have found that for each $j \geqslant 5$ there are precisely 18 solutions $\pmod{3^j}$, of which 12 do not lift to 3^{j+1}, while each of the remaining six lifts to three solutions $\pmod{3^{j+1}}$. These results are depicted in Table 1.

Suppose that $f(a) \equiv 0 \pmod p$, and that $f'(a) \equiv 0 \pmod p$. We wish to know whether a can be lifted to solutions modulo arbitrarily high powers of p. The situation is resolved if we can reach a point at which Theorem 2.24 applies, that is, $j \geqslant 2\tau + 1$. However, there is nothing in our discussion thus far to preclude the possibility that the power of p in f' might steadily increase with that in f, so that Theorem 2.24 might never take effect. In Appendix A.2 we define the *discriminant* $D(f)$ of the polynomial, and show that the critical inequality $j \geqslant 2\tau + 1$ holds whenever j is larger than the power of p in $D(f)$.

PROBLEMS

1. Solve the congruence $x^2 + x + 7 \equiv 0 \pmod{27}$ by using the method of completing the square from elementary algebra, thus $4x^2 + 4x + 28 = (2x + 1)^2 + 27$. Solve this congruence (mod 81) by the same method.

2. Solve $x^5 + x^4 + 1 \equiv 0 \pmod{3^4}$.

3. Solve $x^3 + x + 57 \equiv 0 \pmod{5^3}$.

4. Solve $x^2 + 5x + 24 \equiv 0 \pmod{36}$.

5. Solve $x^3 + 10x^2 + x + 3 \equiv 0 \pmod{3^3}$.

6. Solve $x^3 + x^2 - 4 \equiv 0 \pmod{7^3}$.

7. Solve $x^3 + x^2 - 5 \equiv 0 \pmod{7^3}$.

8. Apply the theory of this section to solve $1000x \equiv 1 \pmod{101^3}$, using a calculator.

9. Suppose that $f(a) \equiv 0 \pmod{p^j}$ and that $f'(a) \not\equiv 0 \pmod{p}$. Let $\overline{f'(a)}$ be an integer chosen so that $f'(a)\overline{f'(a)} \equiv 1 \pmod{p^j}$, and put $b = a - f(a)\overline{f'(a)}$. Show that $f(b) \equiv 0 \pmod{p^{2j}}$.

10. Let p be an odd prime, and suppose that $a \not\equiv 0 \pmod{p}$. Show that if the congruence $x^2 \equiv a \pmod{p^j}$ has a solution when $j = 1$, then it has a solution for all j.

***11.** Let $f(\mathbf{x})$ be a polynomial with integral coefficients in the n variables $x_1, x_2, \cdots, x_n$. Suppose that $f(\mathbf{a}) \equiv 0 \pmod{p}$ where $\mathbf{a} = (a_1, a_2, \cdots, a_n)$, and that $\frac{\partial}{\partial x_i} f(\mathbf{a}) \not\equiv 0 \pmod{p}$ for at least one i. Show that the congruence $f(\mathbf{x}) \equiv 0 \pmod{p^j}$ has a solution for every j.

2.7 PRIME MODULUS

We have now reduced the problem of solving $f(x) \equiv 0 \pmod{m}$ to its last stage, congruences with prime moduli. Although we have no general method for solving such congruences, there are some interesting facts concerning the solutions. A natural question about polynomial congruences of the type $f(x) \equiv 0 \pmod{m}$ is whether there is any analogue to the well-known theorem in algebra that a polynomial equation of degree n whose coefficients are complex numbers has exactly n roots or solutions, allowing for multiple roots. For congruences the situation is more complicated. In the first place, for any modulus $m > 1$, there are polynomial congruences having no solutions. An example of this is given by $x^p - x + 1 \equiv 0 \pmod{m}$, where p is any prime factor of m. This congruence has no solutions because $x^p - x + 1 \equiv 0 \pmod{p}$ has none, by Fermat's theorem.

Moreover, we have already seen that a congruence can have more solutions than its degree, for example, $x^2 - 7x + 2 \equiv 0 \pmod{10}$ with four solutions $x = 3, 4, 8, 9$, and also $x^2 + x + 7 \equiv 0 \pmod{27}$ with three solutions $x = 4, 13, 22$. But if the modulus is a *prime*, a congruence cannot have more solutions than its degree. This is proved in Theorem 2.26 later in the section. It is important here to note carefully the meaning of "degree of congruence," given in Definition 2.5 in Section 2.2. Such a polynomial as $5x^3 + x^2 - x$ has degree 3, but the congruence $5x^3 + x^2 - x \equiv 0 \pmod 5$ has degree 2.

Consider the congruence $5x^2 + 10x + 15 \equiv 0 \pmod 5$, having five solutions $x = 0$, 1, 2, 3, and 4. At first glance, this might appear to be a counterexample to Theorem 2.26. However, by Definition 2.5, this congruence is assigned no degree, so that Theorem 2.26 does not apply.

With this background, we proceed to prove some fundamental results. As before, we write $f(x) = a_n x^n + a_{n-1}x^{n-1} + \cdots + a_0$, and we assume that p is a prime not dividing a_n, so that the congruence $f(x) \equiv 0 \pmod p$ has degree n. In Theorem 2.25, we divide such a polynomial $f(x)$ of degree $n \geqslant p$ by $x^p - x$ to get a quotient and a remainder, both polynomials. This is a limited use of the *division algorithm for polynomials*, which is discussed more fully in Theorem 9.1. By "limited use," we mean that the only idea involved is the division of one polynomial into another, as in elementary algebra. The uniqueness of the quotient and the remainder are not needed.

Theorem 2.25 *If the degree n of $f(x) \equiv 0 \pmod p$ is greater than or equal to p, then either every integer is a solution of $f(x) \equiv 0 \pmod p$ or there is a polynomial $g(x)$ having integral coefficients, with leading coefficient 1, such that $g(x) \equiv 0 \pmod p$ is of degree less than p and the solutions of $g(x) \equiv 0 \pmod p$ are precisely those of $f(x) \equiv 0 \pmod p$.*

Proof Dividing $f(x)$ by $x^p - x$, we get a quotient $q(x)$ and a remainder $r(x)$ such that $f(x) = (x^p - x)q(x) + r(x)$. Here $q(x)$ and $r(x)$ are polynomials with integral coefficients, and $r(x)$ is either zero or a polynomial of degree less than p. Since every integer is a solution of $x^p \equiv x \pmod p$ by Fermat's theorem, we see that the solutions of $f(x) \equiv 0 \pmod p$ are the same as those of $r(x) \equiv 0 \pmod p$. If $r(x) = 0$ or if every coefficient of $r(x)$ is divisible by p, then every integer is a solution of $f(x) \equiv 0 \pmod p$.

On the other hand, if at least one coefficient of $r(x)$ is not divisible by p, then the congruence $r(x) \equiv 0 \pmod p$ has a degree, and that degree is less than p. The polynomial $g(x)$ in the theorem can be obtained from $r(x)$ by getting leading coefficient 1, as follows. We may discard all terms in $r(x)$ whose coefficients are divisible by p, since the congruence properties modulo p are unaltered. Then let bx^m be the term of highest degree in $r(x)$, with $(b, p) = 1$. Choose $\bar{b}$ so that $b\bar{b} \equiv 1 \pmod{p}$, and note that $(\bar{b}, p) = 1$ also. Then the congruence $\bar{b}r(x) \equiv 0 \pmod{p}$ has the same solutions as $r(x) \equiv 0 \pmod{p}$, and so has the same solutions as $f(x) \equiv 0 \pmod{p}$. Define $g(x)$ to be $\bar{b}r(x)$ with its leading coefficient $b\bar{b}$ replaced by 1, that is,

$$g(x) = \bar{b}r(x) - (b\bar{b} - 1)x^m.$$

Theorem 2.26 *The congruence $f(x) \equiv 0 \pmod{p}$ of degree n has at most n solutions.*

Proof The proof is by induction on the degree of $f(x) \equiv 0 \pmod{p}$. If $n = 0$, the polynomial $f(x)$ is just a_0 with $a_0 \not\equiv 0 \pmod{p}$, and hence the congruence has no solution. If $n = 1$, the congruence has exactly one solution by Theorem 2.17. Assuming the truth of the theorem for all congruences of degree $< n$, suppose that there are more than n solutions of the congruence $f(x) \equiv 0 \pmod{p}$ of degree n. Let the leading term of $f(x)$ be $a_n x^n$ and let $u_1, u_2, \cdots, u_n, u_{n+1}$ be solutions of the congruence, with $u_i \not\equiv u_j \pmod{p}$ for $i \neq j$. We define $g(x)$ by the equation

$$g(x) = f(x) - a_n(x - u_1)(x - u_2) \cdots (x - u_n),$$

noting the cancellation of $a_n x^n$ on the right.

Note that $g(x) \equiv 0 \pmod{p}$ has at least n solutions, namely $u_1, u_2, \cdots, u_n$. We consider two cases, first where every coefficient of $g(x)$ is divisible by p, and second where at least one coefficient is not divisible by p. (The first case includes the situation where $g(x)$ is identically zero.) We show that both cases lead to a contradiction. In the first case, every integer is a solution of $g(x) \equiv 0 \pmod{p}$, and since $f(u_{n+1}) \equiv 0 \pmod{p}$ by assumption, it follows that $x = u_{n+1}$ is a solution of

$$a_n(x - u_1)(x - u_2) \cdots (x - u_n) \equiv 0 \pmod{p}.$$

This contradicts Theorem 1.15.

In the second case, we note that the congruence $g(x) \equiv 0 \pmod{p}$ has a degree, and that degree is less than n. By the induction hypothesis, this congruence has fewer than n solutions. This contradicts the earlier observation that this congruence has at least n solutions. Thus the proof is complete.

We have already noted, using the example $5x^2 + 10x + 15 \equiv 0 \pmod{5}$, that the conclusion of Theorem 2.26 need not hold if the assumption is just that the polynomial $f(x)$ has degree n. The following corollary describes the situation.

Corollary 2.27 *If $b_n x^n + b_{n-1}x^{n-1} + \cdots + b_0 \equiv 0 \pmod{p}$ has more than n solutions, then all the coefficients b_j are divisible by p.*

The reason for this is that if some coefficient is not divisible by p, then the polynomial congruence has a degree, and that degree is at most n. Theorem 2.26 implies that the congruence has at most n solutions, and this is a contradiction.

Theorem 2.28 *If $F(x)$ is a function that maps residue classes* (mod p) *to residue classes* (mod p), *then there is a polynomial $f(x)$ with integral coefficients and degree at most $p - 1$ such that $F(x) \equiv f(x) \pmod{p}$ for all residue classes $x \pmod{p}$.*

Proof By Fermat's congruence we see that

$$1 - (x - a)^{p-1} \equiv \begin{cases} 1 \pmod{p} & \text{if } x \equiv a \pmod{p}, \\ 0 \pmod{p} & \text{otherwise.} \end{cases}$$

Hence the polynomial $f(x) = \sum_{i=1}^{p} F(i)(1 - (x - i)^{p-1})$ has the desired properties.

Theorem 2.29 *The congruence $f(x) \equiv 0 \pmod{p}$ of degree n, with leading coefficient $a_n = 1$, has n solutions if and only if $f(x)$ is a factor of $x^p - x$ modulo p, that is, if and only if $x^p - x = f(x)q(x) + ps(x)$, where $q(x)$ and $s(x)$ have integral coefficients, $q(x)$ has degree $p - n$ and leading coefficient 1, and where either $s(x)$ is a polynomial of degree less than n or $s(x)$ is zero.*

Proof First assume that $f(x) \equiv 0 \pmod{p}$ has n solutions. Then $n \leqslant p$, by Definition 2.4 of Section 2.2. Dividing $x^p - x$ by $f(x)$, we get a quotient $q(x)$ and a remainder $r(x)$ satisfying $x^p - x = f(x)q(x) + r(x)$, where $r(x)$ is either identically zero or a polynomial of degree less than n. This equation implies, by application of Fermat's theorem to $x^p - x$, that every solution of $f(x) \equiv 0 \pmod{p}$ is a solution of $r(x) \equiv 0 \pmod{p}$. Thus, $r(x) \equiv 0 \pmod{p}$ has at least n solutions, and by Corollary 2.27, it follows that every coefficient in $r(x)$ is divisible by p, so $r(x) = ps(x)$ as in the theorem.

Conversely, assume that $x^p - x = f(x)q(x) + ps(x)$, as in the statement of the theorem. By Fermat's theorem, the congruence $f(x)q(x) \equiv$

$0 \pmod p$ has p solutions. This congruence has leading term x^p. The leading term of $f(x)$ is x^n by hypothesis, and hence the leading term of $q(x)$ is x^{p-n}. By Theorem 2.26, the congruences $f(x) \equiv 0 \pmod p$ and $q(x) \equiv 0 \pmod p$ have at most n solutions and $p - n$ solutions, respectively. But every one of the p solutions of $f(x)q(x) \equiv 0 \pmod p$ is a solution of at least one of the congruences $f(x) \equiv 0 \pmod p$ and $q(x) \equiv 0 \pmod p$. It follows that these two congruences have exactly n solutions and $p - n$ solutions, respectively.

The restriction $a_n = 1$ in this theorem is needed so that we may divide $x^p - x$ by $f(x)$ and obtain a polynomial $q(x)$ with integral coefficients. However, it is not much of a restriction. We can always find an integer $\overline{a_n}$ such that $a_n\overline{a_n} \equiv 1 \pmod p$. Put $g(x) = \overline{a_n}f(x) - (a_n\overline{a_n} - 1)x^n$. Then $g(x) \equiv 0 \pmod p$ has the same solutions as $f(x) \equiv 0 \pmod p$, and $g(x)$ has leading coefficient 1.

As an example, we see that $x^5 - 5x^3 + 4x \equiv 0 \pmod 5$ has five solutions, and $x^5 - x = (x^5 - 5x^3 + 4x) + (5x^3 - 5x)$. As a second example, we cite $x^3 - x \equiv 0 \pmod 5$ with three solutions, and $x^5 - x = (x^3 - x)(x^2 + 1)$. Theorem 2.29 has many important applications. We now consider one that will be crucial to our discussion of primitive roots in Section 2.8.

Corollary 2.30 *If $d \mid (p - 1)$, then $x^d \equiv 1 \pmod p$ has d solutions.*

Proof Choose e so that $de = p - 1$. Since $(y - 1)(1 + y + \cdots + y^{e-1}) = y^e - 1$, on taking $y = x^d$ we see that $x(x^d - 1)(1 + x^d + \cdots + x^{d(e-1)}) = x^p - x$.

A further application of Theorem 2.29 arises by considering the polynomial

$$f(x) = (x - 1)(x - 2) \cdots (x - p + 1).$$

For convenience we assume that $p > 2$. On expanding, we find that

$$f(x) = x^{p-1} - \sigma_1 x^{p-2} + \sigma_2 x^{p-3} - \cdots + \sigma_{p-1} \tag{2.7}$$

where σ_j is the sum of all products of j distinct members of the set $\{1, 2, \cdots, p - 1\}$. In the two extreme cases we have $\sigma_1 = 1 + 2 + \cdots + (p - 1) = p(p - 1)/2$, and $\sigma_{p-1} = 1 \cdot 2 \cdot \cdots \cdot (p - 1) = (p - 1)!$. The polynomial $f(x)$ has degree $p - 1$ and has the $p - 1$ roots $1, 2, \cdots, p - 1 \pmod p$. Consequently the polynomial $xf(x)$ has degree p and has p roots. By applying Theorem 2.29 to this latter polynomial, we see that there are polynomials $q(x)$ and $s(x)$ such that $x^p - x = xf(x)q(x) + ps(x)$. Since $q(x)$ has degree $p - p = 0$ and leading coefficient 1, we see that $q(x) = 1$. That is, $x^p - x = xf(x) + ps(x)$, which is to say that the coefficients of $x^p - x$ are congruent (mod p) to those of $xf(x)$. On comparing the coefficients of x, we deduce that $\sigma_{p-1} = (p-1)! \equiv -1 \pmod{p}$, which provides a second proof of Wilson's congruence. On comparing the remaining coefficients, we deduce that $\sigma_j \equiv 0 \pmod{p}$ for $1 \leqslant j \leqslant p - 2$. To these useful observations we may add one further remark: if $p \geqslant 5$ then

$$\sigma_{p-2} \equiv 0 \pmod{p^2}. \tag{2.8}$$

This is *Wolstenholme's congruence*. To prove it, we note that $f(p) = (p-1)(p-2)\cdots(p-p+1) = (p-1)!$. On taking $x = p$ in (2.7), we have

$$(p-1)! = p^{p-1} - \sigma_1 p^{p-2} + \cdots + \sigma_{p-3}p^2 - \sigma_{p-2}p + \sigma_{p-1}.$$

We have already observed that $\sigma_{p-1} = (p-1)!$. On subtracting this amount from both sides and dividing through by p, we deduce that

$$p^{p-2} - \sigma_1 p^{p-3} + \cdots + \sigma_{p-3}p - \sigma_{p-2} = 0.$$

All terms except the last two contain visible factors of p^2. Thus $\sigma_{p-3}p \equiv \sigma_{p-2} \pmod{p^2}$. This gives the desired result, since $\sigma_{p-3} \equiv 0 \pmod{p}$.

PROBLEMS

1. Reduce the following congruences to equivalent congruences of degree $\leqslant 6$:
   (*a*) $x^{11} + x^8 + 5 \equiv 0 \pmod{7}$;
   (*b*) $x^{20} + x^{13} + x^7 + x \equiv 2 \pmod{7}$;
   (*c*) $x^{15} - x^{10} + 4x - 3 \equiv 0 \pmod{7}$.
2. Prove that $2x^3 + 5x^2 + 6x + 1 \equiv 0 \pmod{7}$ has three solutions by use of Theorem 2.29.
3. Prove that $x^{14} + 12x^2 \equiv 0 \pmod{13}$ has 13 solutions and so it is an identical congruence.
4. Prove that if $f(x) \equiv 0 \pmod{p}$ has j solutions $x \equiv a_1, x \equiv a_2, \cdots, x \equiv a_j \pmod{p}$, there is a polynomial $q(x)$ such that $f(x) \equiv (x - a_1)(x - a_2)\cdots(x - a_j)q(x) \pmod{p}$. (H)

5. With the assumptions and notation of the preceding problem, prove that if the degree of $f(x)$ is j, then $q(x)$ is a constant and can be taken as the leading coefficient of $f(x)$.

6. Let m be composite. Prove that Theorem 2.26 is false if "mod p" is replaced by "mod m."

7. Show that if the prime number p in Theorem 2.28 is replaced by a composite number m then the statement becomes false.

8. Explain why the proof of Wolstenholme's congruence fails when $p = 3$.

9. For $p = 5$, compute the values of the numbers $\sigma_1, \sigma_2, \sigma_3, \sigma_4$ in (2.7).

10. Write $1/1 + 1/2 + \cdots + 1/(p-1) = a/b$ with $(a, b) = 1$. Show that $p^2 | a$ if $p \geqslant 5$.

***11.** Let p be a prime, $p \geqslant 5$, and suppose that the numbers σ_j are as in (2.7). Show that $\sigma_{p-2} \equiv p\sigma_{p-3} \pmod{p^3}$.

***12.** Show that if $p \geqslant 5$ and m is a positive integer then $\binom{mp-1}{p-1} \equiv 1 \pmod{p^3}$.

***13.** Show that if $p \geqslant 5$ then $(mp)! \equiv m!p!^m \pmod{p^{m+3}}$.

***14.** Suppose that p is an odd prime, and write $1/1 - 1/2 + 1/3 - \cdots - 1/(p-1) = a/(p-1)!$. Show that $a \equiv (2 - 2^p)/p \pmod{p}$.

2.8 PRIMITIVE ROOTS AND POWER RESIDUES

Definition 2.6 *Let m denote a positive integer and a any integer such that $(a, m) = 1$. Let h be the smallest positive integer such that $a^h \equiv 1 \pmod{m}$. We say that the order of a modulo m is h, or that a belongs to the exponent h modulo m.*

The terminology "a belongs to the exponent h" is the classical language of number theory. This language is being replaced more and more in the current literature by "the order of a is h," a usage that is standard in group theory. (In Sections 2.10 and 2.11 we shall explore the relationships between the ideas of number theory and those of group theory.)

Suppose that a has order $h \pmod{m}$. If k is a positive multiple of h, say $k = qh$, then $a^k = a^{qh} = (a^h)^q \equiv 1^q \equiv 1 \pmod{m}$. Conversely, if k is a positive integer such that $a^k \equiv 1 \pmod{m}$, then we apply the division algorithm to obtain integers q and r such that $k = qh + r$, $q \geqslant 0$, and $0 \leqslant r < h$. Thus $1 \equiv a^k = a^{qh+r} = (a^h)^q a^r \equiv 1^q a^r \equiv a^r \pmod{m}$. But $0 \leqslant$

$r < h$ and h is the least positive power of a that is congruent to 1 modulo m, so it follows that $r = 0$. Thus h divides k, and we have proved the following lemma.

Lemma 2.31 *If a has order $h \pmod{m}$, then the positive integers k such that $a^k \equiv 1 \pmod{m}$ are precisely those for which $h \mid k$.*

Corollary 2.32 *If $(a, m) = 1$, then the order of a modulo m divides $\phi(m)$.*

Proof Each reduced residue class a modulo m has finite order, for by Euler's congruence $a^{\phi(m)} \equiv 1 \pmod{m}$. Moreover, if a has order h then by taking $k = \phi(m)$ in the lemma we deduce that $h \mid \phi(m)$.

Lemma 2.33 *If a has order h modulo m, then a^k has order $h/(h, k)$ modulo m.*

Since $h/(h, k) = 1$ if and only if $h \mid k$, we see that Lemma 2.33 contains Lemma 2.31 as a special case.

Proof According to Lemma 2.31, $(a^k)^j \equiv 1 \pmod{m}$ if and only if $h \mid kj$. But $h \mid kj$ if and only if $\{h/(h, k)\} \mid \{k/(h, k)\}j$. As the divisor is relatively prime to the first factor of the dividend, this relation holds if and only if $\{h/(h, k)\} \mid j$. Therefore the least positive integer j such that $(a^k)^j \equiv 1 \pmod{m}$ is $j = h/(h, k)$.

If a has order h and b has order k, both modulo m, then $(ab)^{hk} = (a^h)^k (b^k)^h \equiv 1 \pmod{m}$, and from Lemma 2.31 we deduce that the order of ab is a divisor of hk. If h and k are relatively prime, then we can say more.

Lemma 2.34 *If a has order $h \pmod{m}$, b has order $k \pmod{m}$, and if $(h, k) = 1$, then ab has order $hk \pmod{m}$.*

Proof Let r denote the order of $ab \pmod{m}$. We have shown that $r \mid hk$. To complete the proof it suffices to show that $hk \mid r$. We note that $b^{rh} \equiv (a^h)^r b^{rh} = (ab)^{rh} \equiv 1 \pmod{m}$. Thus $k \mid rh$ by Lemma 2.31. As $(h, k) = 1$, it follows that $k \mid r$. By a similar argument we see that $h \mid r$. Using again the hypothesis $(h, k) = 1$, we conclude that $hk \mid r$.

We have already seen that the order of a modulo m is a divisor of $\phi(m)$. For certain values of m, there are integers a such that the order of a is equal to $\phi(m)$. These cases are of considerable importance, so a special label is used.

Definition 2.7 *If g belongs to the exponent $\phi(m)$ modulo m, then g is called a* primitive root modulo m.

(In algebraic language, this definition can be stated: If the order of g modulo m is $\phi(m)$, then the multiplicative group of reduced residues modulo m is a cyclic group generated by the element g. Readers not too familiar with group theory can find a more detailed explanation of this in Section 2.10.)

In view of Lemma 2.31, the number a is a solution of the congruence $x^k \equiv 1 \pmod{m}$ if and only if the order of $a \pmod{m}$ divides k. In one special case, namely the situation of Corollary 2.30, we have determined the number of solutions of this congruence. That is, if p is prime and $k|(p-1)$, then there are precisely k residue classes $a \pmod{p}$ such that the order of a modulo p is a divisor of k. If k happens to be a prime power, we can then determine the exact number of residues $a \pmod{p}$ of order k.

Lemma 2.35 *Let p and q be primes, and suppose that $q^\alpha|(p-1)$, where $\alpha \geqslant 1$. Then there are precisely $q^\alpha - q^{\alpha-1}$ residue classes $a \pmod{p}$ of order q^α.*

Proof The divisors of q^α are the numbers q^β with $\beta = 0, 1, \cdots, \alpha$. Of these, q^α is the only one that is not a divisor of $q^{\alpha-1}$. There are q^α residues $\pmod{p}$ of order dividing q^α, and among these there are $q^{\alpha-1}$ residues of order dividing $q^{\alpha-1}$. On subtracting we see that there are precisely $q^\alpha - q^{\alpha-1}$ residues a of order $q^\alpha \pmod{p}$.

Theorem 2.36 *If p is a prime then there exist $\phi(p-1)$ primitive roots modulo p.*

Proof We first establish the existence of at least one primitive root. Let $p - 1 = p_1^{\alpha_1} p_2^{\alpha_2} \cdots p_j^{\alpha_j}$ be the canonical factorization of $p-1$. By Lemma 2.35 we may choose numbers $a_i \pmod{p}$ so that a_i has order $p_i^{\alpha_i}$, $i = 1, 2, \cdots, j$. The numbers $p_i^{\alpha_i}$ are pairwise relatively prime, so by repeated use of Lemma 2.34 we see that $g = a_1 a_2 \cdots a_j$ has order $p_1^{\alpha_1} p_2^{\alpha_2} \cdots p_j^{\alpha_j} = p - 1$. That is, g is a primitive root $\pmod{p}$.

To complete the proof, we determine the exact number of primitive roots $\pmod{p}$. Let g be a primitive root $\pmod{p}$. Then the numbers $g, g^2, g^3, \cdots, g^{p-1}$ form a system of reduced residues $\pmod{p}$. By Lemma 2.33 we see that g^k has order $(p-1)/(k, p-1)$. Thus g^k is a primitive root if and only if $(k, p-1) = 1$. By definition of Euler's phi function, there are exactly $\phi(p-1)$ such values of k in the interval $1 \leqslant k \leqslant p-1$.

Remark on Calculation Suppose that we wish to show that a has order $h\,(\text{mod } m)$, where a, h, and m are given. By using the repeated squaring device discussed in Section 2.4, we may quickly verify that $a^h \equiv 1\,(\text{mod } m)$. If h is small, then we simply examine $a, a^2, \cdots, a^{h-1}\,(\text{mod } m)$, but if h is large (e.g., $h = \phi(m)$), then the amount of calculation here would be prohibitively long. Instead, we note by Lemma 2.31 that the order of a must be a divisor of h. If the order of a is a proper divisor of h then the order of a divides h/p for some prime factor p of h. That is, the order of $a\,(\text{mod } m)$ is h if and only if the following two conditions are satisfied: (i) $a^h \equiv 1\,(\text{mod } m)$, and (ii) for each prime factor p of h, $a^{h/p} \not\equiv 1\,(\text{mod } m)$. In case m is prime, we may take $h = m - 1$ in this criterion to determine whether a is a primitive root. To locate a primitive root we simply try $a = 2$, $a = 3, \cdots$, and in general a primitive root is quickly found. For example, to show that 2 is a primitive root (mod 101), we note that 2 and 5 are the primes dividing 100. Then we calculate that $2^{50} \equiv -1 \not\equiv 1\,(\text{mod } 101)$, and that $2^{20} \equiv 95 \not\equiv 1\,(\text{mod } 101)$.

The techniques discussed in Section 2.4 allow us to prove very quickly that a given number m is composite, but they are not so useful in establishing primality. Suppose that a given number p is a strong pseudoprime to several bases, and is therefore expected to be prime. To show that p is prime it suffices to exhibit a number a of order $p - 1\,(\text{mod } p)$, for then $\phi(p) \geqslant p - 1$, and hence p must be prime. Here the hard part is to factor $p - 1$. (If the desired primitive root is elusive, then p is probably composite.) This approach is developed further in Problems 38 and 39 at the end of this Section.

Up to 10^9 or so one may construct primes by sieving. Larger primes (such as those used in public-key cryptography) can be constructed as follows: Multiply several small primes together, add 1 to this product, and call the result p. This number has no greater chance of being prime than a randomly chosen number of the same size, and indeed it is likely that a pseudoprime test will reveal that p is composite (in which case we try again with a new product of small primes). However, if p passes several such tests, then one may proceed as above to show that p is prime, since the factorization of $p - 1$ is known in advance.

Definition 2.8 *If $(a, p) = 1$ and $x^n \equiv a\,(\text{mod } p)$ has a solution, then a is called an n*th power residue *modulo p.*

If $(g, m) = 1$ then the sequence $g, g^2, \cdots\,(\text{mod } m)$ is periodic. If g is a primitive root $(\text{mod } m)$ then the least period of this sequence is $\phi(m)$, and we see that $g, g^2, \cdots, g^{\phi(m)}$ form a system of reduced residues $(\text{mod } m)$. Thus $g^i \equiv g^j\,(\text{mod } m)$ if and only if $i \equiv j\,(\text{mod } \phi(m))$. By expressing numbers as powers of g, we may convert a multiplicative congruence (mod m) to an additive congruence (mod $\phi(m)$), just as we apply logarithms to real numbers. In this way we determine whether a is an nth power residue (mod p).

Theorem 2.37 *If p is a prime and $(a, p) = 1$, then the congruence $x^n \equiv a \pmod{p}$ has $(n, p-1)$ solutions or no solution according as*

$$a^{(p-1)/(n,p-1)} \equiv 1 \pmod{p}$$

or not.

Proof Let g be a primitive root (mod p), and choose i so that $g^i \equiv a \pmod{p}$. If there is an x such that $x^n \equiv a \pmod{p}$ then $(x, p) = 1$, so that $x \equiv g^u \pmod{p}$ for some u. Thus the proposed congruence is $g^{nu} \equiv g^i \pmod{p}$, which is equivalent to $nu \equiv i \pmod{p-1}$. Put $k = (n, p-1)$. By Theorem 2.17, this has k solutions if $k|i$, and no solution if $k \nmid i$. If $k|i$, then $i(p-1)/k \equiv 0 \pmod{p-1}$, so that $a^{(p-1)/k} \equiv g^{i(p-1)/k} = (g^{p-1})^{i/k} \equiv 1 \pmod{p}$. On the other hand, if $k \nmid i$ then $i(p-1)/k \not\equiv 0 \pmod{p-1}$, and hence $a^{(p-1)/k} \equiv g^{i(p-1)/k} \not\equiv 1 \pmod{p}$.

Example 14 Show that the congruence $x^5 \equiv 6 \pmod{101}$ has 5 solutions.

Solution It suffices to verify that $6^{20} \equiv 1 \pmod{101}$. This is easily accomplished using the technique discussed in Section 2.4. Note that we do not need to find a primitive root g, or to find i such that $g^i \equiv 6 \pmod{101}$. The mere fact that $6^{20} \equiv 1 \pmod{101}$ assures us that $5|i$. (With more work one may prove that $g = 2$ is a primitive root (mod 101), and that $2^{70} \equiv 6 \pmod{101}$. Hence the five solutions are $x \equiv 2^{14+20j} \pmod{101}$ where $j = 0, 1, 2, 3, 4$. That is, $x \equiv 22, 70, 85, 96, 30 \pmod{101}$.)

Corollary 2.38 *Euler's criterion. If p is an odd prime and $(a, p) = 1$, then $x^2 \equiv a \pmod{p}$ has two solutions or no solution according as $a^{(p-1)/2} \equiv 1$ or $\equiv -1 \pmod{p}$.*

Proof Put $b = a^{(p-1)/2}$. Thus $b^2 = a^{p-1} \equiv 1 \pmod{p}$ by Fermat's congruence. From Lemma 2.10 it follows that $b \equiv \pm 1 \pmod{p}$. If $b \equiv -1 \pmod{p}$ then the congruence $x^2 \equiv a \pmod{p}$ has no solution, by Theorem 2.37. If $b \equiv 1 \pmod{p}$ then the congruence has exactly two solutions, by Theorem 2.37.

By taking $a = -1$ in Euler's criterion we obtain a second proof of Theorem 2.12. In the next section we give an algorithm for solving the congruence $x^2 \equiv a \pmod{p}$. In Sections 3.1 and 3.2 a quite different approach of Gauss is developed, which offers an alternative to Euler's criterion for determining whether a given number a is a quadratic residue (mod p).

We have seen that primitive roots provide a valuable tool for analyzing certain congruences (mod p). We now investigate the extent to which this can be generalized to other moduli.

Theorem 2.39 *If p is a prime then there exist $\phi(\phi(p^2)) = (p-1)\phi(p-1)$ primitive roots modulo p^2.*

Proof We show that if g is a primitive root (mod p) then $g + tp$ is a primitive root (mod p^2) for exactly $p-1$ values of t (mod p). Let h denote the order of $g + tp$ (mod p^2). (Thus h may depend on t.) Since $(g+tp)^h \equiv 1 \pmod{p^2}$, it follows that $(g+tp)^h \equiv 1 \pmod{p}$, which in turn implies that $g^h \equiv 1 \pmod{p}$, and hence that $(p-1)|h$. On the other hand, by Corollary 2.32 we know that $h|\phi(p^2) = p(p-1)$. Thus $h = p - 1$ or $h = p(p-1)$. In the latter case $g + tp$ is a primitive root (mod p^2), and in the former case it is not. We prove that the former case arises for only one of the p possible values of t. Let $f(x) = x^{p-1} - 1$. In the former case, $g + tp$ is a solution of the congruence $f(x) \equiv 0 \pmod{p^2}$ lying above g (mod p). Since $f'(g) = (p-1)g^{p-2} \not\equiv 0 \pmod{p}$, we know from Hensel's lemma (Theorem 2.23) that g (mod p) lifts to a unique solution $g + tp$ (mod p^2). For all other values of t (mod p), the number $g + tp$ is a primitive root (mod p^2).

Since each of the $\phi(p-1)$ primitive roots (mod p) give rise to exactly $p-1$ primitive roots (mod p^2), we have now shown that there exist at least $(p-1)\phi(p-1)$ primitive roots (mod p^2). To show that there are no other primitive roots (mod p^2), it suffices to argue as in the preceding proof. Let g denote a primitive root (mod p^2), so that the numbers $g, g^2, \cdots, g^{p(p-1)}$ form a system of reduced residues (mod p^2). By Lemma 2.33, we know that g^k is a primitive root if and only if $(k, p(p-1)) = 1$. By the definition of Euler's phi function, there are precisely $\phi(p(p-1))$ such values of k among the numbers $1, 2, \cdots, p(p-1)$. Since $(p, p-1) = 1$, we deduce from Theorem 2.19 that $\phi(p(p-1)) = \phi(p)\phi(p-1) = (p-1)\phi(p-1)$.

Theorem 2.40 *If p is an odd prime and g is a primitive root modulo p^2, then g is a primitive root modulo p^α for $\alpha = 3, 4, 5, \cdots$.*

Proof Suppose that g is a primitive root (mod p^2), and that h is the order of g (mod p^α) where $\alpha > 2$. From the congruence $g^h \equiv 1 \pmod{p^\alpha}$ we deduce that $g^h \equiv 1 \pmod{p^2}$, and hence that $\phi(p^2)|h$. By Corollary 2.32 we also know that $h|\phi(p^\alpha)$. Thus $h = p^\beta(p-1)$ for some β among

$\beta = 1, 2, \cdots$, or $\alpha - 1$. To prove that $\beta = \alpha - 1$, it suffices to show that

$$g^{p^{\alpha-2}(p-1)} \not\equiv 1 \pmod{p^\alpha}. \tag{2.9}$$

We use induction to show that this holds for all $\alpha \geqslant 2$. By hypothesis, the order of $g \pmod{p^2}$ is $\phi(p^2) = p(p-1)$. Hence $g^{p-1} \not\equiv 1 \pmod{p^2}$, and we have (2.9) when $\alpha = 2$. By Fermat's congruence $g^{p-1} \equiv 1 \pmod p$, so we may write $g^{p-1} = 1 + b_1 p$ with $p \nmid b_1$. By the binomial theorem,

$$g^{p(p-1)} = (1 + b_1 p)^p = 1 + \binom{p}{1} b_1 p + \binom{p}{2} b_1^2 p^2 + \cdots.$$

Since $p > 2$ by hypothesis, $\binom{p}{2} = p(p-1)/2 \equiv 0 \pmod p$, and hence the above is $\equiv 1 + b_1 p^2 \pmod{p^3}$. This gives (2.9) when $\alpha = 3$. Thus we may write $g^{p(p-1)} = 1 + b_2 p^2$ with $p \nmid b_2$. We raise both sides of this to the pth power and repeat this procedure to find that $g^{p^2(p-1)} \equiv 1 + b_2 p^3 \pmod{p^4}$, which gives (2.9) for $\alpha = 4$. Continuing in this way, we conclude that (2.9) holds for all $\alpha \geqslant 2$, and the proof is complete.

The prime $p = 2$ must be excluded, for $g = 3$ is a primitive root (mod 4), but not (mod 8). Indeed it is easy to verify that $a^2 \equiv 1 \pmod 8$ for any odd number a. As $\phi(8) = 4$, it follows that there is no primitive root (mod 8). Suppose that a is odd. Since $8 \mid (a^2 - 1)$ and $2 \mid (a^2 + 1)$, it follows that $16 \mid (a^2 - 1)(a^2 + 1) = a^4 - 1$. That is, $a^4 \equiv 1 \pmod{16}$. On repeating this argument we see that $a^8 \equiv 1 \pmod{32}$, and in general that $a^{2^{\alpha-2}} \equiv 1 \pmod{2^\alpha}$ for $\alpha \geqslant 3$. Since $\phi(2^\alpha) = 2^{\alpha-1}$, we conclude that if $\alpha \geqslant 3$ then

$$a^{\phi(2^\alpha)/2} \equiv 1 \pmod{2^\alpha} \tag{2.10}$$

for all odd a, and hence that there is no primitive root $\pmod{2^\alpha}$ for $\alpha = 3, 4, 5, \cdots$.

Suppose that p is an odd prime and that g is a primitive root $\pmod{p^\alpha}$. We may suppose that g is odd, for if g is even then we have only to replace g by $g + p^\alpha$, which is odd. The numbers $g, g^2, \cdots, g^{\phi(p^\alpha)}$ form a reduced residue system $\pmod{p^\alpha}$. Since these numbers are odd, they also form a reduced residue system $\pmod{2p^\alpha}$. Thus g is a primitive root $\pmod{2p^\alpha}$.

We have established that a primitive root exists modulo m when $m = 1, 2, 4, p^\alpha$, or $2p^\alpha$, (p an odd prime), but that there is no primitive root $\pmod{2^\alpha}$ for $\alpha \geqslant 3$. Suppose now that m is not a prime power or twice a prime power. Then m can be expressed as a product, $m = m_1 m_2$

with $(m_1, m_2) = 1$, $m_1 > 2$, $m_2 > 2$. Let $e = \text{l.c.m.}(\phi(m_1), \phi(m_2))$. If $(a, m) = 1$ then $(a, m_1) = 1$, so that $a^{\phi(m_1)} \equiv 1 \pmod{m_1}$, and hence $a^e \equiv 1 \pmod{m_1}$. Similarly $a^e \equiv 1 \pmod{m_2}$, and hence $a^e \equiv 1 \pmod{m}$. Since $2 \mid \phi(n)$ for all $n > 2$, we see that $2 \mid (\phi(m_1), \phi(m_2))$, so that by Theorem 1.13,

$$e = \frac{\phi(m_1)\phi(m_2)}{(\phi(m_1), \phi(m_2))} < \phi(m_1)\phi(m_2) = \phi(m).$$

Thus there is no primitive root in this case. We have now determined precisely which m possess primitive roots.

Theorem 2.41 *There exists a primitive root modulo m if and only if $m = 1, 2, 4, p^\alpha$, or $2p^\alpha$, where p is an odd prime.*

Theorem 2.37 (and its proof) generalizes to any modulus m possessing a primitive root.

Corollary 2.42 *Suppose that $m = 1, 2, 4, p^\alpha$, or $2p^\alpha$, where p is an odd prime. If $(a, m) = 1$ then the congruence $x^n \equiv a \pmod{m}$ has $(n, \phi(m))$ solutions or no solution, according as*

$$a^{\phi(m)/(n, \phi(m))} \equiv 1 \pmod{m} \tag{2.11}$$

or not.

For the general composite m possessing no primitive root, we factor m and apply the above to the prime powers dividing m.

Example 15 Determine the number of solutions of the congruence $x^4 \equiv 61 \pmod{117}$.

Solution We note that $117 = 3^2 \cdot 13$. As $\phi(9)/(4, \phi(9)) = 6/(4, 6) = 3$ and $61^3 \equiv (-2)^3 \equiv 1 \pmod 9$, we deduce that the congruence $x^4 \equiv 61 \pmod 9$ has $(4, \phi(9)) = 2$ solutions. Similarly $\phi(13)/(4, \phi(13)) = 3$ and $61^3 \equiv (-4)^3 \equiv 1 \pmod{13}$, so the congruence $x^4 \equiv 61 \pmod{13}$ has $(4, \phi(13)) = 4$ solutions. Thus by Theorem 2.20, the number of solutions modulo 117 is $2 \cdot 4 = 8$.

This method fails in case the modulus is divisible by 8, as Corollary 2.42 does not apply to the higher powers of 2. In order to establish an analogue of Corollary 2.42 for the higher powers of 2, we first show that 5 is nearly a primitive root $\pmod{2^\alpha}$.

Theorem 2.43 *Suppose that* $\alpha \geqslant 3$. *The order of* $5 \pmod{2^\alpha}$ *is* $2^{\alpha-2}$. *The numbers* $\pm 5, \pm 5^2, \pm 5^3, \cdots, \pm 5^{2^{\alpha-2}}$ *form a system of reduced residues* $\pmod{2^\alpha}$. *If a is odd, then there exist i and j such that* $a \equiv (-1)^i 5^j \pmod{2^\alpha}$. *The values of i and j are uniquely determined* $\pmod 2$ *and* $\pmod{2^{\alpha-2}}$, *respectively*.

Proof We first show that $2^\alpha \| (5^{2^{\alpha-2}} - 1)$ for $\alpha \geqslant 2$. This is clear for $\alpha = 2$. If $a \equiv 1 \pmod 4$ then $2 \| (a+1)$, and hence the power of 2 dividing $a^2 - 1 = (a-1)(a+1)$ is exactly one more than the power of 2 dividing $a - 1$. Taking $a = 5$, we deduce that $2^3 \| (5^2 - 1)$. Taking $a = 5^2$, we then deduce that $2^4 \| (5^4 - 1)$, and so on. Now let h denote the order of $5 \pmod{2^\alpha}$. Since $h | \phi(2^\alpha)$ and $\phi(2^\alpha) = 2^{\alpha-1}$, we know that $h = 2^\beta$ for some β. But the least β for which $5^{2^\beta} \equiv 1 \pmod{2^\alpha}$ is $\beta = \alpha - 2$. Thus 5 has order $2^{\alpha-2} \pmod{2^\alpha}$, so that the numbers $5, 5^2, 5^3, \cdots, 5^{2^{\alpha-2}}$ are mutually incongruent $\pmod{2^\alpha}$. Of the $2^{\alpha-1}$ integers in a reduced residue system $\pmod{2^\alpha}$, half are $\equiv 1 \pmod 4$, and half are $\equiv 3 \pmod 4$. The numbers 5^j are all $\equiv 1 \pmod 4$. Since the powers of 5 lie in $2^{\alpha-2}$ distinct residue classes $\pmod{2^\alpha}$, and since $2^{\alpha-2}$ of the integers $\pmod{2^\alpha}$ are $\equiv 1 \pmod 4$, for any $a \equiv 1 \pmod 4$ there is a j such that $a \equiv 5^j \pmod{2^\alpha}$. For any integer $a \equiv 3 \pmod 4$, we observe that $-a \equiv 1 \pmod 4$, and hence that $-a \equiv 5^j \pmod{2^\alpha}$ for some j.

Corollary 2.44 *Suppose that* $\alpha \geqslant 3$ *and that a is odd. If n is odd, then the congruence* $x^n \equiv a \pmod{2^\alpha}$ *has exactly one solution. If n is even, then choose* β *so that* $(n, 2^{\alpha-2}) = 2^\beta$. *The congruence* $x^n \equiv a \pmod{2^\alpha}$ *has* $2^{\beta+1}$ *solutions or no solution according as* $a \equiv 1 \pmod{2^{\beta+2}}$ *or not*.

Proof Since a is odd, we may choose i and j so that $a \equiv (-1)^i 5^j \pmod{2^\alpha}$. As any x for which $x^n \equiv a \pmod{2^\alpha}$ is necessarily odd, we may suppose that $x \equiv (-1)^u 5^v \pmod{2^\alpha}$. The desired congruence then takes the form $(-1)^{nu} 5^{nv} \equiv (-1)^i 5^j \pmod{2^\alpha}$. By Theorem 2.43, this is equivalent to the pair of congruences $nu \equiv i \pmod 2$, $nv \equiv j \pmod{2^{\alpha-2}}$. If n is odd, then by Theorem 2.17 there exists exactly one $u \pmod 2$ for which the first congruence holds, and exactly one $v \pmod{2^{\alpha-2}}$ for which the second congruence holds, and hence there exists precisely one solution x in this case.

Suppose now that n is even. We apply Theorem 2.17 two more times. If $i \equiv 0 \pmod 2$ then the congruence $nu \equiv i \pmod 2$ has two solutions. Otherwise it has none. If $j \equiv 0 \pmod{2^\beta}$ then the congruence $nv \equiv j \pmod{2^{\alpha-2}}$ has exactly 2^β solutions. Otherwise it has none. Thus the congruence $x^n \equiv a \pmod{2^\alpha}$ has $2^{\beta+1}$ solutions or no solution, according as $a \equiv 5^j \pmod{2^\alpha}$, $j \equiv 0 \pmod{2^\beta}$, or not. From Theorem 2.43 we know that 5 has order $2^\beta \pmod{2^{\beta+2}}$. Thus by Lemma 2.31, $5^j \equiv 1 \pmod{2^{\beta+2}}$ if and only if $2^\beta | j$. Since $2^{\beta+2} | 2^\alpha$, the condition on a is precisely that $a \equiv 1 \pmod{2^{\beta+2}}$.

PROBLEMS

1. Find a primitive root of the prime 3; the prime 5; the prime 7; the prime 11; the prime 13.
2. Find a primitive root of 23.
3. How many primitive roots does the prime 13 have?
4. To what exponents do each of $1, 2, 3, 4, 5, 6$ belong modulo 7? To what exponents do they belong modulo 11?
5. Let p be an odd prime. Prove that a belongs to the exponent 2 modulo p if and only if $a \equiv -1 \pmod{p}$.
6. If a belongs to the exponent h modulo m, prove that no two of $a, a^2, a^3, \cdots, a^h$ are congruent modulo m.
7. If p is an odd prime, how many solutions are there to $x^{p-1} \equiv 1 \pmod{p}$; to $x^{p-1} \equiv 2 \pmod{p}$?
8. Use Theorem 2.37 to determine how many solutions each of the following congruences has:
   (*a*) $x^{12} \equiv 16 \pmod{17}$ (*b*) $x^{48} \equiv 9 \pmod{17}$
   (*c*) $x^{20} \equiv 13 \pmod{17}$ (*d*) $x^{11} \equiv 9 \pmod{17}$.
9. Show that $3^8 \equiv -1 \pmod{17}$. Explain why this implies that 3 is a primitive root of 17.
10. Show that the powers of $3 \pmod{17}$ are 3, 9, 10, 13, 5, 15, 11, 16, 14, 8, 7, 4, 12, 2, 6, 1. Use this information to find the solutions of the congruences in Problem 8.
11. Using the data in the preceding problem, decide which of the congruences $x^2 \equiv 1, x^2 \equiv 2, x^2 \equiv 3, \cdots, x^2 \equiv 16 \pmod{17}$, have solutions.
12. Prove that if p is a prime, $(a, p) = 1$ and $(n, p-1) = 1$, then $x^n \equiv a \pmod{p}$ has exactly one solution.
13. Show that the numbers $1^k, 2^k, \cdots, (p-1)^k$ form a reduced residue system $\pmod{p}$ if and only if $(k, p-1) = 1$.
14. Suppose that a has order $h \pmod{p}$, and that $a\bar{a} \equiv 1 \pmod{p}$. Show that $\bar{a}$ also has order h. Suppose that g is a primitive root $\pmod{p}$, and that $a \equiv g^i \pmod{p}$, $0 \leqslant i < p-1$. Show that $\bar{a} \equiv g^{p-1-i} \pmod{p}$.
15. Prove that if a belongs to the exponent h modulo a prime p, and if h is even, then $a^{h/2} \equiv -1 \pmod{p}$.

16. Let m and n be positive integers. Show that $(2^m - 1, 2^n + 1) = 1$ if m is odd.

17. Show that if $a^k + 1$ is prime, $k > 0$, and $a > 1$ then k is a power of 2. Show that if $p \mid (a^{2n} + 1)$ then $p = 2$ or $p \equiv 1 \pmod{2^{n+1}}$. (H)

18. Show that if g and g' are primitive roots modulo an odd prime p, then gg' is not a primitive root of p.

19. Show that if $a^h \equiv 1 \pmod{p}$ then $a^{ph} \equiv 1 \pmod{p^2}$. Show that if g is a primitive root $\pmod{p^2}$ then it is a primitive root $\pmod{p}$.

20. Of the 101 integers in a complete residue system $\pmod{101^2}$ that are $\equiv 2 \pmod{101}$, which one is not a primitive root $\pmod{101^2}$?

21. Let g be a primitive root of the odd prime p. Show that $-g$ is a primitive root, or not, according as $p \equiv 1 \pmod 4$ or $p \equiv 3 \pmod 4$.

22. Let g be a primitive root $\pmod{p}$. Show that $(p-1)! \equiv g \cdot g^2 \cdot \cdots \cdot g^{p-1} \equiv g^{p(p-1)/2} \pmod{p}$. Use this to give another proof of Wilson's congruence (Theorem 2.11).

***23.** Prove that if a belongs to the exponent 3 modulo a prime p, then $1 + a + a^2 \equiv 0 \pmod{p}$, and $1 + a$ belongs to the exponent 6.

***24.** Let a and $n > 1$ be any integers such that $a^{n-1} \equiv 1 \pmod{n}$ but $a^d \not\equiv 1 \pmod{n}$ for every proper divisor d of $n - 1$. Prove that n is a prime.

***25.** Show that the number of reduced residues $a \pmod{m}$ such that $a^{m-1} \equiv 1 \pmod{m}$ is exactly $\prod_{p \mid m} (p - 1, m - 1)$.

***26.** (Recall that m is a Carmichael number if $a^{m-1} \equiv 1 \pmod{m}$ for all reduced residues $a \pmod{m}$.) Show that m is a Carmichael number if and only if m is square-free and $(p - 1) \mid (m - 1)$ for all primes p dividing m. Deduce that $2821 = 7 \cdot 13 \cdot 31$ is a Carmichael number.

***27.** Show that m is a Carmichael number if and only if m is composite and $a^m \equiv a \pmod{m}$ for all integers a.

***28.** Show that the following are equivalent statements concerning the positive integer n:

(i) n is square-free and $(p - 1) \mid n$ for all primes p dividing n;
(ii) If j and k are positive integers such that $j \equiv k \pmod{n}$, then

$$a^j \equiv a^k \pmod{n} \text{ for all integers } a.$$

(The numbers 1, 2, 6, 42, 1806 have this property, but there are no others. See J. Dyer-Bennet, "A theorem on partitions of the set of positive integers," *Amer. Math. Monthly*, 47 (1940), 152–154.)

***29.** Show that the sequence $1^1, 2^2, 3^3, \cdots$, considered $\pmod{p}$ is periodic with least period $p(p - 1)$.

***30.** Suppose that $(10a, q) = 1$, and that k is the order of $10 \pmod{q}$. Show that the decimal expansion of the rational number a/q is periodic with least period k.

***31.** Show that the decimal expansion of $1/p$ has period $p - 1$ if and only if 10 is a primitive root of p. (It is conjectured that if g is not a square, and if $g \neq -1$, then there are infinitely many primes of which g is a primitive root.)

***32.** Let $r_1, r_2, \cdots, r_n$ be a reduced residue system modulo m ($n = \phi(m)$). Show that the numbers $r_1^k, r_2^k, \cdots, r_n^k$ form a reduced residue system (mod m) if and only if $(k, \phi(m)) = 1$. (H)

33. Let k and a be positive integers, with $a \geqslant 2$. Show that $k | \phi(a^k - 1)$. (H)

34. Show that if $p | \phi(m)$ and $p \nmid m$ then there is at least one prime factor q of m such that $q \equiv 1 \pmod{p}$.

35. Let p be a given prime number. Prove that there exist infinitely many prime numbers $q \equiv 1 \pmod{p}$. (H)

36. Primes $\equiv 1 \pmod{m}$. For any positive integer m, prove that the arithmetic progression

$$1 + m, 1 + 2m, 1 + 3m, \cdots \tag{2.12}$$

contains infinitely many primes. An elementary proof of this is outlined in parts (i) to (vii) below. (The argument follows that of I. Niven and B. Powell, "Primes in certain arithmetic progressions," *Amer. Math. Monthly*, 83 (1976), 467–469, as simplified by R. W. Johnson.)

(i) Prove that it suffices to show that for every positive integer m, the arithmetic progression (2.12) contains at least one prime. Note also that we may suppose that $m \geqslant 3$.

We now show that for any integer $m \geqslant 3$, the number $m^m - 1$ has at least one prime divisor $\equiv 1$ modulo m. We suppose that $m \geqslant 3$ and that $m^m - 1$ has no prime divisor $\equiv 1 \pmod{m}$, and derive a contradiction.

(ii) Let q be any prime divisor of $m^m - 1$, so that $q \not\equiv 1 \pmod{m}$. Let h denote the order of $m \pmod{q}$, so that $m^h \equiv 1 \pmod{q}$, and moreover $m^d \equiv 1 \pmod{q}$ if and only if $h \mid d$, by Lemma 2.31. Verify that $h \mid (q - 1)$ and $h \mid m$. Prove that $h < m$, so that $m = hc$ with $c > 1$. *Suggestion*: From $h = m$ deduce that $m \mid (q - 1)$.

(iii) Let q^r be the highest power of q dividing $m^m - 1$; thus $q^r \| (m^m - 1)$. Prove that $q^r \| (m^h - 1)$, and that $q^r \| (m^d - 1)$ for every integer d such that $h | d$ and $d | m$. *Suggestion*: Verify that it suffices to prove the property for $m^h - 1$, since each of $m^h - 1$,

$m^d - 1$, and $m^m - 1$ is a divisor of the next one. Since $m = hc$, we have $m^m - 1 = (m^h - 1)F(m)$ where

$$F(m) = m^{hc-h} + m^{hc-2h} + m^{hc-3h} + \cdots + m^h + 1.$$

Then $F(m) \equiv 1 + 1 + 1 + \cdots + 1 \equiv c \pmod{q}$. Also $q \mid (m^m - 1)$ implies $q \nmid m$ and $q \nmid c$.

These properties of q hold for any prime divisor of $m^m - 1$. Of course different prime factors may give different values of h, c, and r, because these depend on q. To finish the proof we need one additional concept. Consider the set of integers of the form m/s, where s is any square-free divisor of m, excluding $s = 1$. We partition this set into two disjoint subsets $\mathscr{T}$ and $\mathscr{V}$ according as the number of primes dividing s is odd or even. Put

$$Q = \Big(\prod_{d \in \mathscr{T}} (m^d - 1)\Big)\Big(\prod_{d \in \mathscr{V}} (m^d - 1)\Big)^{-1}.$$

(iv) Let q be the prime factor of $m^m - 1$ under consideration, and let $m^d - 1$ be a factor that occurs in one of the two products displayed. Use (ii) to show that $q \mid (m^d - 1)$ if and only if $s \mid c$.

(v) Let k denote the number of distinct primes dividing c. Show that the number of $d \in \mathscr{T}$ for which $q \mid (m^d - 1)$ is $\binom{k}{1} + \binom{k}{3} + \binom{k}{5} \cdots$, and that this sum is 2^{k-1}. Similarly show that the number of $d \in \mathscr{V}$ for which $q \mid (m^d - 1)$ is $\binom{k}{2} + \binom{k}{4} + \cdots$, and that this is $2^{k-1} - 1$. Use (iii) to show that $q^r \| Q$ for each prime divisor q of $m^m - 1$. Deduce that $Q = m^m - 1$.

(vi) Show that if b is a positive integer and $m \geqslant 3$ then $m^b - 1 \not\equiv \pm 1 \pmod{m^{b+1}}$.

(vii) Prove that the equation $Q = m^m - 1$ is impossible, by writing the equation in the form $(m^m - 1)\prod_{d \in \mathscr{V}} (m^d - 1) = \prod_{d \in \mathscr{T}} (m^d - 1)$, and evaluating both sides $\pmod{m^{b+1}}$ where b is the least integer of the type d that appears in the definition of Q.

***37.** Show that if $n > 1$ then $n \nmid (2^n - 1)$. (H)

***38.** Let m be given, suppose that q is a prime number, $q^\alpha \| (m - 1)$, $\alpha > 0$, and that there is a number a such that $a^{m-1} \equiv 1 \pmod{m}$ but $(a^{(m-1)/q} - 1, m) = 1$. Show that $p \equiv 1 \pmod{q^\alpha}$ for all prime factors p of m.

***39.** Let m be given, and let s be a product of prime powers q^α each having the property described in the preceding problem. Show that if $s > m^{1/2}$ then m is prime.

2.9 CONGRUENCES OF DEGREE TWO, PRIME MODULUS

If $f(x) \equiv 0 \pmod p$ is of degree 2, then $f(x) = ax^2 + bx + c$, and a is relatively prime to p. We shall suppose $p > 2$ since the case $p = 2$ offers no difficulties. Then p is odd, and $4af(x) = (2ax + b)^2 + 4ac - b^2$. Hence u is a solution of $f(x) \equiv 0 \pmod p$ if and only if $2au + b \equiv v \pmod p$, where v is a solution of $v^2 \equiv b^2 - 4ac \pmod p$. Furthermore, since $(2a, p) = 1$, for each solution v there is one, and only one, u modulo p such that $2au + b \equiv v \pmod p$. Clearly different v modulo p yield different u modulo p. Thus the problem of solving the congruence of degree 2 is reduced to that of solving a congruence of the form $v^2 \equiv k \pmod p$. Following some preliminary observations on this congruence, we turn to an algorithm, called RESSOL, for finding its solutions.

If $a \equiv 0 \pmod p$, then this has the sole solution $x \equiv 0 \pmod p$. If $a \not\equiv 0 \pmod p$, then the congruence $x^2 \equiv a \pmod p$ may have no solution, but if x is a solution then $-x$ is also a solution. Since p is odd, $x \not\equiv -x \pmod p$, and thus the congruence has two distinct solutions in this case. It cannot have more than two, by Corollary 2.27.

If p is a small prime then the solutions of the congruence $x^2 \equiv a \pmod p$ may be found by simply trying $x = 0, x = 1, \cdots, x = (p-1)/2$ until one is found. Since this involves $\approx p$ multiplications, for large p it is desirable to have a more efficient procedure. If $p = 2$ then it suffices to take $x = a$. Thus we may suppose that $p > 2$. By Euler's criterion we may suppose that $a^{(p-1)/2} \equiv 1 \pmod p$, for otherwise the congruence has no solution.

Suppose first that $p \equiv 3 \pmod 4$. In this case we can verify that $x \equiv \pm a^{(p+1)/4}$ are the solutions, for

$$\left(\pm a^{(p+1)/4}\right)^2 = a^{(p+1)/2} = a \cdot a^{(p-1)/2} \equiv a \pmod p.$$

Note that it is not necessary to verify in advance that $a^{(p-1)/2} \equiv 1 \pmod p$. It suffices to calculate $x \equiv a^{(p+1)/4} \pmod p$. If $x^2 \equiv a \pmod p$, then the solutions are $\pm x$. Otherwise $x^2 \equiv -a \pmod p$, and we can conclude that a is a quadratic nonresidue. Thus $x = \pm a^{(p+1)/4}$ are the solutions, if the congruence has a solution. This takes care of roughly half the primes. As always with large exponents, the value of $a^{(p+1)/4} \pmod p$ is determined using the repeated squaring device discussed in Section 2.4. Hence the number of congruential multiplications required is only of the order of magnitude $\log p$.

Suppose now that $p \equiv 1 \pmod 4$. We have already considered the special case $x^2 \equiv -1 \pmod p$, and in proving Theorem 2.12 we gave a formula for the solutions, namely $x = \pm((p-1)/2)!$. However, this formula is useless for large p, as it involves $\approx p$ multiplications. On the other hand, if a quadratic nonresidue z is known then we may take $x \equiv \pm z^{(p-1)/4} \pmod p$, since then $x^2 \equiv z^{(p-1)/2} \equiv -1 \pmod p$ by Euler's criterion. Thus in this special case it suffices to find a quadratic nonresidue. We can try small numbers in turn, or use a random number generator to provide "random" residue classes. In either case, since half the reduced residues are quadratic nonresidues, we may expect that the average number of trials is 2. (Here our interest is not in a deterministic algorithm of proven efficiency, but rather a calculational procedure that is quick in practice.)

We now develop these ideas to find the roots of the congruence $x^2 \equiv a \pmod p$ for arbitrary a and p. We begin with a few general observations. Let a and b be relatively prime to m, and suppose that a and b both have order $h \pmod m$. Then $(ab)^h \equiv 1 \pmod m$, and hence the order of ab is a divisor of h. In general nothing more can be said. It may be that b is the inverse of a, so that $ab \equiv 1 \pmod m$, in which case the order of ab is 1. On the other hand, the order of ab may be as large as h. (Consider $3 \pmod{11}$, $5 \pmod{11}$, and $3 \cdot 5 \equiv 4 \pmod{11}$. All three of these numbers have order 5.) Nevertheless there is one particular situation in which a little more can be established.

Theorem 2.45 *If a and b are relatively prime to a prime number p, and if a and b both have order $2^j \pmod p$ with $j > 0$, then ab has order $2^{j'} \pmod p$ for some $j' < j$.*

Proof Since a has order $2^j \pmod p$, it follows that $2^j | (p-1)$, and thus $p > 2$. Put $x = a^{2^{j-1}}$. Then $x \not\equiv 1 \pmod p$ but $x^2 = a^{2^j} \equiv 1 \pmod p$. Thus by Lemma 2.10 it follows that $x \equiv -1 \pmod p$. Similarly, $b^{2^{j-1}} \equiv -1 \pmod p$, and it follows that

$$(ab)^{2^{j-1}} = a^{2^{j-1}} b^{2^{j-1}} \equiv (-1)(-1) \equiv 1 \pmod p.$$

From this and Lemma 2.31 we deduce that the order of ab is a divisor of 2^{j-1}, that is, the order of ab is $2^{j'}$ for some $j' < j$.

Neither Theorem 2.45 nor its proof involves primitive roots, but some further insight can be obtained by interpreting the situation in terms of powers of a given primitive root g. Write $a \equiv g^\alpha \pmod{p}$, where $0 \leqslant \alpha < p - 1$. By Lemma 2.33, the order of g^α is $(p-1)/(p-1, \alpha)$. Write $p - 1 = m2^k$ with m odd. The hypothesis that a has order 2^j is thus equivalent to the relation $(p-1, \alpha) = m2^{k-j}$. That is, $\alpha = \alpha_1 m 2^{k-j}$ with α_1 odd. Similarly, $b \equiv g^\beta \pmod{p}$ with $\beta = \beta_1 m 2^{k-j}$, β_1 odd. But then $ab \equiv g^{\alpha+\beta} \pmod{p}$, and $\alpha + \beta = (\alpha_1 + \beta_1)m2^{k-j}$. Since α_1 and β_1 are both odd, it follows that $\alpha_1 + \beta_1$ is even. Choose i so that $(\alpha_1 + \beta_1, 2^j) = 2^i$. Since $j > 0$ by hypothesis, it follows that $i > 0$. Moreover, the order of ab is 2^{j-i}, so we have $j' = j - i < j$.

With these tools in hand, we describe the algorithm RESSOL (for RESidue SOLver), which locates x such that $x^2 \equiv a \pmod{p}$. We begin by determining the power of 2 in $p - 1$. Thus we find k and m with m odd, so that $p - 1 = 2^k m$. We are supposing that $p > 2$, so that $k > 0$. Set $r \equiv a^{(m+1)/2} \pmod{p}$ and $n \equiv a^m \pmod{p}$. We note that

$$r^2 \equiv an \pmod{p}. \tag{2.13}$$

If $n \equiv 1 \pmod{p}$, then it suffices to take $x \equiv \pm r \pmod{p}$. If $n \not\equiv 1 \pmod{p}$, then we find a quadratic nonresidue z, and put $c \equiv z^m \pmod{p}$. We note that

$$c^{2^k} = z^{2^k m} = z^{p-1} \equiv 1 \pmod{p}.$$

Thus the order of c is a divisor of 2^k. Moreover,

$$c^{2^{k-1}} = z^{2^{k-1}m} = z^{(p-1)/2} \equiv -1 \pmod{p}$$

since z is a quadratic nonresidue. Thus the order of c is exactly 2^k. Similarly,

$$n^{2^k} = a^{2^k m} = a^{p-1} \equiv 1 \pmod{p},$$

so that the order of n divides 2^k. By repeatedly squaring n we determine the exact order of n, say $2^{k'}$. Since

$$n^{2^{k-1}} = a^{2^{k-1}m} = a^{(p-1)/2},$$

we see that a is a quadratic residue $\pmod{p}$ if and only if

$$n^{2^{k-1}} \equiv 1 \pmod{p},$$

which in turn is equivalent to the inequality $k' < k$. It is worth checking that this inequality holds, for otherwise $k' = k$, a is a quadratic nonresidue and the proposed congruence has no solution. At this point of the algorithm, we begin a loop. Set $b \equiv c^{2^{k-k'-1}} \pmod p$. We put $r' \equiv br \pmod p$, $c' \equiv b^2 \pmod p$, $n' \equiv c'n \pmod p$. By multiplying both sides of (2.13) by b^2 we find that

$$r'^2 \equiv an' \pmod p. \tag{2.14}$$

The point of this construction is that c' has order exactly $2^{k'}$. Since $n \not\equiv 1 \pmod p$ in the present case, it follows that $k' > 0$. Thus by Theorem 2.45, the order of $n' \equiv c'n$ is $2^{k''}$ where $k'' < k'$. (We determine the value of k'' by repeated squaring.) If $k'' = 0$, then $n' \equiv 1 \pmod p$, and we see from (2.14) that it suffices to take $x \equiv \pm r' \pmod p$. If $n' \not\equiv 1 \pmod p$, then $k'' > 1$, and the situation is the same as when the loop began, except that the numbers c (of order 2^k) and n (of order $2^{k'}$) with $0 < k' < k$ have been replaced by c' (of order $2^{k'}$) and n' (of order $2^{k''}$) with $0 < k'' < k'$, while r has been replaced by r' and (2.13) has been replaced by (2.14). Since $k'' < k'$, some progress has been made. By executing this loop repeatedly, we eventually arrive at a set of these variables for which $n \equiv 1 \pmod p$, and then $x \equiv \pm r \pmod p$ is the desired solution.

As a numerical example of this algorithm, suppose we wish to find the roots of the congruence $x^2 \equiv 43 \pmod{97}$. Thus $p = 97$, and $p - 1 = 2^5 \cdot 3$. By using the method described in Section 2.4, we find that $r \equiv 43^{(3+1)/2} = 6 \pmod{97}$, and that $n \equiv 43^3 \equiv 64 \pmod{97}$. Thus the congruence (2.13) is $6^2 \equiv 43 \cdot 64 \pmod{97}$. Since $n \not\equiv 1 \pmod{97}$, we must find a quadratic nonresidue. We note that $(p - 1)/2 = 48$, and calculate that $2^{48} \equiv 1 \pmod{97}$. Thus 2 is a quadratic residue, by Euler's criterion. Similarly 3 is a quadratic residue, but 5 is a quadratic nonresidue. We set $z = 5$, $c \equiv 5^3 \equiv 28 \pmod{97}$. Thus c has order $2^5 \pmod{97}$. By repeatedly squaring, we discover that n has order $2^3 \pmod{97}$. That is, $k' = 3$, and we now begin the loop. Since $k - k' - 1 = 1$, we set $b \equiv c^2 \equiv 8 \pmod{97}$, and $c' \equiv b^2 \equiv 64 \pmod{97}$. On multiplying both sides of (2.13) by b^2 we obtain the congruence (2.14) with $r' \equiv 8 \cdot 6 \equiv 48 \pmod{97}$ and $n' \equiv 64 \cdot 64 \equiv 22 \pmod{97}$. That is, $48^2 \equiv 43 \cdot 22 \pmod{97}$. By repeated squaring, we discover that 22 has order $2^2 \pmod{97}$, so we take $k'' = 2$, and we are ready to begin the loop over. With the new values of the parameters, we now have $k - k' - 1 = 0$, so we set $b \equiv c \equiv 64 \pmod{97}$, $c' \equiv 64^2 \equiv 22 \pmod{97}$, and obtain the congruence $65^2 \equiv (64 \cdot 48)^2 \equiv 43 \cdot (22 \cdot 22)^2 \equiv 43 \cdot 96 \pmod{97}$. That is, $r' \equiv 65$, $n' \equiv 96 \pmod{97}$. Here 96 has order 2, so that $k'' = 1$. Since $n' \not\equiv 1 \pmod{97}$, we must execute the loop a third time. As $k - k' - 1 = 0$, we set $b \equiv c \equiv 22 \pmod{97}$, $c' \equiv b^2 \equiv 96 \pmod{97}$, and we obtain the congruence $72^2 \equiv (22 \cdot 65)^2 \equiv 43 \cdot (96 \cdot$

96) $\equiv 43 \pmod{97}$. Thus the solutions are $x \equiv \pm 72 \pmod{97}$. This example of the algorithm is unusually long because $p-1$ is divisible by a high power of 2.

To gain further insight into this algorithm, let g be a primitive root $\pmod p$. Then $z \equiv g^n \pmod p$ for some n, and hence $c \equiv z^m \equiv g^{mn} \pmod p$. But n is odd since z is a quadratic nonresidue, and thus $(mn, p-1) = m$. Consequently by Lemma 2.33 the order of c is 2^k. In general, the order of g^t is a power of 2 if and only if $m|t$. There are precisely 2^k such residue classes, namely $g^m, g^{2m}, g^{3m}, \cdots, g^{2^k m}$. On the other hand, the 2^k residue classes $c, c^2, c^3, \cdots, c^{2^k}$ are distinct, and each one has order a power of 2, so this latter sequence is simply a permutation of the former one. Thus the order of a residue class is a power of 2 if and only if it is a power of c. But $n \equiv a^m \pmod p$ has order that is a power of 2, and hence there is a non-negative integer u such that $n \equiv c^u \pmod p$. A number c^t is a quadratic residue or nonresidue according as t is even or odd. Hence if a is a quadratic residue, then u is even, and the solutions sought are $x \equiv \pm c^{u/2} \pmod p$. Thus it suffices to determine the value of $u \pmod{2^k}$. As it stands, the algorithm does not do this, but it can be slightly modified to yield u. (See Problem 5 below.) If $n \not\equiv 1 \pmod p$, then $u \not\equiv 0 \pmod{2^k}$. Suppose that $0 < u < 2^k$. If the order of n is $2^{k'}$ then $2^{k-k'}|u$ but $2^{k-k'+1} \nmid u$. Thus we obtain some information concerning the binary expansion of u. Repeated iterations of the loop (suitably modified) determine further coefficients in the binary expansion of u, and eventually u is determined. Alternatively, the value of u could be determined by calculating the successive powers of c until n is encountered, but that might require as many as 2^k multiplications. The algorithm given is much faster, as the loop is executed at most k times.

PROBLEMS

1. Reduce the following congruences to the form $(x-r)^2 \equiv k \pmod p$:
   (a) $4x^2 + 2x + 1 \equiv 0 \pmod 5$; (b) $3x^2 - x + 5 \equiv 0 \pmod 7$;
   (c) $2x^2 + 7x - 10 \equiv 0 \pmod{11}$; (d) $x^2 + x - 1 \equiv 0 \pmod{13}$.
2. Suppose that $f(x) = ax^2 + bx + c$, and that $D = b^2 - 4ac$. Show that if p is an odd prime, $p \nmid a, p|D$, then $f(x) \equiv 0 \pmod p$ has exactly one solution. Show that if p is an odd prime, $p \nmid a, p \nmid D$, then the congruence $f(x) \equiv 0 \pmod p$ has either 0 or 2 solutions, and that if x is a solution then $f'(x) \not\equiv 0 \pmod p$.

*3. Let $f(x) = ax^2 + bx + c$, and let p be an odd prime that does not divide all the coefficients a, b, c. Show that the congruence $f(x) \equiv 0 \pmod{p^2}$ has either 0, 1, 2, or p solutions.

4. With the aid of a pocket calculator, use RESSOL to find the solutions of the following congruences:
(*a*) $x^2 \equiv 10 \pmod{13}$; (*b*) $x^2 \equiv 5 \pmod{19}$;
(*c*) $x^2 \equiv 3 \pmod{11}$; (*d*) $x^2 \equiv 7 \pmod{29}$.

***5.** Suppose that p is an odd prime, $p - 1 = m2^k$ with m odd. Let z be a quadratic nonresidue of p, and put $c \equiv z^m \pmod p$. Suppose that $n \not\equiv 1 \pmod p$, and that the order of n is a power of 2, say $2^{k'}$. Let u be chosen, $0 < u < 2^k$, so that $c^u \equiv n \pmod p$. Put $n' \equiv n\bar{c}^{2^{k-k'}} \pmod p$ where $c\bar{c} \equiv 1 \pmod p$. Show that the order of n' is $2^{k''}$ for some $k'' < k'$. If $k'' > 0$, put $n'' \equiv n'\bar{c}^{2^{k-k''}} \pmod p$. Continue in this manner. Show that $2^{k-k'} + 2^{k-k''} + \cdots$ is the binary expansion of u.

6. Suppose that the reduced residue classes a and $b \pmod p$ both have order 3^j. Here $j > 0$ and p is prime. Show that of the two residue classes ab and ab^2, one of them has order 3^j and the other has order $3^{j'}$ for some $j' < j$.

7. Suppose that $(a, p) = 1$ and that p is a prime such that $p \equiv 2 \pmod 3$. Show that the congruence $x^3 \equiv a \pmod p$ has the unique solution $x \equiv a^{(2p-1)/3} \pmod p$.

***8.** Suppose that $p - 1 = m3^k$ with $k > 0$ and $3 \nmid m$. Show that if $(n, p) = 1$ then the order of n is a power of 3 if and only if the congruence $x^m \equiv n \pmod p$ has a solution. If $m \equiv 2 \pmod 3$ then put $r \equiv a^{(m+1)/3}$, $n \equiv a^m \pmod p$. If $m \equiv 1 \pmod 3$, put $r \equiv a^{(2m+1)/3}$, $n \equiv a^{2m} \pmod p$. Show that in either case, $r^3 \equiv an \pmod p$, and that the order of n is a power of 3, say $3^{k'}$. Choose z so that $z^{(p-1)/3} \not\equiv 1 \pmod p$, and set $c \equiv z^m \pmod p$. Show that c has order exactly 3^k, and that there is an integer u, $0 \leqslant u < 3^k$, such that $n \equiv c^u \pmod p$. Show that one of the numbers $nc^{3^{k-k'}}$, $nc^{2 \cdot 3^{k-k'}}$ has order $3^{k'}$, and that the order of the other one is a smaller power of 3, say $3^{k''}$. Let n' denote this number with smaller order. Determine r' so that $r'^3 \equiv an' \pmod p$. Continuing in this manner, construct an algorithm for computing the solutions of the congruence $x^3 \equiv a \pmod p$.

2.10 NUMBER THEORY FROM AN ALGEBRAIC VIEWPOINT

In this section and the next we consider congruences from the perspective of modern algebra. The theory of numbers provides a rich source of examples of the structures of abstract algebra. We shall treat briefly three of these structures: groups, rings, and fields.

Before giving the technical definition of a group, let us explain some of the language used. Operations such as addition and multiplication are called *binary operations* because two elements are added, or multiplied, to produce a third element. The subtraction of pairs of elements, $a - b$, is likewise a binary operation. So also is exponentiation, a^b, in which the element a is raised to the bth power. Now, a group consists of a set of elements together with a binary operation on those elements, such that certain properties hold. The number theoretic groups with which we deal will have integers or sets of integers as elements, and the operation will be either addition or multiplication. However, a general group can have elements of any sort and any kind of binary operation, just as long as it satisfies the conditions that we shall impose shortly.

We begin with a general binary operation denoted by $\oplus$, and we presume that this binary operation is single-valued. This means that for each pair a, b of elements, $a \oplus b$ has a unique value or is not defined. A set of elements is said to be *closed* with respect to an operation $\oplus$, or closed *under* the operation, if $a \oplus b$ is defined and is an element of the set for every pair of elements a, b of the set. For example, the natural numbers $1, 2, 3, \cdots$ are closed under addition but are not closed under subtraction. An element e is said to be an *identity element* of a set with respect to the operation $\oplus$ if the property

$$a \oplus e = e \oplus a = a$$

holds for every element a in the set. In case the elements of the set are numbers, then e is the zero element, $e = 0$, if $\oplus$ is ordinary addition, whereas e is the unity element, $e = 1$, if $\oplus$ is ordinary multiplication. Assuming the existence of an identity element e, an element a is said to have an *inverse*, written a^{-1}, if the property

$$a \oplus a^{-1} = a^{-1} \oplus a = e$$

holds. If the elements are numbers and $\oplus$ is ordinary addition, we usually write $a + b$ for $a \oplus b$ and $-a$ for the inverse a^{-1} because the additive inverse is the negative of the number a. On the other hand, if the operation $\oplus$ is ordinary multiplication, we write $a \cdot b$ for $a \oplus b$. In this case the notation a^{-1} is the customary one in elementary algebra for the multiplicative inverse. Here, and throughout this section, the word "number" means any sort of number, integral, rational, real, or complex.

Definition 2.9 *A group G is a set of elements $a, b, c, \cdots$ together with a single-valued binary operation $\oplus$ such that*

(1) the set is closed under the operation;

(2) the associative law holds, namely,

$$a \oplus (b \oplus c) = (a \oplus b) \oplus c \text{ for all elements } a, b, c \text{ in } G;$$

(3) the set has a unique identity element, e;

(4) each element in G has a unique inverse in G.

A group G is called abelian *or* commutative *if $a \oplus b = b \oplus a$ for every pair of elements a, b, in G. A* finite group *is one with a finite number of elements; otherwise it is an* infinite group. *If a group is finite, the number of its elements is called the* order *of the group.*

Properties 1, 2, 3, and 4 are not the minimum possible postulates for a group. For example, in postulate 4 we could have required merely that each element a have a left inverse, that is an inverse a' such that $a' \oplus a = e$, and then we could prove the other half of postulate 4 as a consequence. However, to avoid too lengthy a discussion of group theory, we leave such refinements to the books on algebra.

The set of all integers $0, \pm 1, \pm 2 \cdots$ is a group under addition; in fact it is an abelian group. But the integers are not a group under multiplication because of the absence of inverses for all elements except ± 1.

Another example of a group is obtained by considering congruences modulo m. In case $m = 6$, to give a concrete example, we are familiar with such simple congruences as

$$3 + 4 \equiv 1 \pmod 6, \qquad 5 + 5 \equiv 4 \pmod 6.$$

We get "the additive group modulo 6" by taking a complete residue system, say $0, 1, 2, 3, 4, 5$ and replacing congruence modulo 6 by equality:

$$3 + 4 = 1, \qquad 5 + 5 = 4.$$

The complete addition table for this system is:

$\oplus$	0	1	2	3	4	5
0	0	1	2	3	4	5
1	1	2	3	4	5	0
2	2	3	4	5	0	1
3	3	4	5	0	1	2
4	4	5	0	1	2	3
5	5	0	1	2	3	4

Of course, any complete residue system modulo 6 would do just as well; thus $1, 2, 3, 4, 5, 6$, or $7, -2, 17, 30, 8, 3$, could serve as the elements, provided we perform additions modulo 6. If we were to use the system $7, -2, 17, 30, 8, 3$, the addition table would look quite different. However, the two groups are essentially the same; we have just renamed the elements: 0 is now called 30, 1 is 7, and so on. We say that the two groups are *isomorphic*, and we do not consider isomorphic groups as being different. Thus we speak of "the" additive group modulo 6, not "an" additive group modulo 6.

Definition 2.10 *Two groups, G with operation $\oplus$ and G' with operation $\odot$, are said to be* isomorphic *if there is a one-to-one correspondence between the elements of G and those of G', such that if a in G corresponds to a' in G', and b in G corresponds to b' in G', then $a \oplus b$ in G corresponds to $a' \odot b'$ in G'. In symbols, $G \cong G'$.*

Another way of thinking of the additive group modulo 6 is in terms of the so-called residue classes. Put two integers a and b into the same residue class modulo 6 if $a \equiv b \pmod 6$, and the result is to separate all integers into six residue classes:

$$C_0: \cdots, -18, -12, -6, 0, \ 6, 12, 18, \cdots$$

$$C_1: \cdots, -17, -11, -5, 1, \ 7, 13, 19, \cdots$$

$$C_2: \cdots, -16, -10, -4, 2, \ 8, 14, 20, \cdots$$

$$C_3: \cdots, -15, \ -9, -3, 3, \ 9, 15, 21, \cdots$$

$$C_4: \cdots, -14, \ -8, -2, 4, 10, 16, 22, \cdots$$

$$C_5: \cdots, -13, \ -7, -1, 5, 11, 17, 23, \cdots$$

If any element in class C_2 is added to any element in class C_3, the sum is an element in class C_5, so it is reasonable to write $C_2 + C_3 = C_5$. Similarly we observe that $C_3 + C_4 = C_1$, $C_5 + C_3 = C_2$, etc., and so we could make up an addition table for these classes. But the addition table so constructed would be simply a repetition of the addition table of the elements $0, 1, 2, 3, 4, 5$ modulo 6. Thus the six classes $C_0, C_1, C_2, C_3, C_4, C_5$ form a group under this addition that is isomorphic to the additive group modulo 6. This residue class formulation of the additive group modulo 6 has the advantage that such a peculiar equation as $5 + 5 = 4$ (in which the symbols have a different meaning than in elementary arithmetic) is replaced by the more reasonable form $C_5 + C_5 = C_4$.

Theorem 2.46 *Any complete residue system modulo m forms a group under addition modulo m. Two complete residue systems modulo m constitute isomorphic groups under addition, and so we speak of "the" additive group modulo m.*

Proof Let us begin with the complete residue system $0, 1, 2, \cdots, m - 1$ modulo m. This system is closed under addition modulo m, and the associative property of addition is inherited from the corresponding property for all integers, that is $a + (b + c) = (a + b) + c$ implies $a + (b + c) \equiv (a + b) + c \pmod{m}$. The identity element is 0, and it is unique. Finally, the additive inverse of 0 is 0, and the additive inverse of any other element a is $m - a$. These inverses are unique.

Passing from the system $0, 1, \cdots, m - 1$ to any complete residue system $r_0, r_1, \cdots, r_{m-1}$, we observe that all the above observations hold with a replaced by r_a, $a = 0, 1, \cdots, m - 1$, so that we have essentially the same group with new notation.

PROBLEMS

1. Which of the following are groups?
   (*a*) the even integers under addition;
   (*b*) the odd integers under addition;
   (*c*) the integers under subtraction;
   (*d*) the even integers under multiplication;
   (*e*) all integers that are multiples of 7, under addition;
   (*f*) all rational numbers under addition (recall that a rational number is one of the form a/b where a and b are integers, with $b \neq 0$);
   (*g*) the same set as in (*f*), but under multiplication;
   (*h*) the set as in (*f*) with the zero element deleted, under multiplication;
   (*i*) all rational numbers a/b having $b = 1$ or $b = 2$, under addition;
   (*j*) all rational numbers a/b having $b = 1$, $b = 2$, or $b = 3$, under addition.
2. Let G have as elements the four pairs $(1, 1), (1, -1), (-1, 1), (-1, -1)$, and let $(a, b) \oplus (c, d) = (ac, bd)$. Prove that G is a group.
3. Using the complete residue system $7, -2, 17, 30, 8, 3$, write out the addition table for the additive group modulo 6. Rewrite this table replacing 7 by 1, 30 by 0, and so on. Verify that this table gives the same values for $a \oplus b$ as the one in the text.

4. Prove that the set of elements e, a, b, c with the following table for the binary operation,

$\oplus$	e	a	b	c
e	e	a	b	c
a	a	e	c	b
b	b	c	a	e
c	c	b	e	a

is a group. Prove that this group is isomorphic to the additive group modulo 4.

5. Prove that the set of elements e, u, v, w, with the following table for the binary operation,

$\oplus$	e	u	v	w
e	e	u	v	w
u	u	e	w	v
v	v	w	e	u
w	w	v	u	e

is a group. Prove that this group is not isomorphic to the additive group modulo 4, but that it is isomorphic to the group described in Problem 2.

6. Prove that the set of elements $1, 2, 3, 4$, under the operation of multiplication modulo 5, is a group that is isomorphic to the group in Problem 4.

7. Prove that the set of complex numbers $+1, -1, +i, -i$, where $i^2 = -1$, is a group under multiplication and that it is isomorphic to the group in Problem 4.

8. Prove that the isomorphism property is *transitive*, that is, if a group G_1 is isomorphic to G_2, and if G_2 is isomorphic to G_3, then G_1 is isomorphic to G_3.

9. Prove that the elements $1, 3, 5, 7$ under multiplication modulo 8 form a group that is isomorphic to the group in Problem 5.

*10. Prove that there are essentially only two groups of order 4, that is that any group of order 4 is isomorphic to one of the groups in Problems 4 and 5.

11. For any positive integer $m > 1$, separate all integers into classes $C_0, C_1, \cdots, C_{m-1}$, putting integers r and s into the same class if $r \equiv s \pmod{m}$, thus

$$C_0: \cdots, -2m, -m, 0, m, 2m, \cdots$$

$$C_1: \cdots, -2m+1, -m+1, 1, m+1, 2m+1, \cdots \text{ etc.}$$

Prove that if any two integers, one from class C_a and one from class C_b, are added, the sum is always an integer in a unique class, namely, either C_{a+b} or C_{a+b-m} according as $a + b < m$ or $a + b \geqslant m$. Define the sum $C_a + C_b = C_{a+b}$ or $C_a + C_b = C_{a+b-m}$ accordingly, and prove that these classes form a group under this addition. Prove that this group is isomorphic to the additive group modulo m.

2.11 GROUPS, RINGS, AND FIELDS

Theorem 2.47 *Let $m > 1$ be a positive integer. Any reduced residue system modulo m is a group under multiplication modulo m. The group is of order $\phi(m)$. Any two such groups are isomorphic, so we speak of the* multiplicative group modulo m, *denoted by R_m.*

Proof Let us consider any reduced residue system $r_1, r_2, \cdots, r_n$ where $n = \phi(m)$. This set is closed under multiplication modulo m by Theorem 1.8. The associative property of multiplication is inherited from the corresponding property for integers, because $a(bc) = (ab)c$ implies that $a(bc) \equiv (ab)c \pmod{m}$. The reduced residue system contains one element, say r_j, such that $r_j \equiv 1 \pmod{m}$, and this is clearly the unique identity element of the group. Finally, for each r_i, the congruence $xr_i \equiv r_j \pmod{m}$ has a solution by Theorem 2.17, and this solution is unique within the reduced residue system $r_1, r_2, \cdots, r_n$. Two different reduced residue systems modulo m are congruent, element by element, modulo m, and so we have an isomorphism between the two groups.

Notation We have been using the symbol $\oplus$ for the binary operation of the group, and we have found that in particular groups $\oplus$ may represent addition or multiplication or some other operation. In dealing with general groups it is convenient to drop the symbol $\oplus$, just as the dot representing ordinary multiplication is usually omitted in algebra. We will write ab for $a \oplus b$, abc for $a \oplus (b \oplus c) = (a \oplus b) \oplus c$, a^2 for $a \oplus a$, a^3 for $a \oplus (a \oplus a)$, and so forth. Also, $abcd$ can be written for $(a \oplus b \oplus c) \oplus d =$

$(a \oplus b) \oplus (c \oplus d)$ and so forth, as can be seen by applying induction to the associative law. We shall even use the word *multiplication* for the operation $\oplus$, but it must be remembered that we do not necessarily mean the ordinary multiplication of arithmetic. In fact, we are dealing with general groups so that a need not be a number, just an abstract element of a group. It is convenient to write a^0 for e, a^{-2} for $(a^{-1})^2$, a^{-3} for $(a^{-1})^3$, and so on. It is not difficult to show that $a^m \cdot a^n = a^{m+n}$ and $(a^m)^n = a^{mn}$ are valid under this definition, for all integers m and n.

Theorem 2.48 *In any group G, $ab = ac$ implies $b = c$, and likewise $ba = ca$ implies $b = c$. If a is any element of a finite group G with identity element e, then there is a unique smallest positive integer r such that $a^r = e$.*

Proof The first part of the theorem is established by multiplying $ab = ac$ on the left by a^{-1}, thus $a^{-1}(ab) = a^{-1}(ac)$, $(a^{-1}a)b = (a^{-1}a)c$, $eb = ec$, $b = c$. To prove the second part, consider the series of elements obtained by repeated multiplication by a,

$$e, a, a^2, a^3, a^4, \cdots.$$

Since the group is finite, and since the members of this series are elements of the group, there must occur a repetition of the form $a^s = a^t$ with, say, $s < t$. But this equation can be written in the form $a^s e = a^s a^{t-s}$, whence $a^{t-s} = e$. Thus there is some positive integer, $t - s$, such that $a^{t-s} = e$ and the smallest positive exponent with this property is the value of r in the theorem.

Definition 2.11 *Let G be any group, finite or infinite, and a an element of G. If $a^s = e$ for some positive integer s, then a is said to be of* finite order. *If a is of finite order, the order of a is the smallest positive integer r such that $a^r = e$. If there is no positive integer s such that $a^s = e$, then a is said to be of infinite order. A group G is said to be* cyclic *if it contains an element a such that the powers of a*

$$\cdots, a^{-3}, a^{-2}, a^{-1}, a^0 = e, a, a^2, a^3, \cdots$$

comprise the whole group; such an element a is said to generate the group and is called a generator.

Consider the multiplicative group R_m of reduced residues (mod m) in Theorem 2.47. For which positive integers m is this a cyclic group? This question is equivalent to asking for the values of m for which a primitive root (mod m) exists, because a primitive root (mod m) can serve as a generator of a cyclic group, and if there is no primitive root, there is no generator. Hence by Theorem 2.41 we conclude that R_m is cyclic if and only if $m = 1, 2, 4, p^\alpha$ or $2p^\alpha$, where p is an odd prime.

Theorem 2.48 shows that all the elements of a finite group are of finite order. Every group, finite or infinite, contains at least the single element e that is of finite order. There are infinite groups consisting entirely of elements of finite order.

If a cyclic group is finite, and has generator a, then the group consists of $e, a, a^2, a^3, \cdots, a^{r-1}$, where r is the order of the element a. All other powers of a are superfluous because they merely repeat these.

Theorem 2.49 *The order of an element of a finite group G is a divisor of the order of the group. If the order of the group is denoted by n, then $a^n = e$ for every element a in the group.*

Proof Let the element a have order r. It is readily seen that

$$e, a, a^2, a^3, \cdots, a^{r-1} \tag{A}$$

are r distinct elements of G. If these r elements do not exhaust the group, there is some other element, say b_2. Then we can prove that

$$b_2, b_2a, b_2a^2, b_2a^3, \cdots, b_2a^{r-1} \tag{B}$$

are r distinct elements, all different from the r elements of A. For in the first place if $b_2a^s = b_2a^t$, then $a^s = a^t$ by Theorem 2.48. And on the other hand, if $b_2a^s = a^t$, then $b_2 = a^{t-s}$, so that b_2 would be among the powers of a.

If G is not exhausted by the sets A and B, then there is another element b_3 that gives rise to r new elements

$$b_3, b_3a, b_3a^2, b_3a^3, \cdots, b_3a^{r-1}$$

all different from the elements in A and B, by a similar argument. This process of obtaining new elements $b_2, b_3, \cdots$ must terminate since G is finite. So if the last batch of new elements is, say

$$b_k, b_ka, b_ka^2, b_ka^3, \cdots, b_ka^{r-1}$$

then the order of the group G is kr, and the first part of the theorem is proved. To prove the second part, we observe that $n = kr$ and $a^r = e$ by Theorem 2.48, whence $a^n = e$.

It can be noted that Theorem 2.49 implies the theorems of Fermat and Euler, where the set of integers relatively prime to the modulus m is taken as the group. In making this implication, you will see the necessity of translating the language and notation of group theory into that of number theory. In the same way we note that the language of Definition 2.7, that "a belongs to the exponent h modulo m," is translated into group theoretic language as "the element a of the multiplicative group modulo m has order h." Also the "primitive root modulo m" of Definition 2.8 is called a "generator" of the multiplicative group modulo m in group theory.

Let G and H be two groups. We may define a multiplication on the ordered pairs (g, h) by setting $(g_1, h_1) \cdot (g_2, h_2) = (g_1 g_2, h_1 h_2)$ where it is assumed that the g_i and h_i lie in G and H, respectively. The ordered pairs, equipped with multiplication in this way, form a group $G \otimes H$, called the *direct product* of G and H. We may similarly form the direct product $G \otimes H \otimes J$ of three groups by considering the ordered triples (g, h, j). It is a general theorem of group theory (which we do not prove here) that any finite abelian group is isomorphic to a direct product of cyclic groups. In the case of the multiplicative group R_m of reduced residues (mod m), we can explicitly determine this decomposition. Let $m = p_1^{\alpha_1} p_2^{\alpha_2} \cdots p_r^{\alpha_r}$ be the canonical factorization of m. By the Chinese Remainder Theorem we see that

$$R_m \cong R_{p_1^{\alpha_1}} \otimes R_{p_2^{\alpha_2}} \otimes \cdots \otimes R_{p_r^{\alpha_r}}.$$

After Definition 2.11 we noted that if p is an odd prime then R_{p^α} is cyclic. It is easy to see that two cyclic groups are isomorphic if and only if they have the same order. Thus we speak of "the" cyclic group of order n, and denote it by C_n. In this notation, we would write $R_{p^\alpha} \cong C_{\phi(p^\alpha)}$ for an odd prime p. For the prime 2 we have $R_2 \cong C_1$, $R_4 \cong C_2$, and by Theorem 2.43 we see that $R_{2^\alpha} \cong C_2 \otimes C_{2^{\alpha-2}}$ for $\alpha \geqslant 3$. The ideas we used to prove Theorem 2.41 can be used to show, more generally, that a direct product $G_1 \otimes G_2 \otimes \cdots \otimes G_r$ of several groups is cyclic if and only if each G_i is cyclic and the orders of the G_i are pairwise relatively prime.

Definition 2.12 *A ring is a set of at least two elements with two binary operations,* $\oplus$ *and* $\odot$, *such that it is a commutative group under* $\oplus$, *is closed under* $\odot$, *and such that* $\odot$ *is associative and distributive with respect to* $\oplus$. *The identity element with respect to* $\oplus$ *is called the* zero *of the ring. If all the elements of a ring, other than the zero, form a commutative group under* $\odot$, *then it is called a* field.

It is customary to call $\oplus$ addition and $\odot$ multiplication and to write $a + b$ for $a \oplus b$, ab for $a \odot b$. The conditions on $\odot$ for a ring are then $a(bc) = (ab)c$, $a(b + c) = ab + ac$, $(b + c)a = ba + ca$. In general, the elements $a, b, c, \cdots$ are not necessarily numbers, and the operations of addition and multiplication need not be the ordinary ones of arithmetic. However, the only rings and fields that will be considered here will have numbers for elements, and the operations will be either ordinary addition and multiplication or addition and multiplication modulo m.

Theorem 2.50 *The set Z_m of elements $0, 1, 2, \cdots, m - 1$, with addition and multiplication defined modulo m, is a ring for any integer $m > 1$. Such a ring is a field if and only if m is a prime.*

Proof We have already seen in Theorem 2.46 that any complete residue system modulo m is a group under addition modulo m. This group is commutative, and the associative and distributive properties of multiplication modulo m are inherited from the corresponding properties for ordinary multiplication. Therefore Z_m is a ring.

Next, by Theorem 2.47 any reduced residue system modulo m is a group under multiplication modulo m. If m is a prime p, the reduced residue system of Z_p is $1, 2, \cdots, p - 1$, that is, all the elements of Z_p other than 0. Since 0 is the zero of the ring, Z_p is a field. On the other hand if m is not a prime, then m is of the form ab with $1 < a \leqslant b < m$. Then the elements of Z_m other than 0 do not form a group under multiplication modulo m because there is no inverse for the element a, no solution of $ax \equiv 1 \pmod{m}$. Thus Z_m is not a field.

Some questions can be settled very readily by using the fields Z_p. For example, consider the following problem: prove that for any prime $p > 3$ the sum

$$\frac{1}{1^2} + \frac{1}{2^2} + \frac{1}{3^2} + \cdots + \frac{1}{(p-1)^2}$$

if written as a rational number a/b has the property that $p|a$. In the field Z_p the term $1/j^2$ in the sum is j^{-2} or x^2 where x is the least positive integer such that $xj \equiv 1 \pmod{p}$. Hence in Z_p the problem can be put in the form, prove that the sum $1^{-2} + 2^{-2} + \cdots + (p - 1)^{-2}$ is the zero element of the field. But the inverses of $1, 2, 3, \cdots, p - 1$ are just the same elements again in some order, so we can write

$$1^{-2} + 2^{-2} + \cdots + (p-1)^{-2} = 1^2 + 2^2 + \cdots + (p-1)^2.$$

For this final sum there is a well-known formula for the sum of the squares of the natural numbers giving $p(p-1)(2p-1)/6$. But this is zero in Z_p, because of the factor p, except in the cases $p = 2$ and $p = 3$ where division by 6 is meaningless.

PROBLEMS

1. Prove that the multiplicative group modulo 9 is isomorphic to the additive group modulo 6.
2. Prove that the additive group modulo m is cyclic with 1 as generator. Prove that any one of $\phi(m)$ elements could serve as generator.
3. Prove that any two cyclic groups of order m are isomorphic.
4. Prove that the group of all integers under addition is an infinite cyclic group.
5. If a is an element of order r of a group G, prove that $a^k = e$ if and only if $r|k$.
6. What is the smallest positive integer m such that the multiplicative group modulo m is not cyclic?
7. A *subgroup* S of a group G is a subset of elements of G that form a group under the same binary operation. If G is finite, prove that the order of a subgroup S is a divisor of the order of G.
8. Prove Theorem 2.49, for the case in which the group is commutative, in a manner analogous to the proof of Theorem 2.8.
9. Prove Theorem 2.8 by the method used in the proof of Theorem 2.49.
10. Let G consist of all possible sequences $(a_1, a_2, a_3, \cdots)$ with each $a_i = 1$ or -1. Let $(a_1, a_2, a_3, \cdots) \oplus (b_1, b_2, b_3, \cdots) = (a_1b_1, a_2b_2, a_3b_3, \cdots)$. Show that G is an infinite group all of whose elements are of finite order.

*11. Let G consist of a, b, c, d, e, f and let $\oplus$ be defined by the following table.

$\oplus$	e	a	b	c	d	f
e	e	a	b	c	d	f
a	a	e	d	f	b	c
b	b	f	e	d	c	a
c	c	d	f	e	a	b
d	d	c	a	b	f	e
f	f	b	c	a	e	d

Show that G is a noncommutative group.

12. Prove that the multiplicative group modulo p is cyclic if p is a prime.

13. Exhibit the addition and multiplication tables for the elements of the field of residues modulo 7.

14. Prove that the set of all integers under ordinary addition and multiplication is a ring but not a field.

15. Prove that the set of all even integers under ordinary addition and multiplication is a ring.

16. Prove that the set $0, 3, 6, 9$ is a ring under addition and multiplication modulo 12.

17. Prove that in any field $a0 = 0a = 0$ for every element a.

***18.** Let a be a divisor of m, say $m = aq$ with $1 < a < m$. Prove that the set of elements $0, a, 2a, 3a, \cdots, (q-1)a$, with addition and multiplication modulo m, forms a ring. Under what circumstances is it a field?

19. Prove that the set of all rational numbers forms a field.

20. An *integral domain* is a ring with the following additional properties: (i) there is a unique identity element with respect to multiplication; (ii) multiplication is commutative; (iii) if $ab = ac$ and $a \neq 0$, then $b = c$. Prove that any field is an integral domain. Which of the following are integral domains?

(a) the set of all integers;

(b) the set Z_m of Theorem 2.50.

21. Let m be a positive integer and consider the set of all the divisors of m. For numbers in this set define two operations $\odot$ and $\oplus$ by $a \odot b = (a, b)$, $a \oplus b = [a, b]$, g.c.d. and l.c.m. Prove that $\odot$ and $\oplus$ are associative and commutative. Prove the distributive law $a \odot (b \oplus c) = (a \odot b) \oplus (a \odot c)$ and its dual $a \oplus (b \odot c) = (a \oplus b) \odot (a \oplus c)$. Show that $a \odot a = a \oplus a = a$. Also prove $1 \odot a = 1$ and $1 \oplus a = a$, so that 1 behaves like an ordinary zero, and $m \odot a = a$, and $m \oplus a = m$. Define a relation $\textcircled{<}$ as $a \textcircled{<} b$ if $a \odot b = a$. Prove $a \textcircled{<} a$, that $\textcircled{<}$ is transitive, and that $a \textcircled{<} b$ if and only if $a \oplus b = b$.

Prove that if m is not divisible by any square other than 1, then corresponding to each divisor a there is a divisor a' such that $a \odot a' = 1$, $a \oplus a' = m$. (These algebras with square-free m are examples of *Boolean algebras*.)

22. Prove that for any prime $p > 2$ the sum

$$\frac{1}{1^3} + \frac{1}{2^3} + \cdots + \frac{1}{(p-1)^3}$$

if written as a rational number a/b, has the property that $p \mid a$. (H)

***23.** Let V_n denote the vector space of dimension n over the field Z_p of integers modulo p. Show that if W is a subspace of V_n of dimension m, then $\operatorname{card}(W) = p^m$. Show that the number of $n \times n$ matrices A with entries considered (mod p) for which $\det(A) \not\equiv 0 \pmod p$ is exactly $(p^n - 1)(p^n - p)(p^n - p^2) \cdots (p^n - p^{n-1})$. (H)

NOTES ON CHAPTER 2

§2.1 It was noted in this section that (i) $a \equiv a \pmod m$, (ii) $a \equiv b \pmod m$ if and only if $b \equiv a \pmod m$, and (iii) $a \equiv b \pmod m$ and $b \equiv c \pmod m$ imply $a \equiv c \pmod m$. Thus the congruence relation has the (i) reflexive property, (ii) the symmetric property, and (iii) the transitive property, and so the congruence relation is a so-called *equivalence relation*. Although the classification of integers by the remainder on division by a fixed modulus goes back at least as far as the ancient Greeks, it was Gauss who introduced the congruence notation.

§2.3 It is often observed of mathematics that there are far more theorems than ideas. The idea used in the proof of Theorem 2.18 is found in many other contexts. For example, Lagrange constructed a polynomial of degree at most n that passes through the $n + 1$ points $(x_0, y_0), (x_1, y_1), \cdots, (x_n, y_n)$ by first constructing the polynomials

$$P_j(x) = \frac{(x - x_0)(x - x_1) \cdots (x - x_{j-1})(x - x_{j+1}) \cdots (x - x_n)}{(x_j - x_0)(x_j - x_1) \cdots (x_j - x_{j-1})(x_j - x_{j+1}) \cdots (x_j - x_n)},$$

which have the property that $P_j(x_j) = 1$, $P_j(x_i) = 0$ for $i \neq j$. Here we are assuming that the x_j are distinct. Then

$$P(x) = \sum_{j=0}^{n} y_j P_j(x)$$

is a polynomial with the desired properties. (This polynomial $P(x)$ is unique. To see this, suppose that $Q(x)$ is another such polynomial. Then the polynomial $R(x) = P(x) - Q(x)$ has degree at most n and vanishes at the $n + 1$ points x_j. But a polynomial that has more zeros than its degree must vanish identically. Thus $P(x)$ and $Q(x)$ are identical.)

The less symmetric procedure applied in Example 4 is similarly analogous to the Hermite formula for polynomial interpolation, by which a polynomial is written in the form

$$P(x) = \sum_{j=1}^{n} c_j \prod_{i=1}^{j-1} (x - x_i).$$

(When $j = 1$ there is no i within the prescribed range, and the resulting empty product is taken to have value 1.) We see that $P(x_1) = c_1$, $P(x_2) = c_1 + c_2(x_2 - x_1)$, $P(x_3) = c_1 + c_2(x_3 - x_1) + c_3(x_3 - x_1)(x_3 - x_2)$, and so on. Thus we may take c_1 so that $P(x_1)$ has the desired value. Having chosen c_1, we may take c_2 so that $P(x_2)$ has the desired value, and so on. This may be compared with Problem 24 at the end of the section.

§2.4 Readers interested in the numerical aspects of number theory may wish to consult the text by Rosen listed in the General References at the end of this book. Number-theoretic algorithms are discussed by D. H. Lehmer, "Computer Technology Applied to the Theory of Numbers," pages 117–151 in the book edited by LeVeque; in Chapter 4 of Volume 2 of Knuth; and in the book edited by Lenstra and Tijdeman. Many of the algorithms that we have discussed can be made more efficient in various ways. For example, when factoring by trial division, one may restrict the trial divisors to prime values.

Methods for factoring numbers have developed considerably during this century. Some of the algorithms employed involve quite sophisticated mathematics, as in the case of the elliptic curve method of Hendrik Lenstra, which we discuss in Section 5.8. During the past decade, the most impressive factorizations were achieved by means of elaborations of the *quadratic sieve* method, proposed by Carl Pomerance in 1982. However, a new strategy called the *general number field sieve* is yielding good results, and offers great promise for the future. Further discussion of factoring techniques can be found in the publications of Pomerance, of Riesel, and of Bressoud, listed in the General References. For the general number field sieve one should see A. K. Lenstra and H. W. Lenstra, Jr., *The development of the number field sieve*, Lecture Notes in Math. 1554, Springer-Verlag (Berlin), 1993.

§2.5 The permutation used here is known as a *trapdoor function* because of the difficulty of computing the inverse permutation. The particular method discussed is known as the RSA method, after Rivest, Shamir, and Adleman, who proposed the method in 1978.

§2.6 In our appeal to Taylor's theorem we have again made a small use of analysis. A more extensive use of analysis is found in Section 8.2, where we investigate arithmetic functions by means of Dirichlet series. Analysis of a somewhat different variety is used in proofs of irrationality or transcendence. A simple example of this is found in our proof that π is irrational, in Section 6.3.

The study of congruences (mod p^k) leads naturally to the theory of p-adic numbers. Solutions of a congruence that lift to arbitrarily high powers of p correspond to the p-adic roots of the equation. The sequence of solutions of the congruence generated by letting n run to infinity form a sequence of approximations to the p-adic root in much the same way that truncations of the decimal expansion of a real number form approximations to the real number being expanded. An attractive introduction to p-adic numbers is found in Chapter 1 of the text by Borevich and Shafarevich.

§2.7 Let $f(x)$ be a fixed polynomial with integral coefficients. The number $N(p)$ of solutions of the congruence $f(x) \equiv 0 \pmod{p}$ fluctuates as p varies, but it can be shown that if f is irreducible then $\Sigma_{p \leqslant x} N(p) \sim x/\log x$ as $x \to \infty$. This is derived from the *prime ideal theorem*, which is a generalization of the prime number theorem to algebraic number fields.

The discussion of the polynomial $f(x)$ in (2.7) can be generalized to composite moduli. This generalization, which is by no means obvious, was discovered by Bauer in 1902. Accounts of Bauer's congruence are found in §§8.5–8.8 of the book by Hardy and Wright, and in articles by Gupta and Wylie, *J. London Math. Soc.*, 14 (1939).

§2.8 In 1769, Lambert stated without proof that every prime number has a primitive root. Euler introduced the term *primitive root*, but his proof of their existence is flawed by gaps and obscurities. Our account, based on Lagrange's result Corollary 2.30, is similar to the method proposed by Legendre in 1785.

For further discussion of methods of proving primality, see the article "Primality testing" in Lenstra and Tijdeman, H. C. Williams, "Primality testing on a computer," *Ars Combinatoria* 5 (1978), 127–185, or Chapter 4 of Riesel. The original account of Atkin's method of proving primality is found in the paper of A. O. L. Atkin and F. Morain, "Elliptic curves and primality proving," *Math. Comp.*, to appear. The method is briefly described in A. K. Lenstra and H. W. Lenstra, Jr., "Algorithms in number theory" in *Handbook of Theoretical Computer Science* (ed. J. van Leeuwen), North-Holland, to appear.

§2.9 The algorithm RESSOL was invented and named by Dan Shanks, "Five number-theoretic algorithms," (*Proc. Second Manitoba Conference on Numerical Mathematics* (1972), 51–70). A similar algorithm for determining u so that $n \equiv c^u \pmod{p}$, had been given in 1891 by Tonelli. D. H. Lehmer ("Computer technology applied to the theory of numbers," *Studies in Number Theory*, (W. J. LeVeque, ed.), *Math. Assoc. Amer.* (1969), 117–151) has given a different algorithm for finding solutions of quadratic congruences.

CHAPTER 3

Quadratic Reciprocity and Quadratic Forms

The purpose of this chapter is to continue the discussion of congruences by means of a remarkable result of Gauss known as the *quadratic reciprocity law*. In the preceding chapter, the problem of solving such a congruence as $x^2 \equiv a \pmod{m}$ was reduced to the case of a prime modulus p. The question remains as to whether $x^2 \equiv a \pmod{p}$ does or does not have a solution. This question can be narrowed to the case $x^2 \equiv q \pmod{p}$, where q is also a prime. The quadratic reciprocity law states that if p and q are distinct odd primes, the two congruences $x^2 \equiv p \pmod{q}$ and $x^2 \equiv q \pmod{p}$ are either both solvable or both not solvable, unless p and q are both of the form $4k + 3$, in which case one of the congruences is solvable and the other is not. This result might appear at first glance to be of very limited use because of the conditional nature of the statement; it is not crisply decisive. However, the result provides a reduction process that enables us to determine very quickly whether $x^2 \equiv q \pmod{p}$ is or is not solvable for any specified primes p and q.

As an example of the remarkable power of the quadratic reciprocity law, consider the question whether $x^2 \equiv 5 \pmod{103}$ has any solutions. Since 5 is not of the form $4k + 3$, the result asserts that $x^2 \equiv 5 \pmod{103}$ and $x^2 \equiv 103 \pmod{5}$ are both solvable or both not. But $x^2 \equiv 103 \pmod{5}$ boils down to $x^2 \equiv 3 \pmod{5}$, which has no solutions. Hence $x^2 \equiv 5 \pmod{103}$ has no solutions.

3.1 QUADRATIC RESIDUES

Definition 3.1 *For all a such that $(a, m) = 1$, a is called a* quadratic residue *modulo m if the congruence $x^2 \equiv a \pmod{m}$ has a solution. If it has no solution, then a is called a* quadratic nonresidue *modulo m.*

Since $a + m$ is a quadratic residue or nonresidue modulo m according as a is or is not, we consider as distinct residues or nonresidues only those that are distinct modulo m. The quadratic residues modulo 5 are 1 and 4, whereas 2 and 3 are the nonresidues.

Definition 3.2 *If p denotes an odd prime, then the* Legendre symbol $\left(\frac{a}{p}\right)$ *is defined to be* 1 *if a is a quadratic residue,* -1 *if a is a quadratic nonresidue modulo p, and* 0 *if $p \mid a$.*

Theorem 3.1 *Let p be an odd prime. Then*

(1) $\left(\frac{a}{p}\right) \equiv a^{(p-1)/2} \pmod{p}$,

(2) $\left(\frac{a}{p}\right)\left(\frac{b}{p}\right) = \left(\frac{ab}{p}\right)$,

(3) $a \equiv b \pmod{p}$ *implies that* $\left(\frac{a}{p}\right) = \left(\frac{b}{p}\right)$,

(4) If $(a, p) = 1$ *then* $\left(\frac{a^2}{p}\right) = 1$, $\left(\frac{a^2b}{p}\right) = \left(\frac{b}{p}\right)$,

(5) $\left(\frac{1}{p}\right) = 1$, $\left(\frac{-1}{p}\right) = (-1)^{(p-1)/2}$.

Remark From our observations in Section 2.9, we see that if p is an odd prime then for any integer a the number of solutions of the congruence $x^2 \equiv a \pmod{p}$ is $1 + \left(\frac{a}{p}\right)$.

Proof If $p \mid a$, then Part 1 of the theorem is obvious. If $(a, p) = 1$ then Part 1 follows from Euler's criterion (Corollary 2.38). The remaining parts are all simple consequences of Part 1.

Part 1 can also be proved without appealing to Euler's criterion, as follows: If $\left(\frac{a}{p}\right) = 1$, then $x^2 \equiv a \pmod{p}$ has a solution, say x_0. Then, by Fermat's congruence (Theorem 2.7), $a^{(p-1)/2} \equiv x_0^{p-1} \equiv 1 \equiv \left(\frac{a}{p}\right) \pmod{p}$. On the other hand, if $\left(\frac{a}{p}\right) = -1$, then $x^2 \equiv a \pmod{p}$ has no solution, and we proceed as in the proof of Wilson's congruence (Theorem 2.11).

To each j satisfying $1 \leqslant j < p$, choose j', $1 \leqslant j' < p$, so that $jj' \equiv a \pmod{p}$. We pair j with j'. We note that $j \not\equiv j' \pmod{p}$, since the congruence $x^2 \equiv a \pmod{p}$ has no solution. The combined contribution of j and j' to $(p-1)!$ is $jj' \equiv a \pmod{p}$. Since there are $(p-1)/2$ pairs j, j', it follows that $a^{(p-1)/2} \equiv (p-1)! \pmod{p}$, and then Wilson's congruence gives Part 1.

The last part of the theorem, which follows immediately from the first part, expresses again the information provided in Theorem 2.12.

Theorem 3.2 *Lemma of Gauss. For any odd prime p let $(a, p) = 1$. Consider the integers $a, 2a, 3a, \cdots, \{(p-1)/2\}a$ and their least positive residues modulo p. If n denotes the number of these residues that exceed $\frac{p}{2}$, then $\left(\frac{a}{p}\right) = (-1)^n$.*

Proof Let $r_1, r_2, \cdots, r_n$ denote the residues that exceed $p/2$, and let $s_1, s_2, \cdots, s_k$ denote the remaining residues. The r_i and s_i are all distinct, and none is zero. Furthermore, $n + k = (p-1)/2$. Now $0 < p - r_i < p/2$, $i = 1, 2, \cdots, n$, and the numbers $p - r_i$ are distinct. Also no $p - r_i$ is an s_j for if $p - r_i = s_j$ then $r_i \equiv \rho a$, $s_j \equiv \sigma a \pmod{p}$ for some ρ, σ, $1 \leqslant \rho \leqslant (p-1)/2$, $1 \leqslant \sigma \leqslant (p-1)/2$, and $p - \rho a \equiv \sigma a \pmod{p}$. Since $(a, p) = 1$ this implies $a(\rho + \sigma) \equiv 0$, $\rho + \sigma \equiv 0 \pmod{p}$, which is impossible. Thus $p - r_1, p - r_2, \cdots, p - r_n, s_1, s_2, \cdots, s_k$ are all distinct, are all at least 1 and less than $p/2$, and they are $n + k = (p-1)/2$ in number. That is, they are just the integers $1, 2, \cdots, (p-1)/2$ in some order. Multiplying them together we have

$$(p - r_1)(p - r_2) \cdots (p - r_n) s_1 s_2 \cdots s_k = 1 \cdot 2 \cdots \frac{p-1}{2}$$

and then

$$(-r_1)(-r_2) \cdots (-r_n) s_1 s_2 \cdots s_k \equiv 1 \cdot 2 \cdots \frac{p-1}{2} \pmod{p},$$

$$(-1)^n r_1 r_2 \cdots r_n s_1 s_2 \cdots s_k \equiv 1 \cdot 2 \cdots \frac{p-1}{2} \pmod{p},$$

$$(-1)^n a \cdot 2a \cdot 3a \cdots \frac{p-1}{2} a \equiv 1 \cdot 2 \cdots \frac{p-1}{2} \pmod{p}.$$

We can cancel the factors $2, 3, \cdots, (p-1)/2$ to obtain $(-1)^n a^{(p-1)/2} \equiv$

$1 \pmod p$ which gives us $(-1)^n \equiv a^{(p-1)/2} \equiv \left(\frac{a}{p}\right) \pmod p$ by Theorem 3.1, part 1.

Definition 3.3 *For real x, the symbol $[x]$ denotes the greatest integer less than or equal to x.*

This is also called the *integral part* of x, and $x - [x]$ is called the *fractional part*. Such an integer as $[1000/23]$ is the quotient when 1000 is divided by 23 and is also the number of positive multiples of 23 less than 1000. On a hand calculator, its value, 43, is immediately obtained by dividing 1000 by 23 and taking the integer part of the answer only. Here are further examples: $[15/2] = 7$, $[-15/2] = -8$, $[-15] = -15$.

Theorem 3.3 *If p is an odd prime and $(a, 2p) = 1$, then*

$$\left(\frac{a}{p}\right) = (-1)^t \quad \textit{where} \quad t = \sum_{j=1}^{(p-1)/2} \left[\frac{ja}{p}\right]; \quad \textit{also} \quad \left(\frac{2}{p}\right) = (-1)^{(p^2-1)/8}.$$

Proof We use the same notation as in the proof of Theorem 3.2. The r_i and s_i are just the least positive remainders obtained on dividing the integers ja by p, $j = 1, 2, \cdots, (p-1)/2$. The quotient in this division is easily seen to be $q = [ja/p]$. Then for $(a, p) = 1$, whether a is odd or even, we have

$$\sum_{j=1}^{(p-1)/2} ja = \sum_{j=1}^{(p-1)/2} p\left[\frac{ja}{p}\right] + \sum_{j=1}^{n} r_j + \sum_{j=1}^{k} s_j$$

and

$$\sum_{j=1}^{(p-1)/2} j = \sum_{j=1}^{n} (p - r_j) + \sum_{j=1}^{k} s_j = np - \sum_{j=1}^{n} r_j + \sum_{j=1}^{k} s_j$$

and hence by subtraction,

$$(a-1) \sum_{j=1}^{(p-1)/2} j = p\left(\sum_{j=1}^{(p-1)/2} \left[\frac{ja}{p}\right] - n \right) + 2 \sum_{j=1}^{n} r_j.$$

But

$$\sum_{j=1}^{(p-1)/2} j = \frac{p^2 - 1}{8}$$

so we have

$$(a-1)\frac{p^2-1}{8} \equiv \sum_{j=1}^{(p-1)/2}\left[\frac{ja}{p}\right] - n \pmod 2.$$

If a is odd, this implies $n \equiv \sum_{j=1}^{(p-1)/2}\left[\frac{ja}{p}\right] \pmod 2$. If $a = 2$ it implies $n \equiv (p^2-1)/8 \pmod 2$ since $[2j/p] = 0$ for $1 \leqslant j \leqslant (p-1)/2$. Our theorem now follows by Theorem 3.2.

Although Theorem 3.2 and the first part of Theorem 3.3 are of considerable importance in theoretical considerations, they are too cumbersome to use for calculations unless p is very small. However, Theorems 3.1 and the other parts of 3.3 are useful in numerical cases. The second part of Theorem 3.3 involves $(-1)^{(p^2-1)/8}$, and this can be easily computed if p is reduced modulo 8. For example, if $p = 59$ then $p \equiv 3 \pmod 8$ and $(-1)^{(p^2-1)/8} = (-1)^{(3^2-1)/8}$. Finally, we point out that the problem of numerical evaluation of $\left(\frac{a}{p}\right)$, apart from the cases $a = \pm 1, \pm 2$, is treated in the next section.

PROBLEMS

1. Find $[3/2]$, $[-3/2]$, $[\pi]$, $[-7]$, and $[x]$ for $0 \leqslant x < 1$.
2. With reference to the notation of Theorem 1.2 prove that $q = [b/a]$.
3. Prove that 3 is a quadratic residue of 13, but a quadratic nonresidue of 7.
4. Find the values of $\left(\frac{a}{p}\right)$ in each of the 12 cases, $a = -1, 2, -2, 3$ and $p = 11, 13, 17$.
5. Prove that the quadratic residues of 11 are $1, 3, 4, 5, 9$, and list all solutions of each of the ten congruences $x^2 \equiv a \pmod{11}$ and $x^2 \equiv a \pmod{11^2}$ where $a = 1, 3, 4, 5, 9$.
6. (*a*) List the quadratic residues of each of the primes $7, 13, 17, 29, 37$.
   (*b*) For any positive integer n, define $F(n)$ to be the minimum value of $|n^2 - 17x|$, where x runs over all integers. Prove that $F(n)$ is either 0 or a power of 2.
7. Which of the following congruences have solutions? How many?
   (*a*) $x^2 \equiv 2 \pmod{61}$ (*b*) $x^2 \equiv 2 \pmod{59}$
   (*c*) $x^2 \equiv -2 \pmod{61}$ (*d*) $x^2 \equiv -2 \pmod{59}$

(*e*) $x^2 \equiv 2 \pmod{122}$ (*f*) $x^2 \equiv 2 \pmod{118}$
(*g*) $x^2 \equiv -2 \pmod{122}$ (*h*) $x^2 \equiv -2 \pmod{118}$.

8. How many solutions are there to each of the congruences?
(*a*) $x^2 \equiv -1 \pmod{61}$ (*b*) $x^2 \equiv -1 \pmod{59}$
(*c*) $x^2 \equiv -1 \pmod{365}$ (*d*) $x^2 \equiv -1 \pmod{3599}$
(*e*) $x^2 \equiv -1 \pmod{122}$ (*f*) $x^2 \equiv -1 \pmod{244}$

9. Let p be a prime, and let $(a, p) = (b, p) = 1$. Prove that if $x^2 \equiv a \pmod{p}$ and $x^2 \equiv b \pmod{p}$ are not solvable, then $x^2 \equiv ab \pmod{p}$ is solvable.

10. Prove that if p is an odd prime then $x^2 \equiv 2 \pmod{p}$ has solutions if and only if $p \equiv 1$ or $7 \pmod{8}$.

11. Let g be a primitive root of an odd prime p. Prove that the quadratic residues modulo p are congruent to $g^2, g^4, g^6, \cdots, g^{p-1}$ and that the nonresidues are congruent to $g, g^3, g^5, \cdots, g^{p-2}$. Thus there are equally many residues and nonresidues for an odd prime.

12. Denote quadratic residues by r, nonresidues by n. Prove that $r_1 r_2$ and $n_1 n_2$ are residues and that rn is a nonresidue for any odd prime p. Give a numerical example to show that the product of two nonresidues is not necessarily a quadratic residue modulo 12.

13. Prove that if r is a quadratic residue modulo $m > 2$, then $r^{\phi(m)/2} \equiv 1 \pmod{m}$. (H)

14. Prove that the quadratic residues modulo p are congruent to $1^2, 2^2, 3^2, \cdots, \{(p-1)/2\}^2$, where p is an odd prime. Hence prove that if $p > 3$, the sum of the quadratic residues is divisible by p. (H)

15. Show that if p is a prime of the form $4k + 1$ then the sum of the quadratic residues $\pmod{p}$ in the interval $[1, p)$ is $p(p-1)/4$.

***16.** Show that if a is a quadratic residue modulo m, and $ab \equiv 1 \pmod{m}$, then b is also a quadratic residue. Then prove that the product of the quadratic residues modulo p is congruent to $+1$ or -1 according as the prime p is of the form $4k + 3$ or $4k + 1$.

***17.** Prove that if p is a prime having the form $4k + 3$, and if m is the number of quadratic residues less than $p/2$, then $1 \cdot 3 \cdot 5 \cdots (p-2) \equiv (-1)^{m+k+1} \pmod{p}$, and $2 \cdot 4 \cdot 6 \cdots (p-1) \equiv (-1)^{m+k} \pmod{p}$. (H)

***18.** For any prime p and any integer a such that $(a, p) = 1$, say that a is a *cubic residue* of p if $x^3 \equiv a \pmod{p}$ has at least one solution. Prove that if p is of the form $3k + 2$, then all integers in a reduced residue system modulo p are cubic residues, whereas if p is of the form $3k + 1$, only one-third of the members of a reduced residue system are cubic residues.

***19.** For all primes p prove that $x^8 \equiv 16 \pmod{p}$ is solvable. (H)

***20.** Let p be an odd prime. Prove that if there is an integer x such that

$$p \mid (x^2 + 1) \text{ then } p \equiv 1 \pmod{4};$$

$$p \mid (x^2 - 2) \text{ then } p \equiv 1 \text{ or } 7 \pmod{8};$$

$$p \mid (x^2 + 2) \text{ then } p \equiv 1 \text{ or } 3 \pmod{8};$$

$$p \mid (x^4 + 1) \text{ then } p \equiv 1 \pmod{8}.$$

Show that there are infinitely many primes of each of the forms $8n + 1, 8n + 3, 8n + 5, 8n + 7$. (H)

***21.** Let p be an odd prime. Prove that every primitive root of p is a quadratic nonresidue. Prove that every quadratic nonresidue is a primitive root if and only if p is of the form $2^{2^n} + 1$ where n is a non-negative integer, that is, if and only if $p = 3$ or p is a Fermat number.

***22.** Show that if p and q are primes, $p = 2q + 1$, and $0 < m < (p + 1)^{1/2}$, then m is a primitive root (mod p) if and only if it is a quadratic nonresidue (mod p).

***23.** Show that if p is an odd prime and $(a, p) = 1$, then $x^2 \equiv a \pmod{p^\alpha}$ has exactly $1 + \left(\frac{a}{p}\right)$ solutions.

***24.** Suppose that m is an odd number. Show that if $(a, p) = 1$ then the number of solutions of the congruence $x^2 \equiv a \pmod{m}$ is $\prod_{p \mid m} \left(1 + \left(\frac{a}{p}\right)\right)$. Show that this holds for all integers a if m is an odd square-free number.

3.2 QUADRATIC RECIPROCITY

Theorem 3.4 *The Gaussian reciprocity law. If p and q are distinct odd primes, then*

$$\left(\frac{p}{q}\right)\left(\frac{q}{p}\right) = (-1)^{\{(p-1)/2\}\{(q-1)/2\}}.$$

Another way to state this is: If p and q are distinct odd primes of the form $4k + 3$, then one of the congruences $x^2 \equiv p \pmod{q}$ and $x^2 \equiv q \pmod{p}$ is solvable and the other is not; but if at least one of the primes is of the form $4k + 1$, then both congruences are solvable or both are not.

Proof Let $\mathscr{S}$ be the set of all pairs of integers (x, y) satisfying $1 \leqslant x \leqslant (p-1)/2$, $1 \leqslant y \leqslant (q-1)/2$. The set $\mathscr{S}$ has $(p-1)(q-1)/4$ members. Separate this set into two mutually exclusive subsets $\mathscr{S}_1$ and $\mathscr{S}_2$ according as $qx > py$ or $qx < py$. Note that there are no pairs (x, y) in $\mathscr{S}$ such that $qx = py$. The set $\mathscr{S}_1$ can be described as the set of all pairs (x, y) such that $1 \leqslant x \leqslant (p-1)/2$, $1 \leqslant y < qx/p$. The number of pairs in $\mathscr{S}_1$ is then seen to be $\Sigma_{x=1}^{(p-1)/2}[qx/p]$. Similarly $\mathscr{S}_2$ consists of the pairs (x, y) such that $1 \leqslant y \leqslant (q-1)/2$, $1 \leqslant x < py/q$, and the number of pairs in $\mathscr{S}_2$ is $\Sigma_{y=1}^{(q-1)/2}[py/q]$. Thus we have

$$\sum_{j=1}^{(p-1)/2}\left[\frac{qj}{p}\right] + \sum_{j=1}^{(q-1)/2}\left[\frac{pj}{q}\right] = \frac{p-1}{2}\frac{q-1}{2}$$

and hence

$$\left(\frac{p}{q}\right)\left(\frac{q}{p}\right) = (-1)^{\{(p-1)/2\}\{(q-1)/2\}}$$

by Theorem 3.3.

This theorem, together with Theorem 3.1 and the second part of Theorem 3.3, makes the computation of $\left(\frac{a}{p}\right)$ fairly simple. For example, we have

$$\left(\frac{-42}{61}\right) = \left(\frac{-1}{61}\right)\left(\frac{2}{61}\right)\left(\frac{3}{61}\right)\left(\frac{7}{61}\right),$$

$$\left(\frac{-1}{61}\right) = (-1)^{60/2} = 1,$$

$$\left(\frac{2}{61}\right) = (-1)^{(61^2-1)/8} = -1,$$

$$\left(\frac{3}{61}\right) = \left(\frac{61}{3}\right)(-1)^{(2/2)(60/2)} = \left(\frac{1}{3}\right) = 1,$$

$$\left(\frac{7}{61}\right) = \left(\frac{61}{7}\right)(-1)^{(6/2)(60/2)} = \left(\frac{5}{7}\right) = \left(\frac{7}{5}\right)(-1)^{(4/2)(6/2)} = \left(\frac{2}{5}\right)$$

$$= (-1)^{24/8} = -1.$$

Hence $\left(\frac{-42}{61}\right) = 1$. This computation demonstrates a number of different sorts of steps; it was chosen for this purpose and is not the shortest possible. A shorter way is

$$\left(\frac{-42}{61}\right) = \left(\frac{19}{61}\right) = \left(\frac{61}{19}\right) \cdot 1 = \left(\frac{4}{19}\right) = 1.$$

One could also obtain the value of $\left(\frac{-42}{61}\right)$ by use of Theorem 3.2 or the first part of Theorem 3.3, but the computation would be considerably longer.

There is another kind of problem that is of some importance. As an example, let us find all odd primes p such that 3 is a quadratic residue modulo p. We have

$$\left(\frac{3}{p}\right) = \left(\frac{p}{3}\right)(-1)^{(p-1)/2},$$

$$\left(\frac{p}{3}\right) = \begin{cases} \left(\frac{1}{3}\right) = 1 & \text{if } \quad p \equiv 1 \pmod 3 \\ \left(\frac{2}{3}\right) = -1 & \text{if } \quad p \equiv 2 \pmod 3, \end{cases}$$

and

$$(-1)^{(p-1)/2} = \begin{cases} 1 & \text{if } \quad p \equiv 1 \pmod 4 \\ -1 & \text{if } \quad p \equiv 3 \pmod 4. \end{cases}$$

Thus $\left(\frac{3}{p}\right) = 1$ if and only if $p \equiv 1 \pmod 3$, $p \equiv 1 \pmod 4$, or $p \equiv 2 \pmod 3$, $p \equiv 3 \pmod 4$; that is $p \equiv 1$ or $11 \pmod{12}$.

Just as we determined which primes have 3 as a quadratic residue, so for any odd prime p we can analyze which primes have p as a quadratic residue. This is done in effect in the following result.

Theorem 3.5 *Let p be an odd prime. For any odd prime $q > p$ let r be determined as follows. First if p is of the form $4n + 1$, define r as the least positive remainder when q is divided by p; thus $q = kp + r$, $0 < r < p$. Next if p is of the form $4n + 3$, there is a unique r defined by the relations $q = 4kp \pm r$, $0 < r < 4p$, $r \equiv 1 \pmod 4$. Then in both cases $\left(\frac{p}{q}\right) = \left(\frac{r}{p}\right)$.*

Proof If $p = 4n + 1$, by Theorems 3.4 and 3.1, part 3, we see that $\left(\frac{p}{q}\right) = \left(\frac{q}{p}\right) = \left(\frac{r}{p}\right)$. In case $p = 4n + 3$, we first prove that r exists to satisfy the conditions stated. Let r_0 be the least positive remainder when q is divided by $4p$, so $0 < r_0 < 4p$. If $r_0 \equiv 1 \pmod 4$, take $r = r_0$; if $r_0 \equiv 3 \pmod 4$ take $r = 4p - r_0$. The uniqueness of r is readily established.

If $q = 4kp + r$, then $q \equiv r \equiv 1 \pmod 4$ and again $\left(\frac{p}{q}\right) = \left(\frac{q}{p}\right) = \left(\frac{r}{p}\right)$. If $q = 4kp - r$, then $q \equiv -r = 3 \pmod 4$ and by Theorems 3.4 and 3.1, Parts 3 and 4, we have

$$\left(\frac{p}{q}\right) = -\left(\frac{q}{p}\right) = -\left(\frac{-r}{p}\right) = -\left(\frac{-1}{p}\right)\left(\frac{r}{p}\right) = \left(\frac{r}{p}\right).$$

For example, suppose we want to determine all odd primes q that have 11 as a quadratic residue. A complete set of quadratic residues r of 11 satisfying $0 < r < 44$ and $r \equiv 1 \pmod 4$ is $1, 5, 9, 25, 37$. Hence by Theorem 3.5 the odd primes q having 11 as a quadratic residue are precisely those primes of the form $44k \pm r$ where $r = 1, 5, 9, 25$, or 37.

PROBLEMS

1. Verify that $x^2 \equiv 10 \pmod{89}$ is solvable.
2. Prove that if p and q are distinct primes of the form $4k + 3$, and if $x^2 \equiv p \pmod q$ has no solution, then $x^2 \equiv q \pmod p$ has two solutions.
3. Prove that if a prime p is a quadratic residue of an odd prime q, and p is of the form $4k + 1$, then q is a quadratic residue of p.
4. Which of the following congruences are solvable?
   (a) $x^2 \equiv 5 \pmod{227}$ (b) $x^2 \equiv 5 \pmod{229}$
   (c) $x^2 \equiv -5 \pmod{227}$ (d) $x^2 \equiv -5 \pmod{229}$
   (e) $x^2 \equiv 7 \pmod{1009}$ (f) $x^2 \equiv -7 \pmod{1009}$
   (Note that 227, 229, and 1009 are primes.)
5. Find the values of $\left(\frac{p}{q}\right)$ in the nine cases obtained from all combinations of $p = 7, 11, 13$ and $q = 227, 229, 1009$.
6. Decide whether $x^2 \equiv 150 \pmod{1009}$ is solvable or not.
7. Find all primes p such that $x^2 \equiv 13 \pmod p$ has a solution.

8. Find all primes p such that $\left(\frac{10}{p}\right) = 1$.

9. Find all primes q such that $\left(\frac{5}{q}\right) = -1$.

10. Of which primes is -2 a quadratic residue?

11. If a is a quadratic nonresidue of each of the odd primes p and q, is $x^2 \equiv a \pmod{pq}$ solvable?

12. In the proof of Theorem 3.4 consider the pairs (x, y) as points in a plane. Let **O, A, B, C** denote the points $(0, 0)$, $(p/2, 0)$, $(p/2, q/2)$, $(0, q/2)$, respectively, and draw the lines **OA, OB, OC, AB**, and **BC**. Repeat the proof of Theorem 3.4 using geometric language—pairs of points, and so forth.

***13.** Prove that there are infinitely many primes of each of the forms $3n + 1$ and $3n - 1$. (H)

14. Let p and q be *twin primes*, that is, primes satisfying $q = p + 2$. Prove that there is an integer a such that $p|(a^2 - q)$ if and only if there is an integer b such that $q|(b^2 - p)$. (There is a famous unsolved problem to prove that the number of pairs of twin primes is infinite. What *is* known is that the sum of the reciprocals of all twin primes is, if not a finite sum, certainly a convergent series; this result can be contrasted with Theorem 1.19. A proof of this result can be found in Chapter 15 of the book by Hans Rademacher, or in Chapter 6 of the 1977 book by W. J. LeVeque listed in the General References.)

***15.** Let $q = 4^n + 1$ where n is a positive integer. Prove that q is a prime if and only if $3^{(q-1)/2} \equiv -1 \pmod{q}$. (In this way it has been shown that $F_{14} = 2^{2^{14}} + 1$ is composite, though no proper divisor of F_{14} is known.)

***16.** Show that if $p = 2^{2^n} + 1$ is prime then 3 is a primitive root $\pmod{p}$ and that 5 and 7 are primitive roots provided that $n > 1$.

***17.** Show that if $19a^2 \equiv b^2 \pmod{7}$ then $19a^2 \equiv b^2 \pmod{7^2}$.

***18.** Given that 1111118111111 is prime, determine whether 1001 is a quadratic residue (mod 1111118111111). (H)

***19.** Show that p is a divisor of numbers of both of the forms $m^2 + 1$, $n^2 + 2$, if and only if it is a divisor of some number of the form $k^4 + 1$.

***20.** Show that $(x^2 - 2)/(2y^2 + 3)$ is never an integer when x and y are integers.

***21.** Show that if x is not divisible by 3 then $4x^2 + 3$ has at least one prime factor of the form $12n + 7$. Deduce that there are infinitely many primes of this sort.

***22.** Suppose that $(ab, p) = 1$ and that $p > 2$. Show that the number of solutions (x, y) of the congruence $ax^2 + by^2 \equiv 1 \pmod p$ is $p - \left(\frac{-ab}{p}\right)$.

***23.** Show that if a and b are positive integers then

$$\sum_{i=1}^{[a/2]} [ib/a] + \sum_{j=1}^{[b/2]} [ja/b] = [a/2][b/2] + [(a,b)/2].$$

***24.** Let p be a prime number of the form $4k + 1$. Show that

$$\sum_{i=1}^{k} \left[\sqrt{ip}\right] = (p^2 - 1)/12.$$

***25.** We call $\mathscr{H}$ a one-half set of reduced residues (mod p) if $\mathscr{H}$ has the property that $h \in \mathscr{H}$ if and only if $-h \notin \mathscr{H}$. Let $\mathscr{H}$ and $\mathscr{K}$ be two complementary one-half sets. Suppose that $(a, p) = 1$. Let ν be the number of $h \in \mathscr{H}$ for which $ah \in \mathscr{K}$. Show that $(-1)^\nu = \left(\frac{a}{p}\right)$. Show that $a\mathscr{H}$ and $a\mathscr{K}$ are complementary one-half sets. Show that

$$\left(\frac{a}{p}\right) = \prod_{h \in \mathscr{H}} \frac{\sin 2\pi ah/p}{\sin 2\pi h/p}.$$

***26.** Let $k > 1$ be given, and suppose that p is a prime such that $k|(p - 1)$. Suppose that a has order k in the multiplicative group of reduced residue classes (mod p). We call $\mathscr{T}$ a *transversal* of the subgroup $(a) = \{1, a, a^2, \cdots, a^{k-1}\}$ if for each reduced residue class $b \pmod p$ there is a unique $t \in \mathscr{T}$ and a unique i, $0 \leqslant i < k$, such that $b \equiv ta^i \pmod p$. Let $\mathscr{T}$ be such a transversal, and let $I(b)$ denote the number i for which $ta^i \equiv b \pmod p$. Show that

$$\prod_{t \in \mathscr{T}} a^{I(bt)} \equiv b^{(p-1)/k} \pmod p.$$

Deduce that b is a kth power residue (mod p) if and only if $\sum_{t \in \mathscr{T}} I(bt) \equiv 0 \pmod k$.

3.3 THE JACOBI SYMBOL

Definition 3.4 *Let Q be positive and odd, so that $Q = q_1 q_2 \cdots q_s$ where the q_i are odd primes, not necessarily distinct. Then the* Jacobi symbol $\left(\frac{P}{Q}\right)$

is defined by

$$\left(\frac{P}{Q}\right) = \prod_{j=1}^{s} \left(\frac{P}{q_j}\right)$$

where $\left(\frac{P}{q_j}\right)$ *is the Legendre symbol.*

If Q is an odd prime, the Jacobi symbol and Legendre symbol are indistinguishable. However, this can cause no confusion since their values are the same in this case. If $(P, Q) > 1$, then $\left(\frac{P}{Q}\right) = 0$, whereas if $(P, Q) = 1$, then $\left(\frac{P}{Q}\right) = \pm 1$. Moreover, if P is a quadratic residue modulo an odd number Q, then P is a quadratic residue modulo each prime q_j dividing Q, so that $\left(\frac{P}{q_j}\right) = 1$ for each j, and hence $\left(\frac{P}{Q}\right) = 1$. However, $\left(\frac{P}{Q}\right) = 1$ *does not* imply that P is a quadratic residue of Q. For example, $\left(\frac{2}{15}\right) = 1$, but $x^2 \equiv 2 \pmod{15}$ has no solution. If Q is odd then a is a quadratic residue $(\bmod\, Q)$ if and only if $\left(\frac{a}{p}\right) = 1$ for every p dividing Q. Let $p_1, p_2, \cdots, p_r$ denote the distinct primes dividing an odd number Q. Then the reduced residue classes modulo Q are partitioned into 2^r subsets of $\phi(Q)/2^r$ classes each, according to the values of $\left(\frac{a}{p_1}\right), \left(\frac{a}{p_2}\right), \cdots, \left(\frac{a}{p_r}\right)$. Of these subsets, the particular one for which $\left(\frac{a}{p_1}\right) = \left(\frac{a}{p_2}\right) = \cdots = \left(\frac{a}{p_r}\right) = 1$ is the set of quadratic residues $(\bmod\, Q)$.

Theorem 3.6 *Suppose that Q and Q' are odd and positive. Then*

(1) $\left(\frac{P}{Q}\right)\left(\frac{P}{Q'}\right) = \left(\frac{P}{QQ'}\right)$,

(2) $\left(\frac{P}{Q}\right)\left(\frac{P'}{Q}\right) = \left(\frac{PP'}{Q}\right)$,

(3) if $(P, Q) = 1$, *then* $\left(\frac{P^2}{Q}\right) = \left(\frac{P}{Q^2}\right) = 1$,

(4) *if* $(PP', QQ') = 1$, *then* $\left(\frac{P'P^2}{Q'Q^2}\right) = \left(\frac{P'}{Q'}\right)$,

(5) $P' \equiv P \pmod{Q}$ *implies* $\left(\frac{P'}{Q}\right) = \left(\frac{P}{Q}\right)$.

Proof Part 1 is obvious from the definition of $\left(\frac{P}{Q}\right)$, and part 2 follows from the definition and Theorem 3.1, part 2. Then part 3 follows from (2) and (1) and so also does (4). To prove part 5, we write $Q = q_1 q_2 \cdots q_s$. Then $P' \equiv P \pmod{q_j}$ so that $\left(\frac{P'}{q_j}\right) = \left(\frac{P}{q_j}\right)$ by Theorem 3.1, part 3, and then we have part 5 from Definition 3.4.

Theorem 3.7 *If Q is odd and $Q > 0$, then*

$$\left(\frac{-1}{Q}\right) = (-1)^{(Q-1)/2} \quad \textit{and} \quad \left(\frac{2}{Q}\right) = (-1)^{(Q^2-1)/8}$$

Proof We have

$$\left(\frac{-1}{Q}\right) = \prod_{j=1}^{s}\left(\frac{-1}{q_j}\right) = \prod_{j=1}^{s}(-1)^{(q_j-1)/2} = (-1)^{\sum_{j=1}^{s}(q_j-1)/2}$$

If a and b are odd, then

$$\frac{ab-1}{2} - \left(\frac{a-1}{2} + \frac{b-1}{2}\right) = \frac{(a-1)(b-1)}{2} \equiv 0 \pmod{2}$$

and hence

$$\frac{a-1}{2} + \frac{b-1}{2} \equiv \frac{ab-1}{2} \pmod{2}.$$

Applying this repeatedly we obtain

$$\sum_{j=1}^{s}\frac{q_j-1}{2} \equiv \frac{1}{2}\left(\prod_{j=1}^{s}q_j - 1\right) \equiv \frac{Q-1}{2} \pmod{2} \tag{3.1}$$

and thus $\left(\frac{-1}{Q}\right) = (-1)^{(Q-1)/2}$.

Similarly, if a and b are odd, then

$$\frac{a^2b^2-1}{8}-\left(\frac{a^2-1}{8}+\frac{b^2-1}{8}\right)=\frac{(a^2-1)(b^2-1)}{8}\equiv 0 \pmod 8$$

so we have

$$\frac{a^2-1}{8}+\frac{b^2-1}{8}\equiv\frac{a^2b^2-1}{8}\pmod 2,$$

$$\sum_{j=1}^{s}\frac{q_j^2-1}{8}\equiv\frac{Q^2-1}{8}\pmod 2$$

and hence,

$$\left(\frac{2}{Q}\right)=\prod_{j=1}^{s}\left(\frac{2}{q_j}\right)=(-1)^{\sum_{j=1}^{s}(q_j^2-1)/8}=(-1)^{(Q^2-1)/8}.$$

Theorem 3.8 *If P and Q are odd and positive and if $(P,Q)=1$, then*

$$\left(\frac{P}{Q}\right)\left(\frac{Q}{P}\right)=(-1)^{\{(P-1)/2\}\{(Q-1)/2\}}.$$

Proof Writing $P=\prod_{i=1}^{r}p_i$ as well as $Q=\prod_{j=1}^{s}q_j$, we have

$$\left(\frac{P}{Q}\right)=\prod_{j=1}^{s}\left(\frac{P}{q_j}\right)=\prod_{j=1}^{s}\prod_{i=1}^{r}\left(\frac{p_i}{q_j}\right)=\prod_{j=1}^{s}\prod_{i=1}^{r}\left(\frac{q_j}{p_i}\right)(-1)^{\{(p_i-1)/2\}\{(q_j-1)/2\}}$$

$$=\left(\frac{Q}{P}\right)(-1)^{\sum_{j=1}^{s}\sum_{i=1}^{r}\{(p_i-1)/2\}\{(q_j-1)/2\}}$$

where we have used Theorem 3.4. But

$$\sum_{j=1}^{s}\sum_{i=1}^{r}\frac{p_i-1}{2}\frac{q_j-1}{2}=\sum_{i=1}^{r}\frac{p_i-1}{2}\sum_{j=1}^{s}\frac{q_j-1}{2}$$

and

$$\sum_{i=1}^{r}\frac{p_i-1}{2}\equiv\frac{P-1}{2},\qquad \sum_{j=1}^{s}\frac{q_j-1}{2}\equiv\frac{Q-1}{2}\pmod 2$$

as in (3.1) in the proof of Theorem 3.7. Therefore we have

$$\left(\frac{P}{Q}\right) = \left(\frac{Q}{P}\right)(-1)^{\{(P-1)/2\}\{(Q-1)/2\}}$$

which proves the theorem.

The theorem we have just proved shows that the Jacobi symbol obeys the law of reciprocity. It is worthwhile to consider what has been done. In this chapter we have been interested in quadratic residues. The definition of the Legendre symbol is a natural one to make. We then proved the useful and celebrated law of reciprocity for this symbol. The Jacobi symbol is an extension of the Legendre symbol, defining $\left(\frac{P}{Q}\right)$ for composite Q. However, at first it might have seemed more natural to define $\left(\frac{P}{Q}\right)$ to be 1 for quadratic residues P and -1 for nonresidues modulo Q. Had this been done, there would have been no reciprocity law ($P = 5$, $Q = 9$ is an example). What we have done is this: We have dropped the connection with quadratic residues in favor of the law of reciprocity. This does not mean that the Jacobi symbol cannot be used in computations like those in Section 3.2. In fact, the Jacobi symbol plays an important role in such calculations. In Section 3.2 we used the reciprocity law to invert the symbol $\left(\frac{p}{q}\right)$ to $\left(\frac{q}{p}\right)$, but we could do it only if q was a prime. In order to compute $\left(\frac{a}{p}\right)$ we had to factor a and consider a product of Legendre symbols. Now however, using Jacobi symbols we do not need to factor a if it is odd and positive. We compute $\left(\frac{a}{p}\right)$ as a Jacobi symbol and then know the quadratic character of a modulo p if p is a prime.

For example,

$$\left(\frac{105}{317}\right) = \left(\frac{317}{105}\right) = \left(\frac{2}{105}\right) = 1$$

and hence 105 is a quadratic residue modulo the prime number 317.

The amount of calculation required to evaluate the Legendre symbol (using the Jacobi symbol and reciprocity) is roughly comparable to the amount required in an application of Euler's criterion (Corollary 2.38). However, the latter method has the disadvantage that it involves multiplying residue classes, a slow process if the modulus is larger than one-half the word length.

The question of how evenly the quadratic residues are distributed in the interval $[1, p]$ is a topic of current research interest. *Vinogradov's hypothesis* asserts that if $\varepsilon > 0$ is given, then there is a $p_0(\varepsilon)$ such that the least positive quadratic nonresidue is less than p^ε provided that $p > p_0(\varepsilon)$. The present status of our knowledge leaves much to be desired, but we now give a simple proof that the least positive quadratic nonresidue cannot be too large.

Theorem 3.9 *Suppose that p is an odd prime. Let n denote the least positive quadratic nonresidue modulo p. Then $n < 1 + \sqrt{p}$.*

Proof Let m be the least positive number for which $mn > p$, so that $(m-1)n \leqslant p$. As $n \geqslant 2$ and p is prime, we have $(m-1)n < p$. Thus $0 < mn - p < n$. As n is the least positive nonresidue (mod p), it follows that $\left(\frac{mn-p}{p}\right) = 1$, and hence that $\left(\frac{m}{p}\right) = -1$. Consequently $m \geqslant n$, so that $(n-1)^2 < (n-1)n \leqslant (m-1)n < p$. Thus $n - 1 < \sqrt{p}$, and we have the stated bound.

In Problem 18 we consider a different kind of question regarding the distribution of the quadratic residues.

PROBLEMS

1. Evaluate: $\left(\frac{-23}{83}\right)$; $\left(\frac{51}{71}\right)$; $\left(\frac{71}{73}\right)$; $\left(\frac{-35}{97}\right)$.

2. Which of the following congruences are solvable?
(*a*) $x^2 \equiv 10 \pmod{127}$
(*b*) $x^2 \equiv 73 \pmod{173}$
(*c*) $x^2 \equiv 137 \pmod{401}$

3. Which of the following congruences are solvable?
(*a*) $x^2 \equiv 11 \pmod{61}$ (*b*) $x^2 \equiv 42 \pmod{97}$
(*c*) $x^2 \equiv -43 \pmod{79}$ (*d*) $x^2 - 31 \equiv 0 \pmod{103}$

4. Determine whether $x^4 \equiv 25 \pmod{1013}$ is solvable, given that 1013 is a prime.

5. Prove that $\sum_{j=1}^{p-1} \left(\frac{j}{p}\right) = 0$, p an odd prime.

6. For any prime p of the form $4k + 3$, prove that $x^2 + (p+1)/4 \equiv 0 \pmod p$ is not solvable.

7. For which primes p do there exist integers x and y with $(x, p) = 1$, $(y, p) = 1$, such that $x^2 + y^2 \equiv 0 \pmod p$?

8. For which prime powers p^a do there exist integers x and y with $(x, p) = 1$, $(y, p) = 1$, such that $x^2 + y^2 \equiv 0 \pmod{p^a}$?

9. For which positive integers n do there exist integers x and y with $(x, n) = 1$, $(y, n) = 1$, such that $x^2 + y^2 \equiv 0 \pmod n$?

10. Let k be odd. Prove that $x^2 \equiv k \pmod 2$ has exactly one solution. Furthermore, $x^2 \equiv k \pmod{2^2}$ is solvable if and only if $k \equiv 1 \pmod 4$, in which case there are two solutions.

11. Let a be odd, and suppose that $\alpha \geqslant 3$. Prove that $x^2 \equiv a \pmod{2^\alpha}$ has 4 solutions or no solution, according as $a \equiv 1 \pmod 8$ or not. Show that if x_0 is one solution, then the other three are $-x_0, x_0 \pm 2^{\alpha-1}$. (H)

12. Consider the congruence $x^2 \equiv a \pmod{p^\alpha}$ with p a prime, $\alpha \geqslant 1$, $a = p^\beta b$, $(b, p) = 1$. Prove that if $\beta \geqslant \alpha$ then the congruence is solvable, and that if $\beta < \alpha$ then the congruence is solvable if and only if β is even and $x^2 \equiv b \pmod{p^{\alpha-\beta}}$ is solvable.

***13.** Let the integers $1, 2, \cdots, p-1$ modulo p, p an odd prime, be divided into two nonempty sets $\mathscr{S}_1$ and $\mathscr{S}_2$ so that the product of two elements in the same set is in $\mathscr{S}_1$, whereas the product of an element of $\mathscr{S}_1$ and an element of $\mathscr{S}_2$ is in $\mathscr{S}_2$. Prove that $\mathscr{S}_1$ consists of the quadratic residues, $\mathscr{S}_2$ of the nonresidues, modulo p. (H)

***14.** Suppose that p is a prime, $p \equiv 1 \pmod 4$, and that $a^2 + b^2 = p$ with a odd and positive. Show that $\left(\frac{a}{p}\right) = 1$.

***15.** Suppose that p is a prime, $p \geqslant 7$. Show that $\left(\frac{n}{p}\right) = \left(\frac{n+1}{p}\right) = 1$ for at least one number n in the set $\{1, 2, \cdots, 9\}$. (H)

***16.** Prove that if $(a, p) = 1$ and p is an odd prime, then $\sum_{n=1}^{p} \left(\frac{an+b}{p}\right) = 0$.

***17.** Let p be an odd prime, and put $s(a, p) = \sum_{n=1}^{p} \left(\frac{n(n+a)}{p}\right)$. Show that $s(0, p) = p - 1$. Show that $\sum_{a=1}^{p} s(a, p) = 0$. Show that if

$(a, p) = 1$ then $s(a, p) = s(1, p)$. Conclude that $s(a, p) = -1$ if $(a, p) = 1$. (H)

***18.** Let p be an odd prime, and let $N_{++}(p)$ denote the number of n, $1 \leqslant n \leqslant p - 2$, such that $\left(\frac{n}{p}\right) = \left(\frac{n+1}{p}\right) = 1$. Show that $N_{++}(p) = (p - \left(\frac{-1}{p}\right) - 4)/4$. Similarly define and evaluate $N_{+-}(p)$, $N_{-+}(p)$, and $N_{--}(p)$. (H)

Remark From a general theorem (the "Riemann hypothesis for curves over finite fields") proved by André Weil in 1948, it can be deduced that if p is an odd prime and k is a positive integer then

$$\left| \sum_{n=1}^{p} \left(\frac{(n+1) \cdots (n+k)}{p} \right) \right| \leqslant 2k\sqrt{p}\,. \qquad (3.2)$$

The technique used in Problem 18 can then be used to show that

$$|N_{\pm\pm\cdots\pm}(p) - p/2^k| \leqslant 3k\sqrt{p}\,.$$

Thus if k is fixed and p is large, the k-tuple of values $\left(\left(\frac{n+1}{p}\right), \left(\frac{n+2}{p}\right), \ldots, \left(\frac{n+k}{p}\right)\right)$ takes on any prescribed set of values ± 1, approximately $p/2^k$ times as n runs from 1 to p.

***19.** Show that if p is an odd prime and h is an integer, $1 \leqslant h \leqslant p$, then

$$\sum_{n=1}^{p} \left(\sum_{m=1}^{h} \left(\frac{m+n}{p} \right) \right)^2 = h(p - h).$$

***20.** Show that if $(a, p) = 1$, p an odd prime, then $\sum_{n=1}^{p} \left(\frac{n^2 + a}{p}\right) = -1$.

***21.** Let m be a positive odd integer, and let $\mathscr{G}$ denote the set of those reduced residue classes $a \pmod m$ such that $a^{(m-1)/2} \equiv \left(\frac{a}{m}\right) \pmod m$. Show that if $a \in \mathscr{G}$ and $b \in \mathscr{G}$, then $ab \in \mathscr{G}$. Show also that if $a \in \mathscr{G}$ and $a\bar{a} \equiv 1 \pmod m$, then $\bar{a} \in \mathscr{G}$. (Thus $\mathscr{G}$ is a subgroup of the multiplicative group of reduced residue classes $\pmod m$.)

22. Find the set $\mathscr{G}$ defined in Problem 21 when $m = 21$.

***23.** Show that if m is an odd composite number then the set $\mathscr{G}$ defined in Problem 21 is a proper subset of the collection of reduced residue classes $\pmod m$. (H)

24. Let m be an odd positive integer, and let $\mathscr{H}$ denote the set of reduced residue classes $a \pmod m$ such that m is a strong probable prime base a (i.e., if $m - 1 = 2^k d$, d odd, then $a^d \equiv 1 \pmod m$ or

$a^{d2^j} \equiv -1$ for some j, $0 \leqslant j < k$). Show that if $m = 65$ then $8 \in \mathscr{H}$ and $18 \in \mathscr{H}$, but that $8 \cdot 18 \equiv 14 \notin \mathscr{H}$. (Thus $\mathscr{H}$ is not a group for this m.)

***25.** Let m be an odd positive integer, and let $\mathscr{G}$ and $\mathscr{H}$ be defined as in Problems 21 and 24. Show that $\mathscr{H} \subseteq \mathscr{G}$.

3.4 BINARY QUADRATIC FORMS

A monomial $ax_1^{k_1}x_2^{k_2} \cdots x_n^{k_n}$ in n variables with coefficient $a \neq 0$ is said to have degree $k_1 + k_2 + \cdots + k_n$. The degree of a polynomial in n variables is the maximum of the degrees of the monomial terms in the polynomial. A polynomial in several variables is called a *form*, or is said to be *homogeneous* if all its monomial terms have the same degree. A form of degree 2 is called a *quadratic* form. Thus the general quadratic form is a sum of the shape

$$\sum_{i=1}^{n} \sum_{j=1}^{n} a_{ij}x_i x_j.$$

A form in two variables is called *binary*. The remainder of this chapter is devoted to the study of binary quadratic forms

$$f(x, y) = ax^2 + bxy + cy^2$$

with integral coefficients. Such forms have many striking number-theoretic properties. In Theorem 2.15 we found that the numbers n represented by the quadratic form $x^2 + y^2$ can be characterized in terms of the prime factors of n. Using quadratic reciprocity, we now investigate the extent to which Theorem 2.15 can be generalized to other quadratic forms.

The *discriminant* of a binary quadratic form is the quantity $d = b^2 - 4ac$. If d is a perfect square (possibly 0), then $f(x, y)$ can be expressed as a product of two linear forms with integral coefficients, as in the cases xy, or $x^2 - y^2 = (x - y)(x + y)$ or $10x^2 - 27xy + 18y^2 = (2x - 3y)(5x - 6y)$, with discriminants $1, 4, 9$, respectively. Conversely, if d is not a perfect square (or 0) then $f(x, y)$ cannot be written as a product of two linear forms with integral coefficients, nor even with rational coefficients. (The proofs of these results are left to the reader in Problems 7–9 at the end of this section.) As the theory develops, we often find it necessary to distinguish between square and nonsquare discriminants.

Theorem 3.10 *Let $f(x, y) = ax^2 + bxy + cy^2$ be a binary quadratic form with integral coefficients and discriminant d. If $d \neq 0$ and d is not a perfect* *square, then* $a \neq 0$, $c \neq 0$, *and the only solution of the equation* $f(x, y) = 0$ *in integers is given by* $x = y = 0$.

Proof We may presume that $a \neq 0$ and $c \neq 0$, for if $a = 0$ or $c = 0$ then $ac = 0$ and $d = b^2 - 4ac = b^2$, a perfect square. Suppose that x_0 and y_0 are integers such that $f(x_0, y_0) = 0$. If $y_0 = 0$ then $ax_0^2 = 0$, and hence $x_0 = 0$ because $a \neq 0$. If $x_0 = 0$, a parallel argument gives $y_0 = 0$. Consequently we take $x_0 \neq 0$ and $y_0 \neq 0$. By completing the square we see that

$$4af(x, y) = (2ax + by)^2 - dy^2 \tag{3.3}$$

and hence $(2ax_0 + by_0)^2 = dy_0^2$ since $f(x_0, y_0) = 0$. But $dy_0^2 \neq 0$, and it follows by unique factorization that d is a perfect square. The proof is now complete.

Definition 3.5 *A form* $f(x, y)$ *is called* indefinite *if it takes on both positive and negative values. The form is called* positive semidefinite (*or* negative semidefinite) *if* $f(x, y) \geqslant 0$ (*or* $f(x, y) \leqslant 0$) *for all integers* x, y. *A semidefinite form is called* definite *if in addition the only integers* x, y *for which* $f(x, y) = 0$ *are* $x = 0$, $y = 0$.

The form $f(x, y) = x^2 - 2y^2$ is indefinite, since $f(1, 0) = 1$ and $f(0, 1) = -2$. The form $f(x, y) = x^2 - 2xy + y^2 = (x - y)^2$ is positive semidefinite, but not definite, because $f(1, 1) = 0$. Finally, $x^2 + y^2$ is an example of a positive definite form. We now show that we may determine whether a quadratic form is definite or indefinite by evaluating its discriminant.

Theorem 3.11 *Let* $f(x, y) = ax^2 + bxy + cy^2$ *be a binary quadratic form with integral coefficients and discriminant* d. *If* $d > 0$ *then* $f(x, y)$ *is indefinite. If* $d = 0$ *then* $f(x, y)$ *is semidefinite but not definite. If* $d < 0$ *then* a *and* c *have the same sign and* $f(x, y)$ *is either positive definite or negative definite according as* $a > 0$ *or* $a < 0$.

Clearly if f is positive definite then $-f$ is negative definite, and conversely. Hence we ignore the negative definite forms, as their properties follow from those of the positive definite forms.

Proof Suppose that $d > 0$. We note that $f(1, 0) = a$, and that $f(b, -2a) = -ad$. These numbers are of opposite sign unless $a = 0$. Similarly, $f(0, 1) = c$ and $f(-2c, b) = -cd$. These numbers are of opposite sign unless $c = 0$. It remains to consider the possibility that $a = c = 0$. Then

$d = b^2 > 0$, so that $b \neq 0$. In this case $f(1, 1) = b$ and $f(1, -1) = -b$, so that f takes values of both signs.

Now suppose that $d = 0$. Consider the possibility that $a \neq 0$. Then from (3.3) we see that the nonzero values of f are all of the same sign as a, so $f(x, y)$ is semidefinite. Moreover, $f(b, -2a) = -ad = 0$. Since $a \neq 0$ in the case under consideration, it follows that f is not definite. Suppose now that $a = 0$. Then $d = b^2$, and hence $b = 0$ since $d = 0$. Thus in this case, $f(x, y) = cy^2$. Here the nonzero values all have the same sign as c, but $f(1, 0) = 0$, so the form is not definite.

Finally, suppose that $d < 0$. From (3.3) and Theorem 3.10 we see that $4af(x, y)$ is positive for all pairs of integers x, y except $0, 0$. Thus f is definite. Since $f(1, 0) = a$ and $f(0, 1) = c$, we deduce in particular that a and c have the same sign, positive for positive definite forms and negative for negative definite forms. (An alternative way to see that a and c have the same sign when $d < 0$ is provided by noting that $4ac = b^2 - d \geqslant -d > 0$, so that $ac > 0$.) This completes the proof.

We now determine which numbers d arise as discriminants of binary quadratic forms.

Theorem 3.12 *Let d be a given integer. There exists at least one binary quadratic form with integral coefficients and discriminant d, if and only if $d \equiv 0$ or $1 \pmod 4$.*

Proof Since $b^2 \equiv 0$ or $1 \pmod 4$ for any integer b, it follows that the discriminant $d = b^2 - 4ac \equiv 0$ or $1 \pmod 4$. For the converse, suppose first that $d \equiv 0 \pmod 4$. Then the form $x^2 - (d/4)y^2$ has discriminant d. Similarly, if $d \equiv 1 \pmod 4$ then the form $x^2 + xy - \left(\frac{d-1}{4}\right)y^2$ has discriminant d, and the proof is complete.

We say that a quadratic form $f(x, y)$ represents an integer n if there exist integers x_0 and y_0 such that $f(x_0, y_0) = n$. Such a representation is called *proper* if g.c.d.$(x_0, y_0) = 1$; otherwise it is *improper*. If $f(x_0, y_0) = n$ and g.c.d.$(x_0, y_0) = g$, then $g^2 | n$, g.c.d.$(x_0/g, y_0/g) = 1$, and $f(x_0/g, y_0/g) = n/g^2$. Thus the representations of n by $f(x, y)$ may be found by determining the proper representations of n/g^2 for those integers g such that $g^2 | n$.

Our object in the remainder of this chapter is to describe those integers n represented, or properly represented, by a particular quadratic form. This aim is only partly achieved, but we can determine whether n is represented by some quadratic form of a prescribed discriminant, as follows.

Theorem 3.13 *Let n and d be given integers with $n \neq 0$. There exists a binary quadratic form of discriminant d that represents n properly if and only if the congruence $x^2 \equiv d \pmod{4|n|}$ has a solution.*

Proof Suppose that b is a solution of the congruence, with $b^2 - d = 4nc$, say. Then the form $f(x, y) = nx^2 + bxy + cy^2$ has integral coefficients and discriminant d. Moreover, $f(1, 0) = n$ is a proper representation of n.

Conversely, suppose we have a proper representation $f(x_0, y_0)$ of n by a form $f(x, y) = ax^2 + bxy + cy^2 = n$ with discriminant $b^2 - 4ac = d$. Since g.c.d.$(x_0, y_0) = 1$, we can choose integers m_1, m_2 such that $m_1 m_2 = 4|n|$, g.c.d.$(m_1, m_2) = 1$, g.c.d.$(m_1, y_0) = 1$ and g.c.d.$(m_2, x_0) = 1$. For example, take m_1 to be the product of those prime-power factors p^α of $4n$ for which $p \mid x_0$, and then put $m_2 = 4|n|/m_1$. From equation (3.3) we see that $4an = (2ax_0 + by_0)^2 - dy_0^2$, and hence $(2ax_0 + by_0)^2 \equiv dy_0^2 \pmod{m_1}$. As $(y_0, m_1) = 1$, there is an integer $\overline{y_0}$ such that $y_0\overline{y_0} \equiv 1 \pmod{m_1}$, and we find that the congruence $u^2 \equiv d \pmod{m_1}$ has a solution, namely $u = u_1 = (2ax_0 + by_0)\overline{y_o}$. We interchange a and c, and also x and y, to see that the parallel congruence $u^2 \equiv d \pmod{m_2}$ also has a solution, say $u = u_2$. Then by the Chinese remainder theorem we find an integer w such that $w \equiv u_1 \pmod{m_1}$ and $w \equiv u_2 \pmod{m_2}$. Thus $w^2 \equiv u_1^2 \equiv d \pmod{m_1}$, and similarly $w^2 \equiv u_2^2 \equiv d \pmod{m_2}$, from which we get $w^2 \equiv d \pmod{m_1 m_2}$. But this last modulus is $4|n|$, so the theorem is proved.

Corollary 3.14 *Suppose that $d \equiv 0$ or $1 \pmod 4$. If p is an odd prime, then there is a binary quadratic form of discriminant d that represents p, if and only if $p \mid d$ or $\left(\frac{d}{p}\right) = 1$.*

Proof Any representation of p must be proper. Hence if p is represented, then it is properly represented, and thus (by the theorem) d must be a square modulo $4p$, so that $\left(\frac{d}{p}\right) = 1$ or 0. Conversely, if $\left(\frac{d}{p}\right) = 1$ or 0, then d is a square modulo p. By hypothesis, d is a square modulo 4. Since p is odd, it follows by the Chinese remainder theorem that d is a square modulo $4p$, and hence (by the theorem) p is properly represented by some form of discriminant d, thus completing the proof.

Let d be given. By quadratic reciprocity we know that the odd primes p for which $\left(\frac{d}{p}\right) = 1$ are precisely the primes lying in certain residue classes modulo $4|d|$. In this way, quadratic reciprocity plays a role in determining which primes are represented by the quadratic forms of a prescribed discriminant.

PROBLEMS

1. For each of the following, determine whether the form is positive definite, negative definite, or indefinite.
   (*a*) $x^2 + y^2$; (*b*) $-x^2 - y^2$; (*c*) $x^2 - 2y^2$;
   (*d*) $10x^2 - 9xy + 8y^2$; (*e*) $x^2 - 3xy + y^2$; (*f*) $17x^2 - 26xy + 10y^2$.
2. Prove that the quadratic form $x^2 - 2xy + y^2$ has discriminant 0. Determine the class of integers represented by this form.
3. If $\mathscr{C}$ is any class of integers, finite or infinite, let $m\mathscr{C}$ denote the class obtained by multiplying each integer of $\mathscr{C}$ by the integer m. Prove that if $\mathscr{C}$ is the class of integers represented by any form f, then $m\mathscr{C}$ is the class of integers represented by mf.
4. Use the binomial theorem to give a formula for positive integers x_k and y_k such that $(3 + 2\sqrt{2})^k = x_k + y_k\sqrt{2}$. Show that $(3 - 2\sqrt{2})^k = x_k - y_k\sqrt{2}$. Deduce that $x_k^2 - 2y_k^2 = 1$ for $k = 1, 2, 3, \cdots$. Show that $(x_k, y_k) = 1$ for each k. Show that $x_{k+1} = 3x_k + 4y_k$ and $y_{k+1} = 2x_k + 3y_k$ for $k = 1, 2, 3 \cdots$. Show that $\{x_k\}$ and $\{y_k\}$ are strictly increasing sequences. Conclude that the number 1 has infinitely many proper representations by the quadratic form $x^2 - 2y^2$.
5. (*a*) Let A and B be real numbers, and put $F(\phi) = A\cos\phi + B\sin\phi$. Using calculus, or otherwise, prove that $\max_{0 \leqslant \phi \leqslant 2\pi} F(\phi) = \sqrt{A^2 + B^2}$, and that $\min_{0 \leqslant \phi \leqslant 2\pi} F(\phi) = -\sqrt{A^2 + B^2}$.
   (*b*) Let $f(x, y)$ denote the quadratic form $ax^2 + bxy + cy^2$. Convert to polar coordinates by writing $x = r\cos\theta$, $y = r\sin\theta$. Show that $f(r\cos\theta, r\sin\theta) = r^2(a + c + (a - c)\cos 2\theta + b\sin 2\theta)2$. Show that if r is fixed and θ runs from 0 to 2π, then the maximum and minimum values of $f(r\cos\theta, r\sin\theta)$ are
   $$r^2\left(a + c \pm \sqrt{(a + c)^2 + d}\right)/2.$$
   (*c*) Let f be a positive definite quadratic form. Prove that there exist positive constants C_1 and C_2 (which may depend on the coefficients of f) such that $C_1(x^2 + y^2) \leqslant f(x, y) \leqslant C_2(x^2 + y^2)$ for all real numbers x and y.
   (*d*) Conclude that if f is a positive definite quadratic form then an integer n has at most a finite number of representations by f.
6. Let d be a perfect square, possibly 0. Show that there is a quadratic form $ax^2 + bxy + cy^2$ of discriminant d for which $a = 0$.

7. Let a, b, and c be integers with $a \neq 0$. Show that if one root of the equation $au^2 + bu + c = 0$ is rational then the other one is, and that $b^2 - 4ac$ is a perfect square, possibly 0. Show also that if $b^2 - 4ac$ is a perfect square, possibly 0, then the roots of the equation $au^2 + bu + c = 0$ are rational.

8. Show that the discriminant of the quadratic form $(h_1x + k_1y)(h_2x + k_2y)$ is the square of the determinant $\begin{vmatrix} h_1 & h_2 \\ k_1 & k_2 \end{vmatrix}$. Deduce that if h_1, h_2, k_1, and k_2 are all integers then the discriminant is a perfect square, possibly 0.

9. Let $f(x, y) = ax^2 + bxy + cy^2$ be a quadratic form with integral coefficients whose discriminant d is a perfect square, possibly 0. Show that there are integers h_1, h_2, k_1, and k_2 such that $f(x, y) = (h_1x + k_1y)(h_2x + k_2y)$. (H)

10. Let $f(x, y) = ax^2 + bxy + cy^2$ be a quadratic form with integral coefficients. Show that there exist integers x_0, y_0, not both 0, such that $f(x_0, y_0) = 0$, if and only if the discriminant d of $f(x, y)$ is a perfect square, possibly 0.

3.5 EQUIVALENCE AND REDUCTION OF BINARY QUADRATIC FORMS

Let $f(x, y) = x^2 + y^2$ and $g(x, y) = x^2 + 2xy + 2y^2$. A quick calculation gives $g(x, y) = f(x + y, y)$ and $f(x, y) = g(x - y, y)$, which implies that these forms represent exactly the same numbers. More precisely, the first identity implies that any number represented by g, such as $34 = g(2, 3)$, is also represented by f, since $f(2 + 3, 3) = g(2, 3) = 34$. Conversely, the second identity implies that any number represented by f is represented by g. For purposes of determining which numbers are represented, these forms may therefore be considered to be equivalent. Here we have used the simple fact that the coordinates of the point (x, y) are integers if and only if the coordinates of the point $(x + y, y)$ are integers. A point whose coordinates are integers is called a *lattice point*. We now determine which linear changes of variable take lattice points to themselves in a one-to-one manner.

Theorem 3.15 *Let* $M = \begin{bmatrix} m_{11} & m_{12} \\ m_{21} & m_{22} \end{bmatrix}$ *be a* 2×2 *matrix with real entries, and put*

$$\begin{bmatrix} u \\ v \end{bmatrix} = M \begin{bmatrix} x \\ y \end{bmatrix}. \tag{3.4}$$

That is, $u = m_{11}x + m_{12}y$, $v = m_{21}x + m_{22}y$. *Then the following two assertions are equivalent*:

(*i*) *the linear transformation* (3.4) *defines a permutation of lattice points* (*i.e., lattice points are mapped to themselves in a one-to-one and onto manner*);

(*ii*) *the matrix M has integral coefficients and* $\det(M) = \pm 1$.

This is analogous to the theorem of linear algebra which asserts that (3.4) defines a permutation of $\mathbb{R}^2$ if and only if $\det(M) \neq 0$.

Proof We first demonstrate that (ii) implies (i). It is clear that if M has integral coefficients then (u, v) is a lattice point whenever (x, y) is a lattice point. For brevity, put $\Delta = \det(M) = m_{11}m_{22} - m_{12}m_{21}$. As $\Delta \neq 0$, the inverse matrix M^{-1} exists, and

$$M^{-1} = \begin{bmatrix} m_{22}/\Delta & -m_{12}/\Delta \\ -m_{21}/\Delta & m_{11}/\Delta \end{bmatrix}.$$

Thus if (ii) holds then M^{-1} also has integral coefficients, and then the inverse map from lattice points (u, v) to lattice points (x, y) is given by matrix multiplication,

$$\begin{bmatrix} x \\ y \end{bmatrix} = M^{-1}\begin{bmatrix} u \\ v \end{bmatrix}.$$

Hence the map is one-to-one and onto (i.e., a permutation).

Suppose now that (i) holds. Taking the lattice point $(x, y) = (1, 0)$, we find that (3.4) gives $(u, v) = (m_{11}, m_{21})$. Since this must be a lattice point, it follows that m_{11} and m_{21} must be integers. Taking $(x, y) = (0, 1)$, we find similarly that m_{12} and m_{22} are integers. It remains to show that $\det(M) = \pm 1$. To this end, consider the lattice point $(u, v) = (1, 0)$. From (i) we know that the map (3.4) is onto. Hence there is a lattice point (x_1, y_1) such that

$$\begin{bmatrix} 1 \\ 0 \end{bmatrix} = M\begin{bmatrix} x_1 \\ y_1 \end{bmatrix}.$$

Similarly, there is a lattice point (x_2, y_2) such that

$$\begin{bmatrix} 0 \\ 1 \end{bmatrix} = M\begin{bmatrix} x_2 \\ y_2 \end{bmatrix}.$$

These two relations may be expressed as a single matrix identity,

$$\begin{bmatrix} 1 & 0 \\ 0 & 1 \end{bmatrix} = M \begin{bmatrix} x_1 & x_2 \\ y_1 & y_2 \end{bmatrix}. \tag{3.5}$$

We now recall from linear algebra that if M and N are two $n \times n$ matrices then

$$\det(MN) = \det(M)\det(N). \tag{3.6}$$

(In the present section we require only the case $n = 2$, which may be verified by checking that $(m_{11}n_{11} + m_{12}n_{21})(m_{21}n_{12} + m_{22}n_{22}) - (m_{21}n_{11} + m_{22}n_{21})(m_{11}n_{12} + m_{12}n_{22}) = (m_{11}m_{22} - m_{21}m_{12})(n_{11}n_{22} - n_{21}n_{12})$ is a valid algebraic identity.) Applying this to (3.5), we find that $1 = \det(M)(x_1y_2 - x_2y_1)$. Here both factors are integers because the matrices on the right side in (3.5) have integral coefficients. Thus $\det(M) | 1$, that is, $\det(M) = \pm 1$, and the proof is complete.

Although Theorem 3.15 allows matrices M with $\det(M) = -1$, we now restrict our attention to matrices with $\det(M) = +1$, as it has been found to lead to a more fruitful theory. We explain this in greater detail in the Notes at the end of this chapter.

Suppose that M and N are 2×2 matrices with integral coefficients. Then the matrix MN is also 2×2, and has integral coefficients. From (3.6) we see that if $\det(M) = \det(N) = 1$ then $\det(MN) = 1$. Moreover, M^{-1} has integral coefficients, and $\det(M^{-1}) = 1$. Thus the set of 2×2 matrices with integral coefficients and determinant 1 form a group.

Definition 3.6 *The group of* 2×2 *matrices with integral elements and determinant* 1 *is denoted by* Γ, *and is called the* modular group.

The modular group is noncommutative. For example, if

$$M = \begin{bmatrix} 0 & 1 \\ -1 & 0 \end{bmatrix} \quad \text{and} \quad N = \begin{bmatrix} 1 & 0 \\ 1 & 1 \end{bmatrix},$$

then

$$MN = \begin{bmatrix} 1 & 1 \\ -1 & 0 \end{bmatrix} \quad \text{but} \quad NM = \begin{bmatrix} 0 & 1 \\ -1 & 1 \end{bmatrix}.$$

Definition 3.7 *The quadratic forms* $f(x, y) = ax^2 + bxy + y^2$ *and* $g(x, y) = Ax^2 + Bxy + Cy^2$ *are equivalent, and we write* $f \sim g$, *if there is an* $M = [m_{ij}] \in \Gamma$ *such that* $g(x, y) = f(m_{11}x + m_{12}y, m_{21}x + m_{22}y)$. *In this case we say that* M *takes* f *to* g.

In this situation, we may calculate the coefficients of g in terms of those of f and of M.

$$A = am_{11}^2 + bm_{11}m_{21} + cm_{21}^2 = f(m_{11}, m_{21}), \tag{3.7a}$$

$$B = 2am_{11}m_{12} + b(m_{11}m_{22} + m_{12}m_{21}) + 2cm_{21}m_{22}, \tag{3.7b}$$

$$C = am_{12}^2 + bm_{12}m_{22} + cm_{22}^2 = f(m_{12}, m_{22}). \tag{3.7c}$$

The effect of this change of variables is made clearer by making systematic use of matrix multiplication. Let

$$F = \begin{bmatrix} a & \frac{1}{2}b \\ \frac{1}{2}b & c \end{bmatrix}, \qquad G = \begin{bmatrix} A & \frac{1}{2}B \\ \frac{1}{2}B & C \end{bmatrix}, \qquad X = \begin{bmatrix} x \\ y \end{bmatrix}.$$

Then $X^tFX = [f(x, y)]$. Here the matrix on the right is a 1×1 matrix, and $X^t = [x \quad y]$ is the transpose of X. Similarly $X^tGX = [g(x, y)]$. Our definition of g states that we obtain g by evaluating f with X replaced by MX. That is, $(MX)^tF(MX) = [g(x, y)]$. Since $(MX)^t = X^tM^t$, this may be written $X^t(M^tFM)X = [g(x, y)]$. The coefficient matrix G of the quadratic form g is uniquely determined by the coefficients of g, so we may conclude that

$$M^tFM = G. \tag{3.8}$$

Indeed, if the matrix multiplications on the left are performed, we discover that this matrix identity is simply a more compact reformulation of the identities (3.7). We now show that the notion of equivalence in Definition 3.7 is an equivalence relation in the usual sense that it is reflexive, symmetric, and transitive.

Theorem 3.16 *Let f, g, and h be binary quadratic forms. Then*

(1) $f \sim f$,

(2) if $f \sim g$, then $g \sim f$,

(3) if $f \sim g$ and $g \sim h$, then $f \sim h$.

Proof We have seen that $f \sim g$ if and only if there is an $M \in \Gamma$ such that (3.8) holds. Take $M = I$, the identity matrix. Since $I \in \Gamma$ and $I^tFI = F$, we conclude that $f \sim f$. Suppose that $f \sim g$. Then we have (3.8) for some $M \in \Gamma$. By multiplying this on the left by $(M^{-1})^t$, and on the right by M^{-1}, we deduce that $F = (M^{-1})^t G M^{-1}$. But Γ is a group, so $M^{-1} \in \Gamma$, and hence $g \sim f$. Suppose finally that $f \sim g$ and $g \sim h$. Then $G = M^{-1}FM$ and $H = N^{-1}GN$ for some matrices M and N in Γ. On substituting the first of these identities in the second, we find that $H = N^{-1}(M^{-1}FM)N = (MN)^{-1}F(MN)$. Since $MN \in \Gamma$, we have established that $f \sim h$.

Since the relation $\sim$ is an equivalence relation, it serves to partition the set of binary quadratic forms into equivalence classes. We now relate this concept to the representability of integers.

Theorem 3.17 *Let f and g be equivalent binary quadratic forms. For any given integer n, the representations of n by f are in one-to-one correspondence with the representations of n by g. Also, the proper representations of n by f are in one-to-one correspondence with the proper representations of n by g. Moreover, the discriminants of f and g are equal.*

Proof The first assertion is immediate from Theorem 3.15 and Definition 3.7. To prove the second assertion, we establish that in this one-to-one correspondence, g.c.d.(x, y) = g.c.d.(u, v) whenever X and U are nonzero lattice points. Let r = g.c.d.(x, y) and s = g.c.d.(u, v). Since $\frac{1}{r}X$ is a lattice point, it follows from Theorem 3.15 that $\frac{1}{r}U = M\left(\frac{1}{r}X\right)$ is a lattice point. That is, $r \mid s$. As it may similarly be shown that $s \mid r$, we conclude that $r = s$.

Let d and D denote the discriminants of f and g, respectively. We note that $\det(F) = -d/4$, $\det(G) = -D/4$. Then from (3.8) and (3.6) we deduce that

$$-D/4 = \det(G) = \det(M^tFM) = \det(M^t)\det(F)\det(M)$$

$$= \det(F) = -d/4.$$

Alternatively, one could establish that $d = D$ by a direct (but less transparent) calculation based on the identities (3.7).

As an aid to determining whether two forms are equivalent, we now identify a special class of forms that we call *reduced* and show how to find a reduced form that is equivalent to any given form.

Definition 3.8 *Let f be a binary quadratic form whose discriminant d is not a perfect square. We call f* reduced *if*

$$-|a| < b \leqslant |a| < |c|$$

or if

$$0 \leqslant b \leqslant |a| = |c|.$$

If the discriminant of f is a square, possibly 0, then we proceed differently; see Problems 7 and 12 at the end of this section.

We now describe two simple transformations that may be used to reduce a given form f. Since the discriminant of f is not a perfect square, we know from Theorem 3.10 that $a \neq 0$ and that $c \neq 0$. If $|c| < |a|$, or if $|a| = |c|$ and $-|a| \leqslant b < 0$, then take $M = \begin{bmatrix} 0 & 1 \\ -1 & 0 \end{bmatrix}$ in (3.3). Thus we see that f is equivalent to the form $g(x, y) = cx^2 - bxy + ay^2$. Alternatively, if b fails to lie in the interval $(-|a|, |a|]$ then we take $M = \begin{bmatrix} 1 & m \\ 0 & 1 \end{bmatrix}$ in (3.8). By (3.7) we see that $A = a$, $B = 2am + b$, and $C = f(m, 1) = am^2 + bm + c$. We take m to be the unique integer for which $-|a| < B \leqslant |a|$. The resulting form may not be reduced, since it may be that $|C| < |A|$. In this case we would apply the first sort of transformation. By alternating between these two transformations, one is eventually led to a reduced form. To see that the process cannot continue indefinitely, note that the absolute value of the coefficient of x^2 is a weakly decreasing sequence, and that this quantity is strictly decreased by the first transformation, unless $|a| = |c|$, in which case the first transformation produces a reduced form. Thus we have proved the following important result.

Theorem 3.18 *Let d be a given integer, which is not a perfect square. Each equivalence class of binary quadratic forms of discriminant d contains at least one reduced form.*

In Section 3.7 we will show that if $d < 0$, then the reduced form in a given equivalence class is unique. For $d > 0$ this is not generally true, but the uniqueness may be recovered by adopting a more elaborate definition of what constitutes a reduced form.

Example 1 Find a reduced form equivalent to the form $133x^2 + 108xy + 22y^2$.

Solution By performing the first transformation, we see that the given form is equivalent to $22x^2 - 108xy + 133y^2$. By performing the second transformation with $m = 2$, we find that this form is equivalent to $22x^2 - 20xy + 5y^2$. By performing the first transformation, we see that this form is equivalent to $5x^2 + 20xy + 22y^2$. By performing the second transformation with $m = -2$, we find that this is equivalent to $5x^2 + 2y^2$. By the first transformation, this is equivalent to $2x^2 + 5y^2$, which is reduced. One may verify that all these quadratic forms have discriminant -40.

Theorem 3.19 *Let f be a reduced binary quadratic form whose discriminant d is not a perfect square. If f is indefinite, then $0 < |a| \leqslant \frac{1}{2}\sqrt{d}$. If f is positive definite then $0 < a \leqslant \sqrt{-d/3}$. In either case, the number of reduced forms of a given nonsquare discriminant d is finite.*

Proof If a and c are of the same sign then $d = b^2 - 4ac = b^2 - 4|ac| \leqslant a^2 - 4|ac| \leqslant a^2 - 4a^2 < 0$. Thus if $d > 0$ then a and c have opposite signs, and $d = b^2 - 4ac = b^2 + 4|ac| \geqslant 4|ac| \geqslant 4a^2$. This gives the bound for $|a|$ in this case. If $d < 0$ then $a > 0$ and $c > 0$, and hence $d = b^2 - 4ac \leqslant a^2 - 4ac \leqslant a^2 - 4a^2 = -3a^2$. This gives the bound for a in this case. In either case, a and b can take only a finite number of values. Once a and b are selected, there exists at most one integer c for which $b^2 - 4ac = d$.

Definition 3.9 *If d is not a perfect square then the number of equivalence classes of binary quadratic forms of discriminant d is called the* class number *of d, denoted $H(d)$.*

Let f be a binary quadratic form whose discriminant d is not a perfect square. In case $H(d) = 1$, we may combine Theorem 3.13 and Theorem 3.17 to determine quite precisely which numbers are representable by f.

Example 2 Show that an odd prime p can be written in the form $p = x^2 - 2y^2$ if and only if $p \equiv \pm 1 \pmod 8$.

Solution We note that the quadratic form $f(x, y) = x^2 - 2y^2$ has discriminant $d = 8$, which is not a perfect square. We first determine all reduced forms of this discriminant. From Theorem 3.19 we have $|a| \leqslant \sqrt{2}$, so that $a = \pm 1$. From Definition 3.8 we deduce that $b = 0$ or 1. But b and d always have the same parity, so we must have $b = 0$. Thus we find that there are precisely two reduced forms of discriminant 8, namely f and $-f$. Let $M = \begin{bmatrix} 1 & -2 \\ 1 & -1 \end{bmatrix}$. We observe that $\det(M) = 1$, so that $M \in \Gamma$. Taking this M in (3.7), we find that $f \sim -f$. Thus $H(8) = 1$. By Corollary 3.14 it follows that p is represented by f if and only $\left(\frac{2}{p}\right) = 1$, and we obtain the stated result by quadratic reciprocity.

It is conjectured that there are infinitely many positive (nonsquare) integers d for which $H(d) = 1$. It is known that for $d < 0$ there are only nine: $d = -3, -4, -7, -8, -11, -19, -43, -67, -163$.

PROBLEMS

1. Find a reduced form that is equivalent to the form $7x^2 + 25xy + 23y^2$.
2. Let G be a group. The set $C = \{c \in G\colon cg = gc \text{ for all } g \in G\}$ is called the *center* of G. Prove that C is a subgroup of G. Prove that the center of the modular group Γ consists of the two elements $I, -I$. (H)
3. Let x and y be integers. Show that there exist integers u and v such that $\begin{bmatrix} x & y \\ u & v \end{bmatrix} \in \Gamma$ if and only if $(x, y) = 1$.
4. Show that a binary quadratic form f properly represents an integer n if and only if there is a form equivalent to f in which the coefficient of x^2 is n. Use this and (3.3) to give a second proof of Theorem 3.13.
5. Show that $x^2 + 5y^2$ and $2x^2 + 2xy + 3y^2$ are the only reduced quadratic forms of discriminant -20. Show that the first of these forms does not represent 2, but that the second one does. Deduce that these forms are inequivalent, and hence that $H(-20) = 2$. Show that an odd prime p is represented by at least one of these forms if and only if $p = 2$, $p = 5$, or $p \equiv 1, 3, 7$, or $9 \pmod{20}$.
6. Let $f(x, y) = ax^2 + bxy + cy^2$ and $g(x, y) = f(-x, y) = ax^2 - bxy + cy^2$. These forms represent precisely the same numbers, but they are not necessarily equivalent (because the determinant of the transformation has determinant -1). Show that $x^2 + xy + 2y^2$ is equivalent to $x^2 - xy + 2y^2$, but that $3x^2 + xy + 4y^2$ and $3x^2 - xy + 4y^2$ are not equivalent.
7. Let $f(x, y)$ be a quadratic form whose discriminant d is a positive perfect square. Show that f is equivalent to a form $ax^2 + bxy + cy^2$ for which $c = 0$ and $0 \leqslant a < |b|$. Deduce that there are only finitely many equivalence classes of forms of this discriminant. (H)
8. Let $f(x, y) = 44x^2 - 97xy + 35y^2$. Show that f is equivalent to the form $g(x, y) = x(47x - 57y)$. Show that n is represented by f if and only if n can be written in the form $n = ab$ where $b \equiv 47a \pmod{57}$. Find the least positive integer n represented by f.
9. Show that if a number n is represented by a quadratic form f of discriminant d, then $4an$ is a square modulo $|d|$. (H)
10. Use the preceding problem to show that if p is represented by the form $x^2 + 5y^2$ then $\left(\frac{p}{5}\right) = 1$ or 0, and that if p is a prime represented by the form $2x^2 + 2xy + 3y^2$ then $\left(\frac{p}{5}\right) = -1$. By combining this information with the result of Problem 4, conclude that an odd prime p is represented by the form $x^2 + 5y^2$ if and only if $p = 5$ or $p \equiv 1$ or $9 \pmod{20}$, and that an odd prime p is represented by the form $2x^2 + 2xy + 3y^2$ if and only if $p = 2$ or $p \equiv 3$ or $7 \pmod{20}$.

11. Suppose that $ax^2 + bxy + cy^2 \sim Ax^2 + Bxy + Cy^2$. Show that g.c.d.$(a, b, c) =$ g.c.d.(A, B, C).

12. Let $f(x, y) = ax^2 + bxy + cy^2$ be a positive semidefinite quadratic form of discriminant 0. Put $g =$ g.c.d.(a, b, c). Show that f is equivalent to the form gx^2.

13. A binary quadratic form $ax^2 + bxy + cy^2$ is called *primitive* if g.c.d.$(a, b, c) = 1$. Prove that if $ax^2 + bxy + cy^2$ is a form of discriminant d and $r =$ g.c.d.(a, b, c) then $(a/r)x^2 + (b/r)xy + (c/r)y^2$ is a primitive form of discriminant d/r^2. If d is not a perfect square, let $h(d)$ denote the number of classes of primitive forms with discriminant d. Prove that $H(d) = \sum h(d/r^2)$ where the sum is over all positive integers r such that $r^2 | d$.

***14.** Show that if f is a primitive form and k is a nonzero integer, then there exists an integer n properly represented by f with the property that $(n, k) = 1$.

***15.** Suppose that $d \equiv 0$ or $1 \pmod 4$ and that d is not a perfect square. Then d is called a *fundamental discriminant* (or *reduced discriminant*) if all binary quadratic forms of discriminant d are primitive. Show that if $d \equiv 1 \pmod 4$ then d is a fundamental discriminant if and only if d is square-free. Show that if $d \equiv 0 \pmod 4$ then d is a fundamental discriminant if and only if $d/4$ is square-free and $d/4 \equiv 2$ or $3 \pmod 4$.

***16.** Let $a_1, a_2, \cdots a_n$ be given integers. Show that there is an $n \times n$ matrix with integral elements and determinant 1 whose first row is $a_1, a_2, \cdots, a_n$, if and only if g.c.d.$(a_1, a_2, \cdots, a_n) = 1$.

3.6 SUMS OF TWO SQUARES

In Theorem 2.15 we characterized those integers n that are represented by the quadratic form $x^2 + y^2$. We now apply the general results obtained in the preceding two sections to give a second proof of this theorem, and we also determine the number of such representations, counted in various ways. For convenient reference, we list four functions that appear repeatedly throughout the section:

$R(n)$: the number of ordered pairs (x, y) of integers such that $x^2 + y^2 = n$;

$r(n)$: the number of ordered pairs (x, y) of integers such that g.c.d.$(x, y) = 1$ and $x^2 + y^2 = n$, that is, the number of proper representations of n;

$P(n)$: the number of proper representations of n by the form $x^2 + y^2$ for which $x > 0$ and $y \geqslant 0$;

$N(n)$: the number of solutions of the congruence $s^2 \equiv -1 \pmod n$.

The form $x^2 + y^2$ has discriminant $d = -4$. Our first task is to construct a list of all reduced quadratic forms of this discriminant. We ignore negative definite forms and restrict our attention to positive forms. As $0 < a \leqslant \sqrt{4/3}$ by Theorem 3.19, we conclude that $a = 1$, and hence from Definition 3.7 that $b = 0$ or 1. But $b = 1$ is impossible since $b^2 - 4ac = -4$, and hence $b = 0$ and $c = 1$. Thus $x^2 + y^2$ is the only positive definite reduced form of discriminant -4. Then by Theorem 3.18 we deduce that all positive definite forms of discriminant -4 are equivalent, that is, $H(-4) = 1$.

From Theorems 3.13 and 3.17 we find that a positive integer n is properly represented by the form $x^2 + y^2$ if and only if -4 is a square modulo $4n$. We observe that -4 is a square modulo 8, but not modulo 16. Thus n may be divisible by 2, but not by 4. If p is an odd prime of the form $4k + 1$, then by Theorem 2.12 we know that -4 is a square (mod p). That is, if $f(x) = x^2 + 4$, then $f(x) \equiv 0 \pmod{p}$ has a solution x_0. Since $f'(x_0) = 2x_0 \not\equiv 0 \pmod{p}$, we deduce by Hensel's lemma (Theorem 2.23) that this solution lifts to a unique solution (mod p^2), and thence to (mod p^3), and so on. Thus we see that n may be divisible by arbitrary powers of primes of the form $4k + 1$. On the other hand, if p is a prime dividing n of the form $4k + 3$, then (by Theorem 2.12) -4 is not a square (mod p) and hence (by Theorem 2.16) -4 is not a square (mod $4n$). Thus we have proved the following theorem.

Theorem 3.20 *A positive integer n is properly representable as a sum of two squares if and only if the prime factors of n are all of the form $4k + 1$, except for the prime 2, which may occur to at most the first power.*

Having described those numbers that are properly represented as a sum of two squares, we may deduce which numbers are represented, properly or otherwise. Suppose that n is positive and that $n = x^2 + y^2$ is an arbitrary representation of n as a sum of two squares. Put $g =$ g.c.d.(x, y). Then $g^2 | n$, and we may write $n = g^2 m$. Since $(x/g, y/g) = 1$, we see that $m = (x/g)^2 + (y/g)^2$ is a proper representation of m. Here g may have some prime factors of the form $4k + 3$, but of course they divide n to an even power. The power of 2 dividing n may be arbitrary, for suppose that $2^\alpha \| n$. If α is even then we take m to be odd, $2^{\alpha/2} \| g$, while if α is odd then we can arrange that $2 \| m$, $2^{(\alpha-1)/2} \| g$. Thus we have a second proof of Theorem 2.15.

Let $R(n)$ denote the number of representations of n as a sum of two squares. That is, $R(n)$ is the number of ordered pairs (x, y) of integers for which $x^2 + y^2 = n$. Let $r(n)$ be the number of such ordered pairs for which g.c.d.$(x, y) = 1$. That is, $r(n)$ is the number of proper representations of n as a sum of two squares. We have determined those n for which $R(n) > 0$, and also those for which $r(n) > 0$. By exercising a little more care, we determine formulae for these functions.

Theorem 3.21 *Suppose that $n > 0$, and let $N(n)$ denote the number of solutions of the congruence $s^2 \equiv -1 \pmod{n}$. Then $r(n) = 4N(n)$, and $R(n) = \sum r(n/d^2)$ where the sum is extended over all those positive d for which $d^2|n$.*

Proof Consider any solution of $x^2 + y^2 = n$, where $n > 0$. Of the four points $(x, y), (-y, x), (-x, -y), (y, -x)$, exactly one of them has positive first coordinate and non-negative second coordinate. Let $P(n)$ denote the number of proper representations $x^2 + y^2 = n$ for which $x > 0$ and $y \geqslant 0$. Then $r(n) = 4P(n)$, and we now prove that $P(n) = N(n)$. Suppose that n is a given positive integer. We shall exhibit a one-to-one correspondence between representations $x^2 + y^2 = n$ with $x > 0$, $y \geqslant 0$, g.c.d.$(x, y) = 1$, and solutions s of the congruence $s^2 \equiv -1 \pmod{n}$. This is accomplished in three steps. First we define a function from the appropriate pairs (x, y) to the appropriate residue classes $s \pmod{n}$. Second, we show that this function is one-to-one. Third, we prove that the function is onto. To define the function, suppose that x and y are integers such that $x^2 + y^2 = n$, $x > 0$, $y \geqslant 0$, and that g.c.d.$(x, y) = 1$. Then g.c.d.$(x, n) = 1$, so there exists a unique $s \pmod{n}$ such that $xs \equiv y \pmod{n}$. More precisely, if $\bar{x}$ is chosen so that $x\bar{x} = 1 \pmod{n}$, then $s \equiv \bar{x}y \pmod{n}$. Since $x^2 \equiv -y^2 \pmod{n}$, on multiplying both sides by $\bar{x}^2$ we deduce that $s^2 \equiv -1 \pmod{n}$.

We now show that our function from the representations counted by $P(n)$ to the residue classes counted by $N(n)$ is one-to-one. To this end, suppose that for $i = 1, 2$ we have $n = x_i^2 + y_i^2$, $x_i > 0$, $y_i \geqslant 0$, g.c.d.$(x_i, y_i) = 1$, and $x_i s_i \equiv y_i \pmod{n}$. We show that if $s_1 \equiv s_2 \pmod{n}$ then $x_1 = x_2$ and $y_1 = y_2$. Suppose that $s_1 \equiv s_2 \pmod{n}$. As $x_1 y_2 s_1 \equiv y_1 y_2 \equiv x_2 y_1 s_2 \pmod{n}$, it follows that $x_1 y_2 \equiv x_2 y_1 \pmod{n}$, since g.c.d.$(s_i, n) = 1$. But $0 < x_i^2 \leqslant n$, so that $0 < x_i \leqslant \sqrt{n}$, and similarly $0 \leqslant y_i < \sqrt{n}$. From these inequalities we deduce that $0 \leqslant x_1 y_2 < n$, and similarly that $0 \leqslant x_2 y_1 < n$. As these two numbers are congruent modulo n and both lie in the interval $[0, n)$, we conclude that $x_1 y_2 = x_2 y_1$. Thus $x_1 | x_2 y_1$. But g.c.d.$(x_1, y_1) = 1$, so it follows that $x_1 | x_2$. Similarly $x_2 | x_1$. As the x_i are positive, we deduce that $x_1 = x_2$, and hence that $y_1 = y_2$. This completes the proof that our function is one-to-one.

To complete the proof that $P(n) = N(n)$, we now show that our function is onto. That is, for each s such that $s^2 \equiv -1 \pmod{n}$, there is a representation $x^2 + y^2 = n$ for which $x > 0$, $y \geqslant 0$, $(x, y) = 1$, and $xs \equiv$

$y \pmod n$. Suppose that such an s is given. Then there is an integer c such that $(2s)^2 - 4nc = -4$. Thus $g(x, y) = nx^2 + 2sxy + cy^2$ is a positive definite binary quadratic form of discriminant -4. In proving Theorem 3.20 we showed that all such forms are equivalent. Thus there is a matrix $M \in \Gamma$ that takes the form $f(x, y) = x^2 + y^2$ to the form g. From (3.7a) we see that $m_{11}^2 + m_{21}^2 = n$. Moreover, g.c.d.$(m_{11}, m_{21}) = 1$ since $\det(M) = m_{11}m_{22} - m_{21}m_{12} = 1$. From (3.7b) we see that $s = m_{11}m_{12} + m_{21}m_{22}$. Hence

$$\begin{aligned} m_{11}s &= m_{11}^2 m_{12} + m_{11}m_{21}m_{22} \\ &\equiv -m_{21}^2 m_{12} + m_{11}m_{21}m_{22} \pmod n \quad \left(\text{since } m_{11}^2 \equiv -m_{21}^2 \pmod n\right) \\ &= -m_{21}^2 m_{12} + m_{21}(1 + m_{21}m_{12}) \quad (\text{since } m_{11}m_{22} - m_{21}m_{12} = 1) \\ &= m_{21}. \end{aligned}$$

If in addition $m_{11} > 0$ and $m_{21} \geqslant 0$, then it suffices to take $x = m_{11}$, $y = m_{21}$. In case these inequalities do not hold, then we take the point (x, y) to be one of the points $(-m_{21}, m_{11}), (-m_{11}, -m_{21}), (m_{21}, m_{11})$. From the congruences $m_{11}s \equiv m_{21} \pmod n$, $s^2 \equiv -1 \pmod n$ we deduce that $(-m_{21})s \equiv m_{11} \pmod n$. Thus $xs \equiv y \pmod n$ in any of these cases. This completes the proof that $r(n) = 4N(n)$.

To prove the last assertion of the theorem we note that if $x^2 + y^2 = n > 0$ and $d =$ g.c.d.(x, y) then $(x/d)^2 + (y/d)^2 = n/d^2$ is a proper representation of n/d^2. Conversely, if $d > 0$, $d^2 | n$, and $u^2 + v^2 = n/d^2$ is a proper representation of n/d^2, then $(du)^2 + (dv)^2 = n$ is a representation of n with g.c.d.$(du, dv) = d$. Thus the representations $x^2 + y^2 = n$ for which g.c.d.$(x, y) = d$ are in one-to-one correspondence with the proper representations of n/d^2, and we have the stated identity expressing $R(n)$ as a sum.

We now apply the methods of Chapter 2 to $N(n)$, and thus determine the precise values of $r(n)$ and $R(n)$.

Theorem 3.22 *Let n be a positive integer, and write $n = 2^\alpha \prod_p p^\beta \prod_q q^\gamma$ where p runs over prime divisors of n of the form $4k + 1$ in the first product, and q runs over prime divisors of n of the form $4k + 3$ in the second. If $\alpha = 0$ or 1 and all the γ are 0, then $r(n) = 2^{t+2}$ where t is the number of primes p of the form $4k + 1$ that divide n. Otherwise $r(n) = 0$. If all the γ are even then $R(n) = 4\prod_p (\beta + 1)$. Otherwise $R(n) = 0$.*

Proof By Theorem 2.20 we know that $N(n) = N(2^\alpha)\prod_p N(p^\beta)\prod_q N(q^\gamma)$. Clearly $N(2) = 1$ and $N(4) = 0$. Thus by Theorem 2.16, $N(2^\alpha) = 0$ for all $\alpha \geqslant 2$. Similarly, $N(q) = 0$, and thus $N(q^\gamma) = 0$ whenever $\gamma > 0$. On the other hand, by Theorem 2.12 and our remarks in Section 2.9 we see that $N(p) = 2$. Then by Hensel's lemma (Theorem 2.23) it follows that $N(p^\beta) = 2$ for all $\beta > 0$. Thus $N(n) = 2^t$ if $\alpha = 0$ or 1 and all the γ vanish, and otherwise $N(n) = 0$.

From Theorem 3.21 we know that $R(n) = 4\Sigma N(n/d^2)$ where d runs over all positive integers for which $d^2|n$. Suppose that $n = m_1 m_2$ where $(m_1, m_2) = 1$. By the unique factorization theorem it is evident that the positive d for which $d^2|n$ are in one-to-one correspondence with pairs (d_1,d_2) of positive numbers for which $d_i^2|m_i$. Thus

$$\sum_{d^2|n} N(n/d^2) = \Bigg(\sum_{d_1^2|m_1} N(m_1/d_1^2)\Bigg)\Bigg(\sum_{d_2^2|m_2} N(m_2/d_2^2)\Bigg).$$

By using this repeatedly we may break n in to prime powers. Thus

$$\sum_{d^2|n} N(n/d^2)$$

$$= \Bigg(\sum_{d^2|2^\alpha} N(2^\alpha/d^2)\Bigg)\prod_p\Bigg(\sum_{d^2|p^\beta} N(p^\beta/d^2)\Bigg)\prod_q\Bigg(\sum_{d^2|q^\gamma} N(q^\gamma/d^2)\Bigg).$$

We evaluate the contributions made by the three types of sums on the right. If α is even, then the only nonzero term in the first factor is obtained by taking $d = 2^{\alpha/2}$. If α is odd, then the only nonzero term is obtained by taking $d = 2^{(\alpha-1)/2}$. Thus the first factor is 1 in any case. If β is even then $N(p^\beta/d^2) = 2$ for $d = 1, p, p^2, \cdots, p^{\beta/2-1}$, and $N(p^\beta/d^2) = 1$ for $d = p^{\beta/2}$. Thus the sum contributed by the prime p is $\beta + 1$ in this case. If β is odd then $N(p^\beta/d^2) = 2$ for $d = 1, p, p^2, \cdots, p^{(\beta-1)/2}$. Thus the sum is $\beta + 1$ in this case also. If γ is odd then $q|q^\gamma/d^2$ for all d in question, and thus all terms vanish. If γ is even then the term arising from $d = q^{\gamma/2}$ is 1 and all other terms vanish. Thus the sum contributed by a prime q is 1 if γ is even, and otherwise vanishes.

Corollary 3.23 *The number of representations of a positive integer n as a sum of two squares is 4 times the excess in the number of divisors of n of the form* $4k + 1$ *over those of the form* $4k + 3$. *That is,* $R(n) = 4\sum\left(\frac{-1}{d}\right)$, *where d runs over the positive odd divisors of n.*

Proof Suppose that $n = m_1 m_2$ with $(m_1, m_2) = 1$. For $d|n$ put $d_i = (d, m_i)$. Then $d_i|m_i$ and $d_1 d_2 = d$. Conversely, if $d_1|m_1$ and $d_2|m_2$, then $d = d_1 d_2|n$, and $d_i = (d, m_i)$. Thus the divisors of n are in one-to-one correspondence with pairs (d_1, d_2) of divisors of m_1 and m_2. Since $\left(\frac{-1}{d}\right) = \left(\frac{-1}{d_1}\right)\left(\frac{-1}{d_2}\right)$ for any odd divisor d of n, it follows that

$$\sum\left(\frac{-1}{d}\right) = \left(\sum\left(\frac{-1}{d_1}\right)\right)\left(\sum\left(\frac{-1}{d_2}\right)\right)$$

where d_i runs over the positive odd divisors of m_i for $i = 1, 2$. By using this repeatedly we may reduce to the case of prime powers. In case the prime is 2, the only nonzero term is obtained by taking $d = 1$. In case of a prime $p \equiv 1 \pmod 4$ each of the $\beta + 1$ summands is 1, so the sum is $\beta + 1$. In case of a prime $q \equiv 3 \pmod 4$, the summands are alternately 1 and -1, so that

$$\sum_{d|q^\gamma}\left(\frac{-1}{d}\right) = \sum_{j=0}^{\gamma}\left(\frac{-1}{q^j}\right) = \sum_{j=0}^{\gamma}(-1)^j = \begin{cases} 1 & \text{if } \gamma \text{ is even,} \\ 0 & \text{if } \gamma \text{ is odd.} \end{cases}$$

Thus the original sum has value $\prod_p (\beta + 1)$ if all the γ are even, and 0 otherwise.

Since $r(n) = 4P(n) = 4N(n)$, it suffices to calculate just $r(n)$ and $R(n)$ if we want the values of the functions R, r, P, N in specific numerical cases. For example, if $n = 1260$ we see that $R(1260) = 0$ by Theorem 3.20, because $7 | 1260$. Hence $r(1260) = 0$, because $R(n) = 0$ implies $r(n) = 0$, by definition. If for a specific value of n we determine that $R(n) > 0$ by Theorem 3.20, we can turn to Theorem 3.22 for formulae that make calculations easy. For $n = 130$, Theorem 3.22 gives $R(130) = 16$ and $r(130) = 16$, and then of course $P(130) = N(130) = 4$. The techniques we have developed may be used to give the representations explicitly.

Example 3 Find integers x and y such that $x^2 + y^2 = p$, where $p = 398417$ is a prime number.

Solution Our first task is to locate a quadratic nonresidue of p. By quadratic reciprocity (or Euler's criterion) we find that $\left(\frac{2}{p}\right) = 1$, but that $\left(\frac{3}{p}\right) = -1$. We let s be the unique integer such that $0 < s < p$ and $s \equiv 3^{(p-1)/4} \pmod p$. By the quick powering method discussed in Section 2.4, we discover that $s = 224149$. By Euler's criterion we know that $s^2 \equiv 3^{(p-1)/2} \equiv -1 \pmod p$, and by direct calculation we verify that $s^2 = kp - 1$ where $k = 126106$. Thus the quadratic form $f(x, y) = px^2 + 2sxy$

$+ ky^2$ has discriminant -4, and $f(1,0) = p$ is a proper representation of p. We now reduce this form, keeping track of the change in x and y as we go. For brevity we let $S = \begin{bmatrix} 0 & 1 \\ -1 & 0 \end{bmatrix}$ and $T = \begin{bmatrix} 1 & 1 \\ 0 & 1 \end{bmatrix}$. By taking $M = S$ or $M = T^m = \begin{bmatrix} 1 & m \\ 0 & 1 \end{bmatrix}$, as appropriate, we eventually locate a reduced form equivalent to f. But we know that there is only one reduced form of discriminant -4, namely $x^2 + y^2$, and the desired representation is achieved.

a	b	c	x	y	Operation
398417	448298	126106	1	0	S
126106	-448298	398417	0	1	T^2
126106	56126	6245	-2	1	S
6245	-56126	126106	-1	-2	T^4
6245	-6166	1522	7	-2	S
1522	6166	6245	2	7	T^{-2}
1522	78	1	16	7	S
1	-78	1522	-7	16	T^{39}
1	0	1	-631	16	

The entry in the last column indicates the operation that will be applied to produce the next row. Thus we conclude that $398417 = (\pm 631)^2 + (\pm 16)^2$.

PROBLEMS

1. Find four consecutive positive integers, each with the property that $r(n) = 0$.
2. What is the maximum value of $R(n)$ for positive $n \leqslant 1000$?
3. What is the maximum value of $r(n)$ for positive $n \leqslant 10000$?
4. Use the method of Example 3 to find integers x and y such that $x^2 + y^2 = 89753$, given that this number is prime.
5. Suppose that n is not a perfect square. Show that the number of ordered pairs (x, y) of positive integers for which $x^2 + y^2 = n$ is $R(n)/4$. Show that if n is a perfect square then the number of such representations of n is $R(n)/4 - 1$.
6. Suppose that $n > 1$. Show that the number of ordered pairs (x, y) of relatively prime positive integers for which $x^2 + y^2 = n$ is $r(n)/4$.

7. Suppose that n is neither a perfect square nor twice a perfect square. Show that the number of ordered pairs (x, y) of integers for which $0 < x < y$ and $x^2 + y^2 = n$ is $R(n)/8$.

8. Prove that if a positive integer n can be expressed as a sum of the squares of two rational numbers then it can be expressed as a sum of the squares of two integers.

9. Suppose that n is a positive integer that can be expressed as a sum of two relatively prime squares. Show that every positive divisor of n must also have this property.

10. Suppose that a matrix M with integral elements and determinant -1 takes a form $f(x, y) = ax^2 + bxy + cy^2$ to $x^2 + y^2$. Prove that f and $x^2 + y^2$ are equivalent by showing that there is another matrix M_1, with integral elements and determinant $+1$, that also takes f to $x^2 + y^2$.

11. Show that if n is a sum of three squares then $n \not\equiv 7 \pmod 8$. Show by example that there exist positive integers m and n, both of which are sums of three squares, but whose product mn is not a sum of three squares.

12. Show that if $x^2 + y^2 + z^2 = n$ and $4|n$, then x, y, and z are even. Deduce that if n is of the form $4^m(8k + 7)$ then n is not the sum of three squares. (Gauss proved that all other positive integers n can be expressed as sums of three squares.)

3.7 POSITIVE DEFINITE BINARY QUADRATIC FORMS

In the further theory of quadratic forms, many differences of detail arise between definite and indefinite forms. As indefinite quadratic forms present greater complications, we now confine our attention to positive definite quadratic forms $f(x, y) = ax^2 + bxy + cy^2$. We have shown that any such form is equivalent to a reduced form, that is, one for which $-a < b \leqslant a < c$ or $0 \leqslant b \leqslant a = c$. We now show that this reduced form is unique. That is, distinct reduced forms are inequivalent, so that the class number $H(d)$ is precisely the number of reduced forms of discriminant d, when $d < 0$. (For $d > 0$ two reduced forms may be equivalent, as we saw in Example 2. To develop a corresponding theory for indefinite forms, one must allow for solutions of the equation $x^2 - dy^2 = \pm 4$. This is a special case of Pell's equation, which we discuss in Section 7.8 as an application of continued fractions.)

Lemma 3.24 *Let $f(x, y) = ax^2 + bxy + cy^2$ be a reduced positive definite form. If for some pair of integers x and y we have* g.c.d.$(x, y) = 1$ *and $f(x, y) \leqslant c$, then $f(x, y) = a$ or c, and the point (x, y) is one of the six* *points* $\pm(1,0)$, $\pm(0,1)$, $\pm(1,-1)$. *Moreover, the number of proper representations of a by f is*

$$\begin{array}{ll} 2 & \text{if } a < c, \\ 4 & \text{if } 0 \leqslant b < a = c, \text{ and} \\ 6 & \text{if } a = b = c. \end{array}$$

Proof Suppose that g.c.d.$(x, y) = 1$. If $y = 0$ then $x = \pm 1$, and we note that $f(\pm 1, 0) = a$. Now suppose that $y = \pm 1$. If $|x| \geqslant 2$ then

$$\begin{aligned} |2ax + by| &\geqslant |2ax| - |by| && \text{(by the triangle inequality)} \\ &\geqslant 4a - |b| \\ &\geqslant 3a && (\text{since } |b| \leqslant a). \end{aligned}$$

Then by (3.3) we deduce that

$$\begin{aligned} 4af(x, y) &= (2ax + by)^2 - dy^2 \\ &\geqslant 9a^2 - dy^2 \\ &= 9a^2 - d \\ &= 9a^2 + 4ac - b^2 \\ &> a^2 - b^2 + 4ac && (\text{since } a > 0) \\ &\geqslant 4ac && (\text{since } |b| \leqslant a). \end{aligned}$$

Thus $f(x, \pm 1) > c$ if $|x| \geqslant 2$. Now suppose that $|y| \geqslant 2$. Then by (3.3) we see that

$$\begin{aligned} 4af(x, y) &= (2ax + by)^2 - dy^2 \\ &\geqslant -dy^2 \\ &\geqslant -4d \\ &= 16ac - 4b^2 \\ &> 8ac - 4b^2 && (\text{since } ac > 0) \\ &\geqslant 4a^2 - 4b^2 + 4ac && (\text{since } 0 < a \leqslant c) \\ &\geqslant 4ac && (\text{since } |b| \leqslant a). \end{aligned}$$

Thus $f(x, y) > c$ if $|y| \geqslant 2$. The only points remaining are $\pm(1, 0)$, $\pm(0, 1)$, $\pm(1, -1)$, and $\pm(1, 1)$. As $b > -a$, we find that $f(1, 1) = a + b + c > c$, so that the proper representations of a and of c are obtained by considering the first three pairs of points.

The last assertion of the lemma now follows on observing that $f(1, 0) = a$, $f(0, 1) = c$, and $f(1, -1) = a - b + c$.

Theorem 3.25 *Let $f(x, y) = ax^2 + bxy + cy^2$ and $g(x, y) = Ax^2 + Bxy + Cy^2$ be reduced positive definite quadratic forms. If $f \sim g$ then $f = g$.*

Proof Suppose that $f \sim g$. By Lemma 3.24, the least positive number properly represented by f is a, and that by g is A. By Theorem 3.17 it follows that $a = A$. We consider first the case $a < c$. Then by Lemma 3.24 there are precisely 2 proper representations of a by f. By Theorem 3.17 it follows that there are precisely 2 proper representations of a by g, and from Lemma 3.24 we deduce that $C > a$. Thus by Lemma 3.24 we see that c is the least number greater than a that is properly represented by f, and C is the least such for g. By Theorem 3.17 it follows that $c = C$. To show that $b = B$, we consider the matrices $M \in \Gamma$ that might take f to g. Since $\det(M) = m_{11}m_{22} - m_{21}m_{12} = 1$, we know that g.c.d.$(m_{11}, m_{21}) = 1$. Thus by (3.7*a*), $f(m_{11}, m_{21}) = a$ is a proper representation of a. By Lemma 3.24 it follows that the first column of M is $\pm\begin{bmatrix}1\\0\end{bmatrix}$. We see similarly that $(m_{12}, m_{22}) = 1$, so that by (3.7*c*), $f(m_{12}, m_{22}) = c$ is a proper representation of c. Hence by Theorem 3.24, the second column of M is $\pm\begin{bmatrix}0\\1\end{bmatrix}$ or $\pm\begin{bmatrix}-1\\1\end{bmatrix}$. Thus we see that the only candidates for M are $\pm I$ and $\pm\begin{bmatrix}1 & -1\\0 & 1\end{bmatrix}$. However, in the latter event (3.7*b*) would give $B = -2a + b$, which is impossible since b and B must both lie in the interval $(-a, a]$. This leaves only $\pm I$, and we see that if $M = \pm I$ then $f = g$.

We now consider the case $a = c$. From Lemma 3.24 we see that a has at least 4 proper representations by f. From Theorem 3.17 it follows that the same is true of g, and then by Lemma 3.24 we deduce that $C = a = c$. Thus by Definition 3.8, $0 \leqslant b \leqslant a = c$ and $0 \leqslant B \leqslant A = C = a$. As $b^2 - 4ac = B^2 - 4AC$, it follows that $b = B$, and hence that $f = g$.

In the case $a < c$ considered, we not only proved that $f = g$, but also established that the only matrices $M \in \Gamma$ that take f to itself are $\pm I$. We now extend this.

Definition 3.10 *Let f be a positive definite binary quadratic form. A matrix $M \in \Gamma$ is called an* automorph *of f if M takes f to itself, that is, if*

$f(m_{11}x + m_{12}y, m_{21}x + m_{22}y) = f(x, y)$. The number of automorphs of f is denoted by $w(f)$.

For example, the matrix $\begin{bmatrix} -1 & -1 \\ 1 & 0 \end{bmatrix}$ is an automorph of $x^2 + xy + y^2$, and of course the identity matrix $I = \begin{bmatrix} 1 & 0 \\ 0 & 1 \end{bmatrix}$ is an automorph of every form.

Theorem 3.26 *Let f and g be equivalent positive definite binary quadratic forms. Then $w(f) = w(g)$, there are exactly $w(f)$ matrices $M \in \Gamma$ that take f to g, and there are exactly $w(f)$ matrices $M \in \Gamma$ that take g to f. Moreover, the only values of $w(f)$ are 2, 4, and 6. If f is reduced then*

$$w(f) = 4 \quad \text{if } a = c \text{ and } b = 0,$$
$$w(f) = 6 \quad \text{if } a = b = c, \text{ and}$$
$$w(f) = 2 \quad \text{otherwise}.$$

Proof Let $A_1, A_2, \cdots, A_r$ be distinct automorphs of f, and let M be a matrix that takes f to g. Then $A_1M, A_2M, \cdots, A_rM$ are distinct members of Γ that take f to g. Conversely, if $M_1, M_2, \cdots, M_s$ are distinct members of Γ that take f to g, then $M_1M_1^{-1}, M_2M_1^{-1}, \cdots, M_sM_1^{-1}$ are distinct automorphs of f. Hence the automorphs of f are in one-to-one correspondence with the matrices M that take f to g. If M takes f to g, then M^{-1} takes g to f, and these matrices M^{-1} are in one-to-one correspondence with the automorphs of g. Thus the automorphs of f are in one-to-one correspondence with those of g, and consequently $w(f) = w(g)$ if either number is finite. But the number is always finite, because any form is equivalent to a reduced form, and in the next paragraph we show that any reduced form has 2, 4, or 6 automorphs.

Suppose that f is reduced. In the course of proving Theorem 3.25, we showed that $w(f) = 2$ if $a < c$, and we saw that $f(m_{11}, m_{21}) = a$ and $f(m_{12}, m_{22}) = c$ are proper representations of a and c. Suppose now that $0 \leqslant b \leqslant a = c$ and that M leaves f invariant (i.e., M takes f to itself). Then by Lemma 3.24 the columns of M lie in the set $\left\{\pm\begin{bmatrix} 1 \\ 0 \end{bmatrix}, \pm\begin{bmatrix} 0 \\ 1 \end{bmatrix}, \pm\begin{bmatrix} 1 \\ -1 \end{bmatrix}\right\}$. Of the 36 such matrices, we need consider only those with determinant 1, and thus we have the six pairs $\pm M_1$, $\pm M_2, \cdots, \pm M_6$ where $M_1 = I$, $M_2 = \begin{bmatrix} 1 & -1 \\ 0 & 1 \end{bmatrix}$, $M_3 = \begin{bmatrix} 1 & 0 \\ -1 & 1 \end{bmatrix}$, $M_4 = \begin{bmatrix} 0 & -1 \\ 1 & 0 \end{bmatrix}$, $M_5 = \begin{bmatrix} 0 & -1 \\ 1 & 1 \end{bmatrix}$, and $M_6 = \begin{bmatrix} 1 & 1 \\ -1 & 0 \end{bmatrix}$. We note that if any one of the four matrices $\pm M^{\pm 1}$ is an automorph, then all four are.

Here M_1 is always an automorph. By (3.7*b*) we see that M_2 takes f to g with $B = b - 2a \neq b$, so that M_2 is never an automorph. As $M_3 = M_2^{-1}$, we deduce that M_3 is likewise never an automorph. As M_4 takes f to $cx^2 - bxy + ay^2$, M_4 is an automorph if and only if $b = 0$ and $a = c$. Since M_5 takes f to $cx^2 + (2c - b)xy + (a - b + c)y^2$, we see that M_5 is an automorph if and only if $a = b = c$. Finally, $M_6 = M_5^{-1}$, so that M_6 is an automorph if and only if $a = b = c$. This gives the stated result.

We now employ our understanding of automorphs to generalize Theorem 3.21 (which was concerned with the particular form $x^2 + y^2$) to arbitrary positive definite binary quadratic forms f of discriminant $d < 0$. Extending the notation of the preceding section, we let $R_f(n)$ denote the number of representations of n by f. Similarly, we let $r_f(n)$ denote the number of these representations that are proper. Finally, let $H_f(n)$ denote the number of integers h, $0 \leqslant h < 2n$, such that $h^2 \equiv d \pmod{4n}$, say $h^2 = d + 4nk$, with the further property that the form $nx^2 + hxy + ky^2$ is equivalent to f.

Theorem 3.27 *Let f be a positive definite binary quadratic form with discriminant $d < 0$. Then for any positive integer n, $r_f(n) = w(f)H_f(n)$, and $R_f(n) = \sum_{m^2|n} r_f(n/m^2)$.*

It may be shown that if a nonzero number n is represented by an indefinite quadratic form whose discriminant is not a perfect square, then n has infinitely many such representations. To construct an analogous theory for indefinite forms one must allow for solutions of Pell's equation $x^2 - dy^2 = \pm 4$.

Proof Let $\mathscr{G}_f(n)$ denote the set of those forms $g(x, y) = nx^2 + hxy + ky^2$ that are equivalent to f, and for which $0 \leqslant h < 2n$. From Theorem 3.17 we know that such a form must have the same discriminant as f, so that $h^2 - 4nk = d$. Thus there are precisely $H_f(n)$ members of the set $\mathscr{G}_f(n)$. If $g \in \mathscr{G}_f(n)$, then g is equivalent to f, which is to say that there is a matrix $M \in \Gamma$ that takes f to g. By Theorem 3.26 it follows that there are precisely $w(f)$ such matrices. Consequently, there are exactly $w(f)H_f(n)$ matrices $M \in \Gamma$ that take f to a member of $\mathscr{G}_f(n)$. We now exhibit a one-to-one correspondence between these matrices M and the proper representations of n.

Suppose that M is of the sort described. Then by (3.7*a*) we see that $f(m_{11}, m_{21}) = n$. As $\det(M) = m_{11}m_{22} - m_{21}m_{12} = 1$, we see that $(m_{11}, m_{21}) = 1$, and thus the representation is proper. Conversely, suppose that $f(x, y) = n$ is a proper representation of n. To recover the matrix M, we take $m_{11} = x$, $m_{21} = y$. It remains to show that m_{12} and m_{21} are uniquely determined. Let u and v be chosen so that $xv - yu = 1$. In order that $\det(M) = 1$, we must have $m_{21} = u + tx$, $m_{22} = v + ty$ for some integer t. For M of this form we see by (3.7b) that

$$\begin{aligned} h &= 2ax(u + tx) + bx(v + ty) + by(y + tx) + 2cy(v + ty) \\ &= (2axu + bxv + byu + 2cyv) + 2nt. \end{aligned}$$

Thus there is a unique t for which $0 \leqslant h < 2n$. This gives a unique matrix M with the desired properties. The first of the asserted identities is thus established.

To establish the second identity, suppose that x and y are integers such that $f(x, y) = n$, and put $m = \text{g.c.d.}(x, y)$. Then $m^2|n$, and indeed $f(x/m, y/m) = n/m^2$ is a proper representation of n/m^2, since $\text{g.c.d.}(x/m, y/m) = 1$. Conversely, if $m^2|n$ and u and v are relatively prime integers such that $f(u, v) = n/m^2$, then $f(mu, mv) = n$ and $\text{g.c.d.}(mu, mv) = m$.

Continuing our quest to generalize Theorem 3.21, we now let $N_d(n)$ denote the number of integers h for which $h^2 \equiv d \pmod{4n}$ and $0 \leqslant h < 2n$. Since h is a solution of the congruence $u^2 \equiv d \pmod{4n}$ if and only if $h + 2n$ is a solution, it follows that $N_d(n)$ is precisely one-half the total number of solutions of the congruence $u^2 \equiv d \pmod{4n}$. Assuming that n is a positive integer, the value of $N_d(n)$ may be determined by applying the tools of Chapter 2, particularly Theorems 2.20 and 2.23. Let $\mathscr{F}$ denote the set of all reduced positive definite binary quadratic forms of discriminant d. If $h^2 \equiv d \pmod{4n}$, say $h^2 = d + 4nk$, and $0 \leqslant h < 2n$, then there is a unique form $f \in \mathscr{F}$ for which $nx^2 + hxy + ky^2 \in \mathscr{G}_f(n)$. Hence

$$\sum_{f \in \mathscr{F}} H_f(n) = N_d(n).$$

For many discriminants d it happens that $w(f)$ is the same for all $f \in \mathscr{F}$. In that case we let w denote the common value. (In this connection, recall Problem 15 in Section 3.5, and see Problem 6 below.) For such d we may multiply both sides by w and appeal to Theorem 3.27 to see that

$$\sum_{f \in \mathscr{F}} r_f(n) = wN_d(n).$$

In this manner we may determine the total number of proper representations of n by reduced forms of discriminant d, but unfortunately it is not always so easy to describe the individual numbers $r_f(n)$.

PROBLEMS

1. Let $f(x, y) = ax^2 + bxy + cy^2$ be a reduced positive definite form. Show that all representations of a by f are proper.

2. Let $f(x, y) = ax^2 + bxy + cy^2$ be a reduced positive definite form. Show that improper representations of c may exist. (H)

3. Show that any positive definite binary quadratic form of discriminant -3 is equivalent to $f(x, y) = x^2 + xy + y^2$. Show that a positive integer n is properly represented by f if and only if n is of the form $n = 3^\alpha \prod p^\beta$, where $\alpha = 0$ or 1 and all the primes p are of the form $3k + 1$. Show that for n of this form, $r_f(n) = 6 \cdot 2^s$, where s is the number of distinct primes $p \equiv 1 \pmod 3$ that divide n.

4. Write the canonical factorization of n in the form $n = 3^\alpha \prod p^\beta \prod q^\gamma$ where the primes p are of the form $3k + 1$ and the primes q are of the form $3k + 2$. Show that n is represented by $f(x, y) = x^2 + xy + y^2$ if and only if all the γ are even. Show that for such n, $R_f(n) = 6\prod_q(\beta + 1)$.

5. Show that for any given $d < 0$, the primitive positive definite quadratic forms of discriminant d all have the same number of automorphs.

6. Show that any positive definite quadratic form of discriminant -23 is equivalent to exactly one of the forms $f_0(x, y) = x^2 + xy + 6y^2$, $f_1(x, y) = 2x^2 + xy + 3y^2$ or $f_2(x, y) = 2x^2 - xy + 3y^2$. Show that if $\left(\frac{-23}{p}\right) = -1$ then p is not represented by any of these forms. Show that if $\left(\frac{-23}{p}\right) = 1$ then p has a total of 4 representations by these forms. Show that in this latter case either p has 4 representations by f_1 or 2 representations apiece by f_1 and f_2. Determine which of these cases applies when $p = 139$. (H)

***7.** Let $f(x, y) = ax^2 + bxy + cy^2$ be a reduced positive definite form. Suppose that g.c.d.$(x, y) = 1$ and that $f(x, y) \leqslant a + |b| + c$. Show that $f(x, y)$ must be one of the numbers $a, c, a - |b| + c$ or $a + |b| + c$.

NOTES ON CHAPTER 3

§3.1, 3.2 Fermat characterized those primes for which 2, -2, 3, and -3 are quadratic residues. His assertions for ± 3 were proved by Euler in 1760, and those for ± 2 by Legendre in 1775. The first part of Theorem 3.1 was proved by Euler in 1755. The last part of Theorem 3.1, first proved by Euler in 1749, is equivalent to Theorem 2.11. We proved Theorem 2.11 by the simpler method discovered by Lagrange in 1773. In 1738 Euler observed that whether the congruence $x^2 \equiv a \pmod{p}$ has a solution or not is determined by the residue class of $p \pmod{4|a|}$. In 1783, Euler gave a faulty proof of an assertion equivalent to the quadratic reciprocity law. (In retrospect, one can see that even much earlier, Euler was just a short step away from having a complete proof of quadratic reciprocity.) In 1785, Legendre introduced his symbol, stated the general case of quadratic reciprocity without using his symbol, introduced the word "reciprocity," and gave an incomplete proof of the law. (In 1859, Kummer noted that the gap in Legendre's proof is easily filled by appealing to Dirichlet's theorem of 1837 concerning primes in arithmetic progressions.) In ignorance of the earlier work of others, Gauss discovered the quadratic reciprocity law just before his eighteenth birthday. After a year of strenuous effort, Gauss found the first proof, in 1795, at the age of nineteen. This was published in 1801. Gauss discovered "Gauss's lemma" (Theorem 3.2) in 1808. Our proof of quadratic reciprocity (Theorem 3.3) follows Gauss's third proof of the theorem, which is considered to have been Gauss's favorite. Eventually Gauss gave eight proofs of quadratic reciprocity, in the hope of finding one that would generalize to give a proof of the quartic reciprocity law that he had empirically discovered.

For an instructive algebraic interpretation of Gauss's lemma, see W. C. Waterhouse, "A tiny note on Gauss's Lemma," *J. Number Theory*, 30 (1988), 105–107.

Theorem 3.5 is a variation of a result by P. Hagis, "A note concerning the law of quadratic reciprocity," *Amer. Math. Monthly*, 77 (1970), 397.

§3.3 In more advanced work, it is useful to extend the Legendre symbol beyond the Jacobi symbol, to the *Kronecker symbol*.

Let $n_2(p)$ denote the least positive quadratic nonresidue of p. Using the inequality (3.2) in a clever way, David Burgess showed that for every $\varepsilon > 0$ there is a $p_0(\varepsilon)$ such that $n_2(p) < p^{c+\varepsilon}$ for $p > p_0(\varepsilon)$, where $c = 1/(4\sqrt{e}) = 0.1516\cdots$.

§3.5 A function $f(z)$ is called a *modular function* if $f\left(\dfrac{az+b}{cz+d}\right) = f(z)$ for every $\begin{bmatrix} a & b \\ c & d \end{bmatrix} \in \Gamma$. The study of modular functions, modular forms, and the more general automorphic functions is an active area of research in advanced number theory. If F is a field, then the $n \times n$ matrices with entries in F and nonzero determinant form a group, known as the *general linear group* of order n over F, and denoted $GL(n, F)$. If R is a commutative ring with identity, then the $n \times n$ matrices with coefficients in R and determinant 1 form a group, known as the *special linear group* of order n over R, denoted $SL(n, R)$. In this notation, the modular group Γ is $SL(2, \mathbb{Z})$.

Two forms $ax^2 + bxy + cy^2$ and $Ax^2 + Bxy + Cy^2$ of discriminant d lie in the same *genus* if aA is a square modulo $|d|$. This defines a new equivalence relation on the forms of discriminant d. Using the observation made in Problem 9, it may be shown that if two forms are equivalent (in the sense of Definition 3.7) then they lie in the same genus. Thus each genus is the union of one or more equivalence classes of forms. The consideration of these genera allows one to refine Corollary 3.14: If p is represented by some form of discriminant d, one may use quadratic reciprocity to determine in which genus this form must lie. An example of this is found in Problem 10, which concerns $d = -20$. In this case it is found that there is only one equivalence class in each genus, and hence we are able to specify precisely which primes are represented by which forms. However, the discriminant $d = -20$ is one of only finitely many discriminants of this sort: If d is large and negative, then each genus contains a large number of equivalence classes of forms.

The problem of finding all negative discriminants d for which $h(d) = 1$ has a long and interesting history, which is recounted in the survey article of D. Goldfeld, "Gauss's class number problem for imaginary quadratic fields," *Bull. Amer. Math. Soc.* 13 (1985), 23–37.

§3.6 Following Fermat, much attention was paid to the problem of giving an explicit formula for the numbers x and y for which $x^2 + y^2 = p$, when p is a prime of the form $4n + 1$. This was first achieved in 1808 by Legendre, using continued fractions. In 1825 Gauss gave a different construction: Since x and y are of opposite parity, we may assume that x is odd. By replacing x by $-x$ if necessary, we may suppose that $x \equiv 1 \pmod 4$. Then x is the unique number for which $|x| < p/2$ and $2x \equiv \binom{2n}{n} \pmod p$ where $p = 4n + 1$. More recently, Jacobsthal discovered that one may express x and y as sums involving the Legendre symbol,

$$x = \frac{1}{2}\sum_{k=1}^{p-1}\left(\frac{k(k^2 - r)}{p}\right), \qquad y = \frac{1}{2}\sum_{k=1}^{p-1}\left(\frac{k(k^2 - n)}{p}\right)$$

where r denotes any quadratic residue of p, and n is any quadratic nonresidue of p. The method of Example 3, though it does not yield an explicit formula for x and y, nevertheless is computationally much more efficient. A similar calculational technique, but using continued fractions instead of the theory of quadratic forms, is found in Problem 6 of Section 7.3.

§3.7 Theorem 3.25 may be proved by considering the action of the modular group Γ on the upper half-plane $\mathcal{H} = \{z \in \mathbb{C}: \mathcal{Im}(z) > 0\}$. A nice account of this is found in Chapter 1 of LeVeque's *Topics*.

It was noted by Gauss that the theory of quadratic forms may be used to provide a method of factoring numbers. An elegant account of this approach has been given by D. H. and E. Lehmer, "A new factorization technique using quadratic forms," *Math. Comp.* 28 (1974), 625–635.

In Chapter 1, our treatment of sums of two squares depended on the identity

$$(x^2 + y^2)(u^2 + v^2) = (xu - yv)^2 + (xv + yu)^2$$

which reflects a familiar property of complex numbers, namely that if $z = x + iy$ and $w = u + iv$, then $|z|\,|w| = |zw|$. This is the first instance of a type of identity known as a *composition formula*. Such formulae exist for forms of other discriminants. For example, the reduced quadratic forms of discriminant -20 are $f_0(x, y) = x^2 + 5y^2$ and $f_1(x, y) = 2x^2 + 2xy + 3y^2$. By Theorems 3.18 and 3.25 it follows that $H(-20) = 2$. Moreover, it is easy to verify that

$$\begin{aligned} f_0(x, y)f_0(u, v) &= f_0(xu - 5yv, xv + yu), \\ f_0(x, y)f_1(u, v) &= f_1(xu - yu - 3yv, xv + 2yu + yv), \\ f_1(x, y)f_1(u, v) &= f_0(2xu + xv + yu - 2yv, xv + yu + yv). \end{aligned}$$

Using these formulae, we see that f_0 and f_1 form a group in which f_0 is the identity. More generally, Gauss proved that if d is not a perfect square then there exist composition formulae relating the various equivalence classes of primitive binary quadratic forms of discriminant d. These formulae cause the equivalence classes of the primitive forms of discriminant d to form an abelian group. Subsequently it was discovered that this corresponds to the ideal class structure in a quadratic field of discriminant d. If in Definition 3.7 we had allowed matrices of determinant -1 then some of our equivalence classes would have been joined, the composition formulae would have become muddled, and the group structure destroyed.

For more extensive treatments of the theory of quadratic forms, one should consult the books of Cassels, Jones, and O'Meara.

CHAPTER 4

Some Functions of Number Theory

4.1 GREATEST INTEGER FUNCTION

The function $[x]$ was introduced in Section 1.2, and again in Definition 3.3 in Section 3.1. It is defined for all real x and it assumes integral values only. Indeed, $[x]$ is the unique integer such that $[x] \leqslant x < [x] + 1$. For brevity it is useful to put $\{x\} = x - [x]$. This is known as the *fractional part of* x. Many of the basic properties of the function $[x]$ are included in the following theorem.

Theorem 4.1 *Let x and y be real numbers. Then we have*

(1) $[x] \leqslant x < [x] + 1,\ x - 1 < [x] \leqslant x,\ 0 \leqslant x - [x] < 1.$

(2) $[x] = \sum_{1 \leqslant i \leqslant x} 1$ *if* $x \geqslant 0$.

(3) $[x + m] = [x] + m$ *if m is an integer.*

(4) $[x] + [y] \leqslant [x + y] \leqslant [x] + [y] + 1.$

(5) $[x] + [-x] = \begin{cases} 0 & \text{if } x \text{ is an integer}, \\ -1 & \text{otherwise}. \end{cases}$

(6) $\left[\dfrac{[x]}{m}\right] = \left[\dfrac{x}{m}\right]$ *if m is a positive integer.*

(7) $-[-x]$ *is the least integer* $\geqslant x$.

(8) $[x + \frac{1}{2}]$ *is the nearest integer to x. If two integers are equally near to x, it is the larger of the two.*

(9) $-[-x + \frac{1}{2}]$ *is the nearest integer to x. If two integers are equally near to x, it is the smaller of the two.*

(10) *If n and a are positive integers,* $[n/a]$ *is the number of integers among* $1, 2, 3, \cdots, n$ *that are divisible by a.*

Proof The first part of (1) is just the definition of $[x]$ in algebraic form. The two other parts are rearrangements of the first part.

In (2) the sum is vacuous if $x < 1$. We adopt the standard convention that a vacuous sum is zero. Then, for $x \geqslant 0$, the sum counts the number of positive integers i that are less than or equal to x. This number is evidently just $[x]$.

Part (3) is obvious from the definition of $[x]$.

To prove (4) we write $x = n + \nu$, $y = m + \mu$, where n and m are integers and $0 \leqslant \nu < 1$, $0 \leqslant \mu < 1$. Then

$$[x] + [y] = n + m \leqslant [n + \nu + m + \mu] = [x + y]$$

$$= n + m + [\nu + \mu] \leqslant n + m + 1 = [x] + [y] + 1.$$

Again writing $x = n + \nu$, we also have $-x = -n - 1 + 1 - \nu$, $0 < 1 - \nu \leqslant 1$. Then

$$[x] + [-x] = n + [-n - 1 + 1 - \nu]$$

$$= n - n - 1 + [1 - \nu] = \begin{cases} 0 & \text{if } \nu = 0 \\ -1 & \text{if } \nu > 0 \end{cases}$$

and we have (5).

To prove (6) we write $x = n + \nu$, $n = qm + r$, $0 \leqslant \nu < 1$, $0 \leqslant r \leqslant m - 1$, and have

$$\left[\frac{x}{m}\right] = \left[\frac{qm + r + \nu}{m}\right] = q + \left[\frac{r + \nu}{m}\right] = q$$

since $0 \leqslant r + \nu < m$. Then (6) follows because

$$\left[\frac{[x]}{m}\right] = \left[\frac{n}{m}\right] = \left[q + \frac{r}{m}\right] = q.$$

Replacing x by $-x$ in (1) we get $-x - 1 < [-x] \leqslant -x$ and hence $x \leqslant -[-x] < x + 1$, which proves (7).

To prove (8) we let n be the nearest integer to x, taking the larger one if two are equally distant. Then $n = x + \theta$, $-\frac{1}{2} < \theta \leqslant \frac{1}{2}$, and $[x + \frac{1}{2}] = n + [-\theta + \frac{1}{2}] = n$, since $0 \leqslant -\theta + \frac{1}{2} < 1$.

The proof of (9) is similar to that of (8).

To prove part (10) we note that if $a, 2a, 3a, \cdots, ja$ are all the positive integers $\leqslant n$ that are divisible by a, then we must prove that $[n/a] = j$. But we see that $(j + 1)a$ exceeds n, so

$$ja \leqslant n < (j + 1)a, \qquad j \leqslant n/a < j + 1, \qquad [n/a] = j.$$

Theorem 4.2 *de Polignac's formula. Let p denote a prime. Then the largest exponent e such that $p^e | n!$ is*

$$e = \sum_{i=1}^{\infty} \left[\frac{n}{p^i} \right].$$

Proof If $p^i > n$, then $[n/p^i] = 0$. Therefore the sum terminates; it is not really an infinite series. The theorem is easily proved by mathematical induction. It is true for 1!. Assume it is true for $(n-1)!$ and let j denote the largest integer such that $p^j | n$. Since $n! = n \cdot (n-1)!$, we must prove that $\Sigma[n/p^i] - \Sigma[(n-1)/p^i] = j$. But

$$\left[\frac{n}{p^i} \right] - \left[\frac{n-1}{p^i} \right] = \begin{cases} 1 & \text{if} \quad p^i | n \\ 0 & \text{if} \quad p^i \nmid n \end{cases}$$

and hence

$$\sum \left[\frac{n}{p^i} \right] - \sum \left[\frac{n-1}{p^i} \right] = j.$$

The preceding proof is short, but it is rather artificial. A different proof can be based on a simple, but interesting, observation. If $a_1, a_2, \cdots, a_n$ are non-negative integers let $f(1)$ denote the number of them that are greater than or equal to 1, $f(2)$ the number greater than or equal to 2, and so on. Then

$$a_1 + a_2 + \cdots + a_n = f(1) + f(2) + f(3) + \cdots$$

since a_i contributes 1 to each of the numbers $f(1), f(2), \cdots, f(a_i)$. For $1 \leqslant j \leqslant n$, let a_j be the largest integer such that $p^{a_j} | j$. Then we see that $e = a_1 + a_2 + \cdots + a_n$. Also $f(1)$ counts the number of integers $\leqslant n$ that are divisible by p, $f(2)$ the number divisible by p^2, and so on. Hence $f(k)$ counts the integers $p^k, 2p^k, 3p^k, \cdots, [n/p^k]p^k$, so that $f(k) = [n/p^k]$. Thus we see that

$$e = a_1 + a_2 + \cdots + a_n = \sum_{i=1}^{\infty} f(i) = \sum_{i=1}^{\infty} \left[\frac{n}{p^i} \right].$$

Formula (6) of Theorem 4.1 shortens the work of computing e in Theorem 4.2. For example, if we wish to find the highest power of 7 that divides 1000! we compute

$$[1000/7] = 142, \qquad [142/7] = 20, \qquad [20/7] = 2, \qquad [2/7] = 0.$$

Adding we find that $7^{164} \mid 1000!$, $7^{165} \nmid 1000!$.

The applications of Theorem 4.2 are not restricted to numerical problems. As an example, let us prove that

$$\frac{n!}{a_1! a_2! \cdots a_r!}$$

is an integer if $a_i \geqslant 0$, $a_1 + a_2 + \cdots + a_r = n$. To do this we merely have to show that every prime divides the numerator to at least as high a power as it divides the denominator. Using Theorem 4.2 we need only prove

$$\sum \left[\frac{n}{p^i}\right] \geqslant \sum \left[\frac{a_i}{p^i}\right] + \sum \left[\frac{a_2}{p^i}\right] + \cdots + \sum \left[\frac{a_r}{p^i}\right].$$

But repeated use of Theorem 4.1, part 4, gives us

$$\left[\frac{a_1}{p^i}\right] + \left[\frac{a_2}{p^i}\right] + \cdots + \left[\frac{a_r}{p^i}\right] \leqslant \left[\frac{a_1 + a_2 + \cdots + a_r}{p^i}\right] = \left[\frac{n}{p^i}\right].$$

Summing this over i we have our desired result.

An alternative way of proving this is that the fraction claimed to be an integer is precisely the number of ways of separating a set of n (distinct) objects into a first set containing a_1 objects, a second set with a_2 objects, $\cdots$, an rth set containing a_r objects. Indeed, the reasoning used to derive Theorem 1.22 can be generalized to yield the *multinomial theorem*, in which it is seen that the quotient in question is the coefficient of $x_1^{a_1} x_2^{a_2} \cdots x_r^{a_r}$ when $(x_1 + x_2 + \cdots + x_r)^n$ is expanded. Similarly, one may use Theorem 4.2 to prove that $\dfrac{(ab)!}{(a!)^b b!}$ is an integer, although it may be simpler to invoke the combinatorial interpretation suggested in Problem 5 in Section 1.4. The advantage offered by Theorem 4.2 is that it supplies a systematic approach that can be used when a combinatorial interpretation is not readily available.

The Day of the Week from the Date The problem is to verify a given formula for calculating the day of the week for any given date. Any date, such as January 1, 2001, defines four integers N, M, C, Y as follows. Let

N be the number of the day in the month, so that $N = 1$ in the example. Let M be the number of the month counting from March, so that $M = 1$ for March, $M = 2$ for April, $\cdots$, $M = 10$ for December, $M = 11$ for January, and $M = 12$ for February. (This peculiar convention arises because the extra leap year day is added at the end of February.) Let C denote the hundreds in the year and Y the rest, so that $C = 20$ and $Y = 01$ for 2001. If d denotes the day of the week, where $d = 0$ for Sunday, $d = 1$ for Monday, $\cdots$, $d = 6$ for Saturday, then

$$d \equiv N + [2.6M - 0.2] + Y + [Y/4] + [C/4]$$
$$- 2C - (1 + L)[M/11] \pmod 7$$

where $L = 1$ for a leap year and $L = 0$ for a nonleap year. For example, in the case of January 1, 2001, we have $L = 0$, so

$$d \equiv 1 + [28.4] + 1 + [1/4][20/4] - 40 - [11/11] \equiv 1 \pmod 7,$$

and hence the first day of 2001 falls on a Monday.

This formula holds for any date after 1582, following the adoption of the Gregorian calendar at that time. The leap years are those divisible by 4, except the years divisible by 100, which are leap years only if divisible by 400. For example, 1984, 2000, 2004, 2400 are leap years, but 1900, 1901, 2100, 2401 are not.

Verify the correctness of the formula by establishing (i) that if it is correct for any date, then it is also correct for the date of the next succeeding day and also the immediately preceding day, and (ii) that it *is* correct for one particular day selected from the current calendar.

PROBLEMS

1. What is the highest power of 2 dividing 533!? The highest power of 3? The highest power of 6? The highest power of 12? The highest power of 70?

2. If 100! were written out in the ordinary decimal notation without the factorial sign, how many zeros would there be in a row at the right end?

3. For what real numbers x is it true that

(*a*) $[x] + [x] = [2x]$?
(*b*) $[x + 3] = 3 + [x]$?
(*c*) $[x + 3] = 3 + x$?

(*d*) $[x + \frac{1}{2}] + [x - \frac{1}{2}] = [2x]$?
(*e*) $[9x] = 9$?

4. Given that $[x + y] = [x] + [y]$ and $[-x - y] = [-x] + [-y]$, prove that x or y is an integer.

5. Find formulas for the highest exponent e of the prime p such that p^e divides (*a*) the product $2 \cdot 4 \cdot 6 \cdots (2n)$ of the first n even numbers; (*b*) the product of the first n odd numbers.

6. For any real number x prove that $[x] + [x + \frac{1}{2}] = [2x]$.

7. For any positive real numbers x and y prove that $[x] \cdot [y] \leqslant [xy]$.

8. For any positive real numbers x and y prove that

$$[x - y] \leqslant [x] - [y] \leqslant [x - y] + 1.$$

***9.** Prove that $(2n)!/(n!)^2$ is even if n is a positive integer.

10. Let m be any real number not zero or a positive integer. Prove that an x exists so that the equation of Theorem 4.1, part 6, is false.

11. If p and q are distinct primes, prove that the divisors of p^2q^3 coincide with the terms of $(1 + p + p^2)(1 + q + q^2 + q^3)$ when the latter is multiplied out.

12. For any integers a and $m \geqslant 2$, prove that $a - m[a/m]$ is the least non-negative residue of a modulo m. Write a similar expression for the least positive residue of a modulo m.

***13.** If a and b are positive integers such that $(a, b) = 1$, and ρ is a real number such that $a\rho$ and $b\rho$ are integers, prove that ρ is an integer. Hence prove that $\rho = n!/(a!b!)$ is an integer if $(a, b) = 1$ and $a + b = n + 1$. Generalize this to prove that

$$\rho = \frac{n!}{a_1!a_2! \cdots a_r!}$$

is an integer if $(a_1, a_2, \cdots, a_r) = 1$ and $a_1 + a_2 + \cdots + a_r = n + 1$. [Note that the first part of this problem implies that the binomial coefficient $\binom{m}{a}$ is divisible by m if $(m, a) = 1$. This follows by writing $n = m - 1$, so that $(a, b) = 1$ is equivalent to $(a, m) = 1$.]

***14.** Consider an integer $n \geqslant 1$ and the integers i, $1 \leqslant i \leqslant n$. For each $k = 0, 1, 2, \cdots$ find the number of i's that are divisible by 2^k but not by 2^{k+1}. Thus prove

$$\sum_{j=1}^{\infty} \left[\frac{n}{2^j} + \frac{1}{2} \right] = n$$

and hence that we get the correct value for the sum $n/2 + n/4 + n/8 + \cdots$ if we replace each term by its nearest integer, using the larger one if two exist.

***15.** If n is any positive integer and ξ any real number, prove that

$$[\xi] + \left[\xi + \frac{1}{n}\right] + \cdots + \left[\xi + \frac{n-1}{n}\right] = [n\xi].$$

16. Prove that $[2\alpha] + [2\beta] \geqslant [\alpha] + [\beta] + [\alpha + \beta]$ holds for every pair of real numbers, but that $[3\alpha] + [3\beta] \geqslant [\alpha] + [\beta] + [2\alpha + 2\beta]$ does not.

***17.** For every positive integer n, prove that $n!(n-1)!$ is a divisor of $(2n-2)!$.

***18.** If $(m, n) = 1$, prove that

$$\sum_{x=1}^{n-1} \left[\frac{mx}{n}\right] = \frac{(m-1)(n-1)}{2}.$$

***19.** If $m \geqslant 1$, prove that $[(1 + \sqrt{3})^{2m+1}]$ is divisible by 2^{m+1} but not by 2^{m+2}.

***20.** Let θ be real, and $0 < \theta < 1$. Define

$$g_n = \begin{cases} 0 & \text{if } [n\theta] = [(n-1)\theta] \\ 1 & \text{otherwise.} \end{cases}$$

Prove that

$$\lim_{n \to \infty} \frac{g_1 + g_2 + \cdots + g_n}{n} = \theta.$$

***21.** Let n be an odd positive integer. If n factors into the product of two integers, $n = uv$, with $u > v$ and $u - v \leqslant \sqrt[4]{64n}$, prove that the roots of $x^2 - 2[\sqrt{n} + 1]x + n = 0$ are integers. (H)

***22.** Let α be a positive irrational number. Prove that the two sequences,

$$[1 + \alpha], [2 + 2\alpha], \cdots, [n + n\alpha], \cdots, \qquad \text{and}$$

$$[1 + \alpha^{-1}], [2 + 2\alpha^{-1}], \cdots, [n + n\alpha^{-1}], \cdots$$

together contain every positive integer exactly once. Prove that this is false if α is rational.

***23.** Let $\mathscr{S}$ be the set of integers given by $[\alpha x]$ and $[\beta x]$ for $x = 1, 2, \cdots$. Prove that $\mathscr{S}$ consists of every positive integer, each appearing exactly once, if and only if α and β are positive irrational numbers such that $\dfrac{1}{\alpha} + \dfrac{1}{\beta} = 1$.

***24.** For positive real numbers α, β, γ define $f(\alpha, \beta, \gamma)$ as the sum of all positive terms of the series

$$\left[\frac{\gamma-\alpha}{\beta}\right]+\left[\frac{\gamma-2\alpha}{\beta}\right]+\left[\frac{\gamma-3\alpha}{\beta}\right]+\left[\frac{\gamma-4\alpha}{\beta}\right]+\cdots.$$

(If there are no positive terms, define $f(\alpha, \beta, \gamma) = 0$.) Prove that $f(\alpha, \beta, \gamma) = f(\beta, \alpha, \gamma)$. (H)

***25.** For any positive integers a, b, n, prove that if n is a divisor of $a^n - b^n$, then n is a divisor of $(a^n - b^n)/(a - b)$. (H)

***26.** Let d be the greatest common divisor of the coefficients of $(x + y)^n$ except the first and last, where n is any positive integer > 1. Prove that $d = p$ if n is a power of a prime p, and that $d = 1$ otherwise.

***27.** Let j and k be positive integers. Prove that

$$[(j+k)\alpha] + [(j+k)\beta] \geqslant [j\alpha] + [j\beta] + [k\alpha + k\beta]$$

for all real numbers α and β if and only if $j = k$. (This is a generalization of Problem 16.) (H)

***28.** Prove that of the two equations

$$\left[\sqrt{n}+\sqrt{n+1}\right]=\left[\sqrt{n}+\sqrt{n+2}\right],$$

$$\left[\sqrt[3]{n}+\sqrt[3]{n+1}\right]=\left[\sqrt[3]{n}+\sqrt[3]{n+2}\right]$$

the first holds for every positive integer n, but the second does not.

***29.** Evaluate the integral $\int_0^1\int_0^1\int_0^1[x+y+z]\,dx\,dy\,dz$ where the square brackets denote the greatest integer function. Generalize to n-dimensions, with an n-fold integral.

30. Show that $(2a)!(2b)!/(a!b!(a+b)!)$ is an integer.

31. Let the positive integer m be written in the base d, so that $m = \sum_i a_i d^i$ with $0 \leqslant a_i < d$ for all i. Prove that $a_i = [m/d^{i-1}] - d[m/d^i]$.

32. Write n in base p, and let $S(n)$ denote the sum of the digits in this representation. Show that $p^e \| n!$ where $e = (n - S(n))/(p - 1)$.

33. Let the positive integers m and n be written in base d, say $m = \sum_i a_i d^i$ and $n = \sum_i b_i d^i$. Show that when m and n are added, that there is a carry in the ith place (the place corresponding to d^i) if and only if $\{m/d^{i+1}\} + \{n/d^{i+1}\} \geqslant 1$.

***34.** Let a and b be positive integers with $a + b = n$. Show that the power of p dividing $\binom{n}{a}$ is exactly the number of carries when a and b are added base p.

***35.** Suppose that $a = \alpha p + a_0$ and that $0 \leqslant a_0 < p$. Show that $a!/(\alpha! p^\alpha) \equiv (-1)^\alpha a_0! \pmod p$. Suppose also that $b = \beta p + b_0$ with $0 \leqslant b_0 < p$. Show that $\binom{a+b}{a} \equiv \binom{\alpha+\beta}{\alpha}\binom{a_0+b_0}{a_0} \pmod p$. Deduce that if $a = \Sigma_i a_i p^i$ and $b = \Sigma_i b_i p^i$ in base p, then

$$\binom{a+b}{a} \equiv \prod_i \binom{a_i+b_i}{a_i} \pmod p.$$

***36.** Show that the least common multiple of the numbers $\binom{n}{1}, \binom{n}{2}, \cdots, \binom{n}{n}$ is l.c.m.$(1, 2, \cdots, n+1)/(n+1)$.

****37.** Show that if x is a real number and n is a positive integer, then $\Sigma_{k=1}^{n}[kx]/k \leqslant [nx]$.

4.2 ARITHMETIC FUNCTIONS

Functions such as $\phi(n)$ of Theorem 2.5 that are defined for all positive integers n are called *arithmetic functions*, or *number theoretic functions*, or *numerical functions*. Specifically, an *arithmetic function* f is one whose domain is the positive integers and whose range is a subset of the complex numbers.

Definition 4.1 *For positive integers n we make the following definitions.*

$d(n)$ is the number of positive divisors of n.
$\sigma(n)$ is the sum of the positive divisors of n.
$\sigma_k(n)$ is the sum of the kth powers of the positive divisors of n.
$\omega(n)$ is the number of distinct primes dividing n.
$\Omega(n)$ is the number of primes dividing n, counting multiplicity.

For example, $d(12) = 6$, $\sigma(12) = 28$, $\sigma_2(12) = 210$, $\omega(12) = 2$, and $\Omega(12) = 3$. These are all arithmetic functions. The value of k can be any real number, positive, negative, or zero. Complex values of k are useful in more advanced investigations. The *divisor function* $d(n)$ is a special case, since $d(n) = \sigma_0(n)$. Similarly, $\sigma(n) = \sigma_1(n)$. It is convenient to use the symbols $\Sigma_{d|n} f(d)$ and $\Pi_{d|n} f(d)$ for the sum and product of $f(d)$ over all positive divisors d of n. Thus we write

$$d(n) = \sum_{d|n} 1, \qquad \sigma(n) = \sum_{d|n} d, \qquad \sigma_k(n) = \sum_{d|n} d^k,$$

and similarly

$$\omega(n) = \sum_{p|n} 1, \qquad \Omega(n) = \sum_{p^\alpha \| n} \alpha = \sum_{p^\beta | n,\, \beta > 0} 1.$$

In the formulae for $\Omega(n)$, the first sum is extended over all prime powers p^α that exactly divide n, while the second sum is over all prime powers p^β dividing n.

Theorem 4.3 *For each positive integer n, $d(n) = \prod_{p^\alpha \| n} (\alpha + 1)$.*

In this notation, $\alpha = \alpha(p)$ depends on the prime being considered, and on n. Those primes p not dividing n may be ignored, since $\alpha = 0$ for such primes, and the factor contributed by such p is 1. If $n = 1$ then this is the case for all p, and we see that this formula gives $d(1) = 1$.

Proof Let $n = \prod p^\alpha$ be the canonical factorization of n. A positive integer $d = \prod p^\beta$ divides n if and only if $0 \leqslant \beta(p) \leqslant \alpha(p)$ for all prime numbers p. Since $\beta(p)$ may take on any one of the values $0, 1, \cdots, \alpha(p)$, there are $\alpha(p) + 1$ possible values for $\beta(p)$, and hence the number of divisors is $\prod_{p^\alpha \| n}(\alpha + 1)$.

From Theorem 4.3 it follows that if $(m, n) = 1$ then $d(mn) = d(m)d(n)$.

Definition 4.2 *If $f(n)$ is an arithmetic function not identically zero such that $f(mn) = f(m)f(n)$ for every pair of positive integers m, n satisfying $(m, n) = 1$, then $f(n)$ is said to be* multiplicative. *If $f(mn) = f(m)f(n)$ whether m and n are relatively prime or not, then $f(n)$ is said to be* totally multiplicative *or* completely multiplicative.

If f is a multiplicative function, $f(n) = f(n)f(1)$ for every positive integer n, and since there is an n for which $f(n) \neq 0$, we see that $f(1) = 1$.

From the definition of a multiplicative function f it follows by mathematical induction that if $m_1, m_2, \cdots, m_r$ are positive integers are relatively prime in pairs, then

$$f(m_1 m_2 \cdots m_r) = f(m_1)f(m_2) \cdots f(m_r).$$

In particular, this result would hold if the integers $m_1, m_2, \cdots, m_r$ are prime powers of distinct primes. Since every positive integer > 1 can be factored into a product of prime powers of distinct primes, it follows that if f is a multiplicative function and we know the value of $f(p^\alpha)$ for every prime p and every positive integer α, then the value of $f(n)$ for every positive integer n can be readily determined by multiplication. For example, $f(3600) = f(2^4)f(3^2)f(5^2)$. Similarly, if g is a totally multiplicative function and we know the value of $g(p)$ for every prime p, then the value of $g(n)$ for every positive integer n can be readily determined. For example $g(3600) = g(2)^4 g(3)^2 g(5)^2$.

These basic properties can be stated in another way. First, if f and g are multiplicative functions such that $f(p^\alpha) = g(p^\alpha)$ for all primes p and all positive integers α, then $f(n) = g(n)$ for all positive integers n, so that $f = g$. Second, if f and g are totally multiplicative functions such that $f(p) = g(p)$ for all primes p, then $f = g$.

Theorem 4.4 *Let $f(n)$ be a multiplicative function and let $F(n) = \sum_{d|n} f(d)$. Then $F(n)$ is multiplicative.*

Proof Suppose that $m = m_1 m_2$ with $(m_1, m_2) = 1$. If $d|m$, then we set $d_1 = (d, m_1)$ and $d_2 = (d, m_2)$. Thus $d = d_1 d_2$, $d_1|m_1$, and $d_2|m_2$. Conversely, if a pair d_1, d_2 of divisors of m_1 and m_2 are given, then $d = d_1 d_2$ is a divisor of m, and $d_1 = (d, m_1)$, $d_2 = (d, m_2)$. Thus we have established a one-to-one correspondence between the positive divisors d of m and pairs d_1, d_2 of positive divisors of m_1 and m_2. Hence

$$F(m) = \sum_{d|m} f(d) = \sum_{d_1|m_1} \sum_{d_2|m_2} f(d_1 d_2)$$

for any arithmetic function f. Since $(d_1, d_2) = 1$, it follows from the hypothesis that f is multiplicative that the right side is

$$\sum_{d_1|m_1} \sum_{d_2|m_2} f(d_1) f(d_2) = \left(\sum_{d_1|m_1} f(d_1) \right) \left(\sum_{d_2|m_2} f(d_2) \right) = F(m_1) F(m_2).$$

We could have used this theorem and Definition 4.1 to prove that $d(n)$ is multiplicative. Since $d(n) = \sum_{d|n} 1$ is of the form $\sum_{d|n} f(d)$, and since the function $f(n) = 1$ is multiplicative, Theorem 4.4 applies, and we see that $d(n)$ is multiplicative. Then Theorem 4.3 would have been easy to prove. If p is a prime, then $d(p^\alpha) = \alpha + 1$, since p^α has the $\alpha + 1$ positive divisors $1, p, p^2, \cdots, p^\alpha$ and no more. Then, since $d(n)$ is multiplicative,

$$d\left(\prod_{p^\alpha \| n} p^\alpha \right) = \prod_{p^\alpha \| n} d(p^\alpha) = \prod_{p^\alpha \| n} (\alpha + 1).$$

This exemplifies a useful method for handling certain arithmetic functions. We shall use it to find a formula for $\sigma(n)$ in the following theorem. However, it should be pointed out that $\sigma(n)$ can also be found quite simply in the same manner as we first obtained the formula for $d(n)$.

Theorem 4.5 *For every positive integer n,* $\sigma(n) = \prod_{p^\alpha \| n} \left(\dfrac{p^{\alpha+1} - 1}{p - 1} \right)$.

In case $n = 1$, $\alpha = 0$ for all primes p, so that each factor in the product is 1, and the formula gives $\sigma(1) = 1$.

Proof By definition $\sigma(n) = \sum_{d|n} d$, so we can apply Theorem 4.4 with $f(n) = n$, $F(n) = \sigma(n)$. Thus $\sigma(n)$ is multiplicative and $\sigma(n) = \prod \sigma(p^\alpha)$. But the positive divisors of p^α are just $1, p, p^2, \cdots, p^\alpha$ whose sum is $(p^{\alpha+1} - 1)/(p - 1)$.

Theorem 4.6 *For every positive integer n,* $\sum_{d|n} \phi(d) = n$.

Proof Let $F(n)$ denote the sum on the left side of the proposed identity. From Theorem 2.19 we see that $\phi(n)$ is multiplicative. Thus $F(n)$ is multiplicative, by Theorem 4.4. Since the right side, n, is also a multiplicative function, to establish that $F(n) = n$ for all n it suffices to prove that $F(p^\alpha) = p^\alpha$ for all prime powers p^α. From Theorem 2.15 we see that if $\beta > 0$ then $\phi(p^\beta) = p^\beta - p^{\beta-1}$. Thus

$$F(p^\alpha) = \sum_{d|p^\alpha} \phi(d) = \sum_{\beta=0}^{\alpha} \phi(p^\beta) = 1 + \sum_{\beta=1}^{\alpha} p^\beta - p^{\beta-1} = p^\alpha.$$

Theorem 4.6 can be proved combinatorially, as follows. Let n be given, and put $\mathscr{S} = \{1, 2, \cdots, n\}$. For each divisor d of n, let $\mathscr{S}_d$ be the subset of those members $k \in \mathscr{S}$ for which $(k, n) = d$. Clearly each member of $\mathscr{S}$ lies in exactly one of the subsets $\mathscr{S}_d$. (In such a situation we say that the subsets *partition* the set.) We note that $k \in \mathscr{S}_d$ if and only if k is of the form $k = jd$ where $(j, n/d) = 1$ and $1 \leqslant j \leqslant n/d$. Thus by Theorem 2.5 we deduce that $\mathscr{S}_d$ contains precisely $\phi(n/d)$ numbers. Since $\mathscr{S}$ contains exactly n numbers, it is now evident that $n = \sum_{d|n} \phi(n/d)$. This is an alternative formulation of the stated identity.

PROBLEMS

1. Find the smallest integer x for which $\phi(x) = 6$.
2. Find the smallest integer x for which $d(x) = 6$.

3. Find the smallest positive integer n so that $\sigma(x) = n$ has no solutions; exactly one solution; exactly two solutions; exactly three solutions.
4. Find the smallest positive integer m for which there is another positive integer $n \neq m$ such that $\sigma(m) = \sigma(n)$.
5. Prove that $\prod_{d|n} d = n^{d(n)/2}$.
6. Prove that $\sum_{d|n} d = \sum_{d|n} n/d$, and more generally that $\sum_{d|n} f(d) = \sum_{d|n} f(n/d)$.
7. Prove that $\sigma_{-k}(n) = n^{-k}\sigma_k(n)$.
8. Find a formula for $\sigma_k(n)$.
9. If $f(n)$ and $g(n)$ are multiplicative functions, and $g(n) \neq 0$ for every n, show that the functions $F(n) = f(n)g(n)$ and $G(n) = f(n)/g(n)$ are also multiplicative.
10. Give an example to show that if $f(n)$ is totally multiplicative, $F(n)$ need not also be totally multiplicative, where $F(n)$ is defined as $\sum_{d|n} f(d)$.
11. Prove that the number of positive irreducible fractions $\leqslant 1$ with denominator $\leqslant n$ is $\phi(1) + \phi(2) + \phi(3) + \cdots + \phi(n)$.
12. Prove that the number of divisors of n is odd if and only if n is a perfect square. If the integer $k \geqslant 1$, prove that $\sigma_k(n)$ is odd if and only if n is a square or double a square.
13. Given any positive integer $n > 1$, prove that there are infinitely many integers x satisfying $d(x) = n$.
14. Given any positive integer n, prove that there is only a finite number of integers x satisfying $\sigma(x) = n$.
15. Prove that if $(a, b) > 1$ then $\sigma_k(ab) < \sigma_k(a)\sigma_k(b)$ and $d(ab) < d(a)d(b)$.
16. We say (following Euclid) that m is a *perfect number* if $\sigma(m) = 2m$, that is, if m is the sum of all its positive divisors other than itself. If $2^n - 1$ is a prime p, prove that $2^{n-1}p$ is a perfect number. Use this result to find three perfect numbers.
17. Prove that an integer q is a prime if and only if $\sigma(q) = q + 1$.
18. Show that if $\sigma(q) = q + k$ where $k|q$ and $k < q$, then $k = 1$.
19. Prove that every even perfect number has the form given in Problem 16. (H)
20. For any positive integer let $\lambda(n) = (-1)^{\Omega(n)}$. This is *Liouville's lambda function*. Prove that $\lambda(n)$ is totally multiplicative, and that

$$\sum_{d|n} \lambda(d) = \begin{cases} 1 & \text{if } n \text{ is a perfect square} \\ 0 & \text{otherwise.} \end{cases}$$

***21.** For any positive integer n prove that $\phi(n) + \sigma(n) \geqslant 2n$, with equality if and only if $n = 1$ or n is a prime.

***22.** (*a*) If $m\phi(m) = n\phi(n)$ for positive integers m and n, prove that $m = n$. (*b*) Given an example to show that this result does not hold if ϕ is replaced by σ. (H)

***23.** Show that the sum of the odd divisors of n is $-\sum_{d|n}(-1)^{n/d}d$, and that this is $\sigma(n) - 2\sigma(n/2)$ where $\sigma(a)$ is defined to be 0 if a is not an integer.

***24.** Show that $\sum_{d|n} d(d)^3 = (\sum_{d|n} d(d))^2$ for all positive integers n.

***25.** Show that for all positive integers n,

$$\sum_{\substack{a=1 \\ (a,n)=1}}^{n} (a-1, n) = d(n)\phi(n).$$

4.3 THE MÖBIUS INVERSION FORMULA

Definition 4.3 *For positive integers n put $\mu(n) = (-1)^{\omega(n)}$ if n is square free, and set $\mu(n) = 0$ otherwise. Then $\mu(n)$ is the* Möbius mu function.

Theorem 4.7 *The function $\mu(n)$ is multiplicative and*

$$\sum_{d|n} \mu(d) = \begin{cases} 1 & \text{if } n - 1 \\ 0 & \text{if } n > 1. \end{cases}$$

Proof It is clear from the definition that $\mu(n)$ is multiplicative. If $F(n) = \sum_{d|n} \mu(d)$, then $F(n)$ is multiplicative by Theorem 4.4. Clearly $F(1) = \mu(1) = 1$. If $n > 1$, then $\alpha > 0$ for some prime p, and in this case $F(p^\alpha) = \sum_{\beta=0}^{\alpha} \mu(p^\beta) = 1 + (-1) = 0$, and we have the desired result.

An alternative formulation of this proof is obtained by considering those square-free divisors d of n with exactly k prime factors. There are $\binom{\omega(n)}{k}$ such divisors, each one contributing $\mu(d) = (-1)^k$. Thus by the binomial theorem, the sum in question is

$$\sum_{k=0}^{\omega(n)} \binom{\omega(n)}{k} (-1)^k = (1-1)^{\omega(n)}.$$

Theorem 4.8 *Möbius inversion formula. If* $F(n) = \sum_{d|n} f(d)$ *for every positive integer* n, *then* $f(n) = \sum_{d|n} \mu(d)F(n/d)$.

Proof We see that

$$\sum_{d|n} \mu(d)F(n/d) = \sum_{d|n} \mu(d) \sum_{k|(n/d)} f(k)$$

$$= \sum_{dk|n} \mu(d)f(k)$$

where the last sum is to be taken over all ordered pairs (d, k) such that $dk|n$. This last formulation suggests that we can reverse the roles of d and k to write the sum in the form

$$\sum_{k|n} f(k) \sum_{d|(n/k)} \mu(d)$$

and this is $f(n)$ by Theorem 4.7.

Theorem 4.9 *If* $f(n) = \sum_{d|n}\mu(d)F(n/d)$ *for every positive integer* n, *then* $F(n) = \sum_{d|n} f(d)$.

Proof First we write

$$\sum_{d|n} f(d) = \sum_{d|n} \sum_{k|d} \mu(k)F(d/k).$$

As k runs through the divisors of d, so does d/k, and hence this sum can be written as

$$\sum_{d|n} \sum_{k|d} \mu(d/k)F(k).$$

In this double sum, $F(k)$ appears for every possible divisor k of n. For each fixed divisor k of n, we collect all the terms involving $F(k)$. The coefficient is the set of all $\mu(d/k)$, where d/k is a divisor of n/k or, more simply, the set of all $\mu(r)$, where r is a divisor of n/k. It follows that the last sum can be rewritten as

$$\sum_{k|n} \sum_{r|(n/k)} \mu(r)F(k).$$

By Theorem 4.7, we see that the coefficient of $F(k)$ here is zero unless $n/k = 1$, so the entire sum reduces to $F(n)$.

It should be noted that Theorem 4.8 and its converse, Theorem 4.9, do not require that $f(n)$ or $F(n)$ be multiplicative.

On inserting the identity of Theorem 4.6 in the inversion formula of Theorem 4.8, we find that

$$\phi(n) = n \sum_{d|n} \mu(d)/d. \tag{4.1}$$

Here the summand is multiplicative, so that by Theorem 4.4 we see once more that $\phi(n)$ is multiplicative. Indeed, if n is a prime power, say $n = p^{\alpha}$, then

$$\sum_{d|p^{\alpha}} \mu(d)/d = \sum_{\beta=0}^{\alpha} \mu(p^{\beta})/p^{\beta} = 1 - 1/p.$$

This, with (4.1), gives again the formula for $\phi(n)$ in Theorem 2.15.

PROBLEMS

1. Find a positive integer n such that $\mu(n) + \mu(n+1) + \mu(n+2) = 3$.

2. Prove that $\mu(n)\mu(n+1)\mu(n+2)\mu(n+3) = 0$ if n is a positive integer.

3. Evaluate $\sum_{j=1}^{\infty} \mu(j!)$.

4. Prove Theorem 4.9 by defining $G(n)$ as $\sum_{d|n} f(d)$, then applying Theorem 4.8 to write $f(n) = \sum_{d|n} \mu(d)G(n/d)$. Thus $\sum_{d|n} \mu(d)G(n/d) = \sum_{d|n} \mu(d)F(n/d)$. Use this to show that $F(1) = G(1)$, $F(2) = G(2)$, $F(3) = G(3)$, and so on.

5. Prove that for every positive integer n, $\sum_{d|n} |\mu(d)| = 2^{\omega(n)}$.

6. If $F(n) = \sum_{d|n} f(d)$ for every positive integer n, prove that $f(n) = \sum_{d|n} \mu(n/d)F(d)$.

7. Prove that for every positive integer n, $\sum_{d|n} \mu(d)d(d) = (-1)^{\omega(n)}$. Similarly, evaluate $\sum_{d|n} \mu(d)\sigma(d)$.

8. If n is any even integer, prove that $\sum_{d|n} \mu(d)\phi(d) = 0$.

9. By use of the algebraic identity $(x+1)^2 - x^2 = 2x + 1$, establish that $(n+1)^2 - 1^2 = \sum_{x=1}^{n}\{(x+1)^2 - x^2\} = \sum_{x=1}^{n}(2x+1)$ and so derive the result $\sum_{x=1}^{n} x = n(n+1)/2$.

10. By use of the algebraic identity $(x+1)^3 - x^3 = 3x^2 - 3x + 1$ establish that $(n+1)^3 - 1^3 = \sum_{x=1}^{n}\{(x+1)^3 - x^3\} = \sum_{x=1}^{n}(3x^2 + 3x + 1)$, and so derive the result $\sum_{x=1}^{n} x^2 = n(n+1)(2n+1)/6$. (The results of this and the preceding problem can be established by other methods, mathematical induction, for example.)

11. Let $S(n)$ denote the sum of the squares of the positive integers $\leqslant n$ and prime to n. Prove that

$$\sum_{j=1}^{n} j^2 = \sum_{d|n} d^2 S\left(\frac{n}{d}\right) = \sum_{d|n} \frac{n^2}{d^2} S(d).$$

(H)

12. Combine the results of the two preceding problems to get

$$\sum_{d|n} \frac{S(d)}{d^2} = \frac{1}{6}\left(2n + 3 + \frac{1}{n}\right).$$

Then apply the Möbius inversion formula to get

$$\frac{S(n)}{n^2} = \sum_{d|n} \frac{1}{6}\mu(d)\left(\frac{2n}{d} + 3 + \frac{d}{n}\right).$$

13. Let $s(n)$ denote the largest square-free divisor of n. That is, $s(n) = \prod_{p|n} p$. Show that $\sum_{d|n} d\mu(d) = (-1)^{\omega(n)}\phi(n)s(n)/n$.

14. In the notation of the two preceding problems, show that $S(n) = n^2\phi(n)/3 + (-1)^{\omega(n)}\phi(n)s(n)/6$ for $n > 1$. (H)

***15.** Given any positive integer k, prove that there exist infinitely many integers n such that

$$\mu(n+1) = \mu(n+2) = \mu(n+3) = \cdots = \mu(n+k).$$

***16.** Let f, g, and h be arithmetic functions such that $h(n) = \sum_{d|n} f(d)g(n/d)$ for all n. Show that if f and g are multiplicative then h is also multiplicative.

***17.** Suppose that $F(n) = \sum_{d|n} f(d)$ for all n. Show that if $F(n)$ is multiplicative then $f(n)$ is multiplicative.

18. Show that for any positive integer n, $\sigma(n) = \sum_{d|n} \phi(d)d(n/d)$.

19. Show that $1/\phi(n) = \dfrac{1}{n}\sum_{d|n} \mu(d)^2/\phi(d)$ for all positive integers n.

20. Let $F(x)$ and $G(x)$ be real-valued functions defined on $[1, \infty)$. Show that $G(x) = \sum_{n \leqslant x} F(x/n)$ for all x if and only if $F(x) = \sum_{n \leqslant x} \mu(n)G(x/n)$ for all x. Here $\sum_{n \leqslant x}$ is a convenient shorthand for $\sum_{n=1}^{[x]}$.

21. Let N be a positive integer, and suppose that f and F are arithmetic functions. Show that the following assertions are equivalent:

(i) $F(n) = \displaystyle\sum_{\substack{m=1 \\ n|m}}^{N} f(m)$ for all n.

(ii) $f(n) = \displaystyle\sum_{\substack{m=1 \\ n|m}}^{N} \mu(m/n)F(m)$ for all n.

*22. For each positive integer n let $\mathscr{F}(n)$ denote the set of those positive integers m such that $\phi(m) = n$. Show that for every positive integer n, $\sum_{m \in \mathscr{F}(n)} \mu(m) = 0$.

*23. Suppose that $f(n)$ is an arithmetic function whose values are all nonzero, and put $F(n) = \prod_{d|n} f(d)$. Show that

$$f(n) = \prod_{d|n} F(n/d)^{\mu(d)}$$

for all positive integers n.

*24. Show that $\prod_{\substack{a=1 \\ (a,n)=1}}^{n} a = n^{\phi(n)} \prod_{d|n} (d!/d^d)^{\mu(n/d)}$.

25. We call a complex number ζ an nth root of unity if $\zeta^n = 1$. Show that ζ is an nth root of unity if and only if ζ is one of the n numbers $e^{2\pi i a/n}$ where $a = 1, 2, \cdots, n$. We call ζ a *primitive* nth *root of unity* if n is the least positive integer such that $\zeta^n = 1$. Show that among the nth roots of unity, $\zeta = e^{2\pi i a/n}$ is a primitive nth root if and only if $(a, n) = 1$.

*26. Let $\Phi_n(x)$ denote the polynomial with leading coefficient 1 and degree $\phi(n)$ whose roots are the $\phi(n)$ different primitive nth roots of unity. Prove that $\prod_{d|n} \Phi_d(x) = x^n - 1$ for all real or complex numbers x. Deduce that $\Phi_n(x) = \prod_{d|n} (x^d - 1)^{\mu(n/d)}$. Show that the coefficients of $\Phi_n(x)$ are integers. This is the *cyclotomic polynomial* of order n.

27. Let $F(n) = \sum_{a=1}^{n} e^{2\pi i a/n}$. Show that $F(1) = 1$, and that $F(n) = 0$ for all $n > 1$. (H)

*28. Show that for each positive integer n, $\sum_{\substack{a=1 \\ (a,n)=1}}^{n} e^{2\pi i a/n} = \mu(n)$.

*29. Let p be prime, and let $\Phi_{p-1}(x)$ denote the cyclotomic polynomial of order $p - 1$. Show that g is a solution of the congruence $\Phi_{p-1}(x) \equiv 0 \pmod{p}$ if and only if g is a primitive root $\pmod{p}$. Slow also that the sum of all the primitive roots $\pmod{p}$ is $\equiv \mu(p-1) \pmod{p}$.

4.4 RECURRENCE FUNCTIONS

We say that the arithmetic function $f(n)$ satisfies a *linear recurrence* (or *recursion*) if $f(n) = af(n-1) + bf(n-2)$ for $n = 2, 3, \cdots$. Here a and b are fixed numbers, which may be real or even complex. For brevity we write u_n for $f(n)$. In this notation the recurrence under consideration is

$$u_n = au_{n-1} + bu_{n-2}. \tag{4.2}$$

Our investigation follows the method used to analyze solutions of the differential equation $y'' = ay' + by$ with constant coefficients, though the details are simpler in the present situation.

Let λ be a root of the polynomial $Q(z) = z^2 - az - b$. Here λ may be complex, even if a and b are real. We note that $\lambda^2 = a\lambda + b$, and on multiplying both sides by λ^{n-2} we see that $\lambda^n = a\lambda^{n-1} + b\lambda^{n-2}$ for all integers $n \geqslant 2$. That is, the sequence $u_n = \lambda^n$ satisfies the recurrence (4.2). If $Q(z)$ has two distinct roots, say λ and μ, then we obtain two different solutions λ^n and μ^n of (4.2).

Suppose that u_n and v_n are two solutions of (4.2), and put $w_n = \alpha u_n + \beta v_n$ where α and β are fixed real or complex numbers. Then

$$\begin{aligned} w_n = \alpha u_n + \beta v_n &= \alpha(a u_{n-1} + b u_{n-2}) + \beta(a v_{n-1} + b v_{n-2}) \\ &= a(\alpha u_{n-1} + \beta v_{n-1}) + b(\alpha v_{n-2} + \beta v_{n-2}) \\ &= a w_{n-1} + b w_{n-2} \end{aligned}$$

for $n \geqslant 2$, and thus w_n is also a solution of (4.2). Hence we see that any linear combination of solutions of (4.2) is again a solution of (4.2). (Consequently the set of solutions forms a vector space in the abstract sense.) In particular, the sequence

$$v_n = \alpha\lambda^n + \beta\mu^n \tag{4.3}$$

is a solution of (4.2), for any values of the constants α and β.

Next we consider the *initial conditions* of our sequence u_n. Suppose we are given two real or complex numbers x_0 and x_1. We note that there is precisely one sequence u_n such that $u_0 = x_0$, $u_1 = x_1$, and which has the property that (4.2) holds for all integers $n \geqslant 2$. If the numbers α and β in (4.3) can be chosen so that

$$\begin{aligned} \alpha + \beta &= x_0 \\ \lambda\alpha + \mu\beta &= x_1 \end{aligned} \tag{4.4}$$

then the sequence v_n given in (4.3) satisfies the initial conditions $v_0 = x_0$, $v_1 = x_1$, and also (4.2), and hence $u_n = v_n$ for all n. The equations (4.4) constitute two simultaneous linear equations in the two variables α and β. The determinant of the coefficient matrix is $\mu - \lambda \neq 0$, and thus the equations (4.4) have a unique solution, for any given values of x_0 and x_1.

In the language of linear algebra, our argument thus far can be expressed succinctly as follows: We observe that the set of solutions of (4.2) form a vector space. If λ is a root of the polynomial $Q(z)$, then the sequence λ^n is a solution. Since a solution is uniquely determined by the values of u_0 and u_1, the space of solutions has dimension 2. If λ and μ are distinct roots of $Q(z)$, then λ^n and μ^n are linearly independent members of the space, and hence they form a basis. Whether we use this terminology or not, we have proved the following theorem.

Theorem 4.10 *Let a, b, x_0, and x_1 be given real or complex numbers, with $b \neq 0$. Suppose that the polynomial $Q(z) = z^2 - az - b$ has two distinct roots, say λ and μ, and let u_n be the unique sequence for which $u_0 = x_0$, $u_1 = x_1$ and for which* (4.2) *holds for all $n \geqslant 2$. Take α and β so that the equations* (4.4) *are satisfied. Then*

$$u_n = \alpha\lambda^n + \beta\mu^n \tag{4.5}$$

for $n = 0, 1, 2, \cdots$.

Conversely, if we begin with a sequence of the form (4.5), then by taking $a = \lambda + \mu$ and $b = -\lambda\mu$, we find that λ and μ are roots of the polynomial $Q(z) = z^2 - az - b$, and hence the sequence (4.5) satisfies (4.2) for this choice of a and b. That is, any sequence of the form (4.5) satisfies a linear recurrence. By excluding the case $b = 0$ we have ensured that $\lambda \neq 0$ and $\mu \neq 0$. Thus there is no difficulty in interpreting (4.5) when $n = 0$. If $b = 0$, then (4.2) defines a geometric progression, and it may be proved by induction that $u_n = u_1 a^{n-1}$ for all $n \geqslant 2$.

The theory thus far is entirely analytic, but a contact is made with number theory when we consider sequences u_n satisfying a linear recurrence, with u_n taking only integer values. For example, the *Fibonacci numbers* $F_0, F_1, \cdots$ are defined by the relations $F_0 = 0$, $F_1 = 1$, and $F_n = F_{n-1} + F_{n-2}$ for $n \geqslant 2$. Thus the first few Fibonacci numbers are $0, 1, 1, 2, 3, 5, 8, 13, 21, 34, 55, 89, 144$. Taking $a = b = 1$, we find that $Q(z)$ has distinct real roots $\lambda = (1 + \sqrt{5})/2$ and $\mu = (1 - \sqrt{5})/2$. The equations (4.4) have the single solution $\alpha = 1/\sqrt{5}$, $\beta = -1/\sqrt{5}$, so we deduce that

$$F_n = \frac{1}{\sqrt{5}}\left(\frac{1 + \sqrt{5}}{2}\right)^n - \frac{1}{\sqrt{5}}\left(\frac{1 - \sqrt{5}}{2}\right)^n \tag{4.6}$$

for $n = 0, 1, 2, \cdots$. In this example, $-1 < \mu < 0$, and the term $\beta\mu^n$ tends to 0 rapidly as n tends to infinity. Thus $\alpha\lambda^n$ is very near an integer for large n. Indeed, F_n is the integer nearest $\alpha\lambda^n$ for all non-negative integers n, and we see that $\alpha\lambda^n$ is slightly larger than F_n if n is even, and that $\alpha\lambda^n$ is slightly smaller than F_n if n is odd.

The Lucas numbers L_n are determined by the relations $L_1 = 1$, $L_2 = 3$, and $L_n = L_{n-1} + L_{n-2}$ for $n > 2$. (The French name Lucas is pronounced "Lu · kah′ ".) By Theorem 4.10 we deduce that

$$L_n = \left(\frac{1+\sqrt{5}}{2}\right)^n + \left(\frac{1-\sqrt{5}}{2}\right)^n \tag{4.7}$$

for $n = 1, 2, 3, \cdots$. We note that the F_n and L_n satisfy the same recurrence, but with different initial conditions.

As another example, we consider the sequence $0, 1, 3, 8, 21, \cdots$, for which $u_0 = 0$, $u_1 = 1$, $a = 3$, and $b = -1$. Then $\lambda = (3 + \sqrt{5})/2$ and $\mu = (3 - \sqrt{5})/2$, and by solving the equations (4.4) we deduce that

$$u_n = \frac{1}{\sqrt{5}}\left(\frac{3+\sqrt{5}}{2}\right)^n - \frac{1}{\sqrt{5}}\left(\frac{3-\sqrt{5}}{2}\right)^n$$

for $n = 0, 1, 2, \cdots$. Here $0 < \mu < 1$ so that $0 < \beta\mu^n < 1$ for all nonnegative n. Hence in this case we may express u_n using the greatest integer notation,

$$u_n = \left[\frac{1}{\sqrt{5}}\left(\frac{3+\sqrt{5}}{2}\right)^n\right].$$

Suppose that a sequence u_n is generated by the recurrence (4.2). We have developed a method by which we may find a formula for u_n, but this method fails if the polynomial $Q(z) = z^2 - az - b$ has a double root λ instead of two distinct roots λ and μ. In the case of a double root, the polynomial $Q(z)$ factors as $Q(z) = (z - \lambda)^2$, and on expanding we find that $a = 2\lambda$ and $b = -\lambda^2$. That is, $a^2 + 4b = 0$ in this case. Conversely, if $a^2 + 4b = 0$, then by the formula for the roots of a quadratic polynomial we see that $Q(z)$ has a double root. We now extend our method to deal with this situation.

Theorem 4.11 *Let a, b, x_0, and x_1 be given real or complex numbers, with $a^2 + 4b = 0$ and $b \neq 0$. Suppose that λ is a root of the polynomial $Q(z) = z^2 - az - b$, and let u_n be the unique sequence for which $u_0 = x_0$, $u_1 = x_1$ and for which (4.2) holds for all $n \geqslant 2$. Take α and β so that*

$$\begin{aligned} \alpha &= x_0, \\ \lambda\alpha + \lambda\beta &= x_1. \end{aligned} \tag{4.8}$$

Then

$$u_n = \alpha\lambda^n + \beta n\lambda^n \tag{4.9}$$

for $n = 0, 1, 2, \cdots$.

Proof The hypothesis $b \neq 0$ ensures that $\lambda \neq 0$, and hence the system (4.8) of linear equations has a unique solution, for any given values of u_0 and u_1. We know that the sequence λ^n satisfies the linear recurrence (4.2). The hypothesis $a^2 + 4b = 0$ implies that $\lambda = a/2$. We multiply both sides of this by $2\lambda^{n-1}$ to see that $2\lambda^n = a\lambda^{n-1}$. On the otherhand, we know that $\lambda^2 = a\lambda + b$. We multiply both sides of this by $(n-2)\lambda^{n-2}$ and add the resulting identity to the preceding equation, to find that $n\lambda^n = a(n-1)\lambda^{n-1} + b(n-2)\lambda^{n-2}$ for all integers $n \geqslant 2$. That is, the sequence $n\lambda^n$ is also a solution of (4.2). A linear combination of solutions of (4.2) is again a solution of (4.2), and thus the expression in (4.9) is a solution of (4.2) for any choice of α and β. To ensure that this expression gives the desired sequence, it suffices to choose α and β so that $u_0 = x_0$ and $u_1 = x_1$. That is, we take α and β so that the equations (4.8) hold.

Remark on Calculation Suppose that the numbers a, b, x_0, and x_1 in Theorem 4.10 are all integers, and let $D = a^2 + 4b$ denote the discriminant of the quadratic polynomial $Q(z)$. Thus $D \neq 0$, since the roots λ and μ are assumed to be distinct. By using (4.2) and mathematical induction, we see that u_n is an integer for all non-negative n. In case D is a perfect square, the value of u_n may be determined quickly from (4.5), but otherwise λ and μ are irrational, and (4.5) is not conducive to calculating exact values. We may use (4.2) instead, but this is slow (involving $\approx n$ arithmetic operations) if n is large. We now develop a method by which u_n may be quickly determined, using only integer arithmetic.

Let a and b be fixed integers. Among the sequences satisfying (4.2), two are especially notable. We denote them by U_n and V_n. They are determined by (4.2) and the initial conditions

$$U_0 = 0, U_1 = 1, V_0 = 2, V_1 = a. \tag{4.10}$$

By Theorem 4.10 it follows that

$$U_n = (\lambda^n - \mu^n)/\sqrt{D}\,, V_n = \lambda^n + \mu^n \tag{4.11}$$

for $n = 0, 1, 2, \cdots$. Alternatively, we could take (4.11) to be the definition and show that the sequences so defined satisfy (4.2) and (4.10). These are the *Lucas functions*, named for the French mathematician who investigated their properties in the late nineteenth century. We assume that the values of a and b are fixed, but whenever their values are at issue we write $U_n(a,b)$, $V_n(a,b)$. Note that $F_n = U_n(1,1)$, and that $L_n = V_n(1,1)$.

From (4.11) it follows that

$$\lambda^n = \left(V_n + U_n\sqrt{D}\right)/2, \quad \mu^n = \left(V_n - U_n\sqrt{D}\right)/2. \tag{4.12}$$

Suppose that u_n is a sequence satisfying (4.2), and that α and β are chosen so that (4.4) holds. Then by (4.12) we have $u_n = \gamma U_n + \delta V_n$ where $\gamma = (\alpha - \beta)/2$ and $\delta = (\alpha + \beta)/2$. Thus to calculate u_n it suffices to calculate U_n and V_n. (In the language of linear algebra, the two sequences U_n, V_n form a basis for the vector space of all solutions of (4.2). The numbers γ and δ are the coordinates of u_n with respect to this basis.)

Using (4.11) and the relations $\lambda + \mu = a$, $\lambda\mu = -b$, we verify by elementary algebra the *duplication formulae*

$$U_{2n} = U_n V_n, \quad V_{2n} = V_n^2 - 2(-b)^n \tag{4.13}$$

and the *sidestep formulae*

$$U_{n+1} = (aU_n + V_n)/2, \quad V_{n+1} = (DU_n + aV_n). \tag{4.14}$$

The identities (4.13) and (4.14) provide a quick means of calculating U_n and V_n when n is large. Suppose that $n = 187$. In binary this is 10111011. We calculate the triple $(U_k, V_k, (-b)^k)$ for the following values of k: 1, 10, 100, 101, 1010, 1011, 10110, 10111, 101110, 1011100, 1011101, 10111010, 10111011 (in binary). Each k in the list is either twice the preceding entry or one more than the preceding entry; we use (4.13) or (4.14) accordingly. The number of steps here is the same as in the procedure we discussed in Section 2.4 to calculate a^n, but now the work is roughly three times greater because we have three entries to calculate at each stage. Nevertheless, the number of steps is $\approx \log n$. Of course U_{187} may be quite large, but this procedure is easily adapted to calculate $U_{187} \pmod{187}$, for example. A somewhat more efficient system of calculation is described in Problem 28 at the end of this section.

The Lucas functions have many interesting congruential properties, of which we give a single example.

Theorem 4.12 *Let a and b be integers, and put $D = a^2 + 4b$. If p is an odd prime such that $\left(\frac{D}{p}\right) = -1$ then $p \mid U_{p+1}$.*

Proof From the binomial theorem we know that

$$\left(\frac{a + \sqrt{D}}{2}\right)^n = 2^{-n} \sum_{k=0}^{n} \binom{n}{k} a^{n-k} \sqrt{D}^k,$$

and similarly

$$\left(\frac{a - \sqrt{D}}{2}\right)^n = 2^{-n} \sum_{k=0}^{n} \binom{n}{k} a^{n-k} (-1)^k \sqrt{D}^k.$$

On inserting these expressions in (4.11), we find that

$$U_n = 2^{-n+1} \sum_{\substack{0 \leqslant k \leqslant n \\ k \text{ odd}}} \binom{n}{k} a^{n-k} D^{(k-1)/2}.$$

Thus we have a formula for U_n that involves only integers and that is amenable to congruential analysis. To simplify matters, we multiply both sides of the identity by 2^{n-1} and then take $n = p + 1$, so that

$$2^p U_{p+1} = \sum_{\substack{0 \leqslant k \leqslant p+1 \\ k \text{ odd}}} \binom{p+1}{k} a^{p+1-k} D^{(k-1)/2}.$$

Here $\binom{p+1}{k} = (p+1)!/(k!(p+1-k)!)$, by Definition 1.6. If $2 \leqslant k \leqslant p-1$, then the denominator is relatively prime to p while the numerator is divisible by p. Hence $\binom{p+1}{k} \equiv 0 \pmod{p}$ for these k, and it follows that

$$2U_{p+1} \equiv \binom{p+1}{1} a^p + \binom{p+1}{p} a D^{(p-1)/2} \equiv a(1 + D^{(p-1)/2}) \pmod{p}.$$

The proof is completed by appealing to Euler's criterion.

Theorem 4.12 can be used to construct a primality test. If n is an odd positive integer, we choose a and b so that $\left(\frac{D}{n}\right) = -1$. This is the Jacobi symbol, calculated as in Section 3.3. If $n \nmid U_{n+1}$, then n must be composite. If $n \mid U_{n+1}$, then we call n a *Lucas probable prime*. A composite Lucas probable prime is called a *Lucas pseudoprime*. In conducting Lucas pseudoprime tests, one should exercise care to avoid those choices of a and b that cause λ and μ to be roots of unity. For example, if $a = 1$ and $b = -1$ then λ and μ are sixth roots of unity, so that any sequence satisfying (4.2) has period 6. In this case $U_{3n} = 0$, and every integer of the form $6k + 5$ is a Lucas probable prime. It may be shown that the pairs (a, b) to avoid are $(\pm 2, -1)$, $(0, \pm 1)$, $(\pm 1, -1)$.

Suppose that a, b, x_0, and x_1 are all integers, and let u_n be the sequence determined by the initial conditions $u_0 = x_0$, $u_1 = x_1$ and the recurrence (4.2). By induction we see that u_n is an integer for all non-negative integers n. The converse is also true, but lies deeper: If u_n is an integer for all non-negative n, then a, b, x_0, and x_1 are all integers. Among the further known properties of linear recurrences, we mention one sample result. Suppose that a and b are given real or complex numbers and that the sequence u_n satisfies (4.2) for all $n \geqslant 2$. If there are at least five different positive integers n for which $u_n = 0$, then there is an arithmetic progression $\mathscr{A}$ such that $u_n = 0$ for all positive integers $n \in \mathscr{A}$. At a more advanced level, the equation (4.2) is called a *linear recurrence of order* 2. In Appendix A.4 we use power series generating functions to develop the analytic theory of linear recurrences of order k. (The use of power series in this context is analogous to the use of the Laplace transform in the study of linear differential equations.)

PROBLEMS

1. Find a formula for u_n if $u_n = 2u_{n-1} - u_{n-2}$, $u_0 = 0$, $u_1 = 1$. Also if $u_0 = 1$ and $u_1 = 1$.

2. Prove that any two consecutive terms of the Fibonacci sequence are relatively prime.

3. Prove that the Fibonacci numbers satisfy the inequalities

$$\left(\frac{1+\sqrt{5}}{2}\right)^{n-1} < F_{n+1} < \left(\frac{1+\sqrt{5}}{2}\right)^{n}$$

if $n > 1$.

4. Prove that for $n \geqslant 2$,

$$F_n = \binom{n-1}{0} + \binom{n-2}{1} + \binom{n-3}{2} + \binom{n-4}{3} + \cdots + \binom{n-j}{j-1}$$

where the sum of the binomial coefficients on the right terminates with the largest j such that $2j \leqslant n + 1$. (H)

5. Prove that $F_1 + F_2 + F_3 + \cdots + F_n = F_{n+2} - 1$.

6. Prove that $F_{n+1}F_{n-1} - F_n^2 = (-1)^n$.

7. Prove that $F_{m+n} = F_{m-1}F_n + F_mF_{n+1}$ for any positive integers m and n. Then prove that $F_m | F_n$ if $m | n$. (H)

8. By induction on n, prove that $L_n = F_{n-1} + F_{n+1}$ for all positive n. Then use (4.6) to give a second proof of (4.7).

9. Let u_0 and u_1 be given, and for $n \geqslant 2$ put $u_n = (u_{n-1} + u_{n-2})/2$. Show that $\lim_{n\to\infty} u_n$ exists, and that it is a certain weighted average of u_0 and u_1.

10. If the Euclidean algorithm is applied to the positive integers b and c, $b \geqslant c$, then $r_j = (b, c)$ for some j, and $r_{j+1} = 0$. Put $E(b, c) = j + 1$, so that $E(b, c)$ is the number of divisions performed in executing the algorithm. Show that $E(F_{n+2}, F_{n+1}) = n$ for all positive integers n.

Prove that $r_j \geqslant F_2$, $r_{j-1} \geqslant F_3$, $r_{j-2} \geqslant F_4, \cdots$, and that $b \geqslant F_{j+3}$. More generally, prove that if $F_{n+2} \geqslant b \geqslant c$, then $E(b, c) \leqslant n$, with equality if and only if $b = F_{n+2}$ and $c = F_{n+1}$. Conclude that if $b \geqslant c$ then $E(b, c) < (\log b)/\log((1+\sqrt{5})/2)$. (This bound was given by Gabriel Lamé in 1845. It was the first occasion in which the worst-case running time of a mathematical algorithm was precisely determined.)

11. Extend the method used to prove Theorem 4.10 to derive a formula for u_n if $u_0 = 1$, $u_1 = 2$, $u_2 = 1$, and $u_n = u_{n-1} + 4u_{n-2} - 4u_{n-3}$ for all integers $n \geqslant 3$.

12. Let $r(n)$ be the number of ways of writing a positive integer n in the form $n = m_1 + m_2 + \cdots + m_k$ where $m_1, m_2, \cdots, m_k$ and k are arbitrary positive integers. Show that $r(n) = 1 + r(1) + r(2) + \cdots + r(n-1)$ for $n \geqslant 2$. Deduce that $r(n) = 2r(n-1)$ for $n \geqslant 2$. Conclude that $r(n) = 2^{n-1}$ for all positive integers n.

13. Show that the number of ways of writing a positive integer n in the form $n = m_1 + m_2 + \cdots + m_k$ where k is an arbitrary positive integer and $m_1, m_2, \cdots, m_k$ are arbitrary odd positive integers is F_n.

***14.** Consider the sequence $1, 2, 3, 5, 8, \cdots = F_2, F_3, F_4, F_5, F_6, \cdots$. Prove that every positive integer has a unique representation as a sum of one or more distinct and non-consecutive terms of this sequence. Here two representations that differ only in the order of the summands are considered to be the same.

***15.** Let $f(n)$ denote the number of sequences $a_1, a_2, \cdots, a_n$ that can be constructed where each a_j is $+1$, -1, or 0, subject to the restrictions that no two consecutive terms can be $+1$, and no two consecutive terms can be -1. Prove that $f(n)$ is the integer nearest to $\frac{1}{2}(1+\sqrt{2})^{n+1}$.

***16.** Let u_0 and u_1 be integers, and for $n \geqslant 2$ let u_n be given by (4.2) where a and b are integers. Let m be a positive integer. Show that the sequence $u_n \pmod{m}$ is eventually periodic, with least period not exceeding $m^2 - 1$.

17. Show that $U_n(ar, br^2) = U_n(a, b)r^{n-1}$ for $n \geqslant 1$, and that $V_n(ar, br^2) = V_n(a, b)r^n$ for $n \geqslant 0$.

***18.** Put $a' = -2 - a^2/b$. Show that $a(-b)^{n-1}U_n(a', -1) = U_{2n}(a, b)$, and that $(-b)^n V_n(a', -1) = V_{2n}(a, b)$.

***19.** Show that if p is an odd prime and $\left(\dfrac{D}{p}\right) = 1$, then $U_{p+1} \equiv a \pmod{p}$.

***20.** Show that if p is an odd prime then $U_p \equiv \left(\dfrac{D}{p}\right) \pmod{p}$.

***21.** Show that if p is odd, $\left(\dfrac{D}{p}\right) = 1$, then $bU_{p-1} \equiv 0 \pmod{p}$.

***22.** Let p be a prime number. Show that $F_p \equiv \left(\frac{p}{5}\right) \pmod p$. Show that $F_{p+1} \equiv 1 \pmod p$ if $p \equiv \pm 1 \pmod 5$, and that $F_{p+1} \equiv 0 \pmod p$ if $p \equiv \pm 2 \pmod 5$. Show that $F_{p-1} \equiv 0 \pmod p$ if $p \equiv \pm 1 \pmod 5$, and that $F_{p-1} \equiv 1 \pmod p$ if $p \equiv \pm 2 \pmod 5$. Conclude that if $p \equiv \pm 1 \pmod 5$ then $p - 1$ is a period of $F_n \pmod p$. (This is not necessarily the least period.) Conclude also that if $p \equiv \pm 2 \pmod 5$ then $2p + 2$ is a period of $F_n \pmod p$.

***23.** Find the most general sequence of real or complex numbers u_n such that for $n \geqslant 2$ (*a*) $u_n = 5u_{n-1} - 6u_{n-2}$, or (*b*) $u_n = 5u_{n-1} - 6u_{n-2} + 1$, or (*c*) $u_n = 5u_{n-1} - 6u_{n-2} + n$.

***24.** Let $f(n)$ be the sum of the first n terms of the sequence $0, 1, 1, 2, 2, 3, 3, 4, 4, \cdots$. Construct a table for $f(n)$. Prove that $f(n) = [n^2/4]$. For integers x and y with $x > y$, prove that $xy = f(x + y) - f(x - y)$. Thus the process of multiplication can be replaced by an addition, a subtraction, looking up two numbers in the table, and subtracting them.

***25.** Show that $[(1 + \sqrt{3})^{2n}] + 1$ and $[(1 + \sqrt{3})^{2n+1}]$ are both divisible by 2^{n+1}. Are they divisible by any higher power of 2?

***26.** Show that if p is an odd prime then $[(2 + \sqrt{5})^p] \equiv 2^{p+1} \pmod{20p}$.

***27.** Let the sequence u_n be determined by the relations $u_1 = 0$, $u_2 = 2$, $u_3 = 3$, and $u_{n+1} = u_{n-1} + u_{n-2}$ for $n \geqslant 3$. Prove that if p is prime then $p | u_p$. (The least composite number with this property is 271,441 $= 521^2$.)

***28.** Show that $V_{2n+1} = V_n V_{n+1} - a(-b)^n$. Explain how this formula and the duplication formula (4.13) can be used to compute the triple $(V_{2n}, V_{2n+1}, (-b)^{2n})$, if the triple $(V_n, V_{n+1}, (-b)^n)$ is known. Similarly, explain how the triple $(V_{2n+1}, V_{2n+2}, (-b)^{2n+1})$ can be computed in terms of the triple $(V_n, V_{n+1}, (-b)^n)$. Explain how this triple can be determined for general n by using these two operations. (This method is not very much more efficient than the method described in the text, but it involves less work in the special case $b = -1$. By constructing congruential analogues of the identities in Problem 18 one may see that for purposes of constructing Lucas pseudoprime tests this does not involve any loss of generality.)

4.5 COMBINATORIAL NUMBER THEORY

Combinatorial mathematics is the study of the arrangements of objects, according to prescribed rules, to count the number of possible arrangements or patterns, to determine whether a pattern of a specified kind exists, and to find methods of constructing arrangements of a given type.

In this section, we treat a few elementary combinatorial problems of number theory, especially those that can be solved by the use of two simple ideas. First, if n sets contain $n + 1$ or more distinct elements in all, at least one of the sets contains two or more elements. This is sometimes familiarly called the *pigeonhole principle*, the idea being that if one places $n + 1$ letters in n slots (called "pigeonholes") then there is a pigeonhole containing more than one letter. The second idea is the one-to-one correspondence procedure, used to pair off elements in a finite set or between two sets to determine the number of elements or to prove the existence of an element of a specified kind.

Arguments of this sort were already used in the earlier parts of this book, such as in Theorem 2.6, where it was proved that the map $x \to ax$ permits residue classes (mod m) if $(a, m) = 1$, and in Fermat's theorem (Lemma 2.13) concerning $p = a^2 + b^2$. The proofs of these theorems reveal that while the two basic arguments outlined in the preceding paragraph are very easy to comprehend, their application to specific problems is another matter. The difficulty lies in determining the set or sets to which these basic arguments should be applied to yield fruitful conclusions. Here are a few illustrations of standard methods.

Example 1 Given any $m + 1$ integers, prove that two can be selected whose difference is divisible by m.

Since there are m residue classes modulo m, two of the integers must be in the same class, and so m is a divisor of their difference.

In this and most other problems in this section, the statement is the best possible of its kind. In Example 1, we could not replace the opening phrase by "Given any m integers," because the integers $1, 2, 3, \cdots, m$ do not have the property that two can be selected whose difference is divisible by m.

Example 2 Given any m integers $a_1, a_2, \cdots, a_m$, prove that a nonempty subset of these can be selected whose sum is a multiple of m.

Solution Consider the $m + 1$ integers

$$0, a_1, a_1 + a_2, a_1 + a_2 + a_3, \cdots, a_1 + a_2 + a_3 + \cdots + a_m$$

consisting of zero and the sums of special subsets of the integers. By Example 1, two of these $m + 1$ integers have a difference that is a multiple of m, and the problem is solved.

Example 3 Let $\mathscr{S}$ be a set of k integers. If $m > 1$ and $2^k > m + 1$, prove that there are two distinct nonempty subsets of $\mathscr{S}$, the sums of whose elements are congruent modulo m. Prove that the conclusion is false if $2^k = m + 1$.

Solution The set $\mathscr{S}$, containing k elements, has 2^k subsets in all, but only $2^k - 1$ nonempty subsets. For each of these nonempty subsets, consider the sum of the elements, so that there are $2^k - 1$ of these sums. Since $2^k - 1 > m$, two of these sums are in the same residue class modulo m, and so are congruent (mod m).

In case $2^k = m + 1$, define $\mathscr{S}$ as the set $\{1, 2, 4, 8, \cdots, 2^{k-1}\}$, with k elements each of a power of 2. It is not difficult to see that the sums of the nonempty subsets of $\mathscr{S}$ are precisely the natural numbers $1, 2, 3, \cdots, 2^k - 1$, each occurring once. One way to see this is to observe that the elements of $\mathscr{S}$, when written to base 2, can be expressed in the form $1, 10, 100, 1000, \cdots, 10^{k-1}$. The sums of the nonempty subsets are then all the integers, in base 2,

$$1, 10, 11, 100, 101, 111, \cdots, 111 \cdots 111$$

where the last integer here contains k digits 1 in a row.

Example 4 If $\mathscr{S}$ is any set of $n + 1$ integers selected from $1, 2, 3, \cdots, 2n$, prove that there are two relatively prime integers in $\mathscr{S}$.

Solution The set $\mathscr{S}$ must contain one of the pairs of consecutive integers 1, 2 or 3, 4 or 5, 6 or $\cdots$ or $2n - 1, 2n$.

Example 5 Find the number of integers in the set $\mathscr{S} = \{1, 2, 3, \cdots, 6300\}$ that are divisible by neither 3 nor 4; also the number divisible by none of 3, 4, or 5.

Solution Of the 6300 integers in $\mathscr{S}$, exactly 2100 are divisible by 3, and 1575 are divisible by 4. The subtraction $6300 - 2100 - 1575$ does not give the correct answer to the first part of the problem, because the sets removed by subtraction are not disjoint. Those integers divisible by 12 have been removed twice. There are 525 such integers, so the answer to the first part of the problem is

$$6300 - 2100 - 1575 + 525 = 3150.$$

Turning to the second part of the problem, we begin by removing from the set $\mathscr{S}$ those integers divisible by 3, in number 2100, those divisible by 4, in number 1575, and those divisible by 5, in number 1260. So we see that

$$6300 - 2100 - 1575 - 1260$$

is a start toward the answer. However, integers divisible by both 3 and 4 have been removed twice; likewise, those divisible by both 3 and 5 and those divisible by both 4 and 5. Hence, we add back in 6300/12 or 525 of the first type, 6300/15 or 420 of the second type, and 6300/20 or 315 of the third type to give

$$6300 - 2100 - 1575 - 1260 + 525 + 420 + 315.$$

This is still not the final answer, because one more adjustment must be made, for the integers $50, 120, 180, \cdots$ that are divisible by 3, 4, and 5. Such integers are counted once in each term of the expression above, and so the net count for each such integer is 1. There are 6300/60 or 105 such integers, so if we subtract this number we get the correct answer,

$$6300 - 2100 - 1575 - 1260 + 525 + 420 + 315 - 105 = 2520.$$

The Inclusion-Exclusion Principle Example 5 illustrates a basic combinatorial argument as follows: Consider a collection of N objects of which $N(\alpha)$ have a certain property α, $N(\beta)$ have property β, and $N(\gamma)$ have property γ. Similarly, let $N(\alpha, \beta)$ be the number having both properties α and β, and $N(\alpha, \beta, \gamma)$ be the number having properties α, β, and γ. Then the number of objects in the collection having none of the properties α, β, γ is

$$\begin{aligned} N - N(\alpha) - N(\beta) - N(\gamma) + N(\alpha, \beta) \\ + N(\alpha, \gamma) + N(\beta, \gamma) - N(\alpha, \beta, \gamma) \end{aligned} \tag{4.15}$$

This is the inclusion-exclusion principle in the case of three properties.

The proof of (4.15) can be given along the same lines as in Example 5: First, that an object having exactly one of the properties, say β, is counted once by N and once by $N(\beta)$ for a net count of $1 - 1$ or 0; that an object having exactly two of the properties has a net count of $1 - 1 - 1 + 1$, again 0; next, that an object with all three properties has a net count of $1 - 1 - 1 - 1 + 1 + 1 + 1 - 1$, again 0. On the other hand, an object having none of the properties is counted by N once in (4.15), and so a net count of 1.

The extension of (4.15) to a collection of N objects having (variously) k properties is very natural. Where (4.15) has three terms of the type $N(\alpha)$, the general formula has k such terms; where (4.15) has three terms of the type $N(\alpha, \beta)$, the general formula has $k(k - 1)/2$ such terms; and so on.

It may be noted that the inclusion-exclusion principle can be used to give an entirely different proof of the formula for the evaluation of the Euler function $\phi(n)$, as set forth in Theorem 2.15. Because that result has been proved in full detail already, we make the argument in the case of an integer n having exactly three distinct prime factors, say p, q, and r. The problem is to determine the number of integers in the set $\mathscr{S} = \{1, 2, 3, \cdots, n\}$ having no prime factor in common with n. Let an integer in the set $\mathscr{S}$ have property α if it is divisible by p, property β if it is divisible by q and property γ if it is divisible by r. A direct application of (4.15) gives

$$n - n/p - n/q - n/r + n/pq + n/pr + n/qr - n/pqr$$
$$= n(1 - 1/p)(1 - 1/q)(1 - 1/r)$$

as the number of integers in the set $\mathscr{S}$ divisible by none of p, q, or r.

PROBLEMS

1. Given any m integers none of which is a multiple of m, prove that two can be selected whose difference is a multiple of m.

***2.** If $\mathscr{S}$ is any set of $n + 1$ integers selected from $1, 2, 3, \cdots, 2n + 1$, prove that $\mathscr{S}$ contains two relatively prime integers. Prove that the result does not hold if $\mathscr{S}$ contains only n integers.

***3.** For any positive integers k and $m > 1$, let $\mathscr{S}$ be a set of k integers none of which is a multiple of m. If $k > m/2$, prove that there are two integers in $\mathscr{S}$ whose sum or whose difference is divisible by m.

***4.** Let the integers $1, 2, \cdots, n$ be placed in any order around the circumference of a circle. For any $k < n$, prove that there are k integers in a consecutive block on the circumference having sum at least $(kn + k)/2$.

***5.** Given any integers a, b, c and any prime p not a divisor of ab, prove that $ax^2 + by^2 \equiv c \pmod{p}$ is solvable.

***6.** Let k and n be integers satisfying $n > k > 1$. Let $\mathscr{S}$ be any set of k integers selected from $1, 2, 3, \cdots, n$. If $2^k > kn$, prove that there exist two distinct nonempty subsets of $\mathscr{S}$ having equal sums of elements.

***7.** Let n and k be positive integers with $n > k$ and $(n, k) = 1$. Prove that if k distinct integers are selected at random from $1, 2, \cdots, n$, the probability that their sum is divisible by n is $1/n$.

***8.** Say that a set $\mathscr{S}$ of positive integers has property M if no element of $\mathscr{S}$ is a multiple of another. (*a*) Prove that there exists a subset $\mathscr{S}$ of $\{1, 2, 3, \cdots, 2n\}$ containing n elements with property M but that no subset of $n + 1$ elements has property M. (*b*) Prove the same results for subsets $\mathscr{S}$ of $\{1, 2, 3, \cdots, 2n - 1\}$. (*c*) How many elements are there in the largest subset $\mathscr{S}$ of $\{1, 3, 5, 7, \cdots, 2n - 1\}$ having property M?

9. Prove that among any ten consecutive positive integers at least one is relatively prime to the product of the others. [*Remark:* if "ten" is replaced by "n", the result is true for every positive integer $n \leqslant 16$, but false for $n > 16$. This is not easy to prove; cf. R. J. Evans, "On blocks of n consecutive integers," *Amer. Math. Monthly*, **76** (1969), 48.]

***10.** Let $a_1, a_2, \cdots, a_n$ be any sequence of positive integers. Let k be the total number of distinct prime factors of the product of the integers. If $n \geqslant 2^k$, prove that there is a consecutive block of integers in the sequence whose product is a perfect square.

***11.** For every positive integer n, construct a minimal set $\mathscr{S}$ of integers having the property that every residue class modulo n occurs at least once among the sums of the elements of the nonempty subsets of $\mathscr{S}$. For example, if $n = 6$, $\mathscr{S} = \{1, 3, 5\}$ will do because every residue class modulo 6 appears among $1, 3, 5, 1 + 3, 3 + 5, 1 + 5, 1 + 3 + 5$.

***12.** Let n and k be positive integers such that $1 \leqslant k \leqslant (n^2 + n)/2$. Prove that there is a subset of the set $\{1, 2, 3, \cdots, n\}$ whose sum is k.

***13.** For any integer $k > 1$, prove that there is exactly one power of 2 having exactly k digits with leading digit 1, when written in standard fashion to base 10. For example $2^4 = 16$, $2^7 = 128$. Prove also that there is exactly one power of 5 having exactly k digits with leading digit *not* equal to 1.

***14.** For any positive integer n, prove that 5^n has leading digit 1 if and only if 2^{n+1} has leading digit 1. Hence, prove that the "probability" that a power of 2 has leading digit 1 is $\log 2/\log 10$ and that this is also the "probability" that a power of 5 has leading digit 1. By "probability," we mean the limit as n tends to infinity of the probability that an arbitrarily selected integer from $2, 2^2, 2^3, \cdots, 2^n$ has leading digit 1, and similarly for powers of 5.

15. Let n be a positive integer having exactly three distinct prime factors p, q and r. Find a formula for the number of positive integers $\leqslant n$ that are divisible by none of pq, pr, or qr.

NOTES ON CHAPTER 4

§4.4 The book by N. J. A. Sloane listed in the General References is very useful in trying to identify or classify a given sequence of integers of an unknown source.

The analytic theory of linear recurrences is developed further in Appendix A.4.

CHAPTER 5

Some Diophantine Equations

We often encounter situations in which we wish to find solutions of an equation with integral values of the variables, or perhaps rational values. Sometimes we seek solutions in non-negative integers. In any of these cases we refer to the equation as a *Diophantine equation*, after the Greek mathematician Diophantus who studied this topic in the third century A.D. We restrict our attention to equations involving polynomials in one or more variables. There is no universal method for determining whether a Diophantine equation has a solution, or for finding them all if solutions exist. However, we are quite successful in dealing with polynomials of low degree, or in a small number of variables. In addition to the material in this chapter, an introductory discussion of $ax + by = c$ is given in Section 1.2, the equation $x^2 + y^2 = n$ is discussed in Sections 2.1 and 3.6, the equation $x^2 + y^2 + z^2 + w^2 = n$ is treated in Section 6.4, Pell's equation, $x^2 - dy^2 = N$, is treated in Chapter 7, by means of continued fractions, and further equations are investigated in Chapter 9, using the arithmetic of quadratic number fields.

5.1 THE EQUATION $ax + by = c$

Any linear equation in two variables having integral coefficients can be put in the form

$$ax + by = c \tag{5.1}$$

where a, b, c are given integers. We consider the problem of identifying all solutions of this equation in which x and y are integers. If $a = b = c = 0$, then every pair (x, y) of integers is a solution of (5.1), whereas if $a = b = 0$ and $c \neq 0$, then (5.1) has no solution. Now suppose that at least one of a

and b is nonzero, and let $g = \text{g.c.d.}(a, b)$. If $g \nmid c$ then (5.1) has no solution, by part (3) of Theorem 1.1. On the other hand, by Theorem 1.3 there exist integers x_0, y_0 such that $ax_0 + by_0 = g$, and hence if $g | c$ then the pair $(cx_0/g, cy_0/g)$ is an integral solution of (5.1). We may find x_0 and y_0 by employing the Euclidean algorithm, as discussed in Section 1.2. Once a single solution is known, say $ax_1 + by_1 = c$, others are given by taking $x = x_1 + kb/g$, $y = y_1 - ka/g$. Here k is an arbitrary integer. Thus (5.1) has infinitely many integral solutions if it has one. We now show that (5.1) has no integral solutions beyond the ones we have already found. For suppose that the pairs $(x_1, y_1), (x, y)$ are integral solutions of (5.1). By subtracting, we find that $a(x - x_1) + b(y - y_1) = 0$. We divide through by g and rearrange, to see that

$$(a/g)(x - x_1) = (b/g)(y_1 - y).$$

That is, a/g divides the product $(b/g)(y_1 - y)$. But $(a/g, b/g) = 1$ by Theorem 1.7, so by Theorem 1.10 it follows that a/g divides $y_1 - y$. That is, $ka/g = y_1 - y$ for some integer k. On substituting this in the equation displayed above, we find that $x - x_1 = kb/g$. Thus we have proved the following theorem.

Theorem 5.1 *Let a, b and c be integers with not both a and b equal to* 0, *and let* $g = \text{g.c.d.}(a, b)$. *If $g \nmid c$ then the equation* (5.1) *has no solution in integers. If $g | c$ then this equation has infinitely many solutions. If the pair (x_1, y_1) is one integral solution, then all others are of the form $x = x_1 + kb/g$, $y = y_1 - ka/g$ where k is an integer.*

The equation (5.1) under consideration is equivalent to the congruence $ax \equiv c \pmod{b}$, whose solutions are described by Theorem 2.17. Indeed Theorem 5.1 is merely a rcformulation of this prior theorem.

Viewed geometrically, the equation (5.1) determines a line in the Euclidean plane. If we hold a and b fixed, and consider different values of c, we obtain a family of mutually parallel lines. Each lattice point in the plane lies on exactly one such line. From Theorem 5.1 we see that the lattice points on such a line (if there are any) form an arithmetic progression and that the common difference between one lattice point on the line and the next is determined by the vector $(b/g, -a/g)$, which is independent of c. If a and b are positive then the line has negative slope, and if in addition c is positive then the line has positive intercepts with the axes. In such a situation, it is interesting to consider the possible existence of solutions of (5.1) in positive integers. From Theorem 5.1 we see that $x > 0$ if and only if $k > -gx_1/b$, and that $y > 0$ if and only if $k < gy_1/a$. Thus the solutions of (5.1) in positive integers are given by those integers k in the open interval $I = (-gx_1/b, gy_1/a)$. Using the fact that the point (x_1, y_1) lies on the line (5.1), we find that the length of I is $gc/(ab)$. Thus if $g|c$ and P denotes the number of solutions of (5.1) in positive integers then $|P - gc/(ab)| \leqslant 1$. In particular, it follows that if $g|c$ and $c > ab/g$ then $P > 0$. Here the hypothesis can not be weakened, for if $c = ab/g$ then the solutions of (5.1) are the points $((k + 1)b/g, -ka/g)$, and we see that there is no integral value of k for which both coordinates are positive. Similarly, the solutions of (5.1) in non-negative integers correspond to integers k lying in the closed interval $J = [-gx_1, gy_1/a]$, so that the total number N of solutions satisfies $|N - gc/(ab)| \leqslant 1$ if $g|c$.

If it is desired to have exact formulae for the numbers P and N defined above, instead of mere approximations, we employ the greatest integer function discussed in Section 4.1. We assume that $g|c$ and that an integral solution (x_1, y_1) of (5.1) is known. The least value of k for which $x_1 + kb/g$ is positive is $[-gx_1/b] + 1$, while the greatest value of k for which $y_1 - ka/g$ is positive is $-[-gy_1/a] - 1$. Thus $P = (-[-gy_1/a] - 1) - ([-gx_1/b] + 1) + 1 = -[-gy_1/a] - [-gx_1/b] - 1$. In terms of the fractional part function $\{x\} = x - [x]$, we deduce that $P = gc/(ab) + \{-gy_1/a\} + \{-gx_1/b\} - 1$.

The methods of Section 1.2 can be used to find integers x_0 and y_0 such that $ax_0 + by_0 = g$, and hence an initial solution x_1, y_1 of (5.1) may be constructed, if $g|c$. In the following numerical examples we tailor those ideas to the present situation.

Example 1 Find all solutions of $999x - 49y = 5000$.

Solution By the division algorithm we observe that $999 = 20 \cdot 49 + 19$. This suggests writing the equation in the form $19x - 49(y - 20x) = 5000$. Putting $x' = x$, $y' = -20x + y$, we find that the original equation is expressed by the condition $19x' - 49y' = 5000$. This is simpler because the coefficients are smaller. Since $49 = 2 \cdot 19 + 11$, we write this equation as $19(x' - 2y') - 11y' = 5000$. That is, $19x'' - 11y'' = 5000$ where $x'' = x' - 2y'$ and $y'' = y'$. Since $19 = 2 \cdot 11 - 3$, we write the equation as $-3x'' - 11(-2x'' + y'') = 5000$. That is, $-3x^{(3)} - 11y^{(3)} = 5000$ where $x^{(3)} = x''$ and $y^{(3)} = -2x'' + y''$. As $11 = 4 \cdot 3 - 1$, we write the equation as $-3(x^{(3)} + 4y^{(3)}) + y^{(3)} = 5000$. That is, $-3x^{(4)} + y^{(4)} = 5000$ where $x^{(4)} = x^{(3)} + 4y^{(3)}$ and $y^{(4)} = y^{(3)}$. Making the further change of variables $x^{(5)} = x^{(4)}$, $y^{(5)} = -3x^{(4)} + y^{(4)}$, we see that the original equation is equivalent to the equation $y^{(5)} = 5000$. Here the value of $y^{(5)}$ is a fixed integer, and $x^{(5)}$ is an arbitrary integer. Since pairs of integers (x, y) are in one-to-one correspondence with pairs of integers $(x^{(5)}, y^{(5)})$, it follows that the original equation has infinitely many solutions in integers. To express x and y explicitly in terms of $x^{(5)}$ and $y^{(5)}$, we first determine x and y in terms of x' and y', then in terms of x'' and y'', and so on. These transformations can be developed at the same time that the original equation is being simplified. We start by writing

$$\begin{aligned} 999x - 49y &= 5000, \\ x \qquad &= x, \\ y &= y. \end{aligned} \tag{5.2}$$

Then we rewrite these equations in the form

$$\begin{aligned} 19x - 49(-20x + y) &= 5000, \\ x \qquad\qquad &= x, \\ 20x + \quad (-20x + y) &= y. \end{aligned}$$

That is,

$$\begin{aligned} 19x' - 49y' &= 5000, \\ x' \qquad &= x, \\ 20x' + \quad y' &= y. \end{aligned} \tag{5.3}$$

We rewrite this as

$$\begin{aligned} 19(x' - 2y') - 11y' &= 5000, \\ x' - 2y' \ + \ 2y' &= x, \\ 20(x' - 2y') + 41y' &= y. \end{aligned}$$

That is,

$$\begin{aligned} 19x'' - 11y'' &= 5000, \\ x'' + \ 2y'' &= x, \\ 20x'' + 41y'' &= y. \end{aligned} \tag{5.4}$$

We rewrite this as

$$\begin{aligned} -3x'' - 11(-2x'' + y'') &= 5000, \\ 5x'' + \ 2(-2x'' + y'') &= x, \\ 102x'' + 41(-2x'' + y'') &= y. \end{aligned}$$

That is,

$$\begin{aligned} -3x^{(3)} - 11y^{(3)} &= 5000, \\ 5x^{(3)} + 2y^{(3)} &= x, \\ 102x^{(3)} + 41y^{(3)} &= y. \end{aligned} \tag{5.5}$$

We rewrite this as

$$\begin{aligned} -3\left(x^{(3)} + 4y^{(3)}\right) + y^{(3)} &= 5000, \\ 5\left(x^{(3)} + 4y^{(3)}\right) - 18y^{(3)} &= x, \\ 102\left(x^{(3)} + 4y^{(3)}\right) - 367y^{(3)} &= y. \end{aligned}$$

That is,

$$\begin{aligned} -3x^{(4)} + y^{(4)} &= 5000, \\ 5x^{(4)} - 18y^{(4)} &= x, \\ 102x^{(4)} - 367y^{(4)} &= y. \end{aligned} \tag{5.6}$$

We rewrite this as

$$\begin{aligned} \left(-3x^{(4)} + y^{(4)}\right) &= 5000, \\ -49x^{(4)} - 18\left(-3x^{(4)} + y^{(4)}\right) &= x, \\ -999x^{(4)} - 367\left(-3x^{(4)} + y^{(4)}\right) &= y. \end{aligned}$$

That is,

$$\begin{aligned} y^{(5)} &= 5000, \\ -49x^{(5)} - 18y^{(5)} &= x, \\ -999x^{(5)} - 367y^{(5)} &= y. \end{aligned} \tag{5.7}$$

Inserting this value of $y^{(5)}$, and writing k in place of $x^{(5)}$, we conclude that the solutions of the proposed equation are given by taking

$$\begin{aligned} x &= -49k - 90000, \\ y &= -999k - 1835000. \end{aligned}$$

This parameterization of the solutions is not unique. For example, we could set $k = -1837 - m$, in which case the equations above would become

$$x = 49m + 13,$$

$$y = 999m + 163.$$

We note that the coefficients in (5.3) are derived from those in (5.2) by subtracting 20 times the second column from the first column. Similarly, the coefficients in (5.4) are obtained from those in (5.3) by adding twice the first column to the second. In (5.4) we add twice the second column to the first to obtain (5.5). In (5.5) we add -4 times the first column to the second column to obtain (5.6), and in (5.6) we add 3 times the second column to the first to obtain (5.7). In general, we may add a multiple of one of the first two columns to the other. In addition we may permute the first two columns or multiply all elements of one of these columns by -1. Thus we may alter the coefficients by means of the following three column operations:

(C1) Add an integral multiple m of one of the first two columns to the other;

(C2) Exchange the first two columns;

(C3) Multiply all elements of one of the first two columns by -1.

These are similar to the elementary column operations of linear algebra, but in linear algebra the multiple in (C1) may be any real number, and in (C3) one may use any nonzero constant in place of -1. In numerical calculations it suffices to manipulate the coefficients according to rules (C1), (C2), and (C3). When applying operation (C1), we are free to take m to be any integer we please, but in practice we choose m so as to reduce the size of a particular coefficient. In particular, we are not confined to the simplest form of the division algorithm—instead we may round to the nearest integer, as we did in passing from (5.4) to (5.5), even though it introduces a negative remainder.

It is not necessary to write out the full set of equations at each stage, as we did in solving Example 1. We now exhibit the method in this more concise format.

Example 2 Find all integers x and y such that $147x + 258y = 369$.

Solution We write

$$\begin{matrix} 147 & 258 & 369 \\ 1 & 0 & \\ 0 & 1 & \end{matrix} \quad \rightarrow \quad \begin{matrix} 147 & 111 & 369 \\ 1 & -1 & \\ 0 & 1 & \end{matrix} \quad \rightarrow \quad \begin{matrix} 36 & 111 & 369 \\ 2 & -1 & \\ -1 & 1 & \end{matrix}$$

$$\rightarrow \quad \begin{matrix} 36 & 3 & 369 \\ 2 & -7 & \\ -1 & 4 & \end{matrix} \quad \rightarrow \quad \begin{matrix} 0 & 3 & 369 \\ 86 & -7 & \\ -49 & 4 & \end{matrix}$$

Let the variables that are implicit in this last array be called u and v. Since $3v = 369$, we deduce that $v = 123$, and that the full set of solutions is given by taking $x = 86u - 861$, $y = -49u + 492$. The variables u and v were obtained from the original variables x and y by a homogeneous change of coordinates. We may reduce the size of the constant term in our answer by introducing an inhomogeneous change of variables. For example, if we put $u = t + 10$, then we find that $x = 86t - 1$, $y = -49t + 2$.

PROBLEMS

1. Prove that all solutions of $3x + 5y = 1$ can be written in the form $x = 2 + 5t$, $y = -1 - 3t$; also in the form $x = 2 - 5t$, $y = -1 + 3t$; also in the form $x = -3 + 5t$, $y = 2 - 3t$. Prove that $x = a + bt$, $y = c + dt$, with t arbitrary, is a form of the general solution if and only if $x = a$, $y = c$ is a solution and either $b = 5$, $d = -3$ or $b = -5$, $d = 3$.
2. Find all solutions of $10x - 7y = 17$.
3. Using a calculator, find all solutions of
   (*a*) $903x + 731y = 2107$;
   (*b*) $903x + 731y = 1106$;
   (*c*) $101x + 99y = 437$.
4. Find all solutions in positive integers:
   (*a*) $5x + 3y = 52$;
   (*b*) $15x + 7y = 111$;
   (*c*) $40x + 63y = 521$;
   (*d*) $123x + 57y = 531$;
   (*e*) $12x + 50y = 1$;
   (*f*) $12x + 510y = 274$;
   (*g*) $97x + 98y = 1000$.
5. Prove that $101x + 37y = 3819$ has a positive solution in integers.
6. Given that $(a, b) = 1$ and that a and b are of opposite sign, prove that $ax + by = c$ has infinitely many positive solutions for any value of c.

7. Let a, b, c be positive integers. Prove that there is no solution of $ax + by = c$ in positive integers if $a + b > c$.

8. If $ax + by = c$ is solvable and $b \neq 0$, prove that it has a solution x_0, y_0 with $0 \leqslant x_0 < |b|$.

9. Prove that $ax + by = a + c$ is solvable if and only if $ax + by = c$ is solvable.

10. Prove that $ax + by = c$ is solvable if and only if $(a, b) = (a, b, c)$.

11. Given that $ax + by = c$ has two solutions, (x_0, y_0) and (x_1, y_1) with $x_1 = 1 + x_0$, and given that $(a, b) = 1$, prove that $b = \pm 1$.

12. A positive integer is called *powerful* if $p^2 | a$ whenever $p | a$. Show that a is powerful if and only if a can be expressed in the form $a = b^2c^3$ where b and c are positive integers.

13. Let a, b, c be positive integers such that $g | c$, where $g =$ g.c.d.(a, b), and let N denote the number of solutions of (5.1) in non-negative integers. Show that $N = [y_1 g/a] + [x_1 g/b] + 1 = gc/(ab) + 1 - \{y_1 g/a\} - \{x_1 g/b\}$.

14. Let a, b, c be positive integers. Assuming that $g | c$ and that $cg/(ab)$ is an integer, prove that $N = 1 + cg/(ab)$, and that $P = -1 + cg/(ab)$.

15. Let a, b, c be positive integers. Assuming that $g | c$ but that $cg/(ab)$ is not an integer, prove that $P = [cg/(ab)]$ or $P = [cg/(ab)] + 1$, and that $N = [cg/(ab)]$ or $N = [cg/(ab)] + 1$. Assuming further that $a | c$, show that $N = [cg/(ab)] + 1$ and that $P = [cg/(ab)]$. (H)

***16.** Let a and b be positive integers with g.c.d.$(a, b) = 1$. Let $\mathscr{S}$ denote the set of all integers that may be expressed in the form $ax + by$ where x and y are non-negative integers. Show that $c = ab - a - b$ is not a member of $\mathscr{S}$, but that every integer larger than c is a member of $\mathscr{S}$.

***17.** Find necessary and sufficient conditions that

$$x + b_1 y + c_1 z = d_1, \qquad x + b_2 y + c_2 z = d_2$$

have at least one simultaneous solution in integers x, y, z, assuming that the coefficients are integers with $b_1 \neq b_2$.

5.2 SIMULTANEOUS LINEAR EQUATIONS

Let $a_1, a_2, \cdots, a_n$ be integers, not all 0, and suppose we wish to find all solutions in integers of the equation

$$a_1x_1 + a_2x_2 + \cdots + a_nx_n = c.$$

As in Theorem 5.1, we may show that such solutions exist if and only if g.c.d.$(a_1, a_2, \cdots, a_n)$ divides c. The numerical technique exposed in the preceding section also extends easily to larger values of n.

Example 3 Find all solutions in integers of $2x + 3y + 4z = 5$.

Solution We write

$$\begin{array}{rrrr} 2 & 3 & 4 & 5 \\ 1 & 0 & 0 & \\ 0 & 1 & 0 & \\ 0 & 0 & 1 & \end{array} \quad \rightarrow \quad \begin{array}{rrrr} 2 & 1 & 0 & 5 \\ 1 & -1 & -2 & \\ 0 & 1 & 0 & \\ 0 & 0 & 1 & \end{array} \quad \rightarrow \quad \begin{array}{rrrr} 0 & 1 & 0 & 5 \\ 3 & -1 & -2 & \\ -2 & 1 & 0 & \\ 0 & 0 & 1 & \end{array}$$

This last array represents simultaneous equations involving three new variables, say t, u, v. The first line gives the condition $u = 5$. On substituting this in the lower lines, we find that every solution of the given equation in integers may be expressed in the form

$$\begin{aligned} x &= 3t - 2v - 5 \\ y &= -2t \qquad + 5 \\ z &= \qquad\quad v \end{aligned}$$

where t and v are integers. From the nature of the changes of variables made, we know that triples (x, y, z) of integers satisfying the given equation are in one-to-one correspondence with triples of integers (t, u, v) for which $u = 5$. Hence each solution of the given equation in integers is given by a unique pair of integers (t, v).

We now consider the problem of treating simultaneous equations. Suppose we have two equations, say

$$\begin{aligned} A &= B, \\ C &= D. \end{aligned} \tag{5.8}$$

By multiplying the first equation by m and adding the result to the second equation, we may obtain a new pair of equations,

$$\begin{aligned} A &= B, \\ C + mA &= D + mB. \end{aligned} \tag{5.9}$$

This pair of equations is equivalent to the original pair (5.8). Here m may be any real number, but since our interest is in equations with integral coefficients, we shall restrict m to be an integer. Similarly, the equation $A = B$ is equivalent to $cA = cB$ provided that $c \neq 0$. Again, since our interest is in equations with integral coefficients, we restrict c to the values $c = \pm 1$. Finally, we may rearrange a collection of equations without altering their significance. Hence we have at our disposal three row operations which we may apply to a system of equations:

(R1) Add an integral multiple m of one equation to another;

(R2) Exchange two equations;

(R3) Multiply both sides of an equation by -1.

By applying these operations in conjunction with the column operations considered in the preceding section, we may determine the integral solutions of a system of linear equations.

Example 4 Find all solutions in integers of the simultaneous equations

$$20x + 44y + 50z = 10,$$

$$17x + 13y + 11z = 19.$$

Solution Among the coefficients of x, y, and z, the coefficient 11 is smallest. Using operation (C1) and the division algorithm (rounding to the nearest integer), reduce the coefficients of x and y in the second row (mod 11):

$$\begin{array}{rrrr} 20 & 44 & 50 & 10 \\ 17 & 13 & 11 & 19 \\ 1 & 0 & 0 & \\ 0 & 1 & 0 & \\ 0 & 0 & 1 & \end{array} \quad \rightarrow \quad \begin{array}{rrrr} -80 & -6 & 50 & 10 \\ -5 & 2 & 11 & 19 \\ 1 & 0 & 0 & \\ 0 & 1 & 0 & \\ -2 & -1 & 1 & \end{array}$$

The coefficient of least absolute value is now in the second row and second column. We use operation (C1) to reduce the other coefficients in the second row (mod 2):

$$\rightarrow \quad \begin{array}{rrrr} -98 & -6 & 80 & 10 \\ 1 & 2 & 1 & 19 \\ 1 & 0 & 0 & \\ 3 & 1 & -5 & \\ -5 & -1 & 6 & \end{array}$$

There are now two coefficients of minimal absolute value. We use the one in the first column as our pivot and use operation (C1) to reduce the other coefficients in the second row:

$$\rightarrow \begin{array}{rrrr} -98 & 190 & 178 & 10 \\ 1 & 0 & 0 & 19 \\ 1 & -2 & -1 & \\ 3 & -5 & -8 & \\ -5 & 9 & 11 & \end{array}$$

The coefficient of least nonzero absolute value is unchanged, so we switch to operation (R1) to reduce the coefficient $-98 \pmod 1$, and then we use (R2) to interchange the two rows:

$$\rightarrow \begin{array}{rrrr} 0 & 190 & 178 & 1872 \\ 1 & 0 & 0 & 19 \\ 1 & -2 & -1 & \\ 3 & -5 & -8 & \\ -5 & 9 & 11 & \end{array} \quad \rightarrow \begin{array}{rrrr} 1 & 0 & 0 & 19 \\ 0 & 190 & 178 & 1872 \\ 1 & -2 & -1 & \\ 3 & -5 & -8 & \\ -5 & 9 & 11 & \end{array}$$

We now ignore the first row and first column. Among the remaining coefficients, the one of least nonzero absolute value is 178. We use operation (C1) to reduce $190 \pmod{178}$, obtaining a remainder 12. Then we use (C1) to reduce $178 \pmod{12}$, obtaining a remainder -2:

$$\rightarrow \begin{array}{rrrr} 1 & 0 & 0 & 19 \\ 0 & 12 & 178 & 1872 \\ 1 & -1 & -1 & \\ 3 & 3 & -8 & \\ -5 & -2 & 11 & \end{array} \quad \rightarrow \begin{array}{rrrr} 1 & 0 & 0 & 19 \\ 0 & 12 & -2 & 1872 \\ 1 & -1 & 14 & \\ 3 & 3 & -53 & \\ -5 & -2 & 41 & \end{array}$$

Next we use (C2) to reduce $12 \pmod 2$. Then we use (C2) to interchange the second and third columns, and finally use (C3) to replace -2 by 2:

$$\rightarrow \begin{array}{rrrr} 1 & 0 & 0 & 19 \\ 0 & 0 & -2 & 1872 \\ 1 & 83 & 14 & \\ 3 & -315 & -53 & \\ -5 & 244 & 41 & \end{array} \quad \rightarrow \begin{array}{rrrr} 1 & 0 & 0 & 19 \\ 0 & 2 & 0 & 1872 \\ 1 & -14 & 83 & \\ 3 & 53 & -315 & \\ -5 & -41 & 244 & \end{array}$$

Let the variables in our new set of equations be called t, u, and v. The two original equations have been replaced by the two new equations $1 \cdot t = 19$ and $2 \cdot u = 1872$. This fixes the values of t and u. Since $1 \mid 19$ and $2 \mid 1872$, these values are integers: $t = 19$, $u = 936$. With these values for t and u, the bottom three rows above give the equations

$$\begin{aligned} x &= t - 14u + 83v = 83v - 13085, \\ y &= 3t + 53u - 315v = -315v + 49665, \\ z &= -5t - 41u + 244v = 244v - 38471. \end{aligned}$$

By making the further change of variable $w = v - 158$ we may adjust the constant terms, so that

$$\begin{aligned} x &= 83w + 29, \\ y &= -315w - 105, \\ z &= 244w + 81. \end{aligned}$$

As integral solutions of the given equations are in one-to-one correspondence with integral values of w, we have achieved our goal.

To demonstrate that this procedure will succeed in general, we describe the strategy more precisely. Suppose we wish to parameterize all integral solutions of a family of m linear equations in n variables,

$$\begin{aligned} a_{11}x_1 + a_{12}x_2 + \cdots + a_{1n}x_n &= b_1, \\ a_{21}x_1 + a_{22}x_2 + \cdots + a_{2n}x_n &= b_2, \\ \vdots \qquad \vdots \qquad\quad \vdots \qquad &\quad \vdots \\ a_{m1}x_1 + a_{m2}x_2 + \cdots + a_{mn}x_n &= b_m. \end{aligned} \tag{5.10}$$

We assume that the a_{ij} and the b_i are integers, with not all $a_{ij} = 0$. Our object is to find an equivalent family of m equations in n equivalent variables that is diagonal, in the sense that the new coefficients a_{ij} vanish whenever $i \neq j$. Let $A = [a_{ij}]$ be the $m \times n$ matrix of given coefficients, let $X = [x_j]$ denote the $n \times 1$ matrix (or column vector) of variables, and let $B = [b_i]$ be the $m \times 1$ matrix (or column vector) of given constant terms. Then the given equations may be expressed as the single matrix equation $AX = B$. If we let $V = [v_{ij}]$ be the $n \times n$ matrix that expresses our original variables in terms of our new variables $Y = [y_j]$, then $VY = X$. Initially, $V = I$, the identity matrix. We describe a reduction step that transforms A into a matrix $A' = [a'_{ij}]$ with the property that $a'_{11} \geqslant 0$, $a'_{1j} = 0$ for $j > 1$, and $a'_{i1} = 0$ for $i > 1$. By repeated use of this reduction step, A is eventually transformed into a diagonal matrix whose diagonal entries are non-negative. As we perform row and column operations on A, we obtain a sequence of coefficient matrices. Let μ denote the minimal absolute value of non-zero elements of the current coefficient matrix. Locating an element of absolute value μ, say in position (i_0, j_0), we use operation (C1) or operation (R1) to reduce the other coefficients in row i_0 or column j_0. This gives rise to a new coefficient matrix with a strictly smaller value of μ, unless all the other coefficients in row i_0 and column

j_0 are 0. Since μ can take on only positive integral values, this latter situation must eventually arise. Then we use operations (R2) and (C2) to move the coefficient from location (i_0, j_0) to $(1, 1)$. If the coefficient is negative, we use (C3) to reverse the sign. Whenever we apply a column operation to the coefficient matrix A, we also apply the same column operation to V, and whenever we apply a row operation to A, we apply the same row operation to B. The reduction procedure will terminate prematurely if in the submatrix that remains to be treated all elements are 0. Thus we obtain a diagonal matrix with positive entries in the first r rows, and 0's elsewhere. In developing standard linear algebra over $\mathbb{R}$ it is found that the rank of a matrix is invariant under row or column operations. Since the row and column operations we are using here are a proper subset of those used in linear algebra over $\mathbb{R}$, the rank is invariant in the present situation, as well. As the rank of a diagonal matrix is simply equal to the number of nonzero elements, we see that the number r is the rank of the matrix A given originally.

Caution At all stages of the reduction process, the column operations must involve only columns 1 through n. Similarly, the row operations must involve only rows 1 through m.

In summary, the change of variables $VY = X$ has the property that n-tuples X of integers are in one-to-one correspondence with n-tuples Y of integers. The m conditions (5.10) on the variables x_j are equivalent to the m conditions

$$d_j y_j = b_j' \qquad (1 \leqslant j \leqslant r), \tag{5.11}$$

$$b_j' = 0 \qquad (r < j \leqslant m). \tag{5.12}$$

Here the d_j are the diagonal entries of the new coefficient matrix, and the b_j' are the new constant terms. In order that integral solutions should exist, it is necessary and sufficient that (5.12) holds, and that

$$d_j | b_j' \qquad (1 \leqslant j \leqslant r). \tag{5.13}$$

If (5.12) holds but (5.13) fails for some $j \leqslant r$, then there exist rational solutions but no integral solution. If (5.12) fails for some $j > r$ then the original equations are inconsistent, and then (5.10) has no solution in real variables. If (5.11) and (5.12) hold and $r = n$, then the integral solution is unique (and indeed this is the unique real solution). If (5.12) and (5.13) hold but $r < n$ then there are infinitely many integral solutions, parameterized by the free integral variables $y_{r+1}, y_{r+2}, \cdots, y_n$.

As we experienced in Example 4, the coefficients encountered during the reduction process may be much larger than the coefficients originally given. (It is not known precisely how much larger, but it is believed that they may be *very much* larger. It is interesting to consider how the reduction process might be modified in order to minimize this phenomenon.) However, this problem does not arise when the method is applied to systems of simultaneous congruences (mod q) instead of simultaneous equations, for then coefficients may be reduced (mod q) during the reduction process. Here q may be any integer > 1, but it is imperative that each congruence involves the same modulus q.

Example 5 Find all solutions of the simultaneous congruences

$$\begin{aligned} 3x \quad\quad + 3z &\equiv 1 \pmod 5, \\ 4x - y + \quad z &\equiv 3 \pmod 5. \end{aligned}$$

Solution We construct an array of coefficients as before. Using operation (C1), we add the third column to both columns 1 and 2.

$$\begin{array}{rrrr} 3 & 0 & 3 & 1 \\ 4 & -1 & 1 & 3 \\ 1 & 0 & 0 & \\ 0 & 1 & 0 & \\ 0 & 0 & 1 & \end{array} \quad \rightarrow \quad \begin{array}{rrrr} 1 & 3 & 3 & 1 \\ 0 & 0 & 1 & 3 \\ 1 & 0 & 0 & \\ 0 & 1 & 0 & \\ 1 & 1 & 1 & \end{array}$$

Using (R1), we multiply the second row by 2 and add the result to the first row. Then we interchange the first and third columns and the first and second rows.

$$\rightarrow \begin{array}{rrrr} 1 & 3 & 0 & 2 \\ 0 & 0 & 1 & 3 \\ 1 & 0 & 0 & \\ 0 & 1 & 0 & \\ 1 & 1 & 1 & \end{array} \quad \rightarrow \quad \begin{array}{rrrr} 1 & 0 & 0 & 3 \\ 0 & 3 & 1 & 2 \\ 0 & 0 & 1 & \\ 0 & 1 & 0 & \\ 1 & 1 & 1 & \end{array}$$

Next we multiply the third column by 2 and add the result to the second column, and then interchange the second and third columns.

$$\rightarrow \begin{array}{rrrr} 1 & 0 & 0 & 3 \\ 0 & 0 & 1 & 2 \\ 0 & 2 & 1 & \\ 0 & 1 & 0 & \\ 1 & 3 & 1 & \end{array} \quad \rightarrow \quad \begin{array}{rrrr} 1 & 0 & 0 & 3 \\ 0 & 1 & 0 & 2 \\ 0 & 1 & 2 & \\ 0 & 0 & 1 & \\ 1 & 1 & 3 & \end{array}$$

Thus we arrive at a new system of congruences, in variables t, u, v, say. We see that $t \equiv 3 \pmod 5$, $u \equiv 2 \pmod 5$, while v can take any value (mod 5). Thus the given system has five solutions, given by

$$\begin{aligned} x &\equiv \quad u + 2v \equiv 2v + 2 \pmod 5, \\ y &\equiv \qquad\quad v \equiv v \qquad \pmod 5, \\ z &\equiv t + u + 3v \equiv 3v \qquad \pmod 5. \end{aligned}$$

In general, the system of simultaneous congruences

$$\begin{aligned} a_{11}x_1 + a_{12}x_2 + \cdots + a_{1n}x_n &\equiv b_1 \pmod q, \\ a_{21}x_1 + a_{22}x_2 + \cdots + a_{2n}x_n &\equiv b_2 \pmod q, \\ \vdots \qquad \vdots \qquad\qquad \vdots \quad &\quad \vdots \quad \vdots \\ a_{m1}x_1 + a_{m2}x_2 + \cdots + a_{mn}x_n &\equiv b_m \pmod q, \end{aligned} \tag{5.14}$$

has a solution (mod q) if and only if

$$\text{g.c.d.}(d_j, q) | b_j' \qquad (1 \leqslant j \leqslant r), \tag{5.15}$$

$$b_j' \equiv 0 \pmod q \qquad (r < j \leqslant m). \tag{5.16}$$

Note that these conditions may hold while (5.12) fails. In such a case the congruences (5.14) have a simultaneous solution even though the equations (5.10) have no real solution. On the other hand, if (5.10) has a real solution then (5.12) holds. If we take q to be a multiple of all of the d_j then the conditions (5.15) are equivalent to (5.13). This gives the following important result.

Theorem 5.2 *If the system of linear equations* (5.10) *has a real solution, and if the system of congruences* (5.14) *has a solution for every modulus* q, *then the equations* (5.10) *have an integral solution.*

We have actually proved more, since we can determine a particular q that suffices. (For a more precise characterization of this q in terms of the original coefficients, see Problem 11 at the end of this section.) The converse of the theorem is obvious, for if a system of equations (even nonlinear equations) has an integral solution then this solution is both a real solution and also a congruential solution for any q. We speak of the congruential and real solutions as "local," while an integral solution is "global." In this parlance, Theorem 5.2 may be expressed by saying that the equations (5.10) have a global solution if they are everywhere locally solvable.

While our main aims in this Section have been achieved, further insights may be gained by making greater use of linear algebra. Suppose that a particular row operation, applied to the $m \times n$ matrix A, gives the matrix A'. Let R denote the matrix obtained by applying this same row operation to the $m \times m$ identity matrix I_m. Then $A' = RA$. We call such a matrix R an *elementary row matrix*. Note that the elementary row matrices here form a proper subset of the elementary row matrices defined in standard linear algebra over $\mathbb{R}$, since we have restricted the row operations that are allowed. Similarly, if a particular column operation takes A to A'' and I_n to C, then $A'' = AC$, and we call C an *elementary column matrix*. Thus the sequence of row and column operations that we have performed in our reduction process may be expressed by matrix multiplication,

$$R_g R_{g-1} \cdots R_2 R_1 A C_1 C_2 \cdots C_{h-1} C_h = D, \tag{5.17}$$

where D is an $m \times n$ diagonal matrix. (Note that a diagonal matrix is not necessarily square.) The matrix V that allows us to express the original variables X in terms of our new variables Y is constructed by applying the same column operations to the identity matrix. That is,

$$V = C_1 C_2 \cdots C_{h-1} C_h. \tag{5.18}$$

Similarly, the new constant terms B' obtained at the end of the reduction process are created by applying the row operations to the original set B of constant terms, so that

$$B' = R_g R_{g-1} \cdots R_2 R_1 B. \tag{5.19}$$

It is useful to characterize those matrices that may be written as products of our elementary row or column matrices.

Definition 5.1 *A square matrix U with integral elements is called* unimodular *if* $\det(U) = \pm 1$.

Theorem 5.3 *Let U be an $m \times m$ matrix with integral elements. Then the following are equivalent:*

(i) U is unimodular;

(ii) The inverse matrix U^{-1} exists and has integral elements;

(iii) U may be expressed as a product of elementary row matrices.

$$U = R_g R_{g-1} \cdots R_2 R_1;$$

(iv) U may be expressed as a product of elementary column matrices,

$$U = C_1 C_2 \cdots C_{h-1} C_h.$$

If U and V are $m \times m$ unimodular matrices, then so also is UV, in view of (3.6). Moreover, U^{-1} is unimodular, by (ii) above. Thus the collection of all $m \times m$ unimodular matrices forms a group.

Proof We first show that (i) implies (ii). From the definition of the adjoint matrix U^{adj} it is evident that if U has integral elements then so does U^{adj}. Since $U^{-1} = U^{\text{adj}}/\det(U)$, it follows that U^{-1} has integral elements if $\det(U) = \pm 1$. We next show that (ii) implies (i). Since $UU^{-1} = I$, it follows by (3.6) that $\det(U)\det(U^{-1}) = \det(I) = 1$. But $\det(U)$ is an integer if U has integral elements, so from (ii) we deduce that both $\det(U)$ and $\det(U^{-1})$ are integers. That is, $\det(U)$ divides 1. As the only divisors of 1 are ± 1, it follows that U is unimodular. Next we show that (iii) implies (i). It is easy to verify that an elementary row matrix is unimodular. From (3.6) it is evident that the product of two unimodular matrices is again unimodular. Thus if $U = R_g R_{g-1} \cdots R_2 R_1$, then U is unimodular.

To show that (i) implies (iii), we first show that if A is an $m \times n$ matrix with integral elements then there exist elementary row matrices such that

$$A = R_1 R_2 \cdots R_{g-1} R_g T \tag{5.20}$$

where T is an upper-triangular $m \times n$ matrix with integral elements. We proceed as in Gaussian elimination in elementary linear algebra, except that we restrict ourselves to the row operations (R1), (R2), and (R3). We apply these row operations to A as follows. In the first column containing nonzero elements, say the first column, we apply the division algorithm and (R1) until only one element in this column is nonzero. By means of (R2) we may place this nonzero entry in the first row. By (R3) we may arrange that this element is positive. We now repeat this process on the columns to the right of the one just considered, but we ignore the first row. Thus the second column operated on may have two nonzero elements, in the first and second rows. Continuing in this manner, we arrive at an upper triangular matrix T. That is, $T = R_g R_{g-1} \cdots R_2 R_1 A$ for suitable elementary row matrices R_i. Hence $A = R_1^{-1} R_2^{-2} \cdots R_{g-1}^{-1} R_g^{-1} T$. Since the inverse of an elementary row matrix is again an elementary row matrix, we have now expressed A in the desired form (5.20).

To complete the proof that (i) implies (iii), we take $A = U$ in (5.20). Applying (3.6), we deduce that $\det(T) = \pm 1$. But since T is upper-triangular, $\det(T)$ is the product of its diagonal elements. As these diagonal elements are non-negative integers, it follows that each diagonal element is 1. With this established, we may now apply the row operation (R1) to T to clear all entries above the diagonal, leaving us with the identity matrix

I_m. That is, T is the product of elementary row matrices, and hence by (5.20), so also is U.

The equivalence of (i) and (iv) may be established similarly. Alternatively, we observe that R is an elementary row matrix if and only if R^t is an elementary column matrix. (Here R^t denotes the transpose of R.) If U is unimodular then U^t is unimodular, and by (iii) we deduce that $U^t = R_g R_{g-1} \cdots R_2 R_1$ for suitable elementary row matrices R_i. Hence $U = R_1^t R_2^t \cdots R_{g-1}^t R_g^t$, a product of column matrices.

We call two $m \times n$ matrices A and A' *equivalent*, and write $A \sim A'$, if there exists an $m \times m$ unimodular matrix U and an $n \times n$ unimodular matrix V such that $A' = UAV$. This is an equivalence relation in the usual sense. With this machinery in hand, we may express (5.17) more compactly by saying that any matrix A is equivalent to a diagonal matrix, say $UAV = D$. Then $A = U^{-1}DV^{-1}$. Writing (5.10) in the form $AX = B$, we deduce that $U^{-1}DV^{-1}X = B$. On putting $Y = V^{-1}X$, $UB = B'$, we are led immediately to the conclusion that (5.10) is equivalent to $DY = B'$, which is precisely the content of (5.11) and (5.12).

Owing to ambiguities in our reduction process, the diagonal matrix D that we have found to be equivalent to A is not uniquely defined. Moreover, two different diagonal matrices may be equivalent, as we see from the example

$$\begin{bmatrix} 1 & 1 \\ -3 & -2 \end{bmatrix} \begin{bmatrix} 2 & 0 \\ 0 & 3 \end{bmatrix} \begin{bmatrix} -1 & -3 \\ 1 & 2 \end{bmatrix} = \begin{bmatrix} 1 & 0 \\ 0 & 6 \end{bmatrix}.$$

However, it is known that among the diagonal matrices equivalent to a given matrix A there is a unique one whose nonzero elements $s_1, s_2, \cdots, s_r$ are positive and satisfy the divisibility relations $s_1 | s_2, s_2 | s_3, \cdots, s_{r-1} | s_r$. This diagonal matrix S is the *Smith normal form of* A, named for the nineteenth-century English mathematician H. J. S. Smith. The numbers s_i, $1 \leqslant i \leqslant r$, are called the *invariant factors* of A. A proof that every $m \times n$ matrix A is equivalent to a unique matrix S in Smith normal form is outlined in Problems 4–9.

PROBLEMS

1. Find all solutions in integers of the system of equations

$$\begin{aligned} x_1 + x_2 + 4x_3 + 2x_4 &= 5, \\ -3x_1 - x_2 \qquad\quad - 6x_4 &= 3, \\ -x_1 - x_2 + 2x_3 - 2x_4 &= 1. \end{aligned}$$

2. For what integers a, b, and c does the system of equations

$$\begin{aligned} x_1 + 2x_2 + 3x_3 + 4x_4 &= a, \\ x_1 + 4x_2 + 9x_3 + 16x_4 &= b, \\ x_1 + 8x_2 + 27x_3 + 64x_4 &= c \end{aligned}$$

have a solution in integers? What are the solutions if $a = b = c = 1$?

3. Suppose that the system of congruences (5.14) has a solution. Show that if q is prime then the number of solutions is a power of q.

***4.** Let a and b be positive integers, and put g = g.c.d. (a, b), h = l.c.m. (a, b). Show that $\begin{bmatrix} a & 0 \\ 0 & b \end{bmatrix} \sim \begin{bmatrix} g & 0 \\ 0 & h \end{bmatrix}$.

***5.** Using the preceding problem, or otherwise, show that if D is a diagonal matrix with integral elements then there is a diagonal matrix S in Smith normal form such that $D \sim S$. Deduce that every $m \times n$ matrix A with integral elements is equivalent to a matrix S in Smith normal form.

***6.** Let A be an $m \times n$ matrix with integral elements, and let r denote the rank of A. For $1 \leqslant k \leqslant r$, let $d_k(A)$ be the greatest common divisor of the determinants of all $k \times k$ minors of A. The numbers $d_k(A)$ are called the *determinantal divisors* of A. Let R be an elementary unimodular row matrix, and put $A' = RA$. Show that A and A' have the same determinantal divisors.

***7.** Use the preceding problem to show that if A and B are equivalent matrices then they have the same determinantal divisors.

***8.** Let S be a matrix in Smith normal form whose positive diagonal elements are $s_1, s_2, \cdots, s_r$. Show that $d_1(S) = s_1$, $d_2(S) = s_1 s_2, \cdots, d_r(S) = s_1 s_2 \cdots s_r$. For convenience, put $d_0(S) = 1$. Deduce that $s_k = d_k(S)/d_{k-1}(S)$ for $1 \leqslant k \leqslant r$.

***9.** Let S and S' be two $m \times n$ matrices in Smith normal form. Using the preceding problems, show that if $S \sim S'$ then $S = S'$. Conclude that the Smith normal form of an $m \times n$ matrix A is unique.

***10.** Show that if two $m \times n$ matrices A and A' have the same rank and the same determinantal divisors then $A \sim A'$.

***11.** Suppose that the system of equations (5.10) has real solutions, and that the system of congruences (5.14) has a solution when $q = d_r(A)/d_{r-1}(A)$. Show that the equations (5.10) have an integral solution. Show also that this is the least integer q for which this conclusion may be drawn.

***12.** Let A be an $n \times n$ matrix with integral elements and nonzero determinant. Then the elements of A^{-1} are rational numbers. Show that the least common denominator of these elements is $d_n(A)/d_{n-1}(A)$.

5.3 PYTHAGOREAN TRIANGLES

We wish to solve the equation $x^2 + y^2 = z^2$ in positive integers. The two most familiar solutions are 3, 4, 5 and 5, 12, 13. We refer to such a triple of positive integers as a *Pythagorean triple* or a *Pythagorean triangle*, since in geometric terms x and y are the legs of a right triangle with hypotenuse z. In view of the algebraic identity

$$(r^2 - s^2)^2 + (2rs)^2 = (r^2 + s^2)^2, \tag{5.21}$$

we may obtain an infinity of Pythagorean triangles by taking

$$\begin{aligned} x &= r^2 - s^2, \\ y &= 2rs, \\ z &= r^2 + s^2 \end{aligned} \tag{5.22}$$

where r and s take integral values with $r > s > 0$. More remarkably, we show that *all* Pythagorean triangles arise in this way.

Since the equation under consideration is homogeneous, if x, y, z is a Pythagorean triple then so also is kx, ky, kz, for any positive integer k. For example, the Pythagorean triangle 3, 4, 5 gives 6, 8, 10 and also 60, 80, 100. Thus any given Pythagorean triangle gives rise to an infinite family of similar triangles. To initiate our analysis, we identify in this family the smallest triangle. Suppose that x, y, and z are given positive integers for which $x^2 + y^2 = z^2$. Let d be a common divisor of x and y. Then $d^2 | x^2$ and $d^2 | y^2$, and hence $d^2 | (x^2 + y^2)$, that is, $d^2 | z^2$. By unique factorization, it follows that $d | z$. Indeed, by further arguments of this sort, we discover that any common factor of two of the numbers x, y, z must divide the third. That is,

$$(x, y) = (y, z) = (z, x) = (x, y, z).$$

Let g denote this common value, and put $x_1 = x/g$, $y_1 = y/g$, $z_1 = z/g$. Then x_1, y_1, z_1 is a Pythagorean triple with $(x_1, y_1) = 1$. We call such a triple *primitive*, since it is not a multiple of a smaller triple. Thus we see that all Pythagorean triangles similar to the given triangle x, y, z are multiples of x_1, y_1, z_1.

We now consider the problem of finding all primitive Pythagorean triples. We note that x and y cannot both be even. They cannot both be odd either, for if they were we would have $x^2 \equiv 1 \pmod 4$, $y^2 \equiv 1 \pmod 4$, and therefore $z^2 \equiv 2 \pmod 4$, which is impossible. Since x and y enter the equation symmetrically, we can now restrict our attention to primitive solutions for which y is even, x and z odd. The equation $x^2 + y^2 = z^2$, being additive, does not seem to offer a line of attack. However, the equation may be expressed in multiplicative form, $(z - x)(z + x) = y^2$. Since the canonical factorization of a perfect square is of a special shape (all the exponents are even), we are now in a position to say something intelligent concerning the prime factorization of $z - x$ and of $z + x$. The key idea here is very simple, but due to its enormous importance in Diophantine analysis, we give it special prominence in the following lemma.

Lemma 5.4 *If u and v are relatively prime positive integers whose product uv is a perfect square, then u and v are both perfect squares.*

Proof Let p be a prime that divides u, and let α be the exact power of p in u. (In symbols, $p^\alpha \| u$.) Since u and v are relatively prime, p does not divide v, and hence $p^\alpha \| uv$. But uv is a perfect square, so α must be even. Since this holds for all primes p dividing u, it follows that u is a perfect square. Similarly, v must be a perfect square.

If x, y, z is a primitive Pythagorean triple with x, z odd, and y even, then $z - x$ and $z + x$ are both even. Accordingly, we divide by 4 and write our equation as

$$\frac{z+x}{2}\,\frac{z-x}{2} = \left(\frac{y}{2}\right)^2.$$

Any common divisor of the two factors on the left divides both their sum, z, and their difference, x. Since $(x, z) = 1$, it follows that the two factors on the left have no common factor. Then by Lemma 5.4 we deduce that $(z + x)/2 = r^2$, $(z - x)/2 = s^2$ and $y/2 = rs$ for some positive integers r, s. We also see that $(r, s) = 1$, and that $r > s$. Also, since z is odd, r and s are of opposite parity (one is even, the other odd). On solving for x, y, and z in terms of r and s, we obtain the equations (5.22) already noted. Thus we have the following result.

Theorem 5.5 *The positive primitive solutions of $x^2 + y^2 = z^2$ with y even are $x = r^2 - s^2$, $y = 2rs$, $z = r^2 + s^2$, where r and s are arbitrary integers of opposite parity with $r > s > 0$ and $(r, s) = 1$.*

The method we have devised here provides a model for attacking many other Diophantine equations. In fact, the approach is so successful that one may go to great lengths in order to make it applicable. For example, the equation $x^2 + 2 = y^3$ does not factor in the field of rational numbers, but we observe that $x^2 + 2 = (x + \sqrt{-2})(x - \sqrt{-2})$. In Section 9.9 we use the arithmetic of the algebraic integers in the field $\mathbb{Q}(\sqrt{-2})$ to treat this equation. In Section 9.10 a similar method is applied to the equation $x^3 + y^3 = z^3$.

PROBLEMS

1. Find all primitive Pythagorean triples for which $0 < z < 30$.
2. Prove that if x, y, z is a Pythagorean triple then at least one of x, y is divisible by 3, and that at least one of x, y, z is divisible by 5.
3. Find all Pythagorean triples whose terms form (a) an arithmetic progression, (b) a geometric progression.
4. Let u and v be positive integers whose product uv is a perfect square, and let $g = (u, v)$. Show that there exist positive integers r, s such that $u = gr^2$ and $v = gs^2$.
5. Let u and v be relatively prime positive integers such that $2uv$ is a perfect square. Show that either (a) $u = 2r^2$, $v = s^2$ or (b) $u = r^2$, $v = 2s^2$, for suitable positive integers r, s.
6. Describe those relatively prime positive integers u and v such that $6uv$ is a perfect square.
7. For which integers n are there solutions to the equation $x^2 - y^2 = n$?
8. If n is any integer $\geqslant 3$, show that there is a Pythagorean triple with n as one of its members.
9. Prove that every integer n can be expressed in the form $n = x^2 + y^2 - z^2$.
10. Prove that $x^2 + y^2 = z^4$ has infinitely many solutions with $(x, y, z) = 1$.
11. Using Theorem 5.5, determine all solutions of the equation $x^2 + y^2 = 2z^2$. (H)
12. Show that if $x = \pm(r^2 - 5s^2)$, $y = 2rs$, $z = r^2 + 5s^2$ then $x^2 + 5y^2 = z^2$. This equation has the solution $x = 2$, $y = 3$, $z = 7$. Show that this solution is not given by any rational values of r, s.
13. Show that all solutions of $x^2 + 2y^2 = z^2$ in positive integers with $(x, y, z) = 1$ are given by $x = |r^2 - 2s^2|$, $y = 2rs$, $z = r^2 + 2s^2$ where r and s are arbitrary positive integers such that r is odd and $(r, s) = 1$. (H)
14. Let x, y, z be positive integers such that $(x, y) = 1$ and $x^2 + 5y^2 = z^2$. Show that if x is odd and y is even then there exist integers r and

s such that x, y, z are given by the equations of Problem 11. Show that if x is even and y is odd then there exist integers r and s such that $x = \pm(2r^2 + 2rs - 2s^2)$, $y = 2rs + s^2$, $z = 2r^2 + 2rs + 3s^2$. (H)

***15.** Prove that no Pythagorean triple of integers belongs to an isosceles right triangle, but that there are infinitely many primitive Pythagorean triples for which the acute angles of the corresponding triangles are, for any given positive ε, within ε of $\pi/4$.

***16.** Find, in the spirit of Theorem 5.5, all primitive triples x, y, z of positive integers such that a triangle with sides x, y, z has an angle of 60°.

***17.** Using the proof of Theorem 5.5 as a model, show that if x and y are integers for which $x^4 - 2y^2 = 1$, then $x = \pm 1$, $y = 0$.

5.4 ASSORTED EXAMPLES

In this section it is not our intent to develop a general theory. Instead, we consider a number of unrelated but instructive examples.

We begin with a very simple remark: If an equation has no solution in real variables, then it cannot have a solution in integers. Thus, for example, the equation $x^2 + y^2 = -1$ has no solution in integers. In most cases we would instantly notice if an equation had no solution in real variables, so this observation is not of much practical value. On the other hand, we may remark similarly that if an equation has a solution in integers, then it has a solution as a congruence (mod m) for every positive integer m. For example, the equation $x^2 + y^2 = 4z + 3$ has no solution in integers, because it has no solution as a congruence (mod 4).

Theorem 5.6 *The equation* $15x^2 - 7y^2 = 9$ *has no solution in integers.*

Proof Since the first and third members are divisible by 3, it follows that $3|7y^2$, and hence $3|y$. Thus the second and third members are divisible by 9, so that $9|15x^2$, and hence $3|x$. Put $x_1 = x/3$, $y_1 = y/3$, so that $15x_1^2 - 7y_1^2 = 1$. This has no solution as a congruence (mod 3).

Let $P(x_1, x_2, \cdots, x_n)$ be a homogeneous polynomial of degree d with integral coefficients. Then the Diophantine equation

$$P(x_1, x_2, \cdots, x_n) = 0 \tag{5.23}$$

has the *trivial solution* $x_1 = x_2 = \cdots = x_n = 0$. If $(x_1, x_2, \cdots, x_n)$ is a nontrivial solution, then we may set $g = \text{g.c.d.}\,(x_1, x_2, \cdots, x_n)$, and divide the equation by g^d to obtain a *primitive solution*, one for which the variables are relatively prime. In general we cannot guarantee that such variables will be pairwise relatively prime. If the homogeneous equation (5.23) has a nontrivial solution in integers, then it has a nontrivial real solution. Moreover, a primitive solution of (5.23) is, for any positive integer m, a solution of the congruence $P(x_1, x_2, \cdots, x_n) \equiv 0 \pmod{m}$ with the property that $\text{g.c.d.}\,(x_1, x_2, \cdots, x_n, m) = 1$. In view of the Chinese Remainder Theorem, it is enough to consider congruences (mod p^j). Thus as a prelude to solving a Diophantine equation, we first ask whether the congruence

$$P(x_1, x_2, \cdots, x_n) \equiv 0 \pmod{p^j} \tag{5.24}$$

has a solution for every prime-power p^j. If P is homogeneous we require that not all the variables be divisible by p.

Theorem 5.7 *The equation $x^3 + 2y^3 + 4z^3 = 9w^3$ has no nontrivial solution.*

Proof We show that the congruence $x^3 + 2y^3 + 4z^3 = 9w^3 \pmod{27}$ has no solution for which $\text{g.c.d.}\,(x, y, z, w, 3) = 1$. We note that for any integer a, $a^3 \equiv 0$ or $\pm 1 \pmod 9$. Thus $x^3 + 2y^3 + 4z^3 \equiv 0 \pmod 9$ implies that $x \equiv y \equiv z \equiv 0 \pmod 3$. But then $x^3 + 2y^3 + 4z^3 \equiv 0 \pmod{27}$, so that $3 \mid w^3$. Hence $3 \mid w$. This contradicts the assumption that $\text{g.c.d.}(x, y, z, w, 3) = 1$.

As we remarked in Section 5.2, we refer to real solutions and congruential solutions as *local*, while a solution in integers is called *global*. In Section 5.2 we established that a system of linear equations is globally solvable if it is everywhere locally solvable. From the work of Hasse and Minkowski it is also known that a single quadratic form, in any number of variables, has a nontrivial solution in integers provided that it has nontrivial solutions everywhere locally. An interesting special case of this is the subject of the next section. Unfortunately, this "Hasse-Minkowski principle" does not hold in general. A counterexample is provided in Problem 11 at the end of this section. A second counterexample, involving a quadratic polynomial in two variables, is indicated in Problem 13 of Section 7.8. In addition, it is known that the equation $3x^3 + 4y^3 + 5z^3 = 0$ has no nontrivial solution in integers, but that it has nontrivial solutions everywhere locally. We now consider a further equation, which can be shown to be solvable everywhere locally, which nevertheless can be treated by congruential considerations.

Theorem 5.8 *The equation $y^2 = x^3 + 7$ has no solution in integers.*

Proof If x is even then the equation is impossible as a congruence (mod 4). Thus in any solution, x must be odd, and hence y must be even. It then follows that $x \equiv 1 \pmod 4$. Since the left side of the equation is non-negative, we deduce that $x \geqslant -1$. We rewrite the equation in the form

$$y^2 + 1 = (x + 2)(x^2 - 2x + 4).$$

Here the left side is odd, and by Lemma 2.14 we know that every prime factor of the left side is $\equiv 1 \pmod 4$. Hence every positive divisor of the left side is $\equiv 1 \pmod 4$. On the other hand, the right side has the positive divisor $x + 2 \equiv 3 \pmod 4$. Thus these two expressions cannot be equal.

In the argument just completed, we discover an inconsistency (mod q) for some prime $q \equiv 3 \pmod 4$, which divides $x + 2$. This q is not fixed, but is instead a function of the hypothetical solution x, y.

Some Diophantine equations can be treated by considering the order of magnitude of the quantities rather than by congruences. Let $f(z)$ be an irreducible polynomial of degree $d > 2$, with integral coefficients, and put $P(x, y) = f(x/y)y^d$. For example, if $f(z) = z^3 - 2$, then $P(x, y) = x^3 - 2y^3$. The coefficients of $P(x, y)$ are the same as those of $f(z)$, but $P(x, y)$ is homogeneous. We note that

$$|cx^jy^k| = |c|\,|x|^j|y|^k \leqslant |c|(\max(|x|, |y|))^{j+k}.$$

By applying this to the various monomial terms of $P(x, y)$ we deduce that

$$|P(x, y)| \leqslant H(P)(\max(|x|, |y|))^d \tag{5.25}$$

where $H(P)$ is the sum of the absolute values of the coefficients of $P(x, y)$, called the *height* of P. For most points (x, y) in the plane, the right side is of the same order of magnitude as the left. However, if the ratio x/y is near a real root of $f(z)$, then $|P(x, y)|$ is smaller, sometimes much smaller. However, it is known that the left side cannot be too much smaller if x and y are integers.

More precisely, if $d > 2$ and $\varepsilon > 0$, then there exists a constant C (depending both on $P(x, y)$ and on ε) such that

$$|P(x, y)| > (\max(|x|, |y|))^{d-2-\varepsilon} \tag{5.26}$$

provided that $\max(|x|, |y|) \geqslant C$. Consequently, if $g(x, y)$ is a polynomial of degree $< d - 2$ then the Diophantine equation

$$P(x, y) = g(x, y)$$

has at most finitely many integral solutions, because the left side has much greater absolute value than the right side, whenever the lattice point (x, y) is far from the origin. In particular, for any given integer c the equation $x^3 - 2y^3 = c$ has at most finitely many integral solutions. The inequality (5.26) is quite deep, but we may nevertheless apply elementary inequalities to certain special types of Diophantine equations.

Theorem 5.9 *The Diophantine equation* $x^4 + x^3 + x^2 + x + 1 = y^2$ *has the integral solutions* $(-1, 1), (0, 1), (3, 11)$, *and no others.*

Proof Put $f(x) = 4x^4 + 4x^3 + 4x^2 + 4x + 4$. Since $f(x) = (2x^2 + x)^2 + 3(x + 2/3)^2 + 8/3$, it follows that $f(x) > (2x^2 + x)^2$ for all real x. On the other hand, $f(x) = (2x^2 + x + 1)^2 - (x + 1)(x - 3)$. Here the last term is positive except for those real numbers x in the interval $I = [-1, 3]$. That is, $f(x) < (2x^2 + x + 1)^2$ provided that $x \notin I$. Thus we see that if x is an integer, $x \notin I$, then $f(x)$ lies between two consecutive perfect squares, namely $(2x^2 + x)^2$ and $(2x^2 + x + 1)^2$. Hence $f(x)$ cannot be a perfect square, except possibly for those integers $x \in I$, which we examine individually.

Theorem 5.10 *The equation*

$$x^4 + y^4 = z^2 \tag{5.27}$$

has no solution in positive integers.

This is one of Fermat's most famous results. From it we see at once that the equation $x^4 + y^4 = z^4$ has no solution in positive integers. Fermat asserted, more generally, that if n is an integer, $n > 2$, then the equation

$$x^n + y^n = z^n \tag{5.28}$$

has no solution in positive integers. This proposition, though still a conjecture for many values of n, is known as *Fermat's last theorem* or *Fermat's big theorem*, as contrasted with Fermat's little theorem (Theorem 2.7). In Section 9.10 we treat the case $n = 3$ using simple ideas in algebraic number theory.

Proof The secret is that one should not consider the given equation in isolation, but rather in tandem with a second equation,

$$a^2 + 4b^4 = c^4. \tag{5.29}$$

We show that if the given equation (5.27) has a solution in positive integers, then so does this second equation, and conversely if (5.29) has a solution in positive integers then so does (5.27). On closer examination we discover that if we start with a solution of (5.27), use it to construct a solution of (5.29), and then use that solution to construct a solution of (5.27), then we do not obtain the original solution of (5.27) that we started with. Instead, the new solution is smaller, in the sense that z is smaller. This allows us to derive a contradiction, since we may assume that our initial solution is minimal. This is Fermat's method of descent.

Let x, y, z be arbitrary positive integers that satisfy (5.27). Set $g =$ g.c.d.(x, y). Since g^4 divides the left side of (5.27), it follows that $g^2|z$. Putting $x_1 = x/g$, $y_1 = y/g$, $z_1 = z/g^2$, we see that x_1, y_1, z_1 are positive integers that satisfy (5.27) and that have the further property that x_1 and y_1 are relatively prime. Thus x_1^2, y_1^2, z_1 is a primitive Pythagorean triple. By interchanging x_1 and y_1, if necessary, we may arrange that x_1 is odd and y_1 is even. Hence by Theorem 5.5 there exist relatively prime positive integers r, s such that

$$x_1^2 = r^2 - s^2, \tag{5.30}$$

$$y_1^2 = 2rs, \tag{5.31}$$

$$z_1 = r^2 + s^2. \tag{5.32}$$

Here r and s are of opposite parity, and to determine which one is odd, we observe from (5.30) that s, x_1, r is a primitive Pythagorean triple. Hence r is odd and s is even. In view of (5.31), we may apply Lemma 5.4 with $u = r$ and $v = 2s$. Thus there exist positive integers b and c such that $r = c^2$, $s = 2b^2$. Taking $a = x_1$, we see from (5.30) that a, b, c is a solution of (5.29) in positive integers. Moreover, using (5.32) we see that

$$c \leqslant c^4 = r^2 < r^2 + s^2 = z_1 \leqslant z. \tag{5.33}$$

Now suppose that a, b, c are positive integers that satisfy (5.29). Put $h =$ g.c.d.(b, c). Then $h^4|a^2$, and hence $h^2|a$. Putting $a_1 = a/h^2$, $b_1 = b/h$, $c_1 = c/h$, we find that a_1, b_1, c_1 are positive integers satisfying (5.29), which have the further property that b_1 and c_1 are relatively prime. Thus $a_1, 2b_1^2, c_1^2$ constitute a primitive Pythagorean triple. Hence by Theorem

5.5 there exist positive relatively prime integers r', s' of opposite parity such that

$$a_1 = r'^2 - s'^2, \tag{5.34}$$

$$b_1^2 = r's', \tag{5.35}$$

$$c_1^2 = r'^2 + s'^2. \tag{5.36}$$

Then by (5.35) and Lemma 5.4 we see that there exist positive integers x' and y' such that $r' = x'^2$, $s' = y'^2$. Setting $z' = c_1$, we conclude by (5.36) that x', y', z' is a solution of (5.27) in positive integers. Here $z' \leqslant c$, and hence by (5.33) we see that $z' < z$. Thus the set of values of z arising in solutions of (5.27) has no least element. Since every nonempty set of positive integers contains a least element, it follows that the set of such z is empty; that is, equation (5.27) has no solution in positive integers.

PROBLEMS

1. Show that the equation $x^2 + y^2 = 9z + 3$ has no integral solution.
2. Show that the equation $x^2 + 2y^2 = 8z + 5$ has no integral solution.
3. Show that the equation $(x^2 + y^2)^2 - 2(3x^2 - 5y^2)^2 = z^2$ has no integral solution. (H)
4. Show that if x, y, z are integers such that $x^2 + y^2 + z^2 = 2xyz$, then $x = y = z = 0$. (H)
5. Show that the equation $x^2 + y^2 = 3(u^2 + v^2)$ has no nontrivial integral solution.
6. Show that if $x^3 + 2y^3 + 4z^3 \equiv 6xyz \pmod 7$ then $x \equiv y \equiv z \equiv 0 \pmod 7$. Deduce that the equation $x^3 + 2y^3 + 4z^3 = 6xyz$ has no nontrivial integral solutions.
7. Let $f(x, y, z) = x^3 + 2y^3 + 4z^3 - 6xyz$. Show that the equation $f(r, s, t) + 7f(u, v, w) + 49f(x, y, z) = 0$ has no nontrivial integral solution.
8. Let $f(\mathbf{x}) = f(x_1, x_2, x_3) = x_1^4 + x_2^4 + x_3^4 - x_1^2x_2^2 - x_2^2x_3^2 - x_3^2x_1^2 - x_1x_2x_3(x_1 + x_2 + x_3)$. Show that $f(\mathbf{x}) \equiv 1 \pmod 4$ unless all three variables are even. Deduce that if $f(\mathbf{x}) + f(\mathbf{y}) + f(\mathbf{z}) = 4(f(\mathbf{u}) + f(\mathbf{v}) + f(\mathbf{w}))$ for integral values of the variables, then all 18 variables are 0.
9. Show that the equation $x^3 + 2y^3 = 7(u^3 + 2v^3)$ has no nontrivial integral solution.

10. Find all integral solutions of the equation $x^4 + 2x^3 + 2x^2 + 2x + 5 = y^2$.

11. Let $F(x) = (x^2 - 17)(x^2 - 19)(x^2 - 323)$. Show that for every integer m, the congruence $F(x) \equiv 0 \pmod{m}$ has a solution. Note that the equation $F(x) = 0$ has no integral solution, nor indeed any rational solution.

12. Show that the equation $x^2 = y^3 + 23$ has no solution in integers. (H)

13. Show that Fermat's equation (5.28) has no solution in positive integers x, y, z if n is a positive integer, $n \equiv 0 \pmod{4}$.

***14.** Construct a descent argument that relates the two equations $x^4 + 4y^4 = z^2$, $a^4 + b^2 = c^4$. Deduce that neither of these equations has a solution in positive integers.

15. Show that there exist no positive integers m and n such that $m^2 + n^2$ and $m^2 - n^2$ are both perfect squares.

***16.** Consider a right triangle the lengths of whose sides are integers. Prove that the area cannot be a perfect square.

17. The preceding problem was asked by Fermat in the following alternative form: If the lengths of the sides of a right triangle are rational numbers, then the area of the triangle cannot be the square of a rational number. Derive this from the former version.

5.5 TERNARY QUADRATIC FORMS

The general ternary quadratic form is a polynomial $f(x, y, z)$ of the sort

$$f(x, y, z) = ax^2 + by^2 + cz^2 + dxy + eyz + fzx.$$

In this section we develop a procedure for determining whether the Diophantine equation $f(x, y, z) = 0$ has a nontrivial solution. In the next section we show how all solutions of this equation may be found, once a single solution has been identified.

A triple (x, y, z) of numbers for which $f(x, y, z) = 0$ is called a *zero* of the form. The solution $(0, 0, 0)$ is the *trivial zero*. If we have a solution in rational numbers, not all zero, then we can construct a primitive solution in integers by multiplying each coordinate by the least common denominator of the three. For example, $(3/5, 4/5, 1)$ is a zero of the form $x^2 + y^2 - z^2$, and hence $(3, 4, 5)$ is a primitive integral solution. Suppose now that $A = [a_{ij}]$ is a 3×3 matrix with rational elements, and put

$$g(x, y, z) = f(a_{11}x + a_{12}y + a_{13}z, a_{21}x + a_{22}y + a_{23}z, a_{31}x + a_{32}y + a_{33}z).$$

Here $g(x, y, z)$ is another ternary quadratic form, whose coefficients are determined in terms of the a_{ij} and the coefficients of $f(x, y, z)$. We assume that the coefficients of $f(x, y, z)$ are rational, so it follows that the coefficients of $g(x, y, z)$ are also rational. If the triple (x_0, y_0, z_0) is a nontrivial rational zero of g, then on inserting these values in the equation we obtain a rational zero of f. To ensure that this zero is nontrivial, we suppose that A is nonsingular. That is, $\det(A) \neq 0$. Let $B = [b_{ij}]$ denote the inverse of A, $B = A^{-1}$, so that

$$f(x, y, z) = g(b_{11}x + b_{12}y + b_{13}z, b_{21}x + b_{22}y + b_{23}z, b_{31}x + b_{32}y + b_{33}z).$$

Thus for our present purposes we may regard g as equivalent to f, since g has a nontrivial rational zero if and only if f does. This is an equivalence relation in the usual sense. The linear transformation

$$\begin{aligned} x' &= a_{11}x + a_{12}y + a_{13}z, \\ y' &= a_{21}x + a_{22}y + a_{23}z, \\ z' &= a_{31}x + a_{32}y + a_{33}z \end{aligned} \tag{5.37}$$

maps $\mathbb{R}^3$ to $\mathbb{R}^3$ in a one-to-one and onto manner, with the origin $(0, 0, 0)$ mapped to itself. Since the elements of A and A^{-1} are rational, this transformation takes $\mathbb{Q}^3$ to $\mathbb{Q}^3$ in the same way. Moreover, if two points have proportional coordinates, say (x_0, y_0, z_0) and $(\alpha x_0, \alpha y_0, \alpha z_0)$, with $\alpha \neq 0$, then their images (x_0', y_0', z_0') and $(\alpha x_0', \alpha y_0', \alpha z_0')$ are proportional.

By making linear changes of variables, we may pass from the given quadratic form f to a new form g of simpler appearance. For example, by completing the square, we may eliminate the coefficients of xy, of yz, and of zx, so that our form is diagonal. By multiplying by a nonzero rational number, if necessary, we may assume that our new coefficients of x^2, y^2, and z^2 are integers with no common factor. That is, through a sequence of changes of variables we reach an equation of the form

$$ax^2 + by^2 + cz^2 = 0 \tag{5.38}$$

with g.c.d.$(a, b, c) = 1$. If $a = 0$ then this equation has nontrivial solutions if and only if $-b/c$ is the square of a rational number. Thus we may confine our attention to the case in which a, b, and c are nonzero. We may write $a = a's^2$ with a' square-free, and thus $ax^2 = a'(sx)^2 = a'x'^2$, say. By making transformations of this kind, we may assume that a, b, and c are square-free. Suppose that $p|a$ and $p|b$. Since a, b, c are relatively prime, it follows that $p \nmid c$, and hence that $p|z$. Writing $z = pz'$, we discover that $ax^2 + by^2 + cz^2 = p((a/p)x^2 + (b/p)y^2 + pcz'^2)$. Here we have passed from a set of coefficients a, b, c of which two are divisible by p, to a new set of coefficients, $a/p, b/p, cp$, only one of which is divisible by p. By making further transformations of this sort, we eventually obtain nonzero square-free coefficients that are pairwise relatively prime. That is, abc is a square-free integer. This situation is very elegantly addressed by the following fundamental theorem of Legendre.

Theorem 5.11 *Let a, b, c be nonzero integers such that the product abc is square-free. Necessary and sufficient conditions that $ax^2 + by^2 + cz^2 = 0$ have a solution in integers x, y, z, not all zero, are that a, b, c do not have the same sign, and that $-bc$, $-ac$, $-ab$ are quadratic residues modulo a, b, c, respectively.*

Before giving the proof of this result we establish two lemmas.

Lemma 5.12 *Let λ, μ, ν be positive real numbers with product $\lambda\mu\nu = m$ an integer. Then any congruence $\alpha x + \beta y + \gamma z \equiv 0 \pmod{m}$ has a solution x, y, z, not all zero, such that $|x| \leqslant \lambda$, $|y| \leqslant \mu$, $|z| \leqslant \nu$.*

Proof Let x range over the values $0, 1, \cdots, [\lambda]$, y over $0, 1, \cdots, [\mu]$, and z over $0, 1, \cdots, [\nu]$. This gives us $(1 + [\lambda])(1 + [\mu])(1 + [\nu])$ different triples x, y, z. Now $(1 + [\lambda])(1 + [\mu])(1 + [\nu]) > \lambda\mu\nu = m$ by Theorem 4.1, part 1, and hence there must be some two triples x_1, y_1, z_1 and x_2, y_2, z_2 such that $\alpha x_1 + \beta y_1 + \gamma z_1 \equiv \alpha x_2 + \beta y_2 + \gamma z_2 \pmod{m}$. Then we have $\alpha(x_1 - x_2) + \beta(y_1 - y_2) + \gamma(z_1 - z_2) \equiv 0 \pmod{m}$, $|x_1 - x_2| \leqslant [\lambda] \leqslant \lambda$, $|y_1 - y_2| \leqslant \mu$, $|z_1 - z_2| \leqslant \nu$.

Lemma 5.13 *Suppose that $ax^2 + by^2 + cz^2$ factors into linear factors modulo m and also modulo n; that is*

$$ax^2 + by^2 + cz^2 \equiv (\alpha_1 x + \beta_1 y + \gamma_1 z)(\alpha_2 x + \beta_2 y + \gamma_2 z) \pmod{m}$$

$$ax^2 + by^2 + cz^2 \equiv (\alpha_3 x + \beta_3 y + \gamma_3 z)(\alpha_4 x + \beta_4 y + \gamma_4 z) \pmod{n}.$$

If $(m, n) = 1$ then $ax^2 + by^2 + cz^2$ factors into linear factors modulo mn.

Proof Using Theorem 2.18, we can choose $\alpha, \beta, \gamma, \alpha', \beta', \gamma'$ to satisfy

$$\alpha \equiv \alpha_1, \beta \equiv \beta_1, \gamma \equiv \gamma_1, \alpha' \equiv \alpha_2, \beta' \equiv \beta_2, \gamma' \equiv \gamma_2 \pmod{m}$$

$$\alpha \equiv \alpha_3, \beta \equiv \beta_3, \gamma \equiv \gamma_3, \alpha' \equiv \alpha_4, \beta' \equiv \beta_4, \gamma' \equiv \gamma_4 \pmod{n}.$$

Then the congruence

$$ax^2 + by^2 + cz^2 \equiv (\alpha x + \beta y + \gamma z)(\alpha' x + \beta' y + \gamma' z)$$

holds modulo m and modulo n, and hence it holds modulo mn.

Proof of Theorem 5.11 If $ax^2 + by^2 + cz^2 = 0$ has a solution x_0, y_0, z_0 not all zero, then a, b, c are not of the same sign. Dividing x_0, y_0, z_0 by (x_0, y_0, z_0) we have a solution x_1, y_1, z_1 with $(x_1, y_1, z_1) = 1$.

Next we prove that $(c, x_1) = 1$. If this were not so there would be a prime p dividing both c and x_1. Then $p \nmid b$ since $p | c$ and abc is square-free. Therefore $p | by_1^2$ and $p \nmid b$, hence $p | y_1^2, p | y_1$, and then $p^2 | (ax_1^2 + by_1^2)$ so that $p^2 | cz_1^2$. But c is square-free so $p | z_1$. We have concluded that p is a factor of x_1, y_1, and z_1 contrary to $(x_1, y_1, z_1) = 1$. Consequently, we have $(c, x_1) = 1$.

Let u be chosen to satisfy $ux_1 \equiv 1 \pmod{c}$. Then the equation $ax_1^2 + by_1^2 + cz_1^2 = 0$ implies $ax_1^2 + by_1^2 \equiv 0 \pmod{c}$, and multiplying this by $u^2 b$ we get $u^2 b^2 y_1^2 \equiv -ab \pmod{c}$. Thus we have established that $-ab$ is a quadratic residue modulo c. A similar proof shows that $-bc$ and $-ac$ are quadratic residues modulo a and b respectively.

Conversely, let us assume that $-bc, -ac, -ab$ are quadratic residues modulo a, b, c respectively. Note that this property does not change if a, b, c are replaced by their negatives. Since a, b, c are not of the same sign, we can change the signs of all of them, if necessary, in order to have one positive and two of them negative. Then, perhaps with a change of notation, we can arrange it so that a is positive and b and c are negative.

Define r as a solution of $r^2 \equiv -ab \pmod{c}$, and a_1 as a solution of $aa_1 \equiv 1 \pmod{c}$. These solutions exist because of our assumptions on a, b, c. Then we can write

$$ax^2 + by^2 \equiv aa_1(ax^2 + by^2) \equiv a_1(a^2x^2 + aby^2) \equiv a_1(a^2x^2 - r^2y^2)$$

$$\equiv a_1(ax - ry)(ax + ry) \equiv (x - a_1 ry)(ax + ry) \pmod{c},$$

$$ax^2 + by^2 + cz^2 \equiv (x - a_1 ry)(ax + ry) \pmod{c}.$$

Thus $ax^2 + by^2 + cz^2$ is the product of two linear factors modulo c, and similarly modulo a and modulo b. Applying Lemma 5.13 twice, we conclude that $ax^2 + by^2 + cz^2$ can be written as the product of two linear factors modulo abc. That is, there exist numbers $\alpha, \beta, \gamma, \alpha', \beta', \gamma'$ such that

$$ax^2 + by^2 + cz^2 \equiv (\alpha x + \beta y + \gamma z)(\alpha' x + \beta' y + \gamma' z) \pmod{abc}. \tag{5.39}$$

Now we apply Lemma 5.12 to the congruence

$$\alpha x + \beta y + \gamma z \equiv 0 \pmod{abc} \tag{5.40}$$

using $\lambda = \sqrt{bc}$, $\mu = \sqrt{|ac|}$, $\nu = \sqrt{|ab|}$. Thus we get a solution x_1, y_1, z_1 of the congruence (5.40) with $|x_1| \leqslant \sqrt{bc}$, $|y_1| \leqslant \sqrt{|ac|}$, $|z_1| \leqslant \sqrt{|ab|}$. But abc is square-free, so $\sqrt{bc}$ is an integer only if it is 1, and similarly for $\sqrt{|ac|}$ and $\sqrt{|ab|}$. Therefore we have

$|x_1| \leqslant \sqrt{bc}$, $\quad x_1^2 \leqslant bc$ with equality possible only if $b = c = -1$

$|y_1| \leqslant \sqrt{|ac|}$, $\quad y_1^2 \leqslant -ac$ with equality possible only if $a = 1$, $c = -1$

$|z_1| \leqslant \sqrt{|ab|}$, $\quad z_1^2 \leqslant -ab$ with equality possible only if $a = 1$, $b = -1$.

Hence, since a is positive and b and c are negative, we have, unless $b = c = -1$,

$$ax_1^2 + by_1^2 + cz_1^2 \leqslant ax_1^2 < abc$$

and

$$ax_1^2 + by_1^2 + cz_1^2 \geqslant by_1^2 + cz_1^2 > b(-ac) + c(-ab) = -2abc.$$

Leaving aside the special case when $b = c = -1$, we have

$$-2abc < ax_1^2 + by_1^2 + cz_1^2 < abc.$$

Now x_1, y_1, z_1 is a solution of (5.40) and so also, because of (5.39), a solution of

$$ax^2 + by^2 + cz^2 \equiv 0 \pmod{abc}.$$

Thus the above inequalities imply that

$$ax_1^2 + by_1^2 + cz_1^2 = 0 \qquad \text{or} \qquad ax_1^2 + by_1^2 + cz_1^2 = -abc.$$

In the first case we have our solution of $ax^2 + by^2 + cz^2 = 0$. In the second case we readily verify that x_2, y_2, z_2, defined by $x_2 = -by_1 + x_1z_1$, $y_2 = ax_1 + y_1z_1$, $z_2 = z_1^2 + ab$, form a solution. In case $x_2 = y_2 = z_2 = 0$ then $z_1^2 + ab = 0$, $z_1^2 = -ab$ and $z_1 = \pm 1$ because ab, like abc, is square-free. Then $a = 1$, $b = -1$, and $x = 1$, $y = 1$. $z = 0$ is a solution.

Finally we must dispose of the special case $b = c = -1$. The conditions on a, b, c now imply that -1 is a quadratic residue modulo a; in other words, that $N(a)$ of Theorem 3.21 is positive. By Theorem 3.21 this implies that $r(a)$ is positive and hence that the equation $y^2 + z^2 = a$ has a solution y_1, z_1. Then $x = 1$, $y = y_1$, $z = z_1$ is a solution of $ax^2 + by^2 + cz^2 = 0$ since $b = c = -1$.

Example 6 Determine whether the Diophantine equation

$$x^2 + 3y^2 + 5z^2 + 7xy + 9yz + 11zx = 0$$

has a nontrivial integral solution.

Solution The given form is

$$\left(x + \frac{7}{2}y + \frac{11}{2}z\right)^2 - \frac{37}{4}y^2 - \frac{101}{4}z^2 - \frac{41}{4}yz = g(x', y', z')$$

where $x' = x + \frac{7}{2}y + \frac{11}{2}z$, $y' = y$, $z' = z$. Thus

$$\begin{aligned} 4g(x, y, z) &= 4x^2 - 37y^2 - 101z^2 - 41yz \\ &= 4x^2 - 37\left(y + \frac{41}{74}z\right)^2 - \frac{13267}{148}z^2 = h(x'', y'', z''). \end{aligned}$$

Here

$$\begin{aligned} 148h(x, y, z) &= 592x^2 - 5476y^2 - 13267z^2 \\ &= 37(4x)^2 - (74y)^2 - 13267z^2, \end{aligned}$$

so we apply Theorem 5.11 to the form $37x^2 - y^2 - 13267z^2$. As 37 and 13267 are prime, and $\left(\frac{37}{13267}\right) = \left(\frac{-13267}{37}\right) = \left(\frac{490879}{1}\right) = 1$, we conclude that the given equation has a nontrivial solution.

Example 7 Determine whether the Diophantine equation $x^2 - 5y^2 - 91z^2 = 0$ has a nontrivial integral solution.

Solution We apply Theorem 5.11. The coefficients are not all of one sign, they are square-free, and $\left(\frac{-455}{1}\right) = \left(\frac{91}{5}\right) = \left(\frac{5}{91}\right) = 1$. However, 91 is not prime, for $91 = 7 \cdot 13$. In order that 5 be a square (mod 91), it is necessary (and sufficient) that 5 be a square (mod 7) and (mod 13). As $\left(\frac{5}{7}\right) = \left(\frac{5}{13}\right) = -1$, we deduce that the proposed equation has no nontrivial solution. Indeed, the equation has no nontrivial solution as a congruence (mod 7) and (mod 91).

It remains to reconcile Theorem 5.11 with our remarks of the preceding section, as it is not obvious that the conditions given for the existence of a nontrivial integral solution guarantee the solvability everywhere locally. In this direction, we note first that if $p|a$ then the congruence

$$ax^2 + by^2 + cz^2 \equiv 0 \pmod{p} \tag{5.41}$$

has the nontrivial solution $x = 1$, $y = z = 0$. However, such a solution does not give rise to a solution of the congruence

$$ax^2 + by^2 + cz^2 \equiv 0 \pmod{p^2},$$

for if $p|y$ and $p|z$, then the above implies that $p^2|ax^2$. But $p^2 \nmid a$, so it follows that $p|x^2$, and hence $p|x$, contrary to the supposition that g.c.d.$(x, y, z, p) = 1$. The hypothesis that $-bc$ be a square modulo a ensures that the congruence (5.41) has a solution for every $p|a$, with the further property that $p \nmid y$, $p \nmid z$. By Hensel's lemma, the congruence is then solvable modulo higher powers of such primes, provided that p is odd. Similar remarks apply to the odd prime divisors of b, and of c. Notably absent from the statement of Theorem 5.11 is any condition modulo primes p not dividing abc. The proof of the theorem establishes, indirectly, that no such condition is needed, but we now demonstrate this more explicitly.

Theorem 5.14 *Let a, b, and c be arbitrary integers. Then the congruence* (5.41) *has a nontrivial solution* (mod p).

This simple result will be useful in Section 6.4, in the proof of Lagrange's theorem concerning representations of n as a sum of four squares.

Proof If p divides one of the coefficients, say $p|a$, then it suffices to take $x = 1$, $y = z = 0$. Suppose that $p \nmid abc$. If $p = 2$ then it suffices to take $x = y = 1$, $z = 0$, so we may suppose that $p > 2$. We put $x = 1$. Let $\mathscr{S} = \{a + by^2\colon\ y = 0, 1, \cdots, (p-1)/2\}$, and let $\mathscr{T} = \{-cz^2\colon\ z = 0, 1, \cdots, (p-1)/2\}$. If $a + by_1^2 \equiv a + by_2^2 \pmod p$, then by Lemma 2.10 we see that $y_1 \equiv \pm y_2 \pmod p$. If such y_1 and y_2 are members of the set $0, 1, \cdots, (p-1)/2$, then it follows that $y_1 = y_2$. Hence the $(p+1)/2$ members of $\mathscr{S}$ lie in distinct residue classes (mod p). Similarly, the $(p+1)/2$ members of $\mathscr{T}$ lie in distinct residue classes (mod p). Since the total number of residue classes is larger than p, by the pigeonhole principle it follows that there is a member of $\mathscr{S}$ that is congruent to a member of $\mathscr{T}$. That is, $a + by^2 + cz^2 \equiv 0 \pmod p$ for some choice of y and z.

If p is odd, $p \nmid abc$, then a nontrivial solution of (5.41) lifts to higher powers of p, by Hensel's lemma. Powers of 2 are another matter. Suppose first that abc is odd. The congruence $ax^2 + by^2 + cz^2 \equiv 0 \pmod 4$ has no solution with g.c.d.$(x, y, z, 2) = 1$ if $a \equiv b \equiv c \pmod 4$. In Theorem 5.11 we find conditions under which the equation has a nontrivial integral solution. These conditions therefore imply that the coefficients do not all lie in the same residue class (mod 4). To demonstrate this more directly, one may note that the hypotheses imply that $\left(\frac{-bc}{p}\right) = 1$ for all prime divisors p of a, and similarly for the prime divisors of b and of c. On multiplying these relations together, we deduce that

$$\prod_{p|a}\left(\frac{-bc}{p}\right)\prod_{p|b}\left(\frac{-ca}{p}\right)\prod_{p|c}\left(\frac{-ab}{p}\right) = 1.$$

By multiplying all three coefficients by -1, and/or permuting them, we may suppose that $a > 0$, $b > 0$, and $c < 0$. Then by quadratic reciprocity the above equation reduces to

$$(-1)^{\frac{b-1}{2}\cdot\frac{c+1}{2} + \frac{c+1}{2}\cdot\frac{a-1}{2} + \frac{a-1}{2}\cdot\frac{b-1}{2} + \frac{c+1}{2}} = 1.$$

It may be seen that this amounts to the assertion that not all of a, b, c lie in the same residue class (mod 4). Once one has a nontrivial solution (mod 4), the solution may be lifted to higher powers of 2. The case in which one of the coefficients is even is a little more complicated, as there are several cases to consider. We omit the details, but remark that the conditions stated in Theorem 5.11 may be shown in a similar manner to imply that the sum of the two odd coefficients is not $\equiv 4 \pmod 8$. This is precisely the condition needed to ensure the existence of a nontrivial solution of the congruence (mod 8), and then such a solution may be lifted to higher powers of 2.

PROBLEMS

1. Use Theorem 5.11 to show that the equation $2x^2 + 5y^2 - 7z^2 = 0$ has a nontrivial integral solution.

2. What is the least positive square-free integer c such that $(c, 105) = 1$, and such that the equation $-7x^2 + 15y^2 + cz^2 = 0$ has a nontrivial integral solution?

3. Determine whether the equation

$$3x^2 + 5y^2 + 7z^2 + 9xy + 11yz + 13zx = 0$$

has a nontrivial integral solution.

4. Determine whether the equation

$$5x^2 + 7y^2 + 9z^2 + 11xy + 13yz + 15zx = 0$$

has a nontrivial integral solution.

5. Determine whether the equation

$$x^2 + 3y^2 + 5z^2 + 2xy + 4yz + 6zx = 0$$

has a nontrivial integral solution.

6. Show that in the proof of Theorem 5.11 we have established more than the theorem stated, that the following stronger result is implied. Let a, b, c be nonzero integers not of the same sign such that the product abc is square-free. Then the following three conditions are equivalent.

(a) $ax^2 + by^2 + cz^2 = 0$ has a solution x, y, z not all zero;

(b) $ax^2 + by^2 + cz^2$ factors into linear factors modulo abc;

(c) $-bc, -ac, -ab$ are quadratic residues modulo a, b, c, respectively.

7. Suppose that a, b, and c are given integers, and let $N(p)$ denote the number of solutions of the congruence (5.41), including the trivial solution. Show that if p divides all the coefficients then $N(p) = p^3$, that if it divides exactly two of the coefficients then $N(p) = p^2$, and that if it divides exactly one of the coefficients then either $N(p) = p$ or $N(p) = 2p^2 - p$, except that $N(2) = 4$.

***8.** Suppose that p divides none of the numbers a, b, c, and let $N(p)$ be defined as in the preceding problem. Show that $N(p) = p^2$. (H)

9. In diagonalizing a quadratic form by repeatedly completing the square, we encounter a problem if $a = b = c = 0$. Show that a quadratic form of the shape $dxy + eyz + fzx$ always takes the value 0 nontrivially. Explain what happens if you put $x = u + v$, $y = u - v$. Similarly, show that any form of the shape $ax^2 + eyz$ takes the value 0 nontrivially.

***10.** Let $Q(x, y, z) = ax^2 + by^2 + cz^2$ where a, b, and c are nonzero integers. Suppose that the Diophantine equation $Q(x, y, z) = 0$ has a nontrivial integral solution. Show that for any rational number g, there exist rational numbers x, y, z such that $Q(x, y, z) = g$.

5.6 RATIONAL POINTS ON CURVES

Let $f(x, y)$ be a polynomial in two variables. The set of points (x, y) in the plane for which $f(x, y) = 0$ constitutes an algebraic curve, which we denote by $\mathscr{C}_f$, or more precisely by $\mathscr{C}_f(\mathbb{R})$, since we are allowing x and y to take real values. A point (x, y) is called a *rational point* if its coordinates are rational numbers. In this section we address the problem of finding the rational points on the curve, that is, the points $\mathscr{C}_f(\mathbb{Q})$. We note that $\mathscr{C}_f(\mathbb{Q}) \subseteq \mathscr{C}_f(\mathbb{R})$. The curve $\mathscr{C}_f(\mathbb{R})$ may be empty, as in the case $f(x, y) = x^2 + y^2 + 1$. Even if the curve $\mathscr{C}_f(\mathbb{R})$ is nonempty, it may contain no rational point. For example, if $f(x, y) = x^2 + y^2 - 3$, then $\mathscr{C}_f(\mathbb{R})$ is the circle of radius $\sqrt{3}$ centered at the origin. The existence of a rational point on this curve is equivalent to the existence of integers X, Y, Z, not all 0, such that $X^2 + Y^2 = 3Z^2$. Since this equation has no nontrivial solution as a congruence (mod 3), we see at once that $\mathscr{C}_f(\mathbb{Q})$ is empty.

The curves we consider all lie in the Euclidean plane $\mathbb{R}^2$ and are consequently called *planar*. The *degree* of the curve $\mathscr{C}_f$ is simply the degree of the polynomial f. If f is of degree 1 we call $\mathscr{C}_f$ a *line*, if f is of degree 2 we call $\mathscr{C}_f$ a *conic* or *quadratic*. A curve of degree three is *cubic*, of degree four *quartic*, and so on. A conic may be an ellipse, parabola, or hyperbola, but as defined here a conic may also be empty ($f(x, y) = x^2 + y^2 + 1$), consist of a single point ($f(x, y) = x^2 + y^2$), two lines ($f(x, y) = (x + y + 1)(2x - y + 3)$) or a double line ($f(x, y) = (x + 5y - 2)^2$).

By considering the intersections of a line with the given curve $\mathscr{C}_f$, we may hope to generate new rational points on $\mathscr{C}_f$ from those already known. This elementary approach succeeds brilliantly when $\mathscr{C}_f$ is a conic,

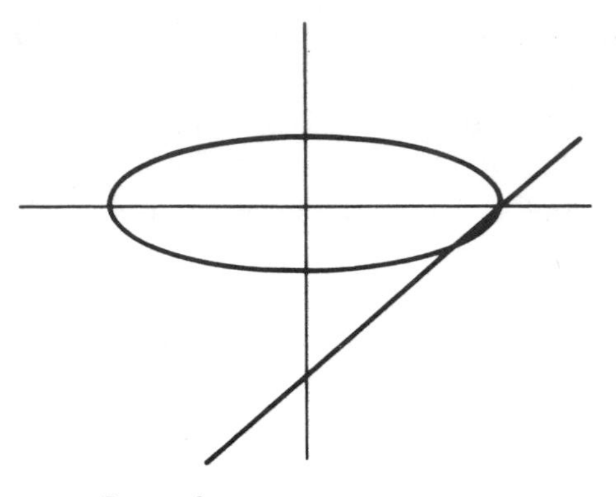

Figure 5.1. The ellipse $x^2 + 5y^2 = 1$. By taking the line through $(1, 0)$ with slope 1, we obtain the second rational point $(2/3, -1/3)$.

and we enjoy some limited success with cubic curves, but curves of degree 4 or larger generally do not surrender to such a simple attack. Before establishing general results, we demonstrate how the method works in practice.

Example 8 Find all rational points on the ellipse $x^2 + 5y^2 = 1$.

Solution We observe that the point $(1, 0)$ is a rational point on this curve. If (x_1, y_1) is a second rational point on this curve, then the slope m of the line joining these two points is a rational number, for $m = y_1/(x_1 - 1)$. Conversely, suppose that m is a rational number, as in Figure 5.1. The line through $(1, 0)$ with slope m has the equation $y = m(x - 1)$. To find the other intersection of this line with the ellipse, we replace y by $m(x - 1)$ in the formula for the ellipse. This gives us a quadratic in x, with one known root, $x = 1$, so we may factor the quadratic to find the other root.

$$
\begin{aligned}
0 = x^2 + 5y^2 - 1 &= x^2 + 5\big(m(x-1)\big)^2 - 1 \\
&= (5m^2 + 1)x^2 - 10m^2x + (5m^2 - 1) \\
&= (x - 1)\big((5m^2 + 1)x - (5m^2 - 1)\big).
\end{aligned}
$$

Thus the second intersection of the line with the ellipse is at a point whose x-coordinate is $x_1 = (5m^2 - 1)/(5m^2 + 1)$. To find the y-coordinate of this point, we use the equation of the line $y_1 = m(x_1 - 1)$. After simplifying, we find that $y_1 = -2m/(5m^2 + 1)$. Since m is assumed to be a rational number, it follows that both x_1 and y_1 are rational. Hence we see that the equations

$$m = y_1/(x_1 - 1), \qquad \begin{aligned} x_1 &= (5m^2 - 1)/(5m^2 + 1), \\ y_1 &= -2m/(5m^2 + 1) \end{aligned} \tag{5.42}$$

determine a one-to-one correspondence between rational numbers m and rational points (x_1, y_1) on the ellipse $x^2 + 5y^2 = 1$, apart from the point $(1, 0)$ that we started with.

The rational number m may be expressed as the quotient of two integers, $m = r/s$, and hence we may write $x_1 = (5r^2 - s^2)/(5r^2 + s^2)$, $y_1 = -2rs/(5r^2 + s^2)$. As a consequence, if a triple (X, Y, Z) of integers is given for which $X^2 + 5Y^2 = Z^2$, then the point $(X/Z, Y/Z)$ lies on our ellipse, and hence there exist integers r and s such that the triple $(5r^2 - s^2, -2rs, 5r^2 + s^2)$ is proportional to the triple (X, Y, Z). We do not necessarily obtain all primitive triples in this way. (Recall Problem 12 at the end of Section 5.3.) Our new method is much more flexible than that used in Section 5.3, but it has the disadvantage of giving less precise information.

In Example 8, our line intersected the ellipse at two points, except for the vertical line ($m = \infty$), which is tangent to the ellipse. More generally, let $f(x, y)$ be a polynomial of degree d with real coefficients, and let $ax + by + c = 0$ be the equation of a line L. Here not both a and b are zero. By interchanging x and y, if necessary, we may assume that $b \neq 0$ (i.e., the line is not "vertical"). Then by a further change of notation, we may write the equation determining the line in the form $y = mx + r$. The x-coordinates of the points of intersection of L with the curve $\mathscr{C}_f(\mathbb{R})$ are the roots of the polynomial

$$p(x) = f(x, mx + r), \tag{5.43}$$

which is of degree at most d. By the fundamental theorem of algebra (discussed in Appendix A.1), we know that the number of complex roots of a polynomial, counting multiplicity, is exactly the degree of the polynomial. Thus $p(x)$ can have at most d distinct real roots, unless $p(x)$ vanishes identically. In the latter case every point of the line is also on the curve $\mathscr{C}_f(\mathbb{R})$, and we say that L is a *component* of $\mathscr{C}_f(\mathbb{R})$. We can actually prove a little more, namely that the polynomial $y - mx - r$ is a factor of the polynomial $f(x, y)$. To see why this is so, let $u = y - mx - r$, so that $f(x, y) = f(x, u + mx + r)$. By multiplying out powers of $u + mx + r$, we see that this new expression is a polynomial in x and u. Each power of u is multiplied by a linear combination of powers of x. That is,

$$f(x, u + mx + r) = f_0(x) + f_1(x)u + f_2(x)u^2 + \cdots + f_d(x)u^d$$

where $f_i(x)$ is a polynomial in x of degree at most $d - i$. Reverting to our original variables, we see that we have shown that any polynomial $f(x, y)$ may be written in the form

$$f(x, y) = \sum_{i=0}^{d} f_i(x)(y - mx - r)^i.$$

From the definition of the polynomial $p(x)$ in (5.43), we see that $p(x) = f_0(x)$. Thus if $p(x)$ vanishes identically, then $f(x, y) = (y - mx - r)k(x, y)$, where $k(x, y) = \sum_{i>0} f_i(x)(y - mx - r)^{i-1}$. We note, moreover, that the coefficients of the $f_i(x)$ are determined by m, r, and the coefficients of $f(x, y)$, using only multiplication and addition. Thus, in particular, if m, r and the coefficients of $f(x, y)$ are all real, then the coefficients of $k(x, y)$ are real, while if m, r and the coefficients of $f(x, y)$ are rational, then the coefficients of $k(x, y)$ are rational. Thus we have proved the following useful result.

Theorem 5.15 *Let $f(x, y)$ be a polynomial with real coefficients and degree d. Let m and r be real numbers, and let L denote the set of points (x, y) for which $y = mx + r$. If the curve $\mathscr{C}_f(\mathbb{R})$ and the line L have strictly more than d distinct points in common, then $L \subseteq \mathscr{C}_f(\mathbb{R})$, and there is a polynomial $k(x, y)$ with real coefficients such that*

$$f(x, y) = (y - mx - r)k(x, y).$$

If m, r, and the coefficients of $f(x, y)$ are all rational, then the coefficients of $k(x, y)$ are also rational.

This may be refined by considering possible multiple roots of $p(x)$. Since the argument hinges on proving that $p(x)$ is identically 0, the conclusion of the theorem may be drawn whenever the total number of roots of $p(x)$, counting multiplicity, is known to be strictly greater than d. If (x_0, y_0) lies on the intersection of $\mathscr{C}_f(\mathbb{R})$ and L, then $p(x)$ has a zero at $x = x_0$. The multiplicity of this zero is called the *intersection multiplicity* of $\mathscr{C}_f$ with L. In general we expect that x_0 is a simple zero of $p(x)$, that is, the intersection multiplicity is 1, but it is important to understand the circumstances in which it is larger. In Example 8 it was of critical importance that the second root of $p(x)$ is distinct from the first, since the approach fails if $p(x)$ has a double root at x_0. To be precise, let us write

$$f(x, y) = a_{00} + a_{10}x + a_{01}y + a_{20}x^2 + a_{11}xy + a_{02}y^2 + \cdots.$$

Since $f(0, 0) = a_{00}$, the origin lies on the curve if and only if $a_{00} = 0$. The further coefficients are related to the partial derivatives of f. If we differentiate i times with respect to x, j times with respect to y, and then set $x = y = 0$, we find that

$$\left(\frac{\partial}{\partial x}\right)^i \left(\frac{\partial}{\partial y}\right)^j f(x, y)\bigg|_{(0,0)} = i!j!a_{ij}.$$

This is a two-variable analogue of Taylor's coefficient formula. For brevity, let $\Delta_{i,j}(0,0)$ denote this partial derivative. In terms of these quantities, we find that

$$f(x, y) = \sum_{n=0}^{d} \frac{1}{n!} \sum_{i=0}^{n} \binom{n}{i} \Delta_{i,n-i}(0,0) x^i y^{n-i}.$$

Here we have sorted monomial terms $x^i y^j$ according to their degree and have put all terms of degree $i + j = n$ in the inner sum. By translating variables, we may expand $f(x, y)$ in powers of $x - x_0$ and $y - y_0$, so that in general

$$f(x, y) = \sum_{n=0}^{d} \frac{1}{n!} \sum_{i=0}^{n} \binom{n}{i} \Delta_{i,n-i}(x_0, y_0)(x - x_0)^i (y - y_0)^{n-i}. \quad (5.44)$$

For a given point $\mathbf{P} = (x_0, y_0)$ in the Euclidean plane $\mathbb{R}^2$, let M be the largest integer so that $\Delta_{i,j}(x_0, y_0) = 0$ whenever $i + j < M$. Then M is called the *multiplicity* of the point $\mathbf{P}$ on $\mathscr{C}_f(\mathbb{R})$. Thus $\mathscr{C}_f(\mathbb{R})$ is precisely the set of points in the plane for which M is positive. A point $\mathbf{P}$ of $\mathscr{C}_f(\mathbb{R})$ is a *simple* point if $M = 1$, and we say that the curve is *smooth* at $\mathbf{P}$. We recall from calculus that the gradient of the function $f(x, y)$ is the vector $(\Delta_{1,0}, \Delta_{0,1})$. This vector points in the direction in which f increases most rapidly. A tangent to the level curve $f(x, y) = 0$ is therefore perpendicular to the gradient. Hence the vector $(-\Delta_{0,1}, \Delta_{1,0})$ is an example of a tangent vector, provided that at least one of these coordinates is nonzero. Thus the tangent vector is well-defined at a simple point of the curve, and the implicit function theorem may be used to show that there is a neighborhood of $\mathbf{P}$ that contains a unique branch of the curve. A point $\mathbf{P}$ of the curve $\mathscr{C}_f(\mathbb{R})$ for which $M > 1$ is called a *singular* point. If all points of a curve are simple, including any points at infinity that may lie on the curve, then the curve is called *nonsingular* or *smooth*. (The idea of points at infinity is clarified in our remarks on projective coordinates at the end of this section.) A point of the curve for which $M = 2$ is a *double* point, $M = 3$ a *triple* point, and so on.

We now relate the multiplicity M of a point on a curve to the intersection multiplicity of a line L passing through a point (x_0, y_0) of the curve. Using the formula (5.44), we may express the polynomial $p(x)$ more explicitly. Substituting $y = y_0 + m(x - x_0)$, we find that

$$p(x) = \sum_{n=0}^{d} \frac{1}{n!}(x - x_0)^n \sum_{i=0}^{n} \binom{n}{i} \Delta_{i,n-i}(x_0, y_0) m^{n-i}.$$

Here the inner sum is a polynomial in m, say $q_n(m)$, of degree at most n. Hence the above may be written more briefly as

$$p(x) = \sum_{n=0}^{d} \frac{1}{n!} q_n(m)(x - x_0)^n.$$

From this formula we see that the intersection multiplicity is the least value of n for which $q_n(m) \neq 0$. In view of the definition of M, if $n < M$ then all coefficients of $q_n(u)$ are 0, and hence $q_n(m) = 0$ for all m. On the other hand, at least one coefficient of $q_M(u)$ is nonzero, so that in general $q_M(m) \neq 0$. Indeed, there can be at most M values of m for which $q_M(m) = 0$. That is, the intersection multiplicity of L with $\mathscr{C}_f(\mathbb{R})$ is at least M for any line through the point (x_0, y_0), but is greater than M for at most M such lines. The case $M = 1$ (i.e., a simple point) is of particular interest. Since $q_1(u) = \Delta_{0,1}u + \Delta_{1,0}$, we see that $q_1(m) = 0$ if and only if $m = -\Delta_{1,0}/\Delta_{0,1}$. But this is the slope of the tangent line to the curve, so we conclude that a line passing through a simple point of the curve has intersection multiplicity 1 unless it is the tangent line, in which case the intersection multiplicity is 2 or greater. Generally it is not greater. If the tangent line at a simple point (x_0, y_0) has intersection multiplicity 3 or more, then the point is called an *inflection point*, or *flex*.

We assume that not all coefficients of the polynomial $f(x, y)$ vanish, for otherwise $\mathscr{C}_f(\mathbb{R})$ consists of the entire plane, and the degree of f is undefined. If d is the degree of f, then at least one coefficient of the polynomial $q_d(u)$ is nonzero, and hence $M \leqslant d$ at any point of the curve. An algebraic curve may consist solely of an isolated point of multiplicity d, as happens with the curve given by $f(x, y) = x^2 + y^2$. If our curve has one point (x_0, y_0) of multiplicity d, and some other point (x_1, y_1) distinct from the first point, then the line through these two points intersects the curve at least d times at the first point, and at least once at the second. Since the sum of the intersection multiplicities is greater than d, the polynomial $p(x)$ has more roots than its degree, and must therefore vanish identically. Thus the line in question is a subset of the curve, and by Theorem 5.15, the linear polynomial defining the line is a factor of $f(x, y)$. Since this argument can be applied to any point (x_1, y_1) of the curve other than (x_0, y_0), we deduce that the curve consists entirely of at most d lines passing through (x_0, y_0). In particular, a conic with a double point that is not isolated is either the union of two distinct lines, $f(x, y) = (a_1x + b_1y + c_1)(a_2x + b_2y + c_2)$, or is a single, doubled line, $f(x, y) = (ax + by + c)^2$. Similarly, a cubic curve may have a triple point, in which case it consists of at most three lines through the point. If a cubic has two distinct singular points, then the line joining them intersects the cubic with multiplicity at least 2 at each point, and therefore the line lies in the cubic and the cubic polynomial has a linear factor.

We are now in a position to demonstrate that the method of Example 8 applies to any nonsingular conic. Let $f(x, y)$ be a quadratic polynomial with rational coefficients, and suppose that the curve $\mathscr{C}_f(\mathbb{R})$ contains a rational point (x_0, y_0). Let m_0 denote the slope of the tangent line to $\mathscr{C}_f$ through (x_0, y_0). Thus m_0 is a rational number. If m is rational, $m \neq m_0$, then the line L through (x_0, y_0) with slope m has intersection multiplicity 1. With $p(x)$ defined by (5.43), we see that the coefficient of x^2 in $p(x)$ is $f(1, m)$. If $f(1, y)$ vanishes identically, then the line $x = 1$ lies in the conic, which is contrary to our supposition that the conic is nonsingular. Since $f(1, y)$ is of degree 2 at most, there may exist one or two rational values of m, say m_1 and m_2, for which $f(1, m) = 0$. For such m, $p(x)$ is linear, and $x = x_0$ is its only root. For all rational m distinct from m_0, m_1, m_2, the polynomial $p(x)$ has rational coordinates, is of degree 2, and has a simple root at the rational number x_0. It must therefore have a second rational root, x_1. Since $y_1 = m(x_1 - x_0)$, the point (x_1, y_1) is a new rational point on the curve, and the method succeeds.

If $p(x) = a_nx^n + a_{n-1}x^{n-1} + \cdots$ is a polynomial, then the sum of the roots of this polynomial is $-a_{n-1}/a_n$. (The reader unfamiliar with this should consult the Appendixes A.1 and A.2.) That is, if $r_1, \cdots, r_n$ denote the roots, then

$$r_1 + r_2 + \cdots + r_n = -a_{n-1}/a_n. \tag{5.45}$$

In general, the roots may be complex, but we see from this identity that if the coefficients of $p(x)$ are rational, and if all but one of the roots is rational, then the last root must also be rational. We have already found this useful when $n = 2$, but we now apply this principle with $n = 3$.

Suppose that $f(x, y)$ is a cubic polynomial with rational coefficients, and suppose that (x_0, y_0) is a double point of the curve $\mathscr{C}_f(\mathbb{R})$. It is known that (x_0, y_0) must be a rational point. (We do not prove this, but note Problems 7 and 8 at the end of this section.) Assuming that it is a rational point, we observe that a line through (x_0, y_0) with rational slope m has intersection multiplicity 2 with the curve, apart from at most two exceptional values of m, say m_1 and m_2. Thus $p(x)$ has three roots, two of which are x_0. The third root must therefore be rational, and we are again able to parameterize the rational points on the curve by means of rational values of m.

Example 9 Find all rational points on the curve $y^2 = x^3 - 3x + 2$.

Solution Put $f(x, y) = y^2 - x^3 + 3x - 2$. To determine whether the curve $\mathscr{C}_f$ has any singular points, we note that $\dfrac{\partial f}{\partial x} = -3x^2 + 3$, which vanishes if and only if $x = \pm 1$. Similarly, $\dfrac{\partial f}{\partial y} = 2y$, which vanishes if and only if $y = 0$. The point $(-1, 0)$ does not lie on the curve, but the point $(1, 0)$ is a singular point on the curve. Since $\dfrac{\partial^2 f}{\partial x^2}(1, 0) = -6$, this point is a double point. Setting $p(x) = f(x, m(x - 1))$, we find by direct calculation that

$$p(x) = -x^3 + m^2x^2 + (3 - 2m^2)x + m^2 - 2.$$

As $x_0 = 1$ is a double root of this polynomial, we deduce from (5.45) that the third root is $x_1 = m^2 - 2$. Hence $y_1 = m(x_1 - 1) = m^3 - 3m$. That is, the equations

$$\begin{aligned} x &= m^2 - 2, \\ y &= m^3 - 3m, \end{aligned} \qquad m = \frac{y}{x - 1}$$

determine a one-to-one correspondence between rational points (x, y) on the curve and rational numbers m, as depicted in Figure 5.2.

Most cubic curves do not have a double point. We now consider the possible existence of rational points on such a nonsingular cubic curve $\mathscr{C}_f$. We observe that if $\mathbf{A} = (x_0, y_0)$ and $\mathbf{B} = (x_1, y_1)$ are rational points on $\mathscr{C}_f$ then the line L through these points has rational slope. If L is a component of $\mathscr{C}_f$ then $\mathscr{C}_f$ is the union of a line and a conic, in which case we have no trouble parameterizing the rational points of $\mathscr{C}_f$. Thus we assume that $\mathscr{C}_f$ contains no line, so that L intersects $\mathscr{C}_f$ at a unique third point of $\mathscr{C}_f$, which we denote $\mathbf{AB}$. In view of (5.45), the point $\mathbf{AB}$ is also a rational point. If $\mathbf{A} \neq \mathbf{B}$, then the line L is called a *chord* of the curve,

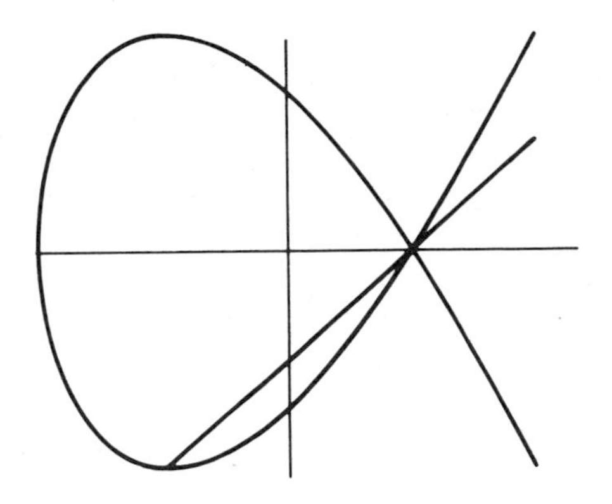

Figure 5.2. The cubic curve $y^2 = x^3 - 3x + 2$ with double point $(1, 0)$. The line through $(1, 0)$ with slope 1 gives the further point $(-1, -2)$.

whereas if $\mathbf{A} = \mathbf{B}$, then L is tangent to the curve. By means of this *chord-and-tangent method* we may construct new rational points from a few known ones. In some cases we obtain only a finite configuration of points, and indeed such a curve may contain only finitely many rational points. In other instances we may use this method to construct infinitely many rational points on the curve, although in general we have no way of knowing whether we have generated all the rational points. In the next section we mention some advanced tools by which one may determine whether a given finite collection of rational points on a nonsingular cubic curve generates infinitely many other rational points, but in many specific cases one may resolve the issue by detailed elementary reasoning. We turn to an example of this type.

Theorem 5.16 *The cubic curve $x^3 + y^3 = 9$ contains infinitely many rational points.*

Proof We define three sequences X_n, Y_n, Z_n of integers by means of the initial conditions $X_0 = 2$, $Y_0 = 1$, $Z_0 = 1$ and the recurrences

$$X_{n+1} = X_n(X_n^3 + 2Y_n^3),$$

$$Y_{n+1} = -Y_n(2X_n^3 + Y_n^3),$$

$$Z_{n+1} = Z_n(X_n^3 - Y_n^3)$$

for $n \geqslant 0$. By induction one may show that $X_n^3 + Y_n^3 = 9Z_n^3$ for all $n \geqslant 0$. Here the basis of the induction is easily verified, and the inductive step is completed by appealing to the recurrences and the inductive hypothesis. Taken out of context, these formulae might seem quite remarkable, but in fact $(X_{n+1}/Z_{n+1}, Y_{n+1}/Z_{n+1})$ is the third point of intersection of the tangent to the curve $x^3 + y^3 = 9$ at the point $(X_n/Z_n, Y_n/Z_n)$. Since each member of the sequence X_n divides the next, it follows that all the X_n are even. By an easy induction we see that the Y_n and Z_n are all odd. It follows that $X_n^3 + 2Y_n^3 \equiv 2 \pmod 4$ for all n. Thus the power of 2 in X_{n+1} is precisely one more than the power of 2 in X_n, so that $2^{n+1} \| X_n$. We have not established that the fractions X_n/Z_n are in lowest terms, but we see in any case that no two of these rational numbers may be equal. Hence we have constructed infinitely many distinct points on the curve.

We conclude this section with a few remarks concerning the properties of algebraic curves. Let $f(x, y)$ have degree d, and $g(x, y)$ have degree e. If $\mathscr{C}_f(\mathbb{R})$ and $\mathscr{C}_g(\mathbb{R})$ have more than de distinct points in common, then they have a common component. More generally, Bézout's theorem asserts that if $\mathscr{C}_f(\mathbb{C})$ and $\mathscr{C}_g(\mathbb{C})$ have no common component then they have exactly de points in common, provided that multiple intersections are counted correctly.

If $f(x, y)$ and $g(x, y)$ are polynomials, then we say that g divides f, and write $g|f$, if there is a polynomial $h(x, y)$ such that we have a polynomial identity $f(x, y) = g(x, y)h(x, y)$. It is not hard to show that if such a polynomial exists, then its coefficients lie in the same field as the field containing the coefficients of f and g. A polynomial with coefficients in a certain field is called *irreducible* over that field if it cannot be written as a product of two polynomials of lower degree, with coefficients in the same field. This is analogous to the definition of a prime number. Although we make no use of the fact, it is nevertheless reassuring to know that the factorization of a polynomial, in any number of variables, and over any given field, is unique, apart from multiplication by constant factors. If $f(x, y)$ and $g(x, y)$ have real coefficients and $g|f$, then $\mathscr{C}_g(\mathbb{R}) \subseteq \mathscr{C}_f(\mathbb{R})$. In Theorem 5.15 we find a converse of this when g is linear, but the converse is false if the degree of g is larger. For example, if $g(x, y) = (x - 1)^2 + y^2$ and $f(x, y) = x + y - 1$, then $\mathscr{C}_g(\mathbb{R}) = \{(1, 0)\} \subseteq \mathscr{C}_f(\mathbb{R})$, but $g \nmid f$. The explanation here is that the field $\mathbb{R}$ of real numbers is not large enough. (In technical language, the field should be algebraically closed.) In the larger field $\mathbb{C}$ of complex numbers, it is true that $\mathscr{C}_g(\mathbb{C}) \subseteq \mathscr{C}_f(\mathbb{C}) \Rightarrow g|f$.

Suppose that $f(x, y)$ is a polynomial with rational coefficients that is irreducible among such polynomials. It may happen that $f(x, y)$ can be factored using polynomials with complex coefficients, but it is known that in such a case the curve $\mathscr{C}_f(\mathbb{R})$ contains at most finitely many points, which may be explicitly determined. We do not prove this in general, but the special case $f(x, y) = x^2 - 2y^2$ is suggestive. This polynomial is irreducible among polynomials with rational coefficients, but may be written as $(x - \sqrt{2}y)(x + \sqrt{2}y)$, using complex numbers. (In this case real numbers are enough.) The curve $\mathscr{C}_f(\mathbb{R})$ is the union of two lines of irrational slope, and the only rational point on these lines is the point $(0, 0)$, at their intersection. If $f(x, y)$ is irreducible over the field $\mathbb{C}$ of complex numbers, then we call f *absolutely irreducible*, and we call the curve $\mathscr{C}_f(\mathbb{C})$ *irreducible*. Thus we see that for purposes of locating rational points on algebraic curves it is enough to consider absolutely irreducible polynomials.

The curves we have considered are called *affine* and may have "points at infinity." Suppose that $f(x, y)$ is a polynomial of degree d. Put $F(X, Y, Z) = f(X/Z, Y/Z)Z^d$. Then $F(X, Y, Z)$ is a homogeneous polynomial of degree d. Consequently, $F(\alpha X, \alpha Y, \alpha Z) = \alpha^d F(X, Y, Z)$ for any values of α, X, Y, Z. In particular, if $\alpha \neq 0$ then $F(X, Y, Z) = 0$ if and only if $F(\alpha X, \alpha Y, \alpha Z) = 0$. On $\mathbb{R}^3 \setminus \{(0, 0, 0)\}$ we define an equivalence relation by saying that two points are equivalent if their coordinates are proportional. That is, there is a nonzero real number α such that $\alpha X_1 = X_2$, $\alpha Y_1 = Y_2$, $\alpha Z_1 = Z_2$. Thus the equivalence classes consist of lines in $\mathbb{R}^3$ passing through the origin, with the origin removed. Such an equivalence class is a point of the *projective plane* $\mathbb{P}_2(\mathbb{R})$. To emphasize that it is the proportion of the coefficients that is significant, we write projective coordinates in the form $X : Y : Z$. Our customary *xy* affine coordinates are embedded in the projective plane by the correspondence $(x, y) \leftrightarrow x : y : 1$, but $\mathbb{P}_2(\mathbb{R})$ includes points of the form $X : Y : 0$, with not both $X = 0$ and $Y = 0$. These are the "points at infinity." For example, the familiar hyperbola $x^2 - y^2 = 1$ becomes $X^2 - Y^2 = Z^2$ in the projective plane. This equation has the solutions $1 : 1 : 0$ and $1 : -1 : 0$, which do not correspond to any finite point in affine *xy* coordinates. One advantage of projective coordinates is that the linear change of variables (5.37) is very natural and can be used to put a given equation into a simpler form. The problem of finding the singularities of a curve is also easier in projective coordinates. For example, the curve $y = x^3 + 1$ has no singularity in the affine *xy*-plane, but if we compute partial derivatives of the homogeneous polynomial $YZ^2 = X^3 + Z^3$ we discover a double point at $0 : 1 : 0$. In affine *xz*-coordinates, this curve is determined by the equation $z^2 = x^3 + z^3$, and the double point at $(0, 0)$ is apparent. A projective line through the point $0 : 1 : 0$ is a line of the form $X = mZ$. In the original *xy*-coordinates, this is the vertical line $x = m$. So once again we have a cubic with a double point, whose rational points are parameterized by lines through the double point, though in this case the final result was obvious at the outset.

PROBLEMS

1. Find a parameterization of the rational points on the hyperbola $x^2 - 2y^2 = 1$, starting from the point $(1, 0)$.
2. Find a parameterization of the rational points on the hyperbola $x^2 - 2y^2 = 1$, starting from the point $(3, 2)$.
3. Apply the analysis in the text to the hyperbola $x^2 - y^2 = 1$ with $(x_0, y_0) = (1, 0)$, and thus find the slope m_0 of the tangent line, and the slopes m_1 and m_2 that give no second intersection.
4. Let $f(x, y)$ be a polynomial of degree d with real coefficients, and set $p(t) = f(2t/(1 + t^2), (1 - t^2)/(1 + t^2))(1 + t^2)^d$. Show that $p(t)$ is a polynomial of degree at most $2d$. Deduce that if $\mathscr{C}_f(\mathbb{R})$ has more than $2d$ distinct points in common with the circle $x^2 + y^2 = 1$ then this circle is a subset of $\mathscr{C}_f(\mathbb{R})$.
5. Show that the curve $y^2 = x^3 + 2x^2$ has a double point. Find all rational points on this curve.
6. Show that the curve $y^2 = x^3 - 3x - 2$ has an isolated double point. Use this double point to parameterize all rational points on the curve.
7. Let $p(x) = ax^3 + bx^2 + cx + d$ where a, b, c, d are real numbers, not all 0. Show that if the curve $y^2 = p(x)$ has a double point, then it must be of the form $(r, 0)$ where r is a double root of $p(x)$.
8. Let $p(x) = ax^3 + bx^2 + cx + d$ where a, b, c, d are rational numbers, not all 0. Show that if r is a double root of $p(x)$, then r is rational.
9. The cubic curve $x^3 + y^3 = 1$ contains the two rational points $(0, 1)$ and $(1, 0)$. Explain why the chord-and-tangent method does not yield any further points on this curve.
10. Show that the cubic curve $y^2 = 4x^3 + x^2 - 2x + 1$ is nonsingular. Note that this curve contains the four rational points $(0, \pm 1), (1, \pm 2)$. Apply the chord-and-tangent method to these points and note the results.
11. Let the triple (X_n, Y_n, Z_n) of integers be defined as in the proof of Theorem 5.16. Show that for $n = 1$ this is $(20, -17, 7)$, and that for $n = 2$ this is $(-36520, 188479, 90391)$. Show also that X_3 is a 21-digit number and that X_4 is an 85-digit number.
12. Let the triple (X_n, Y_n, Z_n) of integers be defined as in the proof of Theorem 5.16. Show that the power of 7 dividing Z_n tends to infinity with n.

13. Let the triple (X_n, Y_n, Z_n) of integers be defined as in the proof of Theorem 5.16, and let $H_n = \max(|X_n|, |Y_n|)$. Show that $H_{n+1} \geqslant \frac{1}{2} H_n^4$. Deduce that $H_n \geqslant 10^{4^{n-2}}$ for $n \geqslant 2$.

14. Apply the tangent method to the curve $x^3 + y^3 = 7$, and thus construct a recurrence that gives a new solution of the equation $X^3 + Y^3 = 7Z^3$ from a known one. Starting from the triple $(2, -1, 1)$, show that this generates infinitely many distinct rational points on the curve $x^3 + y^3 = 7$.

***15.** Let the triple (X_n, Y_n, Z_n) of integers be defined as in the proof of Theorem 5.16. Show that infinitely many of the rational points $(X_n/Z_n, Y_n/Z_n)$ lie in the first quadrant.

5.7 ELLIPTIC CURVES

If the cubic polynomial $f(x, y)$ has rational coefficients, we may use the chord-and-tangent method to produce new rational points on the curve $\mathscr{C}_f(\mathbb{R})$ from a few known ones. As we saw in the preceding section, this sometimes, but not always, produces infinitely many rational points on the curve. We now restrict our attention to those cubic curves such that if **A** and **B** are two points of $\mathscr{C}_f(\mathbb{R})$, then the line L through **A** and **B** intersects the curve at a uniquely defined third point, which we call **AB**. It is understood that if **A** = **B** then the line L is tangent to the curve at this point. Since one or more of the three points **A**, **B**, **AB** may lie at infinity, we consider the curve to be a projective curve in the real projective plane $\mathbb{P}_2(\mathbb{R})$. In order to ensure that **AB** is uniquely defined, two types of cubic curves must be excluded. In the first place, if there is a line L lying within $\mathscr{C}_f(\mathbb{R})$, then **AB** is not uniquely defined when **A** and **B** lie on L. In this case, by Theorem 5.15 the polynomial $f(x, y)$ has a linear factor. In the second place, if **A** is a singular point of $\mathscr{C}_f(\mathbb{R})$ then any line through **A** is tangent to $\mathscr{C}_f(\mathbb{R})$, and hence **AA** is not uniquely defined. This prompts the following definition.

Definition 5.2 *Let $f(x, y)$ be a cubic polynomial with real coefficients. Then $\mathscr{C}_f(\mathbb{R})$ is an* elliptic curve over the field of real numbers *if the polynomial $f(x, y)$ is irreducible over $\mathbb{R}$, and if the curve has no singular point in the real projective plane $\mathbb{P}_2(\mathbb{R})$.*

We define elliptic curves over other fields similarly. We note that if $f(x, y)$ has rational coefficients and if $\mathscr{C}_f(\mathbb{R})$ is an elliptic curve over $\mathbb{R}$, then $\mathscr{C}_f(\mathbb{Q})$ is an elliptic curve over $\mathbb{Q}$. Similarly, if $f(x, y)$ has real coefficients and $\mathscr{C}_f(\mathbb{C})$ is an elliptic curve over $\mathbb{C}$, then $\mathscr{C}_f(\mathbb{R})$ is an elliptic curve over $\mathbb{R}$.

Elliptic curves are precisely those cubic curves for which the binary operation **AB** is well-defined for all pairs of points of the curve. In general the three points **A**, **B**, **AB** are distinct, but they may coincide, as follows:

1. **A** = **B**. In this case L is the tangent to the curve through **A**. If **A** is an inflection point of the curve then **AA** = **A**, but otherwise **AA** ≠ **A**.
2. **A** ≠ **B** but **AB** = **A**. This case arises if the line joining **A** and **B** is tangent to the curve at **A**.
3. **A** ≠ **B** but **AB** = **B**. This case arises if the line joining **A** and **B** is tangent to the curve at **B**.

We note that in any case

$$\mathbf{AB} = \mathbf{BA}, \tag{5.46}$$

and that

$$\mathbf{A}(\mathbf{AB}) = \mathbf{B}. \tag{5.47}$$

When verifying that a particular polynomial $f(x, y)$ defines an elliptic curve, the task of showing that f is irreducible may be tedious. By means of the following result we see that it is enough to demonstrate that $\mathscr{C}_f(\mathbb{C})$ is nonsingular in $\mathbb{P}_2(\mathbb{C})$.

Theorem 5.17 *Let $f(x, y)$ be a cubic polynomial with complex coefficients (which may in fact all be rational or real). If $\mathscr{C}_f(\mathbb{C})$ is nonsingular then $f(x, y)$ is irreducible over $\mathbb{C}$. If $f(x, y)$ is of the special shape $f(x, y) = y^2 - q(x)$ where $q(x)$ is a cubic polynomial, then $\mathscr{C}_f(\mathbb{C})$ is nonsingular if and only if $q(x)$ has no repeated root.*

Proof We show that if $f(x, y)$ is reducible then $\mathscr{C}_f(\mathbb{C})$ has a singular point. Since f is of degree 3, if f is reducible then it can be written as a product of a linear polynomial times a quadratic polynomial. By interchanging x and y, if necessary, and multiplying through by a suitable nonzero constant, we find that $f(x, y)$ may be written in the form $f(x, y) = (y - mx - r)q(x, y)$ where m and r are complex numbers and $q(x, y)$ is a quadratic polynomial with complex coefficients. We pass to projective coordinates, writing $F(X, Y, Z) = f(Z/Z, Y/Z)Z^3$, $L(X, Y, Z) = Y - mX - rZ$, $Q(X, Y, Z) = q(X/Z, Y/Z)Z^2$, so that $F(X, Y, Z) = L(X, Y, Z)Q(X, Y, Z)$. Set $P(X, Z) = Q(X, mX + rZ, Z)$. Then $P(X, Z)$ is a quadratic form in two variables. Any such form is either identically 0, or else factors over $\mathbb{C}$ as the product of two linear forms. In either case there exist complex numbers X_0, Z_0, not both 0, such that $Q(X_0, Z_0) = 0$. Set $Y_0 = mX_0 + rZ_0$. Then $L(X_0, Y_0, Z_0) = Q(X_0, Y_0, Z_0) = 0$. Since

$$\frac{\partial F}{\partial X}(X,Y,Z) = L(X,Y,Z)\frac{\partial Q}{\partial X}(X,Y,Z) + Q(X,Y,Z)\frac{\partial L}{\partial X}(X,Y,Z),$$

we deduce that $\frac{\partial F}{\partial X}(X_0, Y_0, Z_0) = 0$. We argue similarly with the other partial derivatives and conclude that the point $X_0 : Y_0 : Z_0$ is a singularity of the projective curve $F(X,Y,Z) = 0$. If $Z_0 \neq 0$, then the affine curve $\mathscr{C}_f(\mathbb{C})$ has a singularity at $(X_0/Z_0, Y_0/Z_0)$; but if $Z_0 = 0$, then $\mathscr{C}_f(\mathbb{C})$ has a singularity at the point at infinity $X_0 : Y_0 : 0$.

To prove the second assertion, we use projective coordinates and write $F(X,Y,Z) = Y^2Z - aX^3 - bX^2Z - cXZ^2 - dZ^3$. Then we see that

$$\frac{\partial F}{\partial X} = -3aX^2 - 2bXZ - cZ^2, \quad \frac{\partial F}{\partial Y} = 2YZ,$$

$$\frac{\partial F}{\partial Z} = Y^2 - bX^2 - 2cXZ - 3dZ^2.$$

Suppose that $X_0 : Y_0 : Z_0$ is a point of the complex projective plane $\mathbb{P}_2(\mathbb{C})$ such that all three of these expressions vanish. First consider the possibility that $Z_0 = 0$. Then from the first of the above relations we deduce that $-3aX_0^2 = 0$. But $q(x)$ is a cubic polynomial by hypothesis, and hence $a \neq 0$. Thus $X_0 = 0$. Then from the vanishing of the third expression we deduce that $Y_0 = 0$. But $0:0:0$ is not a member of the projective plane, so we conclude that F has no singularity at a point for which $Z = 0$.

Next consider the case $Z_0 \neq 0$. From the vanishing of the second expression displayed above, we deduce that $Y_0 = 0$. Then the identities $F(X_0, 0, Z_0) = \frac{\partial F}{\partial X}(X_0, 0, Z_0) = 0$ are equivalent to the identities $q(X_0/Z_0) = q'(X_0/Z_0) = 0$, which is equivalent to the assertion that $q(x)$ has a repeated root at X_0/Z_0. This completes the proof.

Example 10 Show that the equation $2x(x^2 - 1) = y(y^2 - 1)$ defines an elliptic curve $\mathscr{C}_f(\mathbb{Q})$.

Solution We write $F(X, Y, Z) = 2X^3 - 2XZ^2 - Y^3 + YZ^2$, and find that

$$\frac{\partial F}{\partial X} = 6X^2 - 2Z^2, \quad \frac{\partial F}{\partial Y} = -3Y^2 + Z^2, \quad \frac{\partial F}{\partial Z} = -4XZ + 2YZ.$$

Suppose that $X_0 : Y_0 : Z_0$ is a point of $\mathbb{P}_2(\mathbb{C})$ at which these three expressions vanish. From the first of these relations we deduce that $Z_0 = \pm\sqrt{3}X_0$, and from the second we see that $Z_0 = \pm\sqrt{3}Y_0$, so we deduce that $Y_0 = \pm X_0$. Then the third expression is $X_0Z_0(-4 \pm 2)$, and we see that these three expressions vanish simultaneously only when $X_0 = Y_0 = Z_0 = 0$. But $0:0:0$ is not a member of the projective plane $\mathbb{P}_2(\mathbb{C})$, so we conclude that the curve is nonsingular in $\mathbb{P}_2(\mathbb{C})$. By Theorem 5.17, we deduce that $\mathscr{C}_f(\mathbb{C})$ is an elliptic curve over $\mathbb{C}$. Since the coefficients of f are rational, we conclude that $\mathscr{C}_f(\mathbb{Q})$ is an elliptic curve over $\mathbb{Q}$.

In this example, our work was made no greater by allowing for the possibility that the coordinates of a hypothetical singular point might be complex, not all real. The curve considered here is depicted in Figure 5.5.

The binary operation **AB** on an elliptic curve does not define a group law, because there is no point **0** of the curve with the property that **A0** = **A** for all **A** on the curve. However, we use the point **AB** to construct a further point that we call **A** + **B**, and we show that the points on an elliptic curve $\mathscr{C}_f(\mathbb{R})$ form a group with respect to this addition. When the addition **A** + **B** is defined appropriately, we find that **A** + **B** is a rational point whenever **A** and **B** are rational points, and hence the collection of rational points $\mathscr{C}_f(\mathbb{Q})$ forms a subgroup of $\mathscr{C}_f(\mathbb{R})$ with respect to this addition. By analyzing the structure of these groups, we are led to deeper insights concerning the rational points on an elliptic curve.

To define the addition law for points on an elliptic curve we first choose an arbitrary point of $\mathscr{C}_f(\mathbb{R})$, which we call **0**. Given **A** and **B** on $\mathscr{C}_f(\mathbb{R})$, we construct the point **AB**. Then we construct the line passing through **0** and **AB**, and find the third point **0(AB)** of intersection of this line with the curve $\mathscr{C}_f(\mathbb{R})$. We define **A** + **B** to be this third point. That is, **A** + **B** = **0(AB)**, as depicted in Figure 5.3. This definition of addition depends on the choice of the point **0**, and we explore later how these various additions are related. First we show that the group axioms are satisfied. This is accomplished in several steps.

Lemma 5.18 *Let* **0** *be an arbitrary point of an elliptic curve* $\mathscr{C}_f(\mathbb{R})$, *possibly a point at infinity. Then* **A** + **0** = **A** *for any point* **A** $\in \mathscr{C}_f(\mathbb{R})$. *For any points* **A** *and* **B** *of* $\mathscr{C}_f(\mathbb{R})$, **A** + **B** = **B** + **A**. *Moreover, for any point* **A** $\in \mathscr{C}_f(\mathbb{R})$ *there is a unique point* **B** $\in \mathscr{C}_f(\mathbb{R})$ *such that* **A** + **B** = **0**.

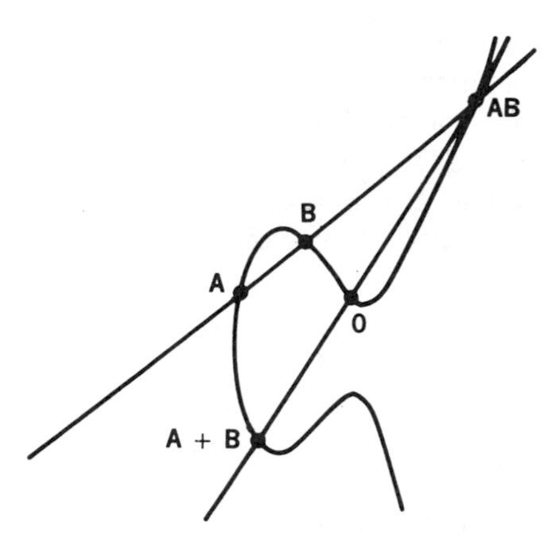

Figure 5.3. Addition of $A = (-3, 2)$ and $B = (-1, 4)$ on the elliptic curve $y^2 = x^3 - 7x + 10$, using $0 = (1, 2)$ as the zero element. Here $AB = (5, 10)$ and $A + B = (-2, -4)$.

Proof Let L denote the line passing through **A** and **0**, which therefore contains the third point **A0**, as in Figure 5.4(a). To find **A** + **0** we consider the line passing through **A0** and **0**. This is the same line L. The third point **0(A0)** on this line is the original point **A**, and thus **A** = **A** + **0**. This argument may be expressed more compactly by noting that the proposed identity **0(A0)** = **0** follows immediately from the general identities (5.46) and (5.47).

Using the definition of the sum of two points, the proposed identity **A** + **B** = **B** + **A** reads **0(AB)** = **0(BA)**. This is immediate from (5.46).

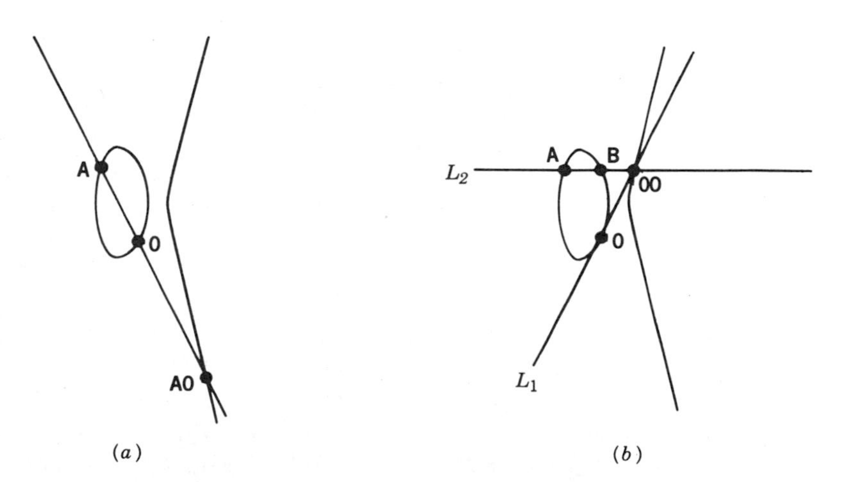

Figure 5.4. The curve $y^2 = 4x^3 - 4x + 1$, with **A** = $(-1, 1)$, **0** = $(1, 1)$. (a) **A0** = $(2, -5)$, **0(A0)** = **A**. (b) **00** = $(1, 1)$, **B** = **A(00)** = $(0, 1)$, **AB** = **00**, **A** + **B** = **0(AB)** = **0(00)** = **0**.

To construct a point $\mathbf{B}$ such that $\mathbf{A} + \mathbf{B} = \mathbf{0}$, let L_1 denote the tangent line to $\mathscr{C}_f(\mathbb{R})$ at $\mathbf{0}$, as in Figure 5.4(b). The further intersection of this line with $\mathscr{C}_f(\mathbb{R})$ is called $\mathbf{00}$. Let L_2 denote the line through $\mathbf{A}$ and $\mathbf{00}$, which intersects $\mathscr{C}_f(\mathbb{R})$ at a third point, $\mathbf{A(00)}$. This is the point $\mathbf{B}$. Then $\mathbf{AB}$ is the point $\mathbf{00}$, and $\mathbf{0(AB)}$ is the point $\mathbf{0}$, so that $\mathbf{A} + \mathbf{B} = \mathbf{0}$, as desired. In algebraic terms, we have $\mathbf{A} + \mathbf{B} = \mathbf{0(AB)}$, by definition. Substituting $\mathbf{B} = \mathbf{A(00)}$, it follows that $\mathbf{A} + \mathbf{B} = \mathbf{0(A(A(00)))}$, and by (5.47) this is $\mathbf{0(00)}$. By a second application of (5.47), this is $\mathbf{0}$. Conversely, if $\mathbf{A} + \mathbf{B} = \mathbf{0}$, then $\mathbf{0(AB)} = \mathbf{0}$, which implies that $\mathbf{00} = \mathbf{0(0(AB))} = \mathbf{AB}$, by (5.47). By (5.47) once more, we find that $\mathbf{A(00)} = \mathbf{A(AB)} = \mathbf{B}$. Thus $\mathbf{B}$ is unique, and the proof is complete.

To prove that the addition of points on an elliptic curve is associative, we first prove the following subsidiary result.

Lemma 5.19 *Let $f(x, y)$ and $g(x, y)$ be cubic polynomials with real coefficients, and suppose that $\mathbf{P}_1, \mathbf{P}_2, \cdots, \mathbf{P}_9$ are nine distinct points in the plane $\mathbb{R}^2$ that are common to the two curves $\mathscr{C}_f(\mathbb{R})$, $\mathscr{C}_g(\mathbb{R})$. Suppose further that the points $\mathbf{P}_1, \mathbf{P}_2, \mathbf{P}_3$ lie on a line L, but that L is not contained in $\mathscr{C}_f(\mathbb{R})$. Then there is a quadratic polynomial $q(x, y)$ such that the six remaining points $\mathbf{P}_4, \mathbf{P}_5, \cdots, \mathbf{P}_9$ all lie on the conic $\mathscr{C}_q(\mathbb{R})$.*

To put this in perspective, note that the general quadratic polynomial $q(x, y)$ in two variables has six coefficients. The condition that $q(x_1, y_1) = 0$ represents a homogeneous linear constraint on these six coefficients. Thus if we choose five distinct points in the plane, the five conditions $q(x_i, y_i)$, $1 \leqslant i \leqslant 5$, give five linear equations in the six unknown coefficients. By a basic theorem of linear algebra, a system of m homogeneous equations in n variables has a nontrivial solution provided that $n > m$. Thus there is a conic passing through any five given points. However, it is known that if six points in the plane are given "in general position," then there is no conic that contains them all. Hence Lemma 5.19 asserts that the six points $\mathbf{P}_4, \mathbf{P}_5, \cdots, \mathbf{P}_9$ are special in some sense.

Proof Since L is not a subset of $\mathscr{C}_f(\mathbb{R})$, we see by Theorem 5.15 that L and $\mathscr{C}_f(\mathbb{R})$ can have at most three distinct points in common. Since three common points are given, the line L and the cubic $\mathscr{C}_f(\mathbb{R})$ can have no further common points. In symbols, $L \cap \mathscr{C}_f(\mathbb{R}) = \{\mathbf{P}_1, \mathbf{P}_2, \mathbf{P}_3\}$. Let $\mathbf{P}_0 = (x_0, y_0)$ be a point on L that is distinct from $\mathbf{P}_1$, $\mathbf{P}_2$, and $\mathbf{P}_3$. Then $f(x_0, y_0) \neq 0$, and we set $\alpha = -g(x_0, y_0)/f(x_0, y_0)$. Let $h(x, y) = \alpha f(x, y) + g(x, y)$. Any point common to $\mathscr{C}_f(\mathbb{R})$ and $\mathscr{C}_g(\mathbb{R})$ will also lie on $\mathscr{C}_h(\mathbb{R})$. Hence the nine given points $\mathbf{P}_1, \mathbf{P}_2, \cdots, \mathbf{P}_9$ all lie on $\mathscr{C}_h(\mathbb{R})$. Moreover, from the choice of α we deduce that $\mathbf{P}_0$ lies on $\mathscr{C}_h(\mathbb{R})$. Since the

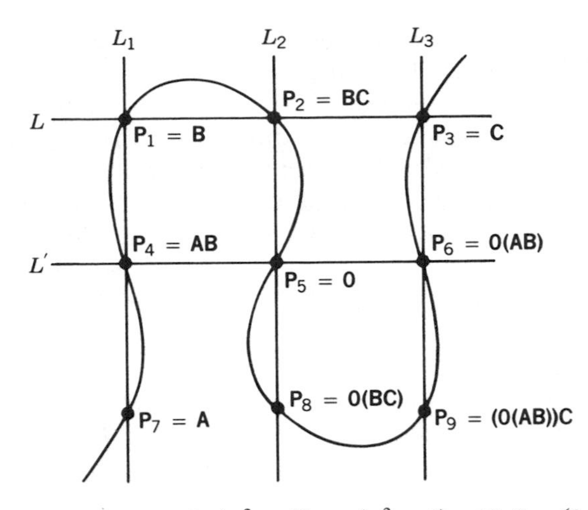

Figure 5.5. The elliptic curve $2x(x^2 - 1) = y(y^2 - 1)$, with $\mathbf{0} = (0, 0)$, $\mathbf{A} = (-1, -1)$, $\mathbf{B} = (-1, 1)$, $\mathbf{C} = (1, 1)$.

cubic $\mathscr{C}_h(\mathbb{R})$ has the four distinct points $\mathbf{P}_0, \mathbf{P}_1, \mathbf{P}_2, \mathbf{P}_3$ in common with L, it follows by Theorem 5.15 that $L \subseteq \mathscr{C}_h(\mathbb{R})$. Not only that, but if $ax + by + c = 0$ defines the line L, then there is a quadratic polynomial $q(x, y)$ such that $h(x, y) = (ax + by + c)q(x, y)$. Hence $\mathscr{C}_h(\mathbb{R}) = L \cup \mathscr{C}_q(\mathbb{R})$. Each of the six points $\mathbf{P}_4, \mathbf{P}_5, \cdots, \mathbf{P}_9$ lies on $\mathscr{C}_h(\mathbb{R})$, but none of them lie on L. Hence they all lie on the conic $\mathscr{C}_q(\mathbb{R})$, and the proof is complete.

Lemma 5.20 *Let* $\mathbf{0}$ *be an arbitrary point of an elliptic curve* $\mathscr{C}_f(\mathbb{R})$, *possibly a point at infinity. Then* $(\mathbf{A} + \mathbf{B}) + \mathbf{C} = \mathbf{A} + (\mathbf{B} + \mathbf{C})$ *for any three points* $\mathbf{A}, \mathbf{B}, \mathbf{C}$ *of* $\mathscr{C}_f(\mathbb{R})$.

Proof Take $\mathbf{P}_1 = \mathbf{B}$, $\mathbf{P}_2 = \mathbf{BC}$, $\mathbf{P}_3 = \mathbf{C}$, $\mathbf{P}_4 = \mathbf{AB}$, $\mathbf{P}_5 = \mathbf{0}$, $\mathbf{P}_6 = \mathbf{0(AB)}$, $\mathbf{P}_7 = \mathbf{A}$, $\mathbf{P}_8 = \mathbf{0(BC)}$, $\mathbf{P}_9 = \mathbf{(0(AB))C}$. We consider first the case in which these nine points are distinct, as depicted in Figure 5.5. Our object is to show that the points $\mathbf{P}_7, \mathbf{P}_8, \mathbf{P}_9$ are collinear, that is, that $\mathbf{A(0(BC))} = \mathbf{(0(AB))C}$. From this it follows immediately that $\mathbf{0(A(0(BC)))} = \mathbf{0((0(AB))C)}$, which is the desired identity.

Let L_1 denote the line determined by the two points $\mathbf{P}_1$ and $\mathbf{P}_7$. From the definition of $\mathbf{P}_4$ we see that $\mathbf{P}_4$ also lies on L_1. Similarly, let L_2 be the line passing through $\mathbf{P}_2$ and $\mathbf{P}_5$, and note that $\mathbf{P}_8$ lies on L_2. Next let L_3 denote the line passing through $\mathbf{P}_3$ and $\mathbf{P}_6$, and note that $\mathbf{P}_9$ lies on L_3. For $i = 1, 2, 3$ let $l_i(x, y) = 0$ be a linear equation defining the line L_i, and put $g(x, y) = l_1(x, y)l_2(x, y)l_3(x, y)$. We now apply Lemma 5.19 to these nine points, which lie on the two cubic curves $\mathscr{C}_f(\mathbb{R})$ and $\mathscr{C}_g(\mathbb{R})$. We note that the line L determined by $\mathbf{P}_1$ and $\mathbf{P}_3$ also passes through $\mathbf{P}_2$, so that Lemma 5.19 applies. Thus the points $\mathbf{P}_4, \mathbf{P}_5, \cdots, \mathbf{P}_9$ all lie on a conic, say $\mathscr{C}_q(\mathbb{R})$. Let L' denote the line passing through $\mathbf{P}_4$ and $\mathbf{P}_5$, and note that $\mathbf{P}_6$ also lies on this line. Since L' has three distinct points in common with the conic $\mathscr{C}_q(\mathbb{R})$, it follows by Theorem 5.15 that $L' \subseteq \mathscr{C}_q(\mathbb{R})$, and moreover that $q(x, y)$ factors into a product of linear functions. That is, $\mathscr{C}_q(\mathbb{R})$ is the union of two lines, L' and L'', say. Each of the remaining points $\mathbf{P}_7, \mathbf{P}_8, \mathbf{P}_9$ lies on $L' \cup L''$. Suppose that one of them, say $\mathbf{P}_7$, were to lie on L'. Then the line L' would have the four distinct points $\mathbf{P}_4, \mathbf{P}_5, \mathbf{P}_6, \mathbf{P}_7$ in common with $\mathscr{C}_f(\mathbb{R})$, which is contrary to the hypothesis that $\mathscr{C}_f(\mathbb{R})$ is an elliptic curve. Thus none of $\mathbf{P}_7, \mathbf{P}_8, \mathbf{P}_9$ lies on L', and hence they must all lie on L''. That is, $\mathbf{P}_7$, $\mathbf{P}_8$, and $\mathbf{P}_9$ are collinear, which is what we set out to prove.

We have proved that $(\mathbf{A} + \mathbf{B}) + \mathbf{C} = \mathbf{A} + (\mathbf{B} + \mathbf{C})$ whenever the nine points $\mathbf{P}_i$ are distinct. We now argue by continuity that this identity still holds even if some of the $\mathbf{P}_i$ coincide. Let $\mathbf{0}', \mathbf{A}', \mathbf{B}', \mathbf{C}'$ be allowed to vary on the elliptic curve $\mathscr{C}_f(\mathbb{R})$, with $\mathbf{0}'$ near $\mathbf{0}$, $\mathbf{A}'$ near $\mathbf{A}$, and so on. We observe that $\mathbf{A}'\mathbf{B}'$ is a continuous function of $\mathbf{A}'$ and $\mathbf{B}'$. Hence $\mathbf{0}'(\mathbf{A}'(\mathbf{0}'(\mathbf{B}'\mathbf{C}')))$ and $\mathbf{0}'(\mathbf{C}'(\mathbf{0}'(\mathbf{A}'\mathbf{B}')))$ are continuous functions of $\mathbf{0}', \mathbf{A}', \mathbf{B}', \mathbf{C}'$. We note that if $\mathbf{B}'$ is fixed and $\mathbf{A}'$ varies, the function $\mathbf{A}'\mathbf{B}'$ never takes the same value twice. That is, the function $\mathbf{A}'\mathbf{B}'$ of $\mathbf{A}'$ with $\mathbf{B}'$ fixed is a one-to-one function of $\mathbf{A}'$. We let the $\mathbf{P}_i$ be as before, but with $\mathbf{0}$ replaced by $\mathbf{0}'$, $\mathbf{A}$ by $\mathbf{A}'$, etc. Thus the $\mathbf{P}_i$ are functions of the four independent variables $\mathbf{0}', \mathbf{A}', \mathbf{B}', \mathbf{C}'$. The original points $\mathbf{P}_i$ are recovered by taking $\mathbf{0}' = \mathbf{0}$, $\mathbf{A}' = \mathbf{A}$, $\mathbf{B}' = \mathbf{B}$, $\mathbf{C}' = \mathbf{C}$. We start with these values, and vary $\mathbf{0}'$ a small amount in such a way that the four points $\mathbf{P}_i$ that depend on $\mathbf{0}'$ (i.e., $\mathbf{P}_5, \mathbf{P}_6, \mathbf{P}_8, \mathbf{P}_9$) are distinct from those points that do not. If we move $\mathbf{0}'$ far enough from $\mathbf{0}$ (always along the curve $\mathscr{C}_f(\mathbb{R})$), two points $\mathbf{P}_i$ which were initially distinct might move together and become coincident. However, this problem does not arise if we keep all the variables $\mathbf{0}', \mathbf{A}', \mathbf{B}', \mathbf{C}'$ sufficiently close to their initial values. Once we have replaced $\mathbf{0}$ by an appropriate $\mathbf{0}'$ near $\mathbf{0}$, we allow $\mathbf{A}'$ to move away from $\mathbf{A}$, to a nearby value chosen so that the $\mathbf{P}_i$ that depend on $\mathbf{A}'$ (i.e., $\mathbf{P}_4, \mathbf{P}_6, \mathbf{P}_7, \mathbf{P}_9$) are distinct from the $\mathbf{P}_i$ that do not depend on $\mathbf{A}'$. Again, by choosing such an $\mathbf{A}'$ sufficiently close to $\mathbf{A}$, we ensure that no new coincidences are introduced among the $\mathbf{P}_i$. Continuing in this manner, we move $\mathbf{B}$ to a point $\mathbf{B}'$ and $\mathbf{C}$ to a point $\mathbf{C}'$. Each $\mathbf{P}_i$ depends on a certain subset of the variables $\mathbf{0}', \mathbf{A}', \mathbf{B}', \mathbf{C}'$, and we note that these subsets are distinct for distinct i. Thus when we replace $\mathbf{0}$ by $\mathbf{0}'$, $\mathbf{A}$ by $\mathbf{A}'$, $\mathbf{B}$ by $\mathbf{B}'$, and $\mathbf{C}$ by $\mathbf{C}'$, the points $\mathbf{P}_i$ move to nearby locations which are all distinct. Thus the argument already given applies to the new $\mathbf{P}_i$, which allows us to deduce that $\mathbf{0}'(\mathbf{A}'(\mathbf{0}'(\mathbf{B}'\mathbf{C}'))) = \mathbf{0}'(\mathbf{C}'(\mathbf{0}'(\mathbf{A}'\mathbf{B}')))$. By continuity, the left side is as close as we like to $\mathbf{A} + (\mathbf{B} + \mathbf{C})$, while the right side is as close as we like to $(\mathbf{A} + \mathbf{B}) + \mathbf{C}$. Since the distance between $\mathbf{A} + (\mathbf{B} + \mathbf{C})$ and $(\mathbf{A} + \mathbf{B}) + \mathbf{C}$ is arbitrarily small, they must be equal, and the proof is complete.

Theorem 5.21 *Let $\mathscr{C}_f(\mathbb{R})$ be an elliptic curve over the field of real numbers, and let* $\mathbf{0}$ *be a point on this curve. Define the sum of two points* $\mathbf{A}$ *and* $\mathbf{B}$ *of* $\mathscr{C}_f(\mathbb{R})$ *to be* $\mathbf{A} + \mathbf{B} = \mathbf{0}(\mathbf{AB})$. *Then the points of* $\mathscr{C}_f(\mathbb{R})$ *form an abelian group with* $\mathbf{0}$ *as the identity. If the coefficients of* $f(x, y)$ *are rational numbers, then the subset* $\mathscr{C}_f(\mathbb{Q})$ *of rational points on* $\mathscr{C}_f(\mathbb{R})$ *form a subgroup if and only if* $\mathbf{0}$ *is a rational point.*

Proof That $\mathscr{C}_f(\mathbb{R})$ satisfies the axioms of an abelian group has been established in Lemma 5.18 and 5.20. Suppose that the coefficients of $f(x, y)$ are rational. In order that $\mathscr{C}_f(\mathbb{Q})$ should be a subgroup, it is necessary that the zero element $\mathbf{0}$ should lie in this subset. That is, $\mathbf{0}$ must be a rational point. Suppose, conversely, that $\mathbf{0}$ is a rational point. We observe that $\mathbf{AB}$ is a rational point whenever $\mathbf{A}$ and $\mathbf{B}$ are rational points on the curve. Hence $\mathbf{A} + \mathbf{B} = \mathbf{0}(\mathbf{AB})$ and $-\mathbf{A} = \mathbf{A}(\mathbf{00})$ are rational points if $\mathbf{A}, \mathbf{B} \in \mathscr{C}_f(\mathbb{Q})$. Since $\mathscr{C}_f(\mathbb{Q})$ is closed under addition and negation, it follows that $\mathscr{C}_f(\mathbb{Q})$ is a subgroup of $\mathscr{C}_f(\mathbb{R})$.

We obtain an infinitude of different addition laws on an elliptic curve $\mathscr{C}_f(\mathbb{R})$ by making different choices of the zero element $\mathbf{0}$. This may be distracting, but in fact these addition laws are all closely related. The elliptic curve is an example of what is called a *homogeneous space*. A more familiar example of such a space is provided by a line L in the plane. We may add two points $\mathbf{A}$ and $\mathbf{B}$ on this line, but we need a point of reference $\mathbf{0}$ on the line, from which to make measurements. Once $\mathbf{0}$ has been chosen, we define $\mathbf{A} + \mathbf{B}$ to be the point on the line that lies to the right (or left) of $\mathbf{A}$ by the same distance that $\mathbf{B}$ lies to the right (or left) of $\mathbf{0}$. If we translate the configuration of points along the line, we replace $\mathbf{0}$ by $\mathbf{0}'$, $\mathbf{A}$ by $\mathbf{A}'$, and so on, but the situation is not changed in any significant way. We now show that any two of our addition laws are related by a similar translation.

Theorem 5.22 *Let* $\mathbf{0}$ *and* $\mathbf{0}'$ *denote two points on an elliptic curve* $\mathscr{C}_f(\mathbb{R})$. *For points* $\mathbf{A}$ *and* $\mathbf{B}$ *on this curve, let* $\mathbf{A} + \mathbf{B}$ *denote the addition defined using* $\mathbf{0}$ *as the zero element, and let* $\mathbf{A} \oplus \mathbf{B}$ *denote the addition defined using* $\mathbf{0}'$. *Then* $\mathbf{A} \oplus \mathbf{B} = \mathbf{A} + \mathbf{B} - \mathbf{0}'$ *for any two points* $\mathbf{A}$ *and* $\mathbf{B}$ *on the curve.*

Proof We show that $\mathbf{0}' + (\mathbf{A} \oplus \mathbf{B}) = \mathbf{A} + \mathbf{B}$. Here $\mathbf{A} \oplus \mathbf{B} = \mathbf{0}'(\mathbf{AB})$, and hence $\mathbf{0}' + (\mathbf{A} \oplus \mathbf{B}) = \mathbf{0}(\mathbf{0}'(\mathbf{0}'(\mathbf{AB})))$. By (5.47) this is $\mathbf{0}(\mathbf{AB}) = \mathbf{A} + \mathbf{B}$.

Let G denote the group $(\mathscr{C}_f(\mathbb{R}), +)$, with $\mathbf{0}$ as the identity, and let H denote the group $(\mathscr{C}_f(\mathbb{R}), \oplus)$ with $\mathbf{0}'$ as identity. We define a map φ: $G \to H$ by the formula $\varphi(\mathbf{A}) = \mathbf{A} + \mathbf{0}'$. We note that $\varphi(\mathbf{A} + \mathbf{B}) = \mathbf{A} + \mathbf{B} + \mathbf{0}' = (\mathbf{A} + \mathbf{0}') + (\mathbf{B} + \mathbf{0}') - \mathbf{0}' = \varphi(\mathbf{A}) + \varphi(\mathbf{B}) - \mathbf{0}' = \varphi(\mathbf{A}) \oplus \varphi(\mathbf{B})$. Thus φ defines a group homomorphism from G to H. We also note that φ is one-to-one and onto. Hence G and H are isomorphic groups, $G \cong H$. Since the group $(\mathscr{C}_f(\mathbb{R}), +)$ is uniquely determined up to isomorphism, we let $E_f(\mathbb{R})$ represent the group of points on $\mathscr{C}_f(\mathbb{R})$, without regard to any particular choice of $\mathbf{0}$. If the coefficients of f are rational, and if $\mathscr{C}_f(\mathbb{Q})$ is nonempty, then we may take $\mathbf{0}$ to be a point of $\mathscr{C}_f(\mathbb{Q})$, and thus the set $\mathscr{C}_f(\mathbb{Q})$ forms a group $E_f(\mathbb{Q})$ that is likewise uniquely determined up to isomorphism. We note that $E_f(\mathbb{Q})$ is a subgroup of $E_f(\mathbb{R})$.

It is instructive to interpret collinearity of points on an elliptic curve as an additive relationship.

Theorem 5.23 *Let* $\mathbf{A}$, $\mathbf{B}$, *and* $\mathbf{C}$ *be three distinct points on an elliptic curve* $\mathscr{C}_f(\mathbb{R})$. *Then these three points are collinear if and only if* $\mathbf{A} + \mathbf{B} + \mathbf{C} = \mathbf{00}$.

Proof We note first that by two applications of (5.47), $\mathbf{A} + (\mathbf{B} + \mathbf{AB}) = \mathbf{0}(\mathbf{A}(\mathbf{0}(\mathbf{BC}))) = \mathbf{0}(\mathbf{A}(\mathbf{0}(\mathbf{B}(\mathbf{AB})))) = \mathbf{0}(\mathbf{A}(\mathbf{0A})) = \mathbf{00}$. But if $\mathbf{C}$ is collinear with $\mathbf{A}$ and $\mathbf{B}$, then $\mathbf{C} = \mathbf{AB}$, and hence $\mathbf{A} + \mathbf{B} + \mathbf{C} = \mathbf{00}$. Suppose, conversely, that this identity holds. The point $\mathbf{C}$ with this property must be unique, and since $\mathbf{AB}$ is such a point, it follows that $\mathbf{C} = \mathbf{AB}$, which implies that $\mathbf{C}$ is collinear with $\mathbf{A}$ and $\mathbf{B}$.

We recall that a point $\mathbf{A}$ of an elliptic curve $\mathscr{C}_f(\mathbb{R})$ is an inflection point if and only if $\mathbf{AA} = \mathbf{A}$. Thus if we choose the point $\mathbf{0}$ to be an inflection point, then we can characterize addition by saying that three points are collinear if and only if they sum to $\mathbf{0}$. On the other hand, it is also important to us that $\mathscr{C}_f(\mathbb{Q})$ be a subgroup of $\mathscr{C}_f(\mathbb{R})$, and for this purpose we require $\mathbf{0}$ to be a rational point. Unfortunately, there exist elliptic curves that possess rational points but no rational inflection point (see Problem 8 at the end of this section for an example), but if the curve has a rational inflection point it is very convenient to take $\mathbf{0}$ to be such a point.

We return to the curve $x^3 + y^3 = 9$, which we considered in Theorem 5.16. This curve has inflection points at $(0, 3^{2/3})$, $(3^{2/3}, 0)$, and at the point at infinity, $1 : -1 : 0$. Since this latter point is a rational point, we take $\mathbf{0}$ to be this point at infinity. We note that the curve has a symmetry about the line $x = y$. Let $\mathbf{A} = (x_0, y_0)$ be a point on this curve, and put $\mathbf{B} = (y_0, x_0)$. The line through these two points has slope -1 and has no further intersection with the curve in the affine plane. Instead, its third intersection with the curve is at the point $1 : -1 : 0$ at infinity. Since $\mathbf{A}$, $\mathbf{B}$ and $\mathbf{0}$ are collinear, it follows by Theorem 5.23 that $\mathbf{A} + \mathbf{B} + \mathbf{0} = \mathbf{00} = \mathbf{0}$. That is, $\mathbf{B} = -\mathbf{A}$. In proving Theorem 5.16 we applied the tangent process to the point $\mathbf{P}_0 = (2, 1)$, to construct the point $\mathbf{P}_0\mathbf{P}_0$. Now $\mathbf{P}_0 + \mathbf{P}_0 = \mathbf{0}(\mathbf{P}_0\mathbf{P}_0)$ is the third point on the curve that passes through $\mathbf{P}_0\mathbf{P}_0$ and $\mathbf{0}$, and hence this third intersection is at $-\mathbf{P}_0\mathbf{P}_0$. That is, $2\mathbf{P}_0 = -\mathbf{P}_0\mathbf{P}_0$, which is to say $\mathbf{P}_0\mathbf{P}_0 = (-2)\mathbf{P}_0$. This is the point $\mathbf{P}_1$ we constructed. Repeating this, we constructed the point $\mathbf{P}_2 = (-2)\mathbf{P}_1 = 4\mathbf{P}_0$. In general, $\mathbf{P}_k = (-2)^k\mathbf{P}_0$. In proving Theorem 5.16 we used only tangents, and we now see that we generated only a small subset of the rational points generated by $\mathbf{P}_0$. If we construct the chord through $\mathbf{P}_0$ and $n\mathbf{P}_0$, the third point of intersection of the chord with the curve is at the point $-(n+1)\mathbf{P}_0$. To obtain $(n+1)\mathbf{P}_0$ from this point we simply interchange coordinates. For example, the chord through $\mathbf{P}_0 = (2, 1)$ and $2\mathbf{P}_0 = (-17/7, 20/7)$ intersects the curve at the third point $(-271/438, 919/438)$, and thus $3\mathbf{P}_0 = (919/438, -271/438)$.

It is easy to construct elliptic curves that contain no rational point. For example, the equation $X^3 + 2Y^3 = 7Z^3$ has no nontrivial solution in integers because the congruence $x^3 + 2y^3 \equiv 7z^3 \pmod{49}$ has no solution for which g.c.d.$(x, y, z, 7) = 1$. Hence the elliptic curve $x^3 + 2y^3 = 7$ contains no rational point. We now consider an example in which we can show that $\mathscr{C}_f(\mathbb{Q})$ consists of precisely four points. The example is carefully selected to take advantage of Theorem 5.10.

Theorem 5.24 *The only rational solutions of the equation* $y^2 = x^3 - 4x$ *are* $(2, 0), (0, 0), (-2, 0)$, *and the point* $0:1:0$ *at infinity.*

In this example, the point $0:1:0$ at infinity is an inflection point of the curve, so it is natural to take $\mathbf{0}$ to be this point. Call the remaining points $\mathbf{A}, \mathbf{B}, \mathbf{C}$. Then $\mathbf{A} + \mathbf{B} + \mathbf{C} = \mathbf{0}$. The tangent lines through these three points are vertical, which is to say that $2\mathbf{A} = 2\mathbf{B} = 2\mathbf{C} = \mathbf{0}$. Thus we see that each of these four points can be written in precisely one way in the form $m\mathbf{A} + n\mathbf{B}$ with $m = 0$ or 1, $n = 0$ or 1, and the group $E_f(\mathbb{Q})$ is isomorphic to $C_2 \oplus C_2$.

Proof We note that the point $(0, 0)$ is a point on this curve. If $\mathbf{P} = (x_0, y_0)$ is a rational point on this curve, then the slope m of the line from $(0, 0)$ to $\mathbf{P}$ has rational slope, $m = y_0/x_0$. Suppose, conversely, that we start with a line L with rational slope m through the point $(0, 0)$. This line intersects the curve at two other points, and we wish to determine those rational values of m for which these further intersections are at rational points. The x-coordinates of the three points of intersection are the roots of a cubic equation with rational coefficients, but since one of these x-coordinates is 0, it follows that the other two x-coordinates are the roots of a quadratic equation with rational coefficients. In the case at hand, we have $f(x, y) = x^3 - y^2 - 4x$, and the x-coordinates in question are the roots of the equation $f(x, mx) = 0$. That is, $x^3 - m^2x^2 - 4x = 0$. After removing the factor x, we see that the x-coordinates of the two remaining points of intersection are the roots of the quadratic $x^2 - m^2x - 4 = 0$. But the roots of a quadratic polynomial with rational coefficients are rational if and only if the discriminant of the polynomial is the square of a rational number. That is, the roots of this equation are either both rational or both irrational, and they are rational if and only if there is a rational number n such that $m^4 + 16 = n^2$. We rewrite this as $(m/2)^4 + 1 = (n/4)^2$. Thus we wish to determine the rational points on the quartic curve $u^4 + 1 = v^2$. Equivalently, in projective coordinates we wish to find all solutions of the equation $U^4 + W^4 = V^2W^2$. Here we may assume that U, V, and W are relatively prime integers. From Theorem 5.10 we deduce that the solutions are $(0, \pm 1, 0), (0, \pm 1, \pm 1)$. Here the first triple represents a point at infinity, which does not correspond to rational values of u and v. Thus it follows that the only rational points on the curve $u^4 + 1 = v^2$ are $(0, \pm 1)$. This gives $m = 0$, $n = \pm 4$ as the only rational solutions of the equation $m^4 + 16 = n^2$, and hence the only line L from $(0, 0)$ that intersects the curve at rational points is the line of slope $m = 0$. The other two points of intersection are therefore $(-2, 0)$ and $(2, 0)$.

When the foregoing approach is analyzed, a marvelous phenomenon emerges. Put $g(u, v) = u^4 + 1 - v^2$, and let $\varphi(x, y) = (u, v)$ be a map from pairs (x, y) of real numbers to pairs of real numbers (u, v) given by the equations

$$u = \frac{y}{2x}, \qquad v = \frac{y^2}{4x^2} - \frac{1}{2}x.$$

Here points of the form $(0, y)$ must be excluded from the domain, in view of the poles that these rational functions have at $x = 0$. Solving for x and y in terms of u and v, we find that

$$x = 2u^2 - 2v, \qquad y = 4u^3 - 4uv.$$

These equations define the inverse map $\vartheta(u, v) = (x, y)$ from pairs of real numbers (u, v) to pairs of real numbers (x, y). By elementary algebra we may verify that

$$-4xg\big(y/(2x), y^2/(2x)^2 - x/2\big) = f(x, y),$$

and that

$$f(2u^2 - 2v, 4u^3 - 4uv) = -8(u^2 - v)g(u, v).$$

Thus if $(x, y) \in \mathscr{C}_f(\mathbb{R})$, $x \neq 0$, then $\varphi(x, y) \in \mathscr{C}_g(\mathbb{R})$, and conversely if $(u, v) \in \mathscr{C}_g(\mathbb{R})$ then $\vartheta(u, v) \in \mathscr{C}_f(\mathbb{R})$. That is, φ: $\mathscr{C}_f(\mathbb{R}) \to \mathscr{C}_g(\mathbb{R})$ and ϑ: $\mathscr{C}_g(\mathbb{R}) \to \mathscr{C}_f(\mathbb{R})$. Moreover, these maps, when restricted to these curves, are inverse to each other, so that the composite map $\vartheta \circ \varphi$ is the identity map on $\mathscr{C}_f(\mathbb{R})$, and $\varphi \circ \vartheta$ is the identity map on $\mathscr{C}_g(\mathbb{R})$. This is an instance of *birational equivalence* of two curves. Since the rational functions employed here have rational coefficients, we say, more precisely, that $\mathscr{C}_f(\mathbb{R})$ and $\mathscr{C}_g(\mathbb{R})$ are $\mathbb{Q}$-birationally equivalent. This equivalence establishes a one-to-one correspondence between the points of $\mathscr{C}_f(\mathbb{R})$ and those of $\mathscr{C}_g(\mathbb{R})$, apart from those points that must be excluded due to poles of the rational functions involved. Since the polynomials $f(x, y)$ and $g(u, v)$ have rational coefficients as well, we find that we have also established a one-to-one correspondence between the rational points $\mathscr{C}_f(\mathbb{Q})$ and the rational points $\mathscr{C}_g(\mathbb{Q})$. By means of this correspondence, we discover that Theorem 5.10 and Theorem 5.24 are equivalent.

The use of $\mathbb{Q}$-birational equivalence is essential to the further study of rational points on algebraic curves. It is known that if $q(x)$ is a polynomial of degree 4 with rational coefficients and distinct roots, then the curve $y^2 = q(x)$ is $\mathbb{Q}$-birationally equivalent to an elliptic curve. Moreover, if $f(x, y)$ has rational coefficients, if the equation $f(x, y) = 0$ determines an elliptic curve, and if this elliptic curve $\mathscr{C}_f(\mathbb{R})$ contains at least one rational point, then this elliptic curve is $\mathbb{Q}$-birationally equivalent to an elliptic curve in *Weierstrass normal form*:

$$y^2 = x^3 - Ax - B. \tag{5.48}$$

One may determine whether the roots of a polynomial $q(x)$ are distinct by calculating the discriminant of the polynomial. (This is discussed in Appendix A.2.) If $q(x)$ is a quadratic polynomial this is simply the familiar quantity $b^2 - 4ac$, but for polynomials of higher degree the discriminant is more complicated. However, for a cubic polynomial in the special shape $x^3 - Ax - B$, the discriminant is the quantity

$$D = 4A^3 - 27B^2. \tag{5.49}$$

Thus by Theorem 5.17 we see that (5.48) defines an elliptic curve if and only if $D \neq 0$. If $D > 0$ then the polynomial $x^3 - Ax - B$ has three distinct real roots, and the elliptic curve $\mathscr{C}_f(\mathbb{R})$ has two connected components, one a closed oval and the other extending to the point $0:1:0$ at infinity. An example of this type is depicted in Figure 5.4. If $D < 0$, then the polynomial $x^3 - Ax - B$ has only one real root, and the elliptic curve $\mathscr{C}_f(\mathbb{R})$ has one connected component, as seen in Figure 5.3.

We now derive explicit formulae for the coefficients of $\mathbf{P}_1 + \mathbf{P}_2$ on an elliptic curve, in terms of the coefficients of $\mathbf{P}_1$ and $\mathbf{P}_2$ and the defining equation of the curve. To provide greater flexibility, we do not restrict ourselves to curves of the form (5.48), but instead consider the more general equation

$$y^2 = x^3 + ax^2 + bx + c. \tag{5.50}$$

In order that this should define an elliptic curve, it is necessary and sufficient that $D \neq 0$ where

$$D = a^2b^2 - 4a^3c - 4b^3 + 18abc - 27c^2 \tag{5.51}$$

is the discriminant of the cubic polynomial in x, discussed in Appendix A.2. Any elliptic curve of the form (5.50) contains the point at infinity $0:1:0$, which is a point of inflection of the curve. Thus it is traditional to take $\mathbf{0}$ to be this point, so that three points on the curve sum to $\mathbf{0}$ if and only if they are collinear. Let $\mathbf{P}_1 = (x_1, y_1)$ and $\mathbf{P}_2 = (x_2, y_2)$ be two points on this curve, and put $\mathbf{P}_3 = \mathbf{P}_1 + \mathbf{P}_2 = (x_3, y_3)$. We assume for the moment that $x_1 \neq x_2$. Let m denote the slope of the line through these points, $m = (y_2 - y_1)/(x_2 - x_1)$. This line intersects the curve at the third point $\mathbf{P}_1\mathbf{P}_2 = -\mathbf{P}_3 = (x_3, -y_3)$. Setting $y = y_1 + m(x - x_1)$ in (5.50), we find that x_1, x_2, and x_3 are the roots of the equation

$$x^3 + (a - m^2)x + (b + 2mx_1 - 2y_1)x^2 + \left(c - (mx_1 - x_2)^2\right) = 0.$$

Hence by (5.45) we see that the sum of the three roots is $m^2 - a$, so that $x_3 = m^2 - a - x_1 - x_2$, and $y_3 = -y_1 - m(x_3 - x_1)$. If $x_1 = x_2$, then either $y_2 = -y_1$, in which case $\mathbf{P}_2 = -\mathbf{P}_1$ and $\mathbf{P}_1 + \mathbf{P}_2 = \mathbf{0}$, or else $y_2 = y_1$, in which case $\mathbf{P}_1 = \mathbf{P}_2$. To find the coordinates of $\mathbf{P}_3 = 2\mathbf{P}_1$, let m denote the slope of the tangent line to the curve through $\mathbf{P}_1$, $m = (3x_1^2 + 2ax_1 + b)/(2y_1)$. If $y_1 = 0$, then this line is vertical, and $2\mathbf{P}_1 = \mathbf{0}$, but otherwise we obtain a finite value for m. Proceeding as before, we find that $x_3 = m^2 - a - 2x_1$, $y_3 = -y_1 - m(x_3 - x_1)$. In summary, we have shown that if $x_1 \neq x_2$, then the coordinates of $\mathbf{P}_1 + \mathbf{P}_2$ are

$$\begin{aligned} x_3 &= \left(\frac{y_2 - y_1}{x_2 - x_1}\right)^2 - a - x_1 - x_2, \\ y_3 &= -y_1 - \left(\frac{y_2 - y_1}{x_2 - x_1}\right)(x_3 - x_1), \end{aligned} \tag{5.52}$$

and that if $y_1 \neq 0$, then the coordinates of $2\mathbf{P}_1$ are

$$\begin{aligned} x_3 &= \left(\frac{3x_1^2 + 2ax_1 + b}{2y_1}\right)^2 - a - 2x_1, \\ y_3 &= -y_1 - \left(\frac{3x_1^2 + 2ax_1 + b}{2y_1}\right)(x_3 - x_1). \end{aligned} \tag{5.53}$$

Using these formulae, it is a simple matter to calculate the coordinates of the sum of two points. For example, we find that the first 10 multiples of the point $(1, 2)$ on the curve $x^3 - 7x + 10 = y^2$ depicted in Figure 5.4 are as follows. We give the coordinates both as decimals and as fractions.

n	x_n	y_n	x_n	y_n
1	1.00000	2.00000	1/1	2/1
2	−1.00000	−4.00000	−1/1	−4/1
3	9.00000	−26.00000	9/1	−26/1
4	2.25000	2.37500	9/4	19/8
5	−3.16000	−0.75200	−79/25	−94/125
6	2.59763	−3.05690	439/169	−6716/2197
7	6.42112	15.15917	4681/729	298378/19683
8	−1.52891	4.13865	−8831/5776	1816769/438976
9	1.24409	−1.79358	364121/292681	−283996102/158340421
10	239.30450	3701.68885	13215591/55225	48040055236/12977875

In these values we note that the denominator of x_n is a perfect square, say w_n^2, and that the denominator of y_n is w_n^3. We now show that this holds for any rational point on this curve.

Theorem 5.25 *Let $\mathscr{C}_f(\mathbb{R})$ be an elliptic curve determined by an equation of the form* (5.50) *with integral coefficients. Let (x_1, y_1) be a rational point on this curve, not at infinity. Then there exist integers u, v, w such that $x_1 = u/w^2$, $y_1 = v/w^3$, and* g.c.d.(u, w) = g.c.d.$(v, w) = 1$.

Proof Let Z be the least common denominator of the rational numbers x_1, y_1, so that $x_1 = X/Z$, $y_1 = Y/Z$ with $Z > 0$, g.c.d.$(X, Y, Z) = 1$. Substituting into (5.50), we find that

$$Y^2Z = X^3 + aX^2Z + bXZ^2 + cZ^3.$$

Put w = g.c.d.(X, Z). Then w^3 divides the right side, and hence $w^3 | Y^2Z$. But g.c.d.$(w, Y) = 1$, since g.c.d.$(X, Y, Z) = 1$. Therefore $w^3 | Z$, say

$Z = tw^3$, $X = uw$. We substitute these new variables in the equation displayed above, and divide both sides by w^3, to find that

$$Y^2t = u^3 + atu^2w^2 + bt^2uw^4 + ct^3w^6.$$

Here t occurs in all terms but one, so we conclude that $t|u^3$. But g.c.d. $(X/w, Z/w) = 1$, which implies that g.c.d. $(u, tw) = 1$. Thus $t = \pm 1$. But $Z > 0$ and $w > 0$, so $t > 0$, and hence $t = 1$. Setting $v = Y$, we have $x_1 = u/w^2$, $y_1 = v/w^3$ with g.c.d. $(u, w) = 1$. To see that g.c.d. $(v, w) = 1$, we note that

$$v^2 = u^3 + au^2w^2 + buw^4 + cw^6, \tag{5.54}$$

so that any common divisor of v and w would also have to divide u^3. This completes the proof.

By manipulating the formulae (5.52) it may be shown that if $\mathbf{P}_1 = (u_1/w_1^2, v_1/w_1^3)$ and $\mathbf{P}_2 = (u_2/w_2^2, v_2/w_2^3)$ are two points of the curve (5.50) with $u_1/w_1^2 \neq u_2/w_2^2$ then we may write $\mathbf{P}_1 + \mathbf{P}_2 = \mathbf{P}_3$ in the form $\mathbf{P}_3 = (u_3/w_3^2, v_3/w_3^3)$ where

$$\begin{aligned} u_3 &= \left(v_2w_1^3 - v_1w_2^3\right)^2 - aw_1^2w_2^2\left(u_2w_1^2 - u_1w_2^2\right)^2 \\ &\quad - \left(u_1w_2^2 + u_2w_1^2\right)\left(u_1w_2^2 - u_2w_1^2\right)^2, \\ v_3 &= -v_1w_2^3\left(u_2w_1^2 - u_1w_2^2\right)^3 - \left(v_2w_1^3 - v_1w_2^3\right)u_3 \\ &\quad + w_2^2\left(u_2w_1^2 - u_1w_2^2\right)^2u_1\left(v_2w_1^3 - v_1w_2^3\right), \\ w_3 &= w_1w_2\left(u_2w_1^2 - u_1w_2^2\right). \end{aligned} \tag{5.55}$$

Similarly, from (5.53) we find that we may write $2\mathbf{P}_1 = (u_3/w_3^2, v_3/w_3^3)$ with

$$\begin{aligned} u_3 &= \left(3u_1^2 + 2au_1w_1^2 + bw_1^4\right)^2 - 4\left(aw_1^2 + 2u_1\right)v_1^2, \\ v_3 &= -8v_1^4 - \left(3u_1^2 + 2au_1w_1^2 + bw_1^4\right)\left(u_3 - 4u_1v_1^2\right) \\ w_3 &= 2v_1w_1. \end{aligned} \tag{5.56}$$

In these formulae, the numbers u_3, v_3, w_3 may have common factors, even if g.c.d.(u_1, w_1) = g.c.d.$(u_2, w_2) = 1$.

We conclude this section with a description of some further properties of the group $E_f(\mathbb{Q})$ of rational points on an elliptic curve. First we introduce some notation taken from the theory of infinite abelian groups. An abelian group G is said to be *finitely generated* if there exists a finite collection $g_1, g_2, \cdots, g_k$ of elements of G such that every element of G can be written in the form $n_1g_1 + n_2g_2 + \cdots + n_kg_k$, where the n_i are integers. If two elements g, g' of infinite order differ by an element of finite order, say $h = g' - g$ has finite order, then we say that g' is a twisted copy of g. The elements h of finite order which produce these twistings form a subgroup H, called the *torsion group* of G. In symbols, $H = \text{tors}(G)$. If G is finitely generated, then $\text{tors}(G)$ is necessarily a finite group. Moreover, it can be shown that if G is a finitely generated abelian group then there exist members $g_1, g_2, \cdots, g_r$ of G such that every element of G is uniquely of the form $h + n_1g_1 + n_2g_2 + \cdots + n_rg_r$. The elements g_i are not uniquely determined, but in any such presentation of the group the number r is the same. This r is called the *rank* of the group. All this is relevant because of the following fundamental result.

Mordell's Theorem *Suppose that the cubic polynomial $f(x, y)$ has rational coefficients, and that the equation $f(x, y) = 0$ defines an elliptic curve $\mathscr{C}_f(\mathbb{R})$. Then the group $E_f(\mathbb{Q})$ of rational points on $\mathscr{C}_f(\mathbb{R})$ is finitely generated.*

In elementary language, this says that on any elliptic curve that contains a rational point, there exists a finite collection of rational points such that all other rational points can be generated by using the chord-and-tangent method. In Theorem 5.16 we proved that the elliptic curve $x^3 + y^3 = 9$ has positive rank, while in Theorem 5.24 we showed that the curve $y^2 = x^3 - 4x$ has rank 0. It is known that the rank of an elliptic curve can be as large as 14, and it is guessed that it can be arbitrarily large. While the rank and generators g_i are known for many particular elliptic curves, we lack a procedure for finding these quantities in general. On the other hand, we have an effective technique for finding all points of finite order (called *torsion points*) on an elliptic curve.

The Lutz-Nagell Theorem *Let $\mathscr{C}_f(\mathbb{R})$ be an elliptic curve given by an equation of the form* (5.50) *with integral coefficients. If (x_0, y_0) is a rational point of finite order on $\mathscr{C}_f(\mathbb{R})$, then x_0 and y_0 are integers. Moreover, either $y_0 = 0$ or y_0^2 divides the discriminant D given in* (5.51).

By applying this theorem we can construct a finite list of integral points on the curve that must include all points of finite order. By examining the multiples of such points, we quickly discover which have finite order, and which not. An elliptic curve may contain other integral points, for which y_0^2 does not divide D, but from a general theorem of Siegel it follows that the number of such points is at most finite. A precise description of groups that can occur as $\operatorname{tors}(E_f(\mathbb{Q}))$ is provided by:

Mazur's Theorem *Let $f(x, y)$ be a nonsingular cubic polynomial with rational coefficients. Then the group* $\operatorname{tors}(E_f(\mathbb{Q}))$ *of points on $\mathscr{C}_f(\mathbb{Q})$ of finite order is isomorphic to one of the following groups: C_n with $n = 1, 2, \cdots, 10$, or $n = 12$, or $C_n \oplus C_2$ with $n = 2, 4, 6$ or 8.*

It is known that each of these groups occurs as the torsion group of $E_f(\mathbb{Q})$ for some elliptic curve $\mathscr{C}_f(\mathbb{Q})$ defined over the rational numbers. From this theorem we see that an elliptic curve can have at most 16 torsion points. Moreover, we see from the foregoing that a rational point **P** is a torsion point if and only if at least one of the points 7**P**, 8**P**, 9**P**, 10**P**, 12**P** is **0**.

PROBLEMS

1. Let $f(x, y) = y^2 - p(x)$, where $p(x)$ is a cubic polynomial with no repeated root. Take the point **0** on $\mathscr{C}_f(\mathbb{R})$ to be the point $0:1:0$ at infinity. Show that $2\mathbf{A} = \mathbf{0}$ if and only if **A** is of the form $\mathbf{A} = (r, 0)$, where r is a root of $p(x)$.
2. Let $\mathscr{C}_f(\mathbb{R})$ be an elliptic curve for which **0** is an inflection point. Show that $3\mathbf{A} = \mathbf{0}$ if and only if **A** is an inflection point. Deduce that if **A** and **B** are inflection points then **AB** is also an inflection point.
3. Show that the general polynomial of degree d in two variables has $\binom{d+2}{2}$ coefficients. Deduce that if $\binom{d+2}{2} - 1$ points in the plane are given, then there exists a curve of degree d that passes through them.
4. Show that if $\binom{d+2}{2} - 1$ rational points are given in the plane, then there exists a polynomial $f(x, y)$ of degree at most d, with integral coefficients, not all 0, such that the given points all lie on the curve $\mathscr{C}_f(\mathbb{Q})$.
5. For what values of c is the curve $cx(x^2 - 1) = y(y^2 - 1)$ not an elliptic curve?
6. Show that the projective curve $X^3 + Y^3 + Z^3 = dXYZ$ is nonsingular if and only if $d^3 \neq 27$. Show that if $d = 3$, then this curve is the union of a line and a conic. Show that if $d^3 \neq 27$, then the points

$1:-1:0$, $0:1:-1$, $-1:0:1$ are inflection points, and that the curve has no other inflection points.

7. Let **A** and **B** be distinct points on an elliptic curve $\mathscr{C}_f(\mathbb{R})$, and suppose that the line through **A** and **B** is tangent to $\mathscr{C}_f(\mathbb{R})$ at **B**. Show that $\mathbf{A} + 2\mathbf{B} = \mathbf{00}$.

8. Let $f(x, y) = x^3 + 2y^3 - 3$. Show that $\mathscr{C}_f(\mathbb{Q})$ is nonempty. Show also that $\mathscr{C}_f(\mathbb{R})$ has three inflection points (including one at infinity), but no inflection point with rational coordinates.

9. Use the method employed to prove Theorem 5.24 to relate the elliptic curve $y^2 = x^3 + x$ to equation (5.29), and thus find all rational points on this elliptic curve.

10. Find all rational points on the elliptic curve $y^2 = x^3 - x$. (H)

11. Find all rational points on the elliptic curve $y^2 = x^3 + 4x$. (H)

12. Suppose that the elliptic curve $\mathscr{C}_f(\mathbb{R})$ is given by (5.50), and that the coefficients a, b, c are integers. Let $\mathbf{P}_1 = (u_1/w_1^2, v_1/w_1^3)$ be a rational point on this curve, and write $2\mathbf{P} = (u_3/w_3^2, v_3/w_3^3)$ with g.c.d.$(u_1, w_1) = 1$. Show that if b is even and u_1 is odd, then u_3 is odd, and the power of 2 in w_3 is greater than the power of 2 in w_1. Deduce that the points $2^k\mathbf{P}_1$ are all distinct, and hence that $\mathbf{P}_1$ has infinite order. In particular, show that the point $(1, 3)$ on the curve $y^2 = x^3 + 6x^2 + 2x$ has infinite order.

13. Show that the formula for u_3 in (5.56) can be rewritten as $u_3 = (u_1^2 - bw_1^4)^2 - 4c(2u_1 + aw_1^2)w_1^6$. Deduce that if the equation (5.50) has integral coefficients, and if **P** is a rational point on $\mathscr{C}_f(\mathbb{R})$, then the x-coordinate of $2\mathbf{P}$ is the square of a rational number if $c = 0$.

14. Let $\mathbf{P}_1 = (x_1, y_1)$ be a point with integral coordinates on the elliptic curve (5.50), where a, b, c are integers. Show that if $2\mathbf{P}_1$ also has integral coordinates then $(2y_1) | (3x_1^2 + 2ax_1 + b)$.

15. Show that $27(x^3 - Ax + B)(x^3 - Ax - B) - (3x^2 - 4A)(3x^2 - A)^2 = 4A^3 - 27B^2$. Deduce that if an elliptic curve is given by (5.48), with A and B integers, and if $\mathbf{P}_1 = (x_1, y_1)$ and $2\mathbf{P}_1$ are points with integral coordinates, $2\mathbf{P}_1 \neq \mathbf{0}$, then $y_1^2 | (4A^3 - 27B^2)$.

16. Suppose that the equation $y^2 = x^3 + ax^2 + bx$ determines an elliptic curve. Suppose also that a and b are integers. Explain why $b \neq 0$. Show that if $(u_1/w_1^2, v_1/w_1^3)$ is a rational point on this curve, with g.c.d. $(u_1, w_1) = 1$, then there exist integers d and s such that $d > 0$, $d | b$, and $u_1 = \pm ds^2$. In the particular case of the curve $y^2 = x^3 + 6x^2 + 2x$, show by congruences (mod 4) that the case $u_1 = -s^2$ yields no solution, and by congruences (mod 8) show that the case $u_1 = -2s^2$ also gives no solution. Deduce that this elliptic curve contains no rational point (x_1, y_1) with $x_1 < 0$.

***17.** Let $f(x, y) = y^2 - x^3 - 6x^2 - 4x$, $g(u, v) = v^2 - u^3 + 12u^2 - 20u$. Let φ take pairs (x, y) to pairs (u, v) by means of the formulae $u = y^2/x^2$, $v = y - 4y/x^2$. Show that if $\mathbf{P} \in \mathscr{C}_f(\mathbb{R})$ then $\varphi(\mathbf{P}) \in \mathscr{C}_g(\mathbb{R})$. Let ϑ take pairs (u, v) to pairs (x, y) by the formulae $x = v^2/(4u^2)$, $y = (1 - 20/u^2)v/8$. Show that if $\mathbf{Q} \in \mathscr{C}_g(\mathbb{R})$ then $\vartheta(\mathbf{Q}) \in \mathscr{C}_f(\mathbb{R})$. Take $\mathbf{P} = (-1, 1)$. Show that $\mathbf{P} \in \mathscr{C}_f(\mathbb{R})$, that $\varphi(\mathbf{P}) = (1, -3) \in \mathscr{C}_g(\mathbb{R})$, and that $\vartheta \circ \varphi(\mathbf{P}) = (9/4, 57/8) = 2\mathbf{P}$. Show, more generally, that if $\mathbf{P} \in \mathscr{C}_f(\mathbb{R})$ then $\vartheta \circ \varphi(\mathbf{P}) = 2\mathbf{P}$.

18. For what values of the constants a and b does the curve

$$axy = (x + 1)(y + 1)(x + y + b) \tag{5.57}$$

contain a line? This curve has three points at infinity. What are they?

***19.** Let b, x_0, x_1 be given real numbers. Generate a sequence of numbers x_n by means of the recursion $x_{n+1} = (x_n + b)/x_{n-1}$ for $n \geqslant 1$. Choose a so that the point (x_0, x_1) lies on the curve (5.57). Show that all further points (x_n, x_{n+1}) lie on the same curve. Show that if $x_0 > 0$, $x_1 > 0$ and $b = 1$, then the sequence x_n has period 5. Show that if x_0, x_1, and b are positive then the sequence x_n is bounded.

***20.** Let $\mathscr{C}_f(\mathbb{R})$ be defined by (5.50) where a, b, c are real numbers, and suppose that the polynomial on the right side of (5.50) has only one real root (so that the curve $\mathscr{C}_f(\mathbb{R})$ lies in one connected component). Show that if $\mathbf{P} \in \mathscr{C}_f(\mathbb{R})$ has infinite order, then the points $n\mathbf{P}$ are dense on $\mathscr{C}_f(\mathbb{R})$.

***21.** Let $\mathscr{C}_f(\mathbb{R})$ be defined by (5.50) where a, b, c are real numbers, and suppose that the polynomial on the right side of (5.50) has three real roots $r_1 < r_2 < r_3$. Let $\mathscr{C}_0$ be the connected component of points $(x, y) \in \mathscr{C}_f(\mathbb{R})$ for which $x \geqslant r_3$, including the point at infinity, and let $\mathscr{C}_1$ denote the connected component of points for which $r_1 \leqslant x \leqslant r_2$. Let $\mathbf{P}$ and $\mathbf{Q}$ be arbitrary points of $\mathscr{C}_f(\mathbb{R})$. Show that $\mathbf{P} + \mathbf{Q}$ lies on $\mathscr{C}_0$, or $\mathscr{C}_1$, according as $\mathbf{P}$ and $\mathbf{Q}$ lie on the same, or different, components. (That is, $\mathscr{C}_0$ is a subgroup of index 2 in $\mathscr{C}_f(\mathbb{R})$.)

***22.** Let $\mathscr{C}_f(\mathbb{R})$ be defined as in the preceding problem. Show that if $\mathbf{P}$ is a point of infinite order, $\mathbf{P} \in \mathscr{C}_0$, then the points $n\mathbf{P}$ form a dense subset of $\mathscr{C}_0$. Show that if $\mathbf{P}$ is of infinite order, $\mathbf{P} \in \mathscr{C}_1$, then the points $n\mathbf{P}$ are dense on $\mathscr{C}_f(\mathbb{R})$.

***23.** Suppose that we have an elliptic curve as described in Problem 20. We construct a function $\mathbf{P}(t)$ from $\mathbb{R}$ to $\mathscr{C}_f(\mathbb{R})$ as follows. The function $\mathbf{P}(t)$ is to have period 1. Put $\mathbf{P}(0) = \mathbf{0}$. Put $\mathbf{P}(1/2) = \mathbf{P}_1 = (r, 0)$ where r is chosen so that $(r, 0) \in \mathscr{C}_f(\mathbb{R})$. Of the two points $\mathbf{P}$ of $\mathscr{C}_f(\mathbb{R})$ for which $2\mathbf{P} = \mathbf{P}_1$, let $\mathbf{P}_2 = (x_2, y_2)$ be the one for which $y_2 < 0$. Put $\mathbf{P}(1/4) = \mathbf{P}_2$. Similarly put $\mathbf{P}(1/8) = \mathbf{P}_3 = (x_3, y_3)$ where $y_3 < 0$ and $2\mathbf{P}_3 = \mathbf{P}_2$, and so on. For k odd, put $\mathbf{P}(k/2^s) = k\mathbf{P}_s$. Thus

$\mathbf{P}(t)$ is defined on a dense subset of $\mathbb{R}$. Extend this to all of $\mathbb{R}$ by continuity. Show that $\mathbf{P}(t_1 + t_2) = \mathbf{P}(t_1) + \mathbf{P}(t_2)$ for arbitrary real numbers t_1, t_2. Show that $\mathbf{P}(t)$ has finite order in $\mathscr{C}_f(\mathbb{R})$ if and only if t is a rational number. Show also that if g.c.d.$(a, b) = 1$ then $\mathbf{P}(a/b)$ has order b. Conclude that $\mathscr{C}_f(\mathbb{R})$ is isomorphic to the additive group $\mathbb{R}/\mathbb{Z}$ of real numbers modulo 1. (This group is called the *circle group*, and is often denoted by $\mathbb{T}$.)

***24.** Let $\mathscr{C}_f(\mathbb{R})$ be an elliptic curve as described in Problem 21. Construct a function from $\mathbb{R}$ to $\mathscr{C}_0$, as in the preceding problem. Show that $\mathscr{C}_0 \cong \mathbb{R}/\mathbb{Z}$, and that $\mathscr{C}_f(\mathbb{R}) \cong \mathbb{R}/\mathbb{Z} \oplus C_2$.

***25.** Let G be a finite subgroup of n points on an elliptic curve $\mathscr{C}_f(\mathbb{R})$ as given by (5.50). (For example, G might be the group tors$(E_f(\mathbb{Q}))$.) Show that if $\mathscr{C}_f(\mathbb{R})$ has one connected component then G is cyclic, and that if $\mathscr{C}_f(\mathbb{R})$ has two connected components then either G is cyclic or is isomorphic to $C_{n/2} \oplus C_2$.

5.8 FACTORIZATION USING ELLIPTIC CURVES

In this section we draw on the ideas of the two preceding sections to devise a factorization strategy called the *Elliptic Curve Method* (abbreviated ECM). When applied to a large composite number, this method can be expected to locate a factor much more rapidly than the methods we discussed in Section 2.4.

The ECM is modeled on the Pollard $p - 1$ method, which we describe first. Let m denote the number we wish to factor. We let $r_1, r_2, \cdots$ be integers greater than 1, and generate a sequence $a_1, a_2, \cdots$ by choosing a to be an arbitrary integer > 1, and then setting $a_1 = a$, $a_{n+1} = a_n^{r_n}$. That is, $a_n = a^{(r_1 r_2 \cdots r_{n-1})}$. Put $g_n = (a_n - 1, m)$. Since $(a_n - 1) | (a_{n+1} - 1)$ for all n, it follows that $g_1 | g_2 | g_3 \cdots$. Our object is to find an n such that $1 < g_n < m$, for then g_n is a proper divisor of m. In practice we do not calculate the exact value of a_n, but only the residue class in which a_n falls modulo m. As $a^{p-1} \equiv 1 \pmod{p}$ by Fermat's congruence, it follows that if p is a prime factor of m for which $(p - 1) | r_1 r_2 \cdots r_{n-1}$, then $a_n \equiv 1 \pmod{p}$, and hence $p | g_n$. The simplest useful choice of the numbers r_n is to take $r_n = n$. A somewhat more efficient choice, but also more complicated, is obtained as follows. Let $q_1 < q_2 < \cdots$ be the sequence of all positive prime powers, and for each n let r_n be the prime of which q_n is a power. Thus the initial q_n's are 2, 3, 4, 5, 7, 9, 11, 13, 16, 17, and the corresponding r_n's are 2, 3, 2, 5, 7, 3, 11, 13, 2, 17. With this determination of the r_n's, we see that the product $r_1 r_2 \cdots r_n$ is the least common multiple of the numbers $q_1, q_2, \cdots, q_n$, which in turn is equal to the least common multiple of all the positive integers not exceeding q_n.

In general, the running time of the Pollard $p - 1$ method is expected to be comparable to the minimum over $p|m$ of the maximum prime divisor of $p - 1$. This is faster than the Pollard rho method for a substantial fraction of m, but on average it is barely faster than trial division. Some choices of a will lead to a proper divisor faster than others, but the ones that yield a substantial savings are comparatively rare and are unlikely to be found by random trials. Thus in practice we simply take $a = 2$. The numbers g_n are calculated by the Euclidean algorithm, but since the g_n form an increasing sequence, it is not necessary to evaluate g_n for every n. Hence some time may be saved by computing g_n for only one n out of 100, say.

The strategy of the Pollard $p - 1$ method is to find the identity in the multiplicative group of reduced residue classes (mod p) by raising a given number to a highly composite power. Here p is a prime divisor of m, and as the value of p is unknown, the calculation is executed modulo m. The method is quick if there is a prime $p|m$ such that the order of the group, $p - 1$, is composed entirely of small primes. We now construct, for each prime p, a large family of additive groups, in which the group addition is calculated using congruence arithmetic (mod p). Since the order of any member g of a finite group G divides the order of the group (recall Theorem 2.49), it follows that $r_1 r_2 \cdots r_n g$ is the identity of the group if the order of the group divides $r_1 r_2 \cdots r_n$. Working modulo m, we calculate a highly composite multiple of some initial element, in order to find the identity in the group. Since this identity is related to the residue class $0 \pmod p$, this reveals the value of p, and a proper divisor of m is located. These groups are of various orders, and we expect that some of them will yield a factorization of m very quickly. We use the same highly composite number $r_1 r_2 \cdots r_n$ as before, but now we limit the size of n. If we are unsuccessful with one group, we start afresh with a different group and continue switching from group to group until a factor is found.

The groups we need are provided by considering elliptic curves modulo p. If $f(x, y)$ is a polynomial with integral coefficients, then the affine curve $\mathscr{C}_f(\mathbb{Z}_p)$ is the collection of pairs (x, y) of integers with $0 \leqslant x < p$, $0 \leqslant y < p$, for which $f(x, y) \equiv 0 \pmod p$. Thus a line (mod p) is the collection of pairs (x, y) satisfying a congruence $ax + by + c \equiv 0 \pmod p$, where p does not divide both a and b. By using Theorem 2.26 we may establish an analogue of Theorem 5.15, and thus show that if a curve $\mathscr{C}_f(\mathbb{Z}_p)$ of degree $d \pmod p$ has more than d points in common with a line $y \equiv mx + r \pmod p$, then there exist polynomials $k(x, y)$ and $q(x, y)$ with integral coefficients such that

$$f(x, y) = (y - mx - r)k(x, y) + pq(x, y).$$

Although the set $\mathscr{C}_f(\mathbb{Z}_p)$ is finite, we may nevertheless define the multiplicity M of a point (x_0, y_0) on this curve to be the largest integer M such that

$$\left(\frac{\partial}{\partial x}\right)^i\left(\frac{\partial}{\partial y}\right)^j f(x_0, y_0) \equiv 0 \pmod{p}$$

whenever $i + j < M$. Continuing in this manner, we may similarly define the intersection multiplicity at (x_0, y_0) of a line $y \equiv y_0 + m(x - x_0) \pmod{p}$ with a curve $f(x, y) \equiv 0 \pmod{p}$ to be the largest integer i such that $f(x, y_0 + m(x - x_0)) = (x - x_0)^i k(x) + pq(x)$. Such a line is tangent if $i > 1$. As an analogue of (5.45), we note that if $p(x) = a_n x^n + a_{n-1}x^{n-1} + \cdots$ is a polynomial with integral coefficients, if $p \nmid a_n$, and if $x_1, x_2, \cdots, x_{n-1}$ are solutions of the congruence $p(x) \equiv 0 \pmod{p}$, then this congruence has an nth solution x_n given by the relation

$$x_1 + x_2 + \cdots + x_n \equiv -a_{n-1}\overline{a_n} \pmod{p} \tag{5.58}$$

where $a_n\overline{a_n} \equiv 1 \pmod{p}$. The x_j may be repeated, provided that the factor $(x - x_j)$ is correspondingly repeated in the factorization of $p(x) \pmod{p}$.

If $f(x, y)$ is a polynomial with integral coefficients, of degree 3 (mod p), if $f(x, y)$ is irreducible (mod p), and *if* the projective curve $\mathscr{C}_f(\mathbb{Z}_p)$ has no singular point, then we call this curve an *elliptic curve modulo p*. If **A** and **B** are any two points of such a curve, we may construct the unique line (mod p) passing through **A** and **B**, and then by (5.58) find the unique third point **AB** of intersection of the line with the curve. If **0** is a further point on the curve, we may define **A** + **B** = **0(AB)**, as in the preceding section. The points $\mathscr{C}_f(\mathbb{Z}_p)$ form a group under this addition, and as with $\mathscr{C}_f(\mathbb{R})$ the hard part of the proof is to verify the associative law. The first part of the proof of Lemma 5.20 carries over to the present situation, but some further work is required to complete the proof. We omit this argument, and take for granted that the points $\mathscr{C}_f(\mathbb{Z}_p)$ form a group, as our only object is to construct a calculational procedure by which a computer might locate a proper divisor of a large composite integer.

Any polynomial of the form $y^2 - x^3 + Ax + B$ is irreducible (mod p), and by calculating partial derivatives we see that the curve $y^2 \equiv x^3 - Ax - B \pmod{p}$ is nonsingular provided that the polynomial $x^3 - Ax - B$ has no repeated root (mod p). Suppose that r is a repeated root of this polynomial. Then by (5.58) the third root is $\equiv -2r \pmod{p}$. Hence the coefficients of the polynomial $(x - r)^2(x + 2r)$ are congruent (mod p) to the coefficients of the given polynomial. Thus $A \equiv 3r^2$, $B \equiv -2r^3 \pmod{p}$, so that $4A^3 - 27B^2 \equiv 108r^6 - 108r^6 \equiv 0 \pmod{p}$. We conclude that the curve

$$y^2 \equiv x^3 - Ax - B \pmod{p} \tag{5.59}$$

is an elliptic curve (mod p) provided that

$$4A^3 - 27B^2 \not\equiv 0 \pmod{p}. \tag{5.60}$$

A projective curve given by (5.59) contains the point $0:1:0$ at infinity, which is an inflection point, and we take this point to be $\mathbf{0}$. The derivation of the formulae for adding two points runs as in the preceding section, and corresponding to (5.55), (5.56), we find that if $\mathbf{P}_1 = (x_1, y_1)$ and $\mathbf{P}_2 = (x_2, y_2)$ are two points of the curve with $x_1 \not\equiv x_2 \pmod{p}$, then $\mathbf{P}_3 = (x_3, y_3) = \mathbf{P}_1 + \mathbf{P}_2$ is given by

$$\begin{aligned} x_3 &\equiv (y_2 - y_1)^2\left(\overline{x_2 - x_1}\right)^2 - x_1 - x_2 \pmod{p}, \\ y_3 &\equiv -y_1 - (y_2 - y_1)\left(\overline{x_2 - x_1}\right)(x_3 - x_1) \pmod{p} \end{aligned} \tag{5.61}$$

where $\overline{x_2 - x_1}$ is chosen so that $(x_2 - x_1)(\overline{x_2 - x_1}) \equiv 1 \pmod{p}$. If $x_1 \equiv x_2 \pmod{p}$, then $y_1 \equiv \pm y_2 \pmod{p}$. If $y_2 \equiv -y_1 \pmod{p}$, then $\mathbf{P}_2 = -\mathbf{P}_1$, and $\mathbf{P}_1 + \mathbf{P}_2 = \mathbf{0}$. If $y_2 \equiv y_1 \pmod{p}$, then $\mathbf{P}_2 = \mathbf{P}_1$, and we find that $2\mathbf{P}_1 = \mathbf{P}_3 = (x_3, y_3)$ is given by the congruences

$$\begin{aligned} x_3 &\equiv \left(3x_1^2 - A\right)^2\left(\overline{2y_1}\right)^2 - 2x_1 \pmod{p}, \\ y_3 &\equiv -y_1 - \left(3x_1^2 - A\right)\left(\overline{2y_1}\right)(x_3 - x_1) \pmod{p} \end{aligned} \tag{5.62}$$

where $\overline{2y_1}$ is an integer chosen so that $2y_1(\overline{2y_1}) \equiv 1 \pmod{p}$.

Example 11 Find the multiples of the point $\mathbf{P}_1 = (3, 2)$ on the curve $y^2 \equiv x^3 - 2x - 3 \pmod{7}$.

Solution From (5.62) we find that $x_3 \equiv (3 \cdot 3^2 - 2)^2(\overline{2 \cdot 2})^2 - 2 \cdot 3 \equiv 4^2 \cdot 2^2 + 1 \equiv 2 \pmod{7}$, and hence that $y_3 \equiv -2 - 4 \cdot 2(2 - 3) \equiv 6 \pmod{7}$. One may verify independently that the point $\mathbf{P}_3 = (2, 6)$ lies on the elliptic curve. We apply (5.61) similarly to see that $3\mathbf{P}_1 = (4, 2)$, $4\mathbf{P}_1 = (0, 5)$, $5\mathbf{P}_1 = (5, 0)$, $6\mathbf{P}_1 = (0, 2)$, $7\mathbf{P}_1 = (4, 5)$, $8\mathbf{P}_1 = (2, 1)$, $9\mathbf{P}_1 = (3, 5)$, $10\mathbf{P}_1 = \mathbf{0}$. By evaluating the Legendre symbol $\left(\dfrac{x^3 - 2x - 3}{7}\right)$ for $x = 0, 1, 2, \cdots, 6$, we discover that $\mathscr{C}_f(\mathbb{Z}_7)$ consists of precisely these 10 points. Hence in this case $E_f(\mathbb{Z}_p)$ is a cyclic group of order 10.

For each x, the congruence (5.59) is satisfied by at most two values of $y \pmod{p}$. Hence the total number of solutions of (5.59) lies between 0 and $2p$. The projective curve $\mathscr{C}_f(\mathbb{Z}_p)$ contains precisely one point at infinity, namely $0:1:0$, so that the group $E_f(\mathbb{Z}_p)$ has order between 1 and $2p+1$. One would expect that the right side of (5.59) is a quadratic residue (mod p) for roughly half the residues x, so that the order of $E_f(\mathbb{Z}_p)$ should be close to p. Indeed, it is known that

$$\left|\left|E_f(\mathbb{Z}_p)\right| - (p+1)\right| < 2\sqrt{p}\,. \tag{5.63}$$

We now apply these groups to factor a composite number m. We calculate multiples of a point $\mathbf{P}_0$ that lies on an elliptic curve. More precisely, we compute $\mathbf{P}_1 = r_1\mathbf{P}_0, \mathbf{P}_2 = r_2\mathbf{P}_1, \mathbf{P}_3 = r_3\mathbf{P}_2, \cdots, \mathbf{P}_N = r_N\mathbf{P}_{N-1}$ where the numbers r_n are the same as in the Pollard $p-1$ method, and N is a parameter at our disposal. Since the prime divisors of m are unknown, we use congruences modulo m.

To calculate the multiple of a point we repeatedly double, in the manner of the repeated squaring technique used in Section 2.4 to compute powers. For example, to compute $101\mathbf{P}$, we double 6 times to compute $2\mathbf{P}, 4\mathbf{P}, 8\mathbf{P}, 16\mathbf{P}, 32\mathbf{P}, 64\mathbf{P}$, and then we perform three additions to compute $\mathbf{P} + 4\mathbf{P} + 32\mathbf{P} + 64\mathbf{P} = 101\mathbf{P}$. Since we intend to perform many such doublings and additions, it is important that these basic manipulations be performed as quickly as possible. Unfortunately, the formulae (5.61) and (5.62) involve inverting a residue class. Even by the Euclidean algorithm, this involves a number of additional manipulations. To avoid this extra burden, we instead use congruential analogues of the formulae (5.55) and (5.56), which involve only addition and multiplication of residue classes.

Example 12 Use the ECM to factor the number $m = 1938796243$.

As with the methods of Pollard, if we apply the ECM to a prime number, then calculations are performed for a very long time, with no definitive outcome. Thus one should only apply these methods to numbers that are already known to be composite. In the present case it is easy to verify that $2^{m-1} \equiv 1334858860 \not\equiv 1 \pmod{m}$, so that m must be composite. Before trying more sophisticated techniques, one should also use trial division to remove any small prime factors, say those not exceeding 10000. In the present case, the trial divisions fail to disclose any factor, so we know that the composite number m is composed entirely of primes larger than 10000.

Solution We use the curves $y^2 \equiv x^3 - Ax + A \pmod{p}$, $A = 1, 2, 3, \cdots$, and take our initial point to be $(1, 1)$. Condition (5.60) fails if $4A \equiv 27 \pmod{p}$, but since we have already determined that m has no prime divisor less than 10000, it follows that g.c.d. $(4A - 27, m) = 1$ for $1 \leqslant A \leqslant$

2500. For a given value of A, we use (5.55) and (5.56) as congruences $(\bmod m)$ and compute triples (u_n, v_n, w_n), which determine the points $\mathbf{P}_n$, where $\mathbf{P}_0$ is given by the triple $(1, 1, 1)$ and $\mathbf{P}_n = r_n \mathbf{P}_{n-1}$. We take $N = 16$. Since $q_{16} = 29$, this amounts to considering those prime powers not exceeding 30. After the triple (u_{16}, v_{16}, w_{16}) determining $\mathbf{P}_{16}$ is calculated, we evaluate g.c.d.(w_{16}, m). For $A = 1, 2, \cdots, 6$ we discover that these numbers are relatively prime, but when we take $A = 7$, we find that g.c.d.$(w_{16}, m) = 37409$. Hence $m = 37409 \cdot 51827$. Since we have already verified that m has no prime divisor less than 10000, it follows that these two factors are prime numbers.

The desired factorization has been achieved, but a few further remarks are in order. By calculating multiples of the point $\mathbf{P}_0 = (1, 1)$ on the curve $y^2 \equiv x^3 - 7x + 7 \pmod{37409}$, we may verify that $(2^4 \cdot 3^3 \cdot 5^2 \cdot 7 \cdot 11 \cdot 13 \cdot 17 \cdot 19 \cdot 23 \cdot 29)\mathbf{P}_0 = \mathbf{0}$. By calculating various multiples of $\mathbf{P}_0$ we may determine that its order is exactly $2 \cdot 3^2 \cdot 5 \cdot 11 \cdot 19$. By a more lengthy calculation based on Problem 6 at the end of this section, we may also show that $|E_f(\mathbb{Z}_p)| = 37620 = 2^2 \cdot 3^2 \cdot 5 \cdot 11 \cdot 19$. Thus the order of $\mathbf{P}_0$ divides the order of the group, as it must by Theorem 2.49.

When using the formulae (5.55) and (5.56) modulo p, we appeal to (5.55) if $u_1 v_2^2 \not\equiv u_2 v_1^2 \pmod p$, and otherwise use (5.56). When using these formulae to factor a number m, we proceed with congruences modulo m, and use (5.55) if $u_1 v_2^2 \not\equiv u_2 v_1^2 \pmod m$. In the course of such calculations we may encounter a situation in which $u_1 v_2^2 - u_2 v_1^2$ is divisible by p, but not by m. In such a case, we use (5.55) whereas the corresponding calculation $(\bmod p)$ would use (5.56) instead. Consequently, further calculations $(\bmod m)$ no longer correspond to the calculation of multiples of $\mathbf{P}_0 \pmod p$. No harm is done, however, for we see from (5.55) that the resulting number w_3 is divisible by p. From the formulae for w_3 in (5.55) and (5.56) we see that all subsequent w's will be divisible by p. Thus the prime p is disclosed when we calculate g.c.d. (w_N, m), even though the triple (u_N, v_N, w_N) may not correspond to the point $\mathbf{P}_N \pmod p$.

One may experiment with various choices of the parameter N, to determine which value minimizes the total mount of calculation. Suppose we wish to find a prime factor p of m, and let $f(u)$ be the function $f(u) = \exp(\sqrt{(\log u)(\log\log u)/2})$. Heuristic arguments indicate that in the limit one should construct multiples of $\mathbf{P}_0$ corresponding to prime-powers q_n not exceeding $f(p)$, and that the number of different values of A that will be treated before finding p may be expected to be comparable to $f(p)$, on average. Thus it is expected that the total number of arithmetic manipulations needed to find p by this method is roughly of the order of magnitude of $f(p)^2 = \exp(\sqrt{2(\log p)(\log\log p)})$. Since the least prime factor of a composite number m is $\leqslant \sqrt{m}$, it follows that one should be able to factor m by performing not too much more than $\exp(\sqrt{(\log m)(\log\log m)})$ arithmetic operations. One advantage of this method is that one may use it to locate the smaller prime factors p of a number m that is much too large to factor completely.

PROBLEMS

1. Show that the number of pairs (A, B) of integers, $0 \leqslant A < p, 0 \leqslant B < p$, for which $4A^3 \not\equiv 27B^2$ is exactly $p^2 - p$. (H)
2. Let $\mathscr{C}_f(\mathbb{Z}_p)$ be an elliptic curve modulo p given by the congruence $y^2 \equiv x^3 - Ax - B \pmod{p}$. Let r be a number such that $(r, p) = 1$, put $A' = r^4A$, $B' = r^6B$, and let $\mathscr{C}_g(\mathbb{Z}_p)$ be the elliptic curve given by $v^2 \equiv u^3 - A'u - B' \pmod{p}$. Show that if $(x, y) \in \mathscr{C}_f(\mathbb{Z}_p)$, then $(r^2x, r^3y) \in \mathscr{C}_g(\mathbb{Z}_p)$, and that this linear map places the points of $\mathscr{C}_f(\mathbb{Z}_p)$ in one-to-one correspondence with those of $\mathscr{C}_g(\mathbb{Z}_p)$. Show that this linear map takes lines to lines, and thus preserves addition. Conclude that $E_f(\mathbb{Z}_p) \cong E_g(\mathbb{Z}_p)$. Call two curves that are related in this way *isomorphic*. Show that isomorphisms among curves define an equivalence relation, and that if $p > 2$ then there are $(p-1)/2$ curves in each equivalence class, and $2p$ equivalence classes. (In addition to these obvious isomorphisms among the groups $E_f(\mathbb{Z}_p)$, there may be other, less obvious ones.)
3. Show that the projective plane $\mathbb{P}_2(\mathbb{Z}_p)$ contains exactly $p^2 + p + 1$ points.
4. Let p be a prime number, $p > 2$, and suppose that x and y are integers such that $x^2 + y^2 \equiv 1 \pmod{p}$, $x \not\equiv 1 \pmod{p}$. Let u be determined by the congruence $(1 - x)u \equiv y \pmod{p}$. Show that $u^2 + 1 \not\equiv 0 \pmod{p}$, and that $x \equiv (u^2 - 1)v$, $y \equiv 2uv \pmod{p}$, where $(u^2 + 1)v \equiv 1 \pmod{p}$. Show, conversely, that if u is an integer such that $u^2 + 1 \not\equiv 0 \pmod{p}$, and if v, x, y are given in terms of u as above, then $x^2 + y^2 \equiv 1 \pmod{p}$ and $x \not\equiv 1 \pmod{p}$. Show that the number of $u \pmod{p}$ that arise in this way is $p - 1 - \left(\frac{-1}{p}\right)$. Deduce that the number of solutions (x, y) of the congruence $x^2 + y^2 \equiv 1 \pmod{p}$ is $p - \left(\frac{-1}{p}\right)$.
5. Show that if $p > 3$, $4A^3 + 27B^2 \equiv 0 \pmod{p}$, $p \nmid A$, then the root r of the congruence $-2Ar \equiv 3B \pmod{p}$ is a repeated root $\pmod{p}$ of the polynomial $x^3 - Ax - B$.
6. Suppose that the polynomial $x^3 + ax^2 + bx + c$ has no repeated root $\pmod{p}$, and put $f(x, y) = y^2 - (x^3 + ax^2 + bx + c)$. Show that the

group of points on the elliptic curve $\mathscr{C}_f(\mathbb{Z}_p)$ has order

$$\left|E_f(\mathbb{Z}_p)\right| = p + 1 + \sum_{x=1}^{p} \left(\frac{f(x)}{p}\right).$$

5.9 CURVES OF GENUS GREATER THAN 1

Let $f(x, y)$ be a polynomial of degree d whose coefficients may be rational, real, or complex. We speak of the set $\mathscr{C}_f(\mathbb{C})$ as a projective curve, though topologically it is a closed oriented surface. As such, it has a topological *genus*, which is a non-negative integer g. It turns out that this genus is of fundamental importance in classifying curves. We do not give a precise definition of the genus of a curve, but we state a few useful rules by which it may be calculated in elementary terms. We suppose that $f(x, y)$ is irreducible over $\mathbb{C}$, so that $\mathscr{C}_f(\mathbb{C})$ is an irreducible curve. If $\mathscr{C}_f(\mathbb{C})$ is nonsingular, then its genus is $g = (d-1)(d-2)/2$. Thus a conic has genus 0, and an elliptic curve has genus 1. It may be shown that an irreducible curve of degree d can have at most $(d-1)(d-2)/2$ singular points. A double point (x_0, y_0) of $\mathscr{C}_f(\mathbb{C})$ is called *ordinary* if the quadratic form

$$\frac{\partial^2 f}{\partial x^2}(x_0, y_0)u^2 + 2\frac{\partial f}{\partial x\,\partial y}(x_0, y_0)uv + \frac{\partial^2 f}{\partial y^2}(x_0, y_0)v^2$$

has distinct roots. At such a double point, the curve crosses itself nontangentially. If $f(x, y)$ is irreducible and the only singular points of $\mathscr{C}_f(\mathbb{C})$ are ordinary double points, of which there are N, then $g = (d-1)(d-2)/2 - N$.

Some care must be exercised in applying these rules to calculate the genus. For example, the quartic curve $y^2 = x^4 + 1$ has no singularity in affine space, but it has a double point at $0:1:0$. Moreover, this double point is not an ordinary double point. In our discussion following the proof of Theorem 5.24, we found that this quartic is birationally equivalent to the elliptic curve $y^2 = x^3 - 4x$. It is known that the genus is invariant under birational transformation, although, as we see in this example, the degree is not. Hence the curve $y^2 = x^4 + 1$ has genus 1. More generally, it is known that any irreducible planar curve is $\mathbb{C}$-birationally equivalent to a planar curve whose only singular points are ordinary double points. In addition, if $p(x)$ is of degree d and has distinct roots, then the curve $y^2 = p(x)$ has genus $g = [(d-1)/2]$.

Suppose now that $f(x, y)$ is an irreducible polynomial with rational coefficients. It is known that if $\mathscr{C}_f(\mathbb{C})$ has genus 0 and if the curve contains at least one rational point (i.e., $\mathscr{C}_f(\mathbb{Q})$ is nonempty), then $\mathscr{C}_f(\mathbb{C})$ is $\mathbb{Q}$-birationally equivalent to a line. Our treatment of conics and of singular cubics are special cases of this. If $\mathscr{C}_f(\mathbb{C})$ has genus 1 and if $\mathscr{C}_f(\mathbb{Q})$ is nonempty, then the curve is $\mathbb{Q}$-birationally equivalent to an elliptic curve.

In 1923, Mordell conjectured that a curve of genus greater than 1 can possess at most finitely many rational points. This conjecture, known as the Mordell Conjecture, was proved in 1983.

Faltings' Theorem *Let $f(x, y)$ be a polynomial with rational coefficients that is irreducible over the field of complex numbers. If the curve $\mathscr{C}_f(\mathbb{C})$ has genus $g > 1$, then the set $\mathscr{C}_f(\mathbb{Q})$ of rational points on the curve is at most finite.*

To see how this might be applied to Diophantine equations, we note that integral solutions of the equation $x^n + y^n = z^n$ with $z \neq 0$ and g.c.d. $(x, y, z) = 1$ are in one-to-one correspondence with rational points on the curve $x^n + y^n = 1$, the so-called Fermat curve. Indeed, in projective coordinates this curve is given by the equation $X^n + Y^n = Z^n$. Taking partial derivatives with respect to X, Y, and Z, we find that all partials vanish only at the origin. Since the origin is not a member of projective space, we conclude that this curve is nonsingular. Hence its genus is $(n - 1)(n - 2)/2$. Thus Faltings' Theorem implies that for each $n > 3$, the equation $x^n + y^n = z^n$ has at most finitely many primitive integral solutions.

Faltings' Theorem does not provide a specific finite upper bound for the number of rational points on the curve, though efforts are being made to strengthen Faltings' theorem in this manner. A more distant goal would be to find an explicit function of the coefficients of $f(x, y)$ that provides an upper bound for the numerators and denominators of the coordinates of the rational points on the curve. Such a bound would have the effect of reducing the problem of finding all rational points to a finite calculation, for any given curve of genus greater than 1.

NOTES ON CHAPTER 5

§5.1 Catalan conjectured that 8 and 9 are the only positive consecutive perfect powers. That is, the only integral solutions of the equation $x^m - y^n = 1$ with $x > 0$, $y > 0$, $m > 1$, $n > 1$ is $3^2 - 2^3 = 1$. Since m and n are variables, this provides a natural example of a Diophantine equation that involves an expression that is not a polynomial. Catalan's conjecture is not fully resolved, but in 1974 Robert Tijdeman applied deep methods of the theory of transcendental numbers to show that there is an effectively computable constant C such that all consecutive perfect powers are less than C. Thus Catalan's question is resolved, apart from a certain finite calculation, which, however, is too long to perform.

§5.2 For further discussion of the equivalence of matrices, additional properties of the Smith canonical form, invariant factors, and determinantal divisors, and for interesting applications of this material, see Chapter 2 of the book by Newman, or Chapter 14 of the book by Hua.

§5.3 The analysis of Pythagorean triples was formerly attributed to the Pythagorean school (ca. 500 B.C.), but it now seems that the full details of Theorem 5.5 were known to the Babylonians as early as 1600 B.C.

§5.4 The Hasse-Minkowski principle for quadratic forms was first proved by Hasse in 1923. The proof proceeds separately for binary, ternary, and quaternary forms, the last case being the most difficult. An easier method gives the result for all quadratic forms in 5 or more variables. Detailed derivations are provided in Borevich and Shafarevich, in Serre, and in Cassels (1978). A more difficult generalized version is found in O'Meara. It seems that the first proof that the Hasse-Minkowski principle does not hold in general was given in 1942 by H. Reichardt, who showed that the equation $x^4 - 17 = 2y^2$ is everywhere locally solvable but has no integral solution.

To solve the congruence (5.24), one must first consider the case $j = 1$. It is known that if P has integral coefficients and is absolutely irreducible (i.e., irreducible over the field $\mathbb{C}$ of complex numbers), then there is a function $p_0(P)$ of P such that the congruence (5.24) is solvable (mod p) for all primes $p > p_0(P)$. Unfortunately, all known proofs of this involve sophisticated techniques of algebraic geometry, although many interesting cases (such as the equation in Theorem 5.8) can be treated by comparatively elementary use of exponential sums. For more details on this, see Chapter 2 of the book by Borevich and Shafarevich. The primes $p \leqslant p_0(P)$ must be considered individually. Once solutions of (5.24) have been found (mod p), one can usually extend the solutions to the moduli p^j by Hensel's lemma, though in some cases one encounters singularities that make this difficult or impossible. In 1884, A. Meyer proved that any quadratic form with integral coefficients in 5 or more variables has a nontrivial zero (mod p^j) for all p and all j. Here the number 5 cannot be reduced. Indeed, it is not difficult to find a form of degree d in d^2 variables that for some suitable p has no nontrivial zero (mod p^d), the example in Problem 7 being typical. In the 1930s, E. Artin conjectured that any form of degree d in at least $d^2 + 1$ variables has a nontrivial zero (mod p^j) for every p and every j. In 1944, R. Brauer proved a weak form of this, namely that there is a number $n_0(d)$ such that every form of degree d in at least $n_0(d)$ variables has a nontrivial zero (mod p^j) for every p and every j. In 1951, D. J. Lewis proved Artin's conjecture for

$d = 3$, but in 1966 Terjanian found a form (see Problem 8) of degree 4 in 18 variables with no nontrivial zero (mod 16). It is now known that $n_0(d)$ grows very rapidly with d. Nevertheless, in 1965 Ax and Kochen used tools of mathematical logic to show that for every d there is a set $\mathscr{P}_d$ of primes, which may be empty but in any case is at most finite, such that if P is homogeneous of degree d in more than d^2 variables, then (5.24) has a nontrivial solution for all $j \geqslant 1$, if $p \notin \mathscr{P}_d$. For $d > 2$ the set $\mathscr{P}_d$ of exceptional primes has not been precisely determined.

Many particular Diophantine equations have been treated by means of special methods. Such techniques are sometimes exceedingly ingenious, as in the proof of Theorem 5.8, given first by V. A. Lebesgue in 1869. In sharp contrast to the special methods used in proving Theorem 5.8 and 5.9, we note that the powerful inequality (5.26) enables us to treat a wide class of Diophantine equations. The first nontrivial estimate in the direction of (5.26) was established in 1909 by Axel Thue. The estimate was improved, first by C. L. Siegel, then by Freeman Dyson, and finally K. F. Roth proved (5.26) in 1955. The estimate is best-possible, though the problem of determining explicitly the dependence of C on ε and P remains unsolved.

The equation in Theorem 5.8 is a special case of *Bachet's equation*, $x^3 + k = y^2$. We treat $k = -1$ and $k = -2$ in Section 9.9, by means of the arithmetic of quadratic number fields. In 1917, Thue used his weak form of (5.26) to show that for any given nonzero integer k, this equation has at most finitely many solutions. Using deep estimates from the theory of transcendental numbers, for each $k \neq 0$ one can give a bound for the size of x and y, and hence reduce the problem of finding all solutions to a finite calculation.

The proof of Theorem 5.10 offers a good example of Fermat's "method of infinite descent." In this application, the argument raises more questions than it answers, concerning the nature of the mysterious connection between the two equations (5.27) and (5.29). One may note that our method constructs a rational transformation from the curve $x^4 + 1 = y^2$ to the curve $x^4 - 4 = y^2$, and a second rational transformation that takes the second curve back to the first. These curves have genus 1, and descent is very effective when applied to such curves, but a full explanation of the reasons for this involves a sophisticated discussion of cohomology and two-coverings of elliptic curves, as in the paper of J. W. S. Cassels, "Diophantine equations with special reference to elliptic curves," *J. London Math. Soc.*, 41 (1966), 193–291. In other situations descent may be used to generate new solutions from a given one, or to show that all solutions are generated from some initial solution.

By Theorem 5.10 we see that Fermat's last theorem is settled when $4|n$. This much was done by Fermat. All other n have an odd prime divisor, and thus to settle the problem completely it suffices to show that for each prime $p > 2$, the equation $x^p + y^p = z^p$ has no solution in positive integers. Euler settled the case $p = 3$ in 1770, and Dirichlet and Legendre proved the result for $p = 5$ in 1825, but the greatest contributions were made by E. E. Kummer in the mid-nineteenth century. To describe Kummer's approach, suppose that p is prime, $p > 2$, and let ζ denote a primitive pth root of unity, say $\zeta = e^{2\pi i/p}$. Using this complex number, we see that

$$x^p + y^p = (x + y)(x + \zeta y) \cdots (x + \zeta^{p-1}y).$$

Kummer used this factorization in the same manner as in the argument in Section 5.3. If these factors have no common divisors, then one would think that each factor must be a pth power, but unique factorization fails in this ring when p is large. Kummer discovered that the unique factorization of these numbers is restored if one works in a still larger algebraic number field obtained by adjoining certain further algebraic numbers. Kummer called these numbers "ideal elements," but it was later found that the same effect can be achieved by manipulating certain sets of numbers within the original algebraic number field. Since these sets replace Kummer's ideal elements, they were called "ideals." It can be shown that the ideals in an algebraic number field factor uniquely into prime ideals, even though the integers in the field may not. Kummer developed the arithmetic of integers in algebraic number fields and formulated a criterion, which if satisfied, guarantees that Fermat's equation has no solution. In this way, Kummer was able to settle the problem for many exponents p. Kummer's criterion has since been greatly strengthened. In 1954, D. H. Lehmer, E. Lehmer, and H. S. Vandiver, "An application of high-speed computing to Fermat's last theorem," *Proc. Nat. Acad. Sci. USA*, 40, (1954), 25–33, 732–735, gave a powerful criterion involving only integer arithmetic, which is not known to fail for any prime p, although it is still not known that the criterion is satisfied for infinitely many primes. J. Tanner and S. Wagstaff, "New congruences for the Bernoulli numbers," *Math. Comp.* 48 (1987), 341–350, verified that the criterion holds for all $p < 150{,}000$, and thus Fermat's last theorem is settled for these exponents. Using somewhat different methods, which go back to work of Sophie Germain in the nineteenth century, together with deep estimates from the analytic theory of prime numbers, in 1985 Adleman, Heath-Brown, and Fouvry proved that there are infinitely many primes p such that the equation $x^p + y^p = z^p$ has no solution for which p divides none of the variables. For a detailed account of the history and mathematics surrounding Fermat's last theorem see the books by Paulo Ribenboim listed in the General References, as well as his more recent journal article "Recent results about Fermat's last theorem," *Expos. Math.*, 5 (1987), 75–90. In

1986, G. Frey proposed that Fermat's last theorem might be approached by considering the elliptic curve $y^2 = x(x - a^p)(x - c^p)$. K. Ribet, "On modular representations of Gal$(\overline{Q}/Q)$ arising from modular forms," *Invent. Math.* **100** (1990), 431–476, has confirmed this by showing that if a and c are nonzero rational numbers such that $a^p + b^p = c^p$ for some nonzero rational number b, then the curve violates the Weil-Taniyama conjecture concerning elliptic curves.

§5.5 To diagonalize a quadratic form by linear transformations with rational coefficients, one may use the Gram-Schmidt process, as discussed in many texts on linear algebra. The proof of Theorem 5.11 follows an account devised by L. J. Mordell, "On the equation $ax^2 + by^2 - cz^2 = 0$," *Monats. Math.*, **55** (1951) 323–327. This proof is quite different from that given by Legendre, but like Legendre's proof, does not use quadratic reciprocity. Indeed, Legendre deduced some special cases of quadratic reciprocity from this result. In 1950, L. Holzer showed that if a, b, c are as described in Theorem 5.11, then not only does a nontrivial integral solution exist, but there is such a solution for which $|x| \leqslant \sqrt{|bc|}$, $|y| \leqslant \sqrt{|ac|}$, $|z| \leqslant \sqrt{|ab|}$. An elementary proof of Holzer's theorem has been given by L. J. Mordell, "On the magnitude of integer solutions of the equation $ax^2 + by^2 + cz^2 = 0$," *J. Number Theory*, 1 (1969), 1–3.

Theorem 5.14 is a special case of a theorem of Chevalley and Warning which asserts that if $f(\mathbf{x})$ is a homogeneous polynomial of degree d in n variables, then the congruence $f(\mathbf{x}) \equiv 0 \pmod p$ has a nontrivial solution provided that $n > d$. An account of this is found in Section 1.1 of Borevich and Shafarevich.

§5.6 The use of the tangent line to generate a point on a cubic curve from a given point is found in Diophantus, and this method was used extensively by Bachet and Fermat. However, it seems that the use of a chord to generate a new point from two given points occurs first in a manuscript of Newton.

§5.7 The definition of the sum of two points on an elliptic curve was given first by Cauchy in 1835, but the further observation that this defines a group seems to have been made first by Poincaré in 1901. Poincaré tacitly assumed that the group is finitely generated, and it was only in 1921 that this was proved by Mordell. André Weil, in his doctoral thesis of 1928, gave not only a new proof of Mordell's theorem, but extended it to algebraic number fields and generalized it to abelian varieties of higher dimension.

It is perhaps not immediately evident why the nonsingular cubic curve is termed "elliptic." To establish the connection, we remark that it is natural to express the arc length of an ellipse as an integral involving the square root of a quartic polynomial. By making a rational change of variables, this may be reduced to an integral involving the square root of a cubic polynomial. In general, an integral involving the square root of a quartic or cubic polynomial is called an *elliptic integral*. Such integrals were extensively studied in the eighteenth and nineteenth centuries, and methods were developed to reduce them to integrals of a few standard forms. An indefinite elliptic integral is not an elementary function, but it can be represented by introducing a new transcendental function, the *Weierstrass $\wp$-function*, which satisfies the differential equation $\wp'^2 = \wp^3 - A\wp - B$. Consequently, the change of variables $x = \wp(t)$ gives

$$\int_a^b (x^3 - Ax - B)^{-1/2}\, dx = \wp^{-1}(b) - \wp^{-1}(a).$$

This is analogous to the observation that $\sin x$ is a solution of the differential equation $y^2 + y'^2 = 1$, so that the change of variables $x = \sin t$ gives $\int_a^b (1 - x^2)^{-1/2}\, dx = \arcsin b - \arcsin a$. In the same way that we parameterize the unit circle as $(\cos t, \sin t)$, we may parameterize the elliptic curve $\mathscr{C}_f(\mathbb{C})$ as $(\wp(t), \wp'(t))$. Moreover, it may be shown that if $\mathbf{A} = (\wp(t), \wp'(t))$ and $\mathbf{B} = (\wp(u), \wp'(u))$, then $\mathbf{A} + \mathbf{B}$ is given by $(\wp(t+u), \wp'(t+u))$. Thus the addition of points on $\mathscr{C}_f(\mathbb{C})$ corresponds to the addition of complex numbers. When approached in this manner, it is immediately evident that this addition of points on an elliptic curve yields a group. A development of the subject along these lines is given in Koblitz (1984). The geometric approach we adopted is easily transferred to elliptic curves over other fields, as arises in Section 5.8. A different proof of Theorem 5.21, but in the same spirit, is found in Reid. A similar proof, accompanied by a more detailed development of the properties of algebraic curves, is found in Husemöller. For a complete account of intersection theory and Bézout's theorem, see the book of Fulton or of Walker. A charming introduction to elliptic curves, at a somewhat more advanced level, is found in Chahal. The graduate text by Silverman is more demanding.

The description of points of finite order on an elliptic curve was given independently by Elizabeth Lutz in 1937 and Trygve Nagell in 1935. The theorem of Siegel, proved in 1929, states that a curve of positive genus contains at most finitely many integral points. Mazur's theorem was first conjectured by Andrew Ogg, and then proved by Barry Mazur in 1977.

By combining the results of Problems 12, 16, 21, 22 one obtains an example of an elliptic curve with two real components, with rational points dense on one component but absent from the other component. This example is due to A. Bremner.

§5.8 The $p-1$ method was proposed by J. M. Pollard, "Theorems on factorization and primality testing," *Proc. Camb. Philos. Soc.*, 76 (1974), 521–528. A corresponding $p+1$ method, using Lucas sequences, has been investigated by H. C. Williams, "A $p+1$ method of factoring,"

Math. Comp., 39 (1982) 225–234. The elliptic curve method of factorization was invented by H. W. Lenstra Jr., "Factoring integers with elliptic curves," *Annals of Math.*, 126 (1987), 649–673. D. V. and G. V. Chudnovsky, "Sequences of numbers generated by addition in formal groups and new primality and factorization tests," *Advances in Applied Math.*, 7 (1986), 385–434 have discussed various formulae that may be used to implement the ECM. Since most machines perform multiplication much more slowly than addition, a rough measure of the time required to evaluate a typical expression is obtained by simply counting the number of multiplications involved. If a point P is given and we use the formulae (5.55) and (5.56) to calculate rP, on average we require about $27 \log r$ multiplications. A more efficient system of formulae, requiring only about $16 \log r$ multiplications, has been found by P. L. Montgomery, "Speeding the Pollard and elliptic curve methods of factorization," *Math. Comp.*, 48 (1987), 243–264, who also describes a number of ways of enhancing the method. A further method, the Quadratic Sieve (abbreviated QS) was invented in 1983 by Carl Pomerance (see his article in the book edited by Lenstra and Tijdeman in the General References). The Quadratic Sieve is also subject to a number of refinements and modifications. Although the description of the QS is more intricate than with ECM, the mathematics involved is more elementary. The running times of the two methods are thought to be roughly the same, but the QS seems to hold an advantage when applied to composite numbers m composed of two large primes, especially on large machines. Much of the calculation performed in executing the QS is single-precision, whereas the operations involved in the ECM are likely to involve multiple-precision arithmetic. In addition, the QS lends itself to parallel processing, even to the extent that several machines, connected only by electronic mail, may share in the task. The disadvantage of the QS is that it is memory-intense, so that it is unsuitable for use on a pocket calculator. On the other hand, the ECM makes very little use of memory and runs very well on small machines.

One way to complete the proof of the associativity of addition of points on an elliptic curve (mod p) involves observing that the field $\mathbb{Z}_p$ is contained in its algebraic closure $\overline{\mathbb{Z}_p}$, which is an infinite field. One may define what it means for two points of an elliptic curve over $\overline{\mathbb{Z}_p}$ to be "close," and thus one may complete the proof of associativity by a continuity argument, as in the proof of Lemma 5.20.

The reduction of an elliptic curve to Weierstrass normal form cannot always be carried out for elliptic curves (mod p), but one can reduce the general elliptic curve (mod p) to the shape $y^2 + axy + by \equiv x^3 + cx^2 + dx + e \pmod{p}$. To finish the reduction, one would want to complete the square, writing the left side as $(y + \overline{2}ax + \overline{2}b)^2 + \cdots$. However, this can be done only if $p \neq 2$. As for the right side of the congruence, we would want to complete the cube, writing $(x + \bar{3}c)^3 + \cdots$. This can be done, provided that $p \neq 3$. Thus one can reduce to Weierstrass form for $p > 3$, but a more general form is required if one is to capture all elliptic curves modulo 2 or 3.

The inequality (5.58) was proved in 1931 by H. Hasse. In 1948, A. Weil proved a similar inequality pertaining to irreducible curves (mod p) of arbitrary degree, and a more complicated generalization to varieties of higher dimension was established in 1973 by P. Deligne.

§5.9 The application of techniques of algebraic geometry to Diophantine equations has given rise to a subdiscipline called *Diophantine geometry*. This area traces its roots to a time just a century ago when the properties of $\mathbb{Q}$-birational equivalence were first investigated by Hilbert, Hurwitz, and Poincaré.

In a front page story on July 19, 1983, *The New York Times* announced that the Mordell Conjecture had been settled by the German mathematician Gerd Faltings. Within a few weeks, Faltings' theorem was hailed as "The theorem of the century." Faltings' paper, "Endlichkeitssätze für abelsche Varietäten über Zahlkörpern," *Invent. Math.*, 73 (1983), 349–366, is quite technical, but a useful perspective is provided by the account of D. Harris, "The Mordell conjecture," *Notices of the AMS*, 33 (1986), 443–449. Our formulation of Faltings' theorem is somewhat weakened. In its full strength, it is not restricted to plane curves, and it applies to points whose coordinates lie in any fixed algebraic number field.

CHAPTER 6

Farey Fractions and Irrational Numbers

A *rational* number is one that is expressible as the quotient of two integers. Real numbers that are not rational are said to be *irrational*. In this chapter the Farey fractions are presented; they give a useful classification of the rational numbers. Some results on irrational numbers are given in Section 6.3, and this material can be read independently of the first two sections. The discussion of irrational numbers is limited to number theoretic considerations, with no attention given to questions that belong more properly to analysis or the foundations of mathematics.

A rational number a/b with g.c.d.$(a, b) = 1$ is said to be in *reduced form*, or in *lowest terms*.

6.1 FAREY SEQUENCES

Let us construct a table in the following way. In the first row we write $0/1$ and $1/1$. For $n = 2, 3, \cdots$ we use the rule: Form the nth row by copying the $(n - 1)$st in order, but insert the fraction $(a + a')/(b + b')$ between the consecutive fractions a/b and a'/b' of the $(n - 1)$st row if $b + b' \leqslant n$. Thus, since $1 + 1 \leqslant 2$ we insert $(0 + 1)/(1 + 1)$ between $0/1$ and $1/1$ and obtain $0/1$, $1/2$, $1/1$, for the second row. The third row is $0/1$, $1/3$, $1/2$, $2/3$, $1/1$. To obtain the fourth row we insert $(0 + 1)/(1 + 3)$ and $(2 + 1)/(3 + 1)$ but not $(1 + 1)/(3 + 2)$ and $(1 + 2)/(2 + 3)$. The first five rows of the table are:

$$\begin{array}{ccccccccccc}
\frac{0}{1} & & & & & & & & & & \frac{1}{1} \\[6pt]
\frac{0}{1} & & & & & \frac{1}{2} & & & & & \frac{1}{1} \\[6pt]
\frac{0}{1} & & & \frac{1}{3} & & \frac{1}{2} & & \frac{2}{3} & & & \frac{1}{1} \\[6pt]
\frac{0}{1} & & \frac{1}{4} & \frac{1}{3} & & \frac{1}{2} & & \frac{2}{3} & \frac{3}{4} & & \frac{1}{1} \\[6pt]
\frac{0}{1} & \frac{1}{5} & \frac{1}{4} & \frac{1}{3} & \frac{2}{5} & \frac{1}{2} & \frac{3}{5} & \frac{2}{3} & \frac{3}{4} & \frac{4}{5} & \frac{1}{1}
\end{array}$$

Up to this row, at least, the table has a number of interesting properties. All the fractions that appear are in reduced form; all reduced fractions a/b such that $0 \leqslant a/b \leqslant 1$ and $b \leqslant n$ appear in the nth row; if a/b and a'/b' are consecutive fractions in the nth row, then $a'b - ab' = 1$ and $b + b' > n$. We shall prove all these properties for the entire table.

Theorem 6.1 *If a/b and a'/b' are consecutive fractions in the nth row, say with a/b to the left of a'/b', then $a'b - ab' = 1$.*

Proof It is true for $n = 1$. Suppose it is true for the $(n - 1)$st row. Any consecutive fractions in the nth row will be either a/b, a'/b' or a/b, $(a + a')/(b + b')$, or $(a + a')/(b + b')$, a'/b' where a/b and a'/b' are consecutive fractions in the $(n - 1)$st row. But then we have $a'b - ab' = 1$, $(a + a') - a(b + b') = a'b - ab' = 1$, $a'(b + b') - (a + a')b' = a'b - ab' = 1$, and the theorem is proved by mathematical induction.

Corollary 6.2 *Every a/b in the table is in reduced form, that is, $(a, b) = 1$.*

Corollary 6.3 *The fractions in each row are listed in order of their size.*

Theorem 6.4 *If a/b and a'/b' are consecutive fractions in any row, then among all rational fractions with values between these two, $(a + a')/(b + b')$ is the unique fraction with smallest denominator.*

Proof In the first place, the fraction $(a + a')/(b + b')$ will be the first fraction to be inserted between a/b and a'/b' as we continue to further rows of the table. It will first appear in the $(b + b')$th row. Therefore we have

$$\frac{a}{b} < \frac{a + a'}{b + b'} < \frac{a'}{b'}$$

by Corollary 6.3.

Now consider any fraction x/y between a/b and a'/b' so that $a/b < x/y < a'/b'$. Then

$$\frac{a'}{b'} - \frac{a}{b} = \left(\frac{a'}{b'} - \frac{x}{y}\right) + \left(\frac{x}{y} - \frac{a}{b}\right)$$

$$= \frac{a'y - b'x}{b'y} + \frac{bx - ay}{by} \geqslant \frac{1}{b'y} + \frac{1}{by} = \frac{b + b'}{bb'y}, \qquad (6.1)$$

and therefore

$$\frac{b + b'}{bb'y} \leqslant \frac{a'b - ab'}{bb'} = \frac{1}{bb'},$$

which implies $y \geqslant b + b'$. If $y > b + b'$ then x/y does not have least denominator among fractions between a/b and a'/b'. If $y = b + b'$, then the inequality in (6.1) must become equality and we have $a'y - b'x = 1$ and $bx - ay = 1$. Solving, we find $x = a + a'$, $y = b + b'$, and hence $(a + a')/(b + b')$ is the unique rational fraction lying between a/b and a'/b' with denominator $b + b'$.

Theorem 6.5 *If* $0 \leqslant x \leqslant y$, $(x, y) = 1$, *then the fraction* x/y *appears in the* yth *and all later rows.*

Proof This is obvious if $y = 1$. Suppose it is true for $y = y_0 - 1$, with $y_0 > 1$. Then if $y = y_0$, the fraction x/y cannot be in the $(y - 1)$st row by definition and so it must lie in value between two consecutive fractions a/b and a'/b' of the $(y - 1)$st row. Thus $a/b < x/y < a'/b'$. Since

$$\frac{a}{b} < \frac{a + a'}{b + b'} < \frac{a'}{b'}$$

and a/b, a'/b' are consecutive, the fraction $(a + a')/(b + b')$ is not in the $(y - 1)$st row and hence $b + b' > y - 1$ by our induction hypothesis. But $y \geqslant b + b'$ by Theorem 6.4, so we have $y = b + b'$. Then the uniqueness part of Theorem 6.4 shows that $x = a + a'$. Therefore $x/y = (a + a')/(b + b')$ enters in the yth row, and it is then in all later rows.

Corollary 6.6 *The* nth *row consists of all reduced rational fractions* a/b *such that* $0 \leqslant a/b \leqslant 1$ *and* $0 < b \leqslant n$. *The fractions are listed in order of their size.*

Definition 6.1 *The sequence of all reduced fractions with denominators not exceeding* n, *listed in order of their size, is called the* Farey sequence of order n.

The nth row of our table gives that part of the Farey sequence of order n that lies between 0 and 1, and so the entire Farey sequence of order n can be obtained from the nth row by adding and subtracting integers. For example, the Farey sequence of order 2 is

$$\ldots, \frac{-3}{1}, \frac{-5}{2}, \frac{-2}{1}, \frac{-3}{2}, \frac{-1}{1}, \frac{-1}{2}, \frac{0}{1}, \frac{1}{2}, \frac{1}{1}, \frac{3}{2}, \frac{2}{1}, \frac{5}{2}, \frac{3}{1}, \ldots$$

This definition of the Farey sequences seems to be the most convenient. However, some authors prefer to restrict the fractions to the interval from 0 to 1; they define the Farey sequences to be just the rows of our table.

Any reduced fraction with positive denominator $\leqslant n$ is a member of the Farey sequence of order n and can be called a *Farey fraction* of order n. Note that consecutive fractions a/b and a'/b' in the Farey sequence of order n satisfy the equality of Theorem 6.1 and also the inequality $b + b' > n$.

PROBLEMS

1. Let a/b and a'/b' be the fractions immediately to the left and the right of the fraction $1/2$ in the Farey sequence of order n. Prove that $b = b' = 1 + 2[(n-1)/2]$, that is, b is the greatest odd integer $\leqslant n$. Also prove that $a + a' = b$.
2. Prove that the number of Farey fractions a/b of order n satisfying the inequalities $0 \leqslant a/b \leqslant 1$ is $1 + \sum_{j=1}^{n} \phi(j)$, and that their sum is exactly half this value.
3. Let $a/b, a'/b', a''/b''$ be any three consecutive fractions in the Farey sequence of order n. Prove that $a'/b' = (a + a'')/(b + b'')$.

4. Let a/b and a'/b' run through all pairs of adjacent fractions in the Farey sequence of order $n > 1$. Prove that

$$\min\left(\frac{a'}{b'} - \frac{a}{b}\right) = \frac{1}{n(n-1)} \qquad \text{and} \qquad \max\left(\frac{a'}{b'} - \frac{a}{b}\right) = \frac{1}{n}$$

5. Consider two rational numbers a/b and c/d such that $ad - bc = 1$, $b > 0$, $d > 0$. Define n as $\max(b, d)$, and prove that a/b and c/d are adjacent fractions in the Farey sequence of order n.

6. Prove that the two fractions described in the preceding problem are not necessarily adjacent in the Farey sequence of order $n + 1$.

7. Consider the fractions from $0/1$ to $1/1$ inclusive in the Farey sequence of order n. Reading from left to right, let the denominators of these fractions be $b_1, b_2, \cdots, b_k$ so that $b_1 = 1$ and $b_k = 1$. Prove that $\sum_{j=1}^{k-1}(b_j b_{j+1})^{-1} = 1$.

8. Show that if n is a positive integer then $\Sigma(bb')^{-1} = 1$ where the sum is over all pairs (b, b') of integers for which $1 \leqslant b \leqslant n$, $1 \leqslant b' \leqslant n$, g.c.d. $(b, b') = 1$, and $b + b' > n$.

9. For each Farey fraction a/b let $\mathscr{C}(a/b)$ denote the circle in the plane of radius $(2b^2)^{-1}$ and center $(a/b, (2b^2)^{-1})$. These circles, called the *Ford circles*, lie in the half-plane $y \geqslant 0$ and are tangent to the x-axis at the point a/b. Show that the interior of a Ford circle contains no point of any other Ford circle, and that two Ford circles $\mathscr{C}(a/b)$, $\mathscr{C}(a'/b')$ are tangent if and only if a/b and a'/b' are adjacent Farey fractions of some order.

6.2 RATIONAL APPROXIMATIONS

Theorem 6.7 *If a/b and c/d are Farey fractions of order n such that no other Farey fraction of order n lies between them, then*

$$\left|\frac{a}{b} - \frac{a+c}{b+d}\right| = \frac{1}{b(b+d)} \leqslant \frac{1}{b(n+1)}$$

and

$$\left|\frac{c}{d} - \frac{a+c}{b+d}\right| = \frac{1}{d(b+d)} \leqslant \frac{1}{d(n+1)}.$$

Proof We have

$$\left|\frac{a}{b} - \frac{a+c}{b+d}\right| = \frac{|ad-bc|}{b(b+d)} = \frac{1}{b(b+d)} \leqslant \frac{1}{b(n+1)}$$

by Theorem 6.1 and the fact that $b + d \geqslant n + 1$. The second formula is proved in a similar way.

Theorem 6.8 *If n is a positive integer and x is real, there is a rational number a/b such that $0 < b \leqslant n$ and*

$$\left|x - \frac{a}{b}\right| \leqslant \frac{1}{b(n+1)}.$$

Proof Consider the set of all Farey fractions of order n and all the fractions $(a+c)/(b+d)$ as described in Theorem 6.7. For some Farey fractions a/b and c/d, the number x will lie between or on, and so by interchanging a/b and c/d if necessary, we can say that x lies in the closed interval between a/b and $(a+c)/(b+d)$. Then, by Theorem 6.7,

$$\left|x - \frac{a}{b}\right| \leqslant \left|\frac{a}{b} - \frac{a+c}{b+d}\right| \leqslant \frac{1}{b(n+1)}.$$

Theorem 6.9 *If ξ is real and irrational, there are infinitely many distinct rational numbers a/b such that*

$$\left|\xi - \frac{a}{b}\right| < \frac{1}{b^2}.$$

Proof For each $n = 1, 2, \cdots$ we can find an a_n and a b_n by Theorem 6.8 such that $0 < b_n \leqslant n$ and

$$\left|\xi - \frac{a_n}{b_n}\right| \leqslant \frac{1}{b_n(n+1)} < \frac{1}{b_n^2}.$$

Many of the a_n/b_n may be equal to each other, but there will be infinitely many distinct ones. For if there were not infinitely many distinct ones, there would be only a finite number of distinct values taken by $|\xi - a_n/b_n|$, $n = 1, 2, 3, \cdots$. Then there would be a least one among these values, and it would be the value of $|\xi - a_n/b_n|$ for some n, say $n = k$. We would have $|\xi - a_n/b_n| \geqslant |\xi - a_k/b_k|$ for all $n = 1, 2, 3, \cdots$. But $|\xi - a_k/b_k| > 0$ since ξ is irrational, and we can find an n sufficiently large that

$$\frac{1}{n+1} < \left|\xi - \frac{a_k}{b_k}\right|.$$

This leads to a contradiction since we would now have

$$\left|\xi - \frac{a_k}{b_k}\right| \leqslant \left|\xi - \frac{a_n}{b_n}\right| \leqslant \frac{1}{b_n(n+1)} \leqslant \frac{1}{n+1} < \left|\xi - \frac{a_k}{b_k}\right|.$$

The condition that ξ be irrational is necessary in the theorem. For if x is any rational number, we can write $x = r/s$, $s > 0$. Then if a/b is any fraction such that $a/b \neq r/s$, $b > s$, we have

$$\left|\frac{r}{s} - \frac{a}{b}\right| = \frac{|rb - as|}{sb} \geqslant \frac{1}{sb} > \frac{1}{b^2}.$$

Hence all fractions a/b, $b > 0$, satisfying $|x - a/b| < 1/b^2$ have denominators $b \leqslant s$, and there can only be a finite number of such fractions.

The result of Theorem 6.9 can be improved, as Theorem 6.11 will show. Different proofs of Theorems 6.11 and 6.12 are given in Section 7.6.

Lemma 6.10 *If x and y are positive integers then not both of the inequalities*

$$\frac{1}{xy} \geqslant \frac{1}{\sqrt{5}}\left(\frac{1}{x^2} + \frac{1}{y^2}\right) \quad \text{and} \quad \frac{1}{x(x+y)} \geqslant \frac{1}{\sqrt{5}}\left(\frac{1}{x^2} + \frac{1}{(x+y)^2}\right)$$

can hold.

Proof The two inequalities can be written as

$$\sqrt{5}\,xy \geqslant y^2 + x^2, \qquad \sqrt{5}\,x(x+y) \geqslant (x+y)^2 + x^2.$$

Adding these inequalities, we get $\sqrt{5}(x^2 + 2xy) \geqslant 3x^2 + 2xy + 2y^2$, hence $2y^2 - 2(\sqrt{5} - 1)xy + (3 - \sqrt{5})x^2 \leqslant 0$. Multiplying this by 2 we put it in the form $4y^2 - 4(\sqrt{5} - 1)xy + (5 - 2\sqrt{5} + 1)x^2 \leqslant 0$, $(2y - (\sqrt{5} - 1)x)^2 \leqslant 0$. This is impossible for positive integers x and y because $\sqrt{5}$ is irrational.

Theorem 6.11 *Hurwitz. Given any irrational number ξ, there exist infinitely many different rational numbers h/k such that*

$$\left|\xi - \frac{h}{k}\right| < \frac{1}{\sqrt{5}\,k^2}. \tag{6.2}$$

Proof Let n be a positive integer. There exist two consecutive fractions a/b and c/d in the Farey sequence of order n, such that $a/b < \xi < c/d$. We prove that at least one of the three fractions a/b, c/d, $(a+c)/(b+d)$ can serve as h/k in (6.2). Suppose this is not so. Either $\xi < (a+c)/(b+d)$ or $\xi > (a+c)/(b+d)$.

Case I. $\xi < (a+c)/(b+d)$. Suppose that

$$\xi - \frac{a}{b} \geqslant \frac{1}{b^2\sqrt{5}}, \quad \frac{a+c}{b+d} - \xi \geqslant \frac{1}{(b+d)^2\sqrt{5}}, \quad \frac{c}{d} - \xi \geqslant \frac{1}{d^2\sqrt{5}}.$$

Adding inequalities we obtain

$$\frac{c}{d} - \frac{a}{b} \geqslant \frac{1}{d^2\sqrt{5}} + \frac{1}{b^2\sqrt{5}}, \quad \frac{a+c}{b+d} - \frac{a}{b} \geqslant \frac{1}{(b+d)^2\sqrt{5}} + \frac{1}{b^2\sqrt{5}}$$

hence

$$\frac{1}{bd} = \frac{cb - ad}{bd} = \frac{c}{d} - \frac{a}{b} \geqslant \frac{1}{\sqrt{5}}\left(\frac{1}{b^2} + \frac{1}{d^2}\right)$$

and

$$\frac{1}{b(b+d)} = \frac{(a+c)b - (b+d)a}{b(b+d)} \geqslant \frac{1}{\sqrt{5}}\left(\frac{1}{b^2} + \frac{1}{(b+d)^2}\right).$$

These two inequalities contradict Lemma 6.10. Therefore at least one of a/b, c/d, $(a+c)/(b+d)$ will serve as h/k in this case.

Case II. $\xi > (a+c)/(b+d)$. Suppose that

$$\xi - \frac{a}{b} \geqslant \frac{1}{b^2\sqrt{5}}, \quad \xi - \frac{a+c}{b+d} \geqslant \frac{1}{(b+d)^2\sqrt{5}}, \quad \frac{c}{d} - \xi \geqslant \frac{1}{d^2\sqrt{5}}.$$

Adding as before, we obtain

$$\frac{c}{d} - \frac{a}{b} \geqslant \frac{1}{d^2\sqrt{5}} + \frac{1}{b^2\sqrt{5}}, \quad \frac{c}{d} - \frac{a+c}{b+d} \geqslant \frac{1}{d^2\sqrt{5}} + \frac{1}{(b+d)^2\sqrt{5}}$$

hence

$$\frac{1}{bd} \geqslant \frac{1}{\sqrt{5}}\left(\frac{1}{d^2} + \frac{1}{b^2}\right), \qquad \frac{1}{d(b+d)} \geqslant \frac{1}{\sqrt{5}}\left(\frac{1}{(b+d)^2} + \frac{1}{d^2}\right)$$

which also contradicts Lemma 6.10. Again at least one of a/b, c/d, $(a+c)/(b+d)$ will serve as h/k.

We have shown the existence of some h/k that satisfies (6.2). This h/k depends on our choice of n. In fact h/k is either a/b, c/d, or $(a+c)/(b+d)$, where a/b and c/d are consecutive fractions in the Farey sequence of order n, and $a/b < \xi < c/d$. Using Theorem 6.7 we see that

$$\left|\xi - \frac{h}{k}\right| < \left|\frac{c}{d} - \frac{a}{b}\right| = \left|\frac{c}{d} - \frac{a+c}{b+d}\right| + \left|\frac{a+c}{b+d} - \frac{a}{b}\right|$$

$$\leqslant \frac{1}{d(n+1)} + \frac{1}{b(n+1)} \leqslant \frac{2}{n+1}.$$

We want to establish that there are infinitely many h/k that satisfy (6.2). Suppose that we have any h_1/k_1 that satisfies (6.2). Then $\left|\xi - \frac{h_1}{k_1}\right|$ is positive, and we can choose $n > 2/\left|\xi - \frac{h_1}{k_1}\right|$. The Farey sequence of order n then yields an h/k that satisfies (6.2) and such that

$$\left|\xi - \frac{h}{k}\right| \leqslant \frac{2}{n+1} < \left|\xi - \frac{h_1}{k_1}\right|.$$

This shows that there exist infinitely many rational numbers h/k that satisfy (6.2) since, given any rational number, we can find another that is closer to ξ.

Theorem 6.12 *The constant* $\sqrt{5}$ *in Theorem* 6.11 *is the best possible. In other words Theorem* 6.11 *does not hold if* $\sqrt{5}$ *is replaced by any larger value.*

Proof We need only exhibit one ξ for which $\sqrt{5}$ cannot be replaced by a larger value. Let us take $\xi = (1+\sqrt{5})/2$. Then

$$(x - \xi)\left(x - \frac{1-\sqrt{5}}{2}\right) = x^2 - x - 1.$$

For integers h, k with $k > 0$, we then have

$$\left|\frac{h}{k} - \xi\right|\left|\frac{h}{k} - \xi + \sqrt{5}\right| = \left|\left(\frac{h}{k} - \xi\right)\left(\frac{h}{k} - \frac{1-\sqrt{5}}{2}\right)\right|$$

$$= \left|\frac{h^2}{k^2} - \frac{h}{k} - 1\right| = \frac{1}{k^2}|h^2 - hk - k^2|. \quad (6.3)$$

The expression on the left in (6.3) is not zero because both ξ and $\sqrt{5} - \xi$ are irrational. The expression $|h^2 - hk - k^2|$ is a non-negative integer. Therefore $|h^2 - hk - k^2| \geqslant 1$ and we have

$$\left|\frac{h}{k} - \xi\right|\left|\frac{h}{k} - \xi + \sqrt{5}\right| \geqslant \frac{1}{k^2}. \quad (6.4)$$

Now suppose we have an infinite sequence of rational numbers h_j/k_j, $k_j > 0$, and a positive real number m such that

$$\left|\frac{h_j}{k_j} - \xi\right| < \frac{1}{mk_j^2}. \quad (6.5)$$

Then $k_j\xi - \dfrac{1}{mk_j} < h_j < k_j\xi + \dfrac{1}{mk_j}$, and this implies that there are only a finite number of h_j corresponding to each value of k_j. Therefore we have $k_j \to \infty$ as $j \to \infty$. Also, by (6.4), (6.5), and the triangle inequality we have

$$\frac{1}{k_j^2} \leqslant \left|\frac{h_j}{k_j} - \xi\right|\left|\frac{h_j}{k_j} - \xi + \sqrt{5}\right| < \frac{1}{mk_j^2}\left(\frac{1}{mk_j^2} + \sqrt{5}\right)$$

hence

$$m < \frac{1}{mk_j^2} + \sqrt{5}$$

and therefore

$$m \leqslant \lim_{j \to \infty}\left(\frac{1}{mk_j^2} + \sqrt{5}\right) = \sqrt{5}\,.$$

PROBLEMS

1. Prove that for every real number x there are infinitely many pairs of integers a, b, with b positive such that $|bx - a| < (\sqrt{5}\,b)^{-1}$.
2. Take $\xi = (1 + \sqrt{5})/2$. Let $\lambda > 0$ and $\alpha > 2$ be real numbers. Prove that there are only finitely many rationals h/k satisfying

$$\left|\xi - \frac{h}{k}\right| < \frac{1}{\lambda k^{\alpha}}.$$

3. Suppose $h = a$, $k = b$ is a solution of the inequality (6.2) for some irrational ξ. Prove that only a finite number of pairs h, k in the set $\{h = ma, k = mb;\ m = 1, 2, 3, \cdots\}$ satisfy (6.2).
4. Let $\alpha > 1$ be a real number. Suppose that for some real number β there are infinitely many rational numbers h/k such that $|\beta - h/k| < k^{-\alpha}$. Prove that β is irrational.
5. Prove that the following are irrational: $\sum_{j=1}^{\infty} 2^{-3^j}, \sum_{j=1}^{\infty} 2^{-j!}$.
6. If an irrational number θ lies between two consecutive terms a/b and c/d of the Farey sequence of order n, prove that at least one of the following inequalities holds:

$$|\theta - a/b| < 1/2b^2, \quad |\theta - c/d| < 1/2d^2.$$

6.3 IRRATIONAL NUMBERS

That $\sqrt{2}$ is irrational can be concluded at once from the unique factorization theorem. For if $\sqrt{2}$ could be represented in the form a/b, it would follow that $a^2 = 2b^2$. But this is impossible with integers a and b because the highest power of 2 that divides a^2 is an even power, whereas the highest power of 2 that divides $2b^2$ is an odd power, by the unique factorization theorem. A more general argument for deducing irrationality is formulated next.

Theorem 6.13 *If a polynomial equation with integral coefficients*

$$c_n x^n + c_{n-1}x^{n-1} + \cdots c_2 x^2 + c_1 x + c_0 = 0, \quad c_n \neq 0 \qquad (6.6)$$

has a nonzero rational solution a/b where the integers a and b are relatively prime, then $a \mid c_0$ and $b \mid c_n$.

Proof Replacing x by a/b in (6.6) and multiplying by b^{n-1}, we note that $c_n a^n/b$ is an integer, and hence $b \mid c_n$ since $(a, b) = 1$. On the other hand, replacing x by a/b in (6.6) and multiplying by b^n/a, we observe that $c_0 b^n/a$ is an integer, so $a \mid c_0$.

Corollary 6.14 *If a polynomial equation* (6.6) *with* $c_n = \pm 1$ *has a nonzero rational solution, that solution is an integer dividing* c_0.

Corollary 6.15 *For any integers* c *and* $n > 0$, *the only rational solutions, if any, of* $x^n = c$ *are integers. Thus* $x^n = c$ *has rational solutions if and only if* c *is the* nth *power of an integer.*

It follows at once that such numbers as $\sqrt{2}, \sqrt{3}, \sqrt{5}$ are irrational because there are no integral solutions of $x^2 = 2$, $x^2 = 3$, and $x^3 = 5$.

Another application of Theorem 6.13 can be made to certain values of the trigonometric functions, as follows.

Theorem 6.16 *Let* θ *be a rational multiple of* π; *thus,* $\theta = r\pi$ *where* r *is rational. Then* $\cos\theta, \sin\theta, \tan\theta$ *are irrational numbers apart from the cases where* $\tan\theta$ *is undefined, and the exceptions*

$$\cos\theta = 0, \pm 1/2, \pm 1; \qquad \sin\theta = 0, \pm 1/2, \pm 1; \qquad \tan\theta = 0, \pm 1.$$

Proof Let n be any positive integer. First we prove by mathematical induction that there is a polynomial $f_n(x)$ of degree n with integral coefficients and leading coefficient 1 such that $2\cos n\theta = f_n(2\cos\theta)$ holds for all real numbers θ. We note that $f_1(x) = x$, and $f_2(x) = x^2 - 2$ because of the well-known identity $2\cos 2\theta = (2\cos\theta)^2 - 2$. The identity

$$2\cos(n+1)\theta = (2\cos\theta)(2\cos n\theta) - 2\cos(n-1)\theta$$

is easily established by elementary trigonometry, and this reveals that $f_{n+1}(x) = xf_n(x) - f_{n-1}(x)$ which completes the proof by induction.

Next, let the positive integer n be chosen so that nr is also an integer. With $\theta = r\pi$ it follows that

$$f_n(2\cos\theta) = 2\cos n\theta = 2\cos nr\pi = \pm 2$$

where the plus sign holds if nr is even, the minus sign if odd. Thus $2\cos\theta$ is a solution of $f_n(x) = \pm 2$. Setting aside the cases where $\cos\theta = 0$, we apply Corollary 6.14 to conclude that $2\cos\theta$, if rational, is a nonzero integer. But $-1 \leqslant \cos\theta \leqslant 1$, so the only possible values of $2\cos\theta$, apart from 0, are ± 1 and ± 2. So Theorem 6.16 has been established in the case of $\cos\theta$.

As to $\sin\theta$, if θ is a rational multiple of π so is $\pi/2 - \theta$, and from the identity $\sin\theta = \cos(\pi/2 - \theta)$ we arrive at the conclusion stated in the theorem.

Finally, the identity $\cos 2\theta = (1 - \tan^2\theta)/(1 + \tan^2\theta)$ reveals that if $\tan\theta$ is rational so is $\cos 2\theta$. In view of what was just proved about the cosine function, we need look only at the possibilities $\cos 2\theta = 0$, $\pm 1/2$, ± 1. When $\cos 2\theta = 0$ it is readily calculated that $\tan\theta = \pm 1$; when $\cos 2\theta = +1$, $\tan\theta = 0$; when $\cos 2\theta = -1$, $\tan\theta$ is undefined; when $\cos 2\theta = \pm 1/2$, $\tan\theta$ is one of the irrational values $\pm\sqrt{3}$, $\pm 1/\sqrt{3}$. This completes the proof of Theorem 6.16.

The logarithm of any positive rational number to a positive rational base is easily classified as rational or irrational. Consider, for example, $\log_6 9$. If this were a rational number a/b, where a and b are positive integers, this would imply that $9 = 6^{a/b}$ or $9^b = 6^a$. The unique factorization theorem can be applied to separate the primes 2 and 3 to give $9^b = 3^a$ and $1 = 2^a$. These equations imply that $a = b = 0$, and so we conclude that $\log_6 9$ is irrational.

The basic mathematical constants π and e are irrational. A proof of this for e is sufficiently simple that we leave it to the reader in Problems 7 and 8 at the end of this section. For π the matter is not quite so easy, so we precede the proof with a lemma.

Lemma 6.17 *If n is any positive integer, and $g(x)$ any polynomial with integral coefficients, then $x^n g(x)$ and all its derivatives, evaluated at $x = 0$, are integers divisible by $n!$.*

Proof Any term in $g(x)$ is of the form cx^j where c and j are integers with $c \neq 0$ and $j \geqslant 0$. The corresponding term in $x^n g(x)$ is cx^{j+n}; if we prove the lemma for this single term, the entire lemma will follow because the derivative of a finite sum is the sum of the derivatives.

At $x = 0$, it is readily seen that cx^{j+n} and all its derivatives are zero, with one exception, namely the $(j+n)$th derivative. The $(j+n)$th derivative is $c\{(j+n)!\}$, and since $j \geqslant 0$, this is divisible by $n!$

Theorem 6.18 *π is irrational.*

Proof Suppose that $\pi = a/b$, where a and b are positive integers. Define the polynomial

$$f(x) = x^n(a - bx)^n/n! = b^n x^n(\pi - x)^n/n!, \tag{6.7}$$

where the second form of $f(x)$ stems from the first by simple algebra. The integer n will be specified later. We apply Lemma 6.17 with $g(x)$ in the form $(a - bx)^n$ to conclude that $x^n(a - bx)^n$ and all its derivatives, evaluated at $x = 0$, are integers divisible by $n!$. Dividing by $n!$, we see that $f(x)$ and all its derivatives, evaluated at $x = 0$, are integers. Denoting the

jth derivative of $f(x)$ by $f^{(j)}(x)$, and writing $f(x) = f^{(0)}(x)$, we can state that $f^{(j)}(0)$ is an integer for every $j = 0, 1, 2, 3, \cdots$.

By the second part of (6.7) we find that $f(\pi - x) = f(x)$, and taking derivatives we get $-f'(\pi - x) = f'(x), f^{(2)}(\pi - x) = f^{(2)}(x)$, and in general $(-1)^j f^{(j)}(\pi - x) = f^{(j)}(x)$. Letting $x = 0$ we obtain the result that $f^{(j)}(\pi)$ is an integer for every $j = 0, 1, 2, 3, \cdots$.

Next the polynomial $F(x)$ is defined by

$$F(x) = f(x) - f^{(2)}(x) + f^{(4)}(x) - f^{(6)}(x) + \cdots + (-1)^n f^{(2n)}(x).$$

Now if this equation is differentiated twice the result is

$$F^{(2)}(x) = f^{(2)}(x) - f^{(4)}(x) + f^{(6)}(x)$$

$$- f^{(8)}(x) + \cdots + (-1)^{n-1} f^{(2n)}(x) + 0$$

because $f^{(2n+2)}(x) = 0$ since $f(x)$ is a polynomial of degree $2n$. Adding these equations we get $F(x) + F^{(2)}(x) = f(x)$. Also, by the preceding paragraphs we observe that $F(0)$ and $F(\pi)$ are integers, because they are sums and differences of integers.

Now by elementary calculus it is seen that

$$\frac{d}{dx}\{F'(x)\sin x - F(x)\cos x\}$$

$$= F''(x)\sin x + F(x)\sin x = f(x)\sin x.$$

Thus we are able to integrate $f(x)\sin x$, to get

$$\int_0^\pi f(x)\sin x\,dx = [F'(x)\sin x - F(x)\cos x]_0^\pi = F(\pi) + F(0). \quad (6.8)$$

A contradiction arises from this equation, because whereas $F(\pi) + F(0)$ is an integer, we demonstrate that the integer n can be chosen sufficiently large in the definition of $f(x)$ in (6.7) that the integral in (6.8) lies strictly between 0 and 1.

From (6.7) we see that from $x = 0$ to $x = \pi$,

$$f(x) < \frac{\pi^n a^n}{n!} \quad \text{and} \quad f(x)\sin x < \frac{\pi^n a^n}{n!}.$$

Also $f(x) \sin x > 0$ in the open interval $0 < x < \pi$, and hence

$$0 < \int_0^{\pi} f(x) \sin x \, dx < \frac{\pi^n a^n}{n!} \cdot \pi$$

because the interval of integration is of length π. From elementary calculus it is well known that for any constant such as πa, the limit of $(\pi a)^n/n!$ is zero as n tends to infinity. Hence we can choose n sufficiently large that the integral in (6.8) lies strictly between 0 and 1, and we have obtained the contradiction stated above. It follows that π is irrational.

PROBLEMS

1. Prove that the irrational numbers are not closed under addition, subtraction, multiplication, or division.
2. Prove that the sum, difference, product, and quotient of two numbers, one irrational and the other a nonzero rational, are irrational.
3. Prove that $\sqrt{2} + \sqrt{3}$ is a root of $x^4 - 10x^2 + 1 = 0$, and hence establish that it is irrational.
4. (*a*) For any positive integer h, note that h^2 ends in an even number of zeros whereas $10h^2$ ends in an odd number of zeros in the ordinary base ten notation. Use this to prove that $\sqrt{10}$ is irrational, by assuming $\sqrt{10} = h/k$ so that $h^2 = 10k^2$. (*b*) Extend this argument to $\sqrt[3]{10}$. (*c*) Extend the argument to prove that $\sqrt{n}$ is irrational, where n is a positive integer not a perfect square, by taking n as the base of the number system instead of ten.
5. (i) Verify the details of the following sketch of an argument that $\sqrt{77}$ is irrational. Suppose that $\sqrt{77}$ is rational, and among its rational representations let a/b be that one having the smallest positive integer denominator b, where a is also an integer. Prove that another rational representation of $\sqrt{77}$ is $(77b - 8a)/(a - 8b)$. Prove that $a - 8b$ is a smaller positive integer than b, which is a contradiction. (ii) Generalize this argument to prove that $\sqrt{n}$ is irrational if n is a positive integer not a perfect square, by assuming $n = a/b$ and then getting another rational representation of n with denominator $a - kb$ where $k = [\sqrt{n}]$, the greatest integer less than $\sqrt{n}$. (An interesting aspect of this problem is that it establishes irrationality by use of the idea that every nonempty set of positive integers has a least member, not by use of the unique factorization theorem.)

6. Let a/b be a positive rational number with $a > 0$, $b > 0$, g.c.d. $(a, b) = 1$. Generalize Corollary 6.15 by proving that for any integer $n > 1$ the equation $x^n = a/b$ has a rational solution if and only if both a and b are nth powers of integers. (H)

7. Prove that a number α is rational if and only if there exists a positive integer k such that $[k\alpha] = k\alpha$. Prove that a number α is rational if and only if there exists a positive integer k such that $[(k!)\alpha] = (k!)\alpha$.

8. Recalling that the mathematical constant e has value $\sum_{j=0}^{\infty} 1/j!$, prove that

$$[(k!)e] = k! \sum_{j=0}^{k} 1/j! < (k!)e$$

Hence prove that e is irrational.

9. Prove that $\cos 1$ is irrational, where "1" is in radian measure. (H)

10. Prove that $(\log 3)/\log 2$ is irrational.

***11.** Prove that no n points with rational coordinates (x, y) can be chosen in the Euclidean plane to form the vertices of a regular polygon with n sides, except in the case $n = 4$. (H)

6.4 THE GEOMETRY OF NUMBERS

In this section we consider sets $\mathscr{S}$ that lie in real n-dimensional space $\mathbb{R}^n$ and find conditions which ensure that $\mathscr{S}$ contains a point whose coordinates are integers, that is, a point of $\mathbb{Z}^n$. If $\mathbf{v}$ is a point (or "vector") of $\mathbb{R}^n$ and c is a real number, then $c\mathbf{v}$ denotes the scalar multiple of $\mathbf{v}$. If $\mathbf{v}$ and $\mathbf{w}$ are two points of $\mathbb{R}^n$, then $\mathbf{v} + \mathbf{w}$ is the vector sum of $\mathbf{v}$ and $\mathbf{w}$. Similarly, if $\mathscr{S} \subseteq \mathbb{R}^n$, then we let $c\mathscr{S}$ denote the set $\mathscr{S}$ dilated by the factor c, that is, $c\mathscr{S} = \{c\mathbf{s} \in \mathbb{R}^n : \mathbf{s} \in \mathscr{S}\}$. In the same way, we define $\mathbf{v} + \mathscr{S}$ to be the set $\mathscr{S}$ translated by $\mathbf{v}$, so that $\mathbf{v} + \mathscr{S} = \{\mathbf{v} + \mathbf{s} \in \mathbb{R}^n : \mathbf{s} \in \mathscr{S}\}$. These definitions apply to arbitrary sets in $\mathbb{R}^n$, but we restrict our attention to those sets $\mathscr{S}$ for which the volume $v(\mathscr{S})$ is defined by multiple Riemann integrals.

Theorem 6.19 *Blichfeldt's principle. Let $\mathscr{S}$ be a set in $\mathbb{R}^n$ with volume $v(\mathscr{S}) > 1$. Then there exist two distinct points $\mathbf{s}' \in \mathscr{S}$ and $\mathbf{s}'' \in \mathscr{S}$ such that $\mathbf{s}' - \mathbf{s}''$ has integral coordinates.*

The analogue of this for sets of integers is obvious by the pigeonhole principle: If $\mathscr{S}$ is a set of more than m integers then there exist two distinct members of $\mathscr{S}$ that are congruent modulo m.

Proof To simplify notation and also to make geometric visualization easier, we suppose that $n = 2$, though the proof is perfectly general. By considering only those points $\mathbf{s} \in \mathscr{S}$ that lie in the disk $|\mathbf{s}| \leqslant R$, with R suitably large, we may suppose that $\mathscr{S}$ is bounded. For each point $\mathbf{k} = (k_1, k_2)$ with integral coordinates we let $\mathscr{U}(\mathbf{k})$ be the unit square consisting of those points $\mathbf{v} = (v_1, v_2)$ for which $k_1 \leqslant v_1 < k_1 + 1$, $k_2 \leqslant v_2 < k_2 + 1$. That is, $[v_1] = k_1$, $[v_2] = k_2$. Since each point $\mathbf{v}$ in the plane $\mathbb{R}^2$ lies in exactly one such square, these squares form a partitioning of $\mathbb{R}^2$. For each integral point $\mathbf{k}$ we let $\mathscr{S}(\mathbf{k})$ denote that part of $\mathscr{S}$ that lies in $\mathscr{U}(\mathbf{k})$. In symbols, $\mathscr{S}(\mathbf{k}) = \mathscr{S} \cap \mathscr{U}(\mathbf{k})$. Thus the subsets $\mathscr{S}(\mathbf{k})$ partition $\mathscr{S}$, and consequently

$$\sum_{\mathbf{k} \in \mathbb{Z}^2} v(\mathscr{S}(\mathbf{k})) = v(\mathscr{S}).$$

Put $\mathscr{T}(\mathbf{k}) = -\mathbf{k} + \mathscr{S}(\mathbf{k})$, so that $\mathscr{T}(\mathbf{k})$ is a translate of $\mathscr{S}(\mathbf{k})$ and $\mathscr{T}(\mathbf{k}) \subseteq \mathscr{U}(0)$. Since translation does not disturb the volume of a set, we have $v(\mathscr{T}(\mathbf{k})) = v(\mathscr{S}(\mathbf{k}))$. On inserting this in the identity above and appealing to our hypothesis that $v(\mathscr{S}) > 1$, we deduce that

$$\sum_{\mathbf{k} \in \mathbb{Z}^2} v(\mathscr{T}(\mathbf{k})) > 1.$$

Here only finitely many of the sets $\mathscr{T}(\mathbf{k})$ are nonempty, since $\mathscr{S}$ is a bounded set. The sets $\mathscr{T}(\mathbf{k})$ lie in the unit square $\mathscr{U}(\mathbf{0})$ whose volume is 1. Since the volumes of these sets sum to more than 1, they cannot all be disjoint. Thus there exist two distinct integral points, say $\mathbf{k}'$ and $\mathbf{k}''$ such that $\mathscr{T}(\mathbf{k}')$ and $\mathscr{T}(\mathbf{k}'')$ have a point $\mathbf{v}$ in common. Put $\mathbf{s}' = \mathbf{k}' + \mathbf{v}$, $\mathbf{s}'' = \mathbf{k}'' + \mathbf{v}$. Then $\mathbf{s}' \in \mathscr{S}(\mathbf{k}')$, $\mathbf{s}'' \in \mathscr{S}(\mathbf{k}'')$, so that $\mathbf{s}'$ and $\mathbf{s}''$ are members of $\mathscr{S}$, and $\mathbf{s}' - \mathbf{s}'' = \mathbf{k}' - \mathbf{k}''$ is a nonzero integral point. This completes the proof.

If $\mathbf{v}$ and $\mathbf{w}$ are points of $\mathbb{R}^n$, then the line segment joining them consists of the points $t\mathbf{v} + (1 - t)\mathbf{w}$, where $0 \leqslant t \leqslant 1$. A set $\mathscr{C}$ in $\mathbb{R}^n$ is said to be *convex* if for any two points $\mathbf{v}, \mathbf{w}$ of $\mathscr{C}$, the line segment joining them is contained in $\mathscr{C}$. A set $\mathscr{S}$ in $\mathbb{R}^n$ that has the property that $\mathbf{s} \in \mathscr{S}$ if and only if $-\mathbf{s} \in \mathscr{S}$ is said to be *symmetric about* $\mathbf{0}$.

Theorem 6.20 *Minkowski's Convex Body Theorem. Let $\mathscr{C}$ be a convex subset of $\mathbb{R}^n$. If $\mathscr{C}$ is convex, symmetric about* $\mathbf{0}$, *and has volume $v(\mathscr{C}) > 2^n$, then $\mathscr{C}$ contains a point* $\mathbf{c}$ *whose coordinates are integers, not all of them* 0.

Proof Let $\mathscr{S} = \frac{1}{2}\mathscr{C}$. Then $v(\mathscr{S}) = (\frac{1}{2})^n v(\mathscr{C}) > 1$. By Blichfeldt's principle (Theorem 6.19) there must exist points $\mathbf{s}'$ and $\mathbf{s}''$ of $\mathscr{S}$ such that $\mathbf{s}' \neq \mathbf{s}''$, $\mathbf{s}' - \mathbf{s}'' \in \mathbb{Z}^n$. We note that $2\mathbf{s}' \in \mathscr{C}$, $2\mathbf{s}'' \in \mathscr{C}$. Since $\mathscr{C}$ is symmetric about 0, it follows that $-2\mathbf{s}'' \in \mathscr{C}$. Since $\mathscr{C}$ is convex, the line segment joining $2\mathbf{s}'$ to $-2\mathbf{s}''$ lies in $\mathscr{C}$. In particular, $\mathscr{C}$ contains the midpoint of this segment, namely the point $\mathbf{s}' - \mathbf{s}''$. This is the point desired, as it has integral coordinates, not all 0.

Let A be an $n \times n$ matrix with real elements. Then A is nonsingular (i.e., the inverse matrix A^{-1} exists) if and only if $\det(A) \neq 0$. For such A, the linear transformation $\mathbf{y} = A\mathbf{x}$ from $\mathbb{R}^n$ into itself is both one-to-one and onto. We now consider how Theorem 6.20 is altered if we apply such a linear transformation. If $\mathscr{S}$ is a set in $\mathbb{R}^n$, then we let $A\mathscr{S}$ denote the image of $\mathscr{S}$ under this linear transformation. That is, $A\mathscr{S} = \{A\mathbf{s} \in \mathbb{R}^n : \mathbf{s} \in \mathscr{S}\}$. In particular, let $\Lambda = A\mathbb{Z}^n$. If $\det(A) \neq 0$, then we call the set Λ a *lattice*. Members of Λ are called *lattice points*. By taking A to be the identity matrix I, we see that $I\mathbb{Z}^n = \mathbb{Z}^n$ is itself a lattice, called the *lattice of integral points*. If A is a nonsingular matrix with columns $\mathbf{a}_1, \mathbf{a}_2, \cdots, \mathbf{a}_n$, and if $\mathbf{x}$ is a column vector with real coordinates $x_1, x_2, \cdots, x_n$, then $A\mathbf{x} = x_1\mathbf{a}_1 + x_2\mathbf{a}_2 + \cdots + x_n\mathbf{a}_n$. Here the $\mathbf{a}_j$ form a basis for $\mathbb{R}^n$, so that every point of $\mathbb{R}^n$ is uniquely of this form. Such a point is a member of Λ if and only if all the x_j are integers. That is, Λ is the set of all vectors $\mathbf{v}$ of the form

$$\mathbf{v} = k_1\mathbf{a}_1 + k_2\mathbf{a}_2 + \cdots + k_n\mathbf{a}_n \tag{6.9}$$

where the k_j are integers. For each such lattice point $\mathbf{v}$, the set of coordinates $k_1, k_2, \cdots, k_n$ is unique, and we say that $\mathbf{a}_1, \mathbf{a}_2, \cdots, \mathbf{a}_n$ form a basis for Λ.

Since a linear transformation takes lines to lines, we see that if $\mathscr{C}$ is a convex set in $\mathbb{R}^n$, then $A\mathscr{C}$ is also convex. Similarly, if $\mathscr{S}$ is a set in $\mathbb{R}^n$ that is symmetric about $\mathbf{0}$, then $A\mathscr{S}$ also has this property. Let $\mathbf{e}_1, \mathbf{e}_2, \cdots, \mathbf{e}_n$ denote the columns of I. These *elementary unit vectors* determine the edges of the unit cube $\mathscr{U}(\mathbf{0})$, whose volume is 1. Under the linear transformation $\mathbf{y} = A\mathbf{x}$, the vectors $\mathbf{e}_1, \mathbf{e}_2, \cdots, \mathbf{e}_n$ are mapped to $\mathbf{a}_1, \mathbf{a}_2, \cdots, \mathbf{a}_n$, which determine the edges of the parallelepiped $A\mathscr{U}(\mathbf{0})$ of volume $|\det(A)|$. This number is called the *determinant* of Λ, and is denoted $d(\Lambda)$. Suppose that $\mathscr{S}$ is a set in $\mathbb{R}^n$ with volume $v(\mathscr{S})$. To estimate $v(\mathscr{S})$ we cut $\mathbb{R}^n$ into small cubes, and sum the volumes of those cubes that lie in $\mathscr{S}$. Under the linear transformation, each such cube is mapped to a parallelepiped whose volume is the volume of the original cube multiplied by $|\det(A)|$. Thus we see that $v(A\mathscr{S}) = v(\mathscr{S})|\det(A)|$ for any set $\mathscr{S}$ for which volume is defined. We are now in a position to extend Theorem 6.20 to arbitrary lattices.

Theorem 6.21 *Minkowski's Convex Body Theorem for general lattices. Let A be a nonsingular $n \times n$ matrix with real elements, and let $\Lambda = A\mathbb{Z}^n$. If $\mathscr{C}$ is a set in $\mathbb{R}^n$ that is convex, symmetric about $\mathbf{0}$, and if $v(\mathscr{C}) > 2^n d(\Lambda)$, then there exists a lattice point $\mathbf{x} \in \Lambda$ such that $\mathbf{x} \neq 0$ and $\mathbf{x} \in \mathscr{C}$.*

Proof Let $\mathscr{C}' = A^{-1}\mathscr{C}$. Then $\mathscr{C}'$ is convex and symmetric about $\mathbf{0}$. Since $\det(A^{-1}) = 1/\det(A)$, it follows that $v(\mathscr{C}') = v(\mathscr{C})/|\det(A)| = v(\mathscr{C})/d(\Lambda) > 2^n$. Thus by Theorem 6.19, there exists a point $\mathbf{c} \in \mathscr{C}'$ such that $\mathbf{c} \neq \mathbf{0}$, $\mathbf{c} \in \mathbb{Z}^n$. Put $\mathbf{x} = A\mathbf{c}$. Then $\mathbf{x}$ has the desired properties.

By introducing a limiting argument we now show that the strict inequality $v(\mathscr{C}) > 2^n d(\Lambda)$ may be replaced by the weak inequality, provided that we place further restrictions on the set $\mathscr{C}$.

Corollary 6.22 *Let A be a nonsingular $n \times n$ matrix with real elements, and let $\Lambda = \mathbb{A}\mathbb{Z}^n$. If $\mathscr{C}$ is a set in $\mathbb{R}^n$ that is closed, bounded, convex, symmetric about $\mathbf{0}$, and if $v(\mathscr{C}) \geqslant 2^n d(\Lambda)$, then there exists a lattice point $\mathbf{x} \in \Lambda$ such that $\mathbf{x} \neq \mathbf{0}$ and $\mathbf{x} \in \mathscr{C}$.*

Proof For $k = 1, 2, 3, \cdots$ let $\mathscr{C}_k = (1 + 1/k)\mathscr{C}$. Then $v(\mathscr{C}_k) = (1 + 1/k)^n v(\mathscr{C}) > 2^n d(\Lambda)$, so that Theorem 6.21 applies to $\mathscr{C}_k$. Let $\mathbf{x}_k$ denote a nonzero member of Λ that lies in $\mathscr{C}_k$. Since each point in the sequence $\{\mathbf{x}_k\}$ lies in the bounded set $2\mathscr{C}$, there must be a nonzero point $\mathbf{x}_0$ of Λ such that $\mathbf{x}_k = \mathbf{x}_0$ for infinitely many k. Since $\mathbf{x}_0 \in \mathscr{C}_k$ for infinitely many k, and since $\mathscr{C}$ is closed, it follows that $\mathbf{x}_0 \in \mathscr{C}$.

Theorem 6.23 *Let A and B be nonsingular $n \times n$ matrices, and put $\Lambda_1 = A\mathbb{Z}^n$, $\Lambda_2 = B\mathbb{Z}^n$. Then $\Lambda_2 \subseteq \Lambda_1$ if and only if B is of the form $B = AK$, where K has integral elements.*

Proof Put $K = A^{-1}B$, and suppose that K has integral elements. If $\mathbf{x}$ has integral coordinates then so also does $K\mathbf{x}$. That is, $K\mathbb{Z}^n \subseteq \mathbb{Z}^n$, and hence $B\mathbb{Z}^n = (AK)\mathbb{Z}^n = A(K\mathbb{Z}^n) \subseteq A\mathbb{Z}^n$.

Suppose, conversely, that $\Lambda_2 \subseteq \Lambda_1$. Let $\mathbf{a}_1, \mathbf{a}_2, \cdots, \mathbf{a}_n$ be the columns of A, and $\mathbf{b}_1, \mathbf{b}_2, \cdots, \mathbf{b}_n$ be the columns of B. Choose j, $1 \leqslant j \leqslant n$. Since $\mathbf{b}_j \in \Lambda_2$ and $\Lambda_2 \subseteq \Lambda_1$, it follows from (6.9) that there exist integers $k_{1j}, k_{2j}, \cdots, k_{nj}$ such that $\mathbf{b}_j = k_{1j}\mathbf{a}_1 + k_{2j}\mathbf{a}_2 + \cdots + k_{nj}\mathbf{a}_n$. Let $K = [k_{ij}]$. Then $B = AK$, and K has integral elements. This completes the proof.

Corollary 6.24 *Let A and B be nonsingular $n \times n$ matrices, and put $\Lambda_1 = A\mathbb{Z}^n$, $\Lambda_2 = B\mathbb{Z}^n$. Then $\Lambda_1 = \Lambda_2$ if and only if there is a unimodular matrix U such that $B = AU$.*

Proof Put $K = A^{-1}B$. Since $\Lambda_2 \subseteq \Lambda_1$, it follows from Theorem 6.23 that K has integral elements. Similarly, the relation $\Lambda_1 \subseteq \Lambda_2$ implies that $B^{-1}A = K^{-1}$ has integral elements. Thus by Theorem 5.3, K is a unimodular matrix.

In the situation of Corollary 6.24, we have a lattice with two different bases. However, as $\det(B) = \det(AU) = \det(A)\det(U) = \pm\det(A)$, we see that the determinant $d(\Lambda)$ is independent of the choice of the basis.

At the end of Section 5.2 we observed that a matrix K with integral elements may be written in the form $K = UDV$, where U and V are unimodular and D is diagonal with non-negative integral elements. This has an interesting application to the situation of Theorem 6.23. We suppose that $B = AK$, $K = UDV$, and let $F = AU$, and $G = AUD$. Then by two applications of Corollary 6.24 we see that $\Lambda_1 = F\mathbb{Z}^n$, $\Lambda_2 = G\mathbb{Z}^n$. Let the columns of F be $\mathbf{f}_1, \mathbf{f}_2, \cdots, \mathbf{f}_n$, and let the diagonal elements of D be the integers $d_1, d_2, \cdots, d_n$. Then the columns of $G = FD$ are $d_1\mathbf{f}_1, d_2\mathbf{f}_2, \cdots, d_n\mathbf{f}_n$. Moreover, since $\det(K) \neq 0$ it follows that none of the d_i vanish, and hence the d_i are positive integers. That is, for any sublattice Λ_2 of a lattice Λ_1, there is a basis $\mathbf{f}_1, \mathbf{f}_2, \cdots, \mathbf{f}_n$ of Λ_1 and positive integers $d_1, d_2, \cdots, d_n$ such that $d_1\mathbf{f}_1, d_2\mathbf{f}_2, \cdots, d_n\mathbf{f}_n$ is a basis for Λ_2.

The geometry of numbers has many applications concerning rational approximations to real numbers (called *Diophantine approximations*), to quadratic forms, and to the theory of algebraic numbers. Although we have established only the first results in an extensive theory, we are already in a position to make some interesting applications. We begin by extending Theorem 6.8 to simultaneous approximation.

Theorem 6.25 *Let $x_1, x_2, \cdots, x_k$ be arbitrary real numbers, and let n be a positive integer. Then there exist integers $a_1, a_2, \cdots, a_k$ and an integer b, $0 < b \leqslant n$, such that $|x_i - a_i/b| < 1/(bn^{1/k})$ for $i = 1, 2, \cdots, k$.*

Proof Let $\mathscr{C}$ be the parallelepiped in $\mathbb{R}^{k+1}$ that consists of those points $(u_0, u_1, \cdots, u_k)$ for which

$$|u_0| < n + 1 \tag{6.10}$$

and

$$|x_i u_0 - u_i| < n^{-1/k} \tag{6.11}$$

for $1 \leqslant i \leqslant k$. Thus $\mathscr{C}$ is convex and symmetric about $\mathbf{0}$. To calculate the volume of $\mathscr{C}$ we observe that u_0 lies in an interval of length $2(n + 1)$, and that for each given value of u_0, the other variables u_j lie in intervals of length $2/n^{1/k}$. Thus the volume of $\mathscr{C}$ is

$$v(\mathscr{C}) = 2(n+1)(2/n^{1/k})^k = 2^{k+1}(n+1)/n > 2^{k+1}.$$

(An alternative method for evaluating this volume is indicated in Problem 10 at the end of this section.) Thus by Theorem 6.20 there exist integers $u_0, u_1, \cdots, u_k$, not all 0, such that the inequalities (6.10), (6.11) hold. For such integers we note that $u_0 \neq 0$, for if $u_0 = 0$ then (6.11) gives $|u_i| < n^{-1/k} \leqslant 1$, which implies that $u_i = 0$ for $i = 1, 2, \cdots, k$, and then all the u_i would be 0. If $u_0 < 0$, then we multiply all the u_i by -1. Thus we may assume that $u_0 > 0$. The desired result now follows by taking $b = u_0$, $a_i = u_i$ for $1 \leqslant i \leqslant k$.

Theorem 6.26 *Lagrange. Every positive integer n can be expressed as the sum of four squares, $n = x_1^2 + x_2^2 + x_3^2 + x_4^2$, where the x_i are non-negative integers.*

Fewer than four squares does not suffice, for if $n \equiv 7 \pmod 8$ then the congruence $n \equiv x_1^2 + x_2^2 + x_3^2 \pmod 8$ has no solution.

Proof In view of the algebraic identity

$$(x_1^2 + x_2^2 + x_3^2 + x_4^2)(y_1^2 + y_2^2 + y_3^2 + y_4^2)$$

$$= (x_1y_1 + x_2y_2 + x_3y_3 + x_4y_4)^2 + (x_1y_2 - x_2y_1 + x_3y_4 - x_4y_3)^2$$

$$+ (x_1y_3 - x_2y_4 - x_3y_1 + x_4y_2)^2 + (x_1y_4 + x_2y_3 - x_3y_2 - x_4y_1)^2$$

we see that if m and n are sums of four squares then so also is mn. Thus it suffices to show that each prime number p is a sum of four squares. To this end let

$$A = \begin{bmatrix} p & 0 & r & s \\ 0 & p & s & -r \\ 0 & 0 & 1 & 0 \\ 0 & 0 & 0 & 1 \end{bmatrix}$$

where r and s are chosen so that $r^2 + s^2 + 1 \equiv 0 \pmod p$. The existence of such integers is assured by Theorem 5.14. Let $\Lambda = A\mathbb{Z}^4$, and suppose that $\mathbf{x} = A\mathbf{t}$ is a point of Λ. Writing $\mathbf{t} = (t_1, t_2, t_3, t_4)$, $\mathbf{x} = (x_1, x_2, x_3, x_4)$, we see that if $\mathbf{x} \in \Lambda$ then

$$\begin{aligned} x_1^2 + x_2^2 + x_3^2 + x_4^2 &= (pt_1 + rt_3 + st_4)^2 + (pt_2 + st_3 - rt_4)^2 + t_3^2 + t_4^2 \\ &\equiv (1 + r^2 + s^2)(t_3^2 + t_4^2) \\ &\equiv 0 \pmod{p}. \end{aligned}$$

We observe that $d(\Lambda) = p^2$. Let $\mathscr{C}$ be the ball in $\mathbb{R}^4$ consisting of those points (x_1, x_2, x_3, x_4) such that $x_1^2 + x_2^2 + x_3^2 + x_4^2 < 2p$. Thus $\mathscr{C}$ is convex and symmetric about $\mathbf{0}$. A ball of radius R in $\mathbb{R}^4$ has volume $\frac{1}{2}\pi^2 R^4$. An unimaginative proof of this may be given by using rectangular coordinates to express the volume as an iterated integral,

$$\int_{-R}^{R} \int_{-\sqrt{R^2-x_1^2}}^{\sqrt{R^2-x_1^2}} \int_{-\sqrt{R^2-x_1^2-x_2^2}}^{\sqrt{R^2-x_1^2-x_2^2}} \int_{-\sqrt{R^2-x_1^2-x_2^2-x_3^2}}^{\sqrt{R^2-x_1^2-x_2^2-x_3^2}} 1 \, dx_4 \, dx_3 \, dx_2 \, dx_1,$$

and then evaluating this using standard techniques. An elegant, but less obvious, method of determining the volumes and surface areas of balls is sketched in Problem 23 at the end of this section. Taking $R = \sqrt{2p}$, we see that $v(\mathscr{C}) = \frac{1}{2}\pi^2(2p)^2 = 2\pi^2 p^2 > 2^4 p^2$. Thus by Theorem 6.21 there is a point $\mathbf{x} = (x_1, x_2, x_3, x_4)$ such that $\mathbf{x} \neq 0$ and $\mathbf{x} \in \mathscr{C}$. Then $0 < x_1^2 + x_2^2 + x_3^2 + x_4^2 < 2p$, and $x_1^2 + x_2^2 + x_3^2 + x_4^2 \equiv 0 \pmod{p}$, and hence $x_1^2 + x_2^2 + x_3^2 + x_4^2 = p$. This completes the proof.

In Theorem 6.26, some of the squares used to represent a positive integer n may be 0. In case it is desired to express n as a sum of positive squares, we have the following result.

Corollary 6.27 *There exist infinitely many positive integers that cannot be written as a sum of four positive perfect squares, but every integer $n > 169$ is a sum of five positive perfect squares.*

Proof We first note that we may restrict our attention to representations $n = x_1^2 + x_2^2 + x_3^2 + x_4^2$ for which

$$x_1 \geq x_2 \geq x_3 \geq x_4 \geq 0. \tag{6.12}$$

Next we observe that if $x_1^2 + x_2^2 + x_3^2 + x_4^2 \equiv 0 \pmod 8$ then all the x_i are even. Hence if $8n = x_1^2 + x_2^2 + x_3^2 + x_4^2$ then $2n = (x_1/2)^2 + (x_2/2)^2 +$

$(x_3/2)^2 + (x_4/2)^2$. Conversely, if $2n = x_1^2 + x_2^2 + x_3^2 + x_4^2$, then $8n = (2x_1)^2 + (2x_2)^2 + (2x_3)^2 + (2x_4)^2$. Thus the representations of $2n$ and of $8n$ are in one-to-one correspondence. The only representation of 2 as a sum of four squares subject to (6.12) is $2 = 1^2 + 1^2 + 0^2 + 0^2$. Hence the only representation of 8 as a sum of four squares, subject to (6.12), is $8 = 2^2 + 2^2 + 0^2 + 0^2$. Relating the representations of 8 as a sum of four squares to those of 32, we deduce that the only representation of 32 as a sum of four squares is $32 = 4^2 + 4^2 + 0^2 + 0^2$. Continuing in this manner, we find that the only representation of 2^{2r+1} as a sum of four squares, subject to (6.12), is $2^{2r+1} = (2^r)^2 + (2^r)^2 + 0^2 + 0^2$. Hence 2^{2r+1} cannot be written as a sum of four positive perfect squares.

Suppose now that $n > 169$. We write $n - 169$ as a sum of four squares, and suppose that (6.12) holds, so that

$$n = 169 + x_1^2 + x_2^2 + x_3^2 + x_4^2.$$

If the x_j are all positive then we write $169 = 13^2$, and then n is the sum of five positive perfect squares. If $x_1 \geqslant x_2 \geqslant x_3 > 0$, $x_4 = 0$ then we write $169 = 5^2 + 12^2$, so that $n = 5^2 + 12^2 + x_1^2 + x_2^2 + x_3^2$. If $x_1 \geqslant x_2 > 0$ but $x_3 = x_4 = 0$, then we write $169 = 12^2 + 4^2 + 3^2$, so that $n = 12^2 + 4^2 + 3^2 + x_1^2 + x_2^2$. If $x_1 > 0$ but $x_2 = x_3 = x_4 = 0$ then we write $169 = 10^2 + 8^2 + 2^2 + 1^2$, so that $n = 10^2 + 8^2 + 2^2 + 1^2 + x_1^2$. In each of these cases, n is represented as a sum of five positive perfect squares.

PROBLEMS

1. Construct a set $\mathscr{C}$ in the plane that is convex, symmetric about $\mathbf{0}$, and has area 4, but contains no nonzero integral point.
2. Construct a set $\mathscr{C}$ in the plane that is convex and has infinite area, but contains no integral point.
3. Let $\mathscr{H}$ be the set of points (x_1, x_2) in the plane such that $|x_1^2 - 2x_2^2| < 1$. Sketch $\mathscr{H}$. Show that $\mathscr{H}$ is symmetric about $\mathbf{0}$, that $\mathscr{H}$ has infinite area, but that $\mathscr{H}$ contains no nonzero integral point.
4. Let x_1 and x_2 be arbitrary real numbers, and let n be an arbitrary positive integer. Show that there exist integers a_1, a_2, b such that $0 < b \leqslant n$ and $(x_1 - a_1/b)^2 + (x_2 - a_2/b)^2 \leqslant 4/(\pi n b^2)$.
5. Let x_1 and x_2 be arbitrary real numbers, and let n be an arbitrary positive integer. Show that there exist integers a_1, a_2, b such that $|a_1| \leqslant n$, $|a_2| \leqslant n$, and $|a_1x_1 + a_2x_2 + b| < 1/n^2$, with not both $a_1 = 0$ and $a_2 = 0$.

6. Show that if $a \not\equiv 0 \pmod p$, p prime, and if n is any positive integer, then there exist integers x and y such that $x \equiv ay \pmod p$, $0 < x \leqslant n$, $|y| \leqslant p/n$.

7. A point $\mathbf{x}$ of a set $\mathscr{C}$ in the plane is called an *interior point* of $\mathscr{C}$ if there is an $r > 0$ such that $\mathscr{C}$ contains all points within a distance r of $\mathbf{x}$. Show that if $\mathscr{C}$ is convex and contains no interior point then $\mathscr{C}$ is a subset of a line.

8. Show that if $\mathscr{C}$ is convex, symmetric about $\mathbf{0}$, and if $\mathscr{C}$ contains an interior point, then $\mathbf{0}$ is an interior point of $\mathscr{C}$.

9. Show that if $\mathscr{C}$ is convex, unbounded, and contains an interior point, then $v(\mathscr{C}) = +\infty$. (Thus the hypothesis in Corollary 6.22 that $\mathscr{C}$ be bounded is superfluous.)

10. Let $\mathscr{C}$ be as in the proof of Theorem 6.25. Construct a $(k+1) \times (k+1)$ matrix A such that $\mathscr{C} = A\mathscr{C}'$ where $\mathscr{C}'$ is the cube consisting of those points $(t_0, t_1, \cdots, t_k)$ of $\mathbb{R}^{k+1}$ for which $|t_i| < 1$. Calculate $v(\mathscr{C}')$ and $\det(A)$, and thus give a second derivation of the value of $v(\mathscr{C})$.

11. Let p be a prime number, $p \equiv 1 \pmod 4$, and choose a so that $a^2 \equiv -1 \pmod p$. Put $\Lambda = A\mathbb{Z}^2$ where $A = \begin{bmatrix} p & a \\ 0 & 1 \end{bmatrix}$. Show that if (x, y) is a point of Λ then $x^2 + y^2 \equiv 0 \pmod p$. Show that Λ contains a nonzero point (x, y) for which $x^2 + y^2 < 2p$. Deduce that p can be represented as a sum of two squares. (This provides an alternative proof of Lemma 2.13.)

12. Let a, b, c be real numbers with $a > 0$. Put $d = b^2 - 4ac$, and suppose that $d < 0$. Show that there exist integers x, y, not both 0, such that $|ax^2 + bxy + cy^2| \leqslant \dfrac{2}{\pi}\sqrt{-d}$.

13. Show that any lattice Λ in the plane contains a nonzero point (x, y) such that $x^2 + y^2 \leqslant \dfrac{4}{\pi} d(\Lambda)$.

14. Show that any lattice Λ in the plane contains a nonzero point (x, y) such that $|xy| \leqslant \frac{1}{2} d(\Lambda)$. (H)

15. Let $\mathscr{S}$ be a set in $\mathbb{R}^n$ with volume $v(\mathscr{S})$. For each $\mathbf{x} \in \mathscr{U}(0)$, let $f(\mathbf{x})$ denote the number of $\mathbf{k} \in \mathbb{Z}^n$ for which $\mathbf{k} + \mathbf{x} \in \mathscr{S}$. Show that

$$\int_{\mathscr{U}(\mathbf{0})} f(\mathbf{x})\, d\mathbf{x} = v(\mathscr{S}).$$

16. Let r be a positive integer, and suppose that $\mathscr{S}$ is a set in $\mathbb{R}^n$ for which $v(\mathscr{S}) > r$. Show that there exist $r + 1$ distinct points $\mathbf{s}_0, \mathbf{s}_1, \cdots, \mathbf{s}_r$ of $\mathscr{S}$ such that $\mathbf{s}_i - \mathbf{s}_j \in \mathbb{Z}^n$ for $0 \leqslant i \leqslant j \leqslant r$.

17. Let $\mathbf{c} = (c_1, c_2, \cdots, c_n)$ be a row vector with integral coordinates, and put $g = \text{g.c.d.}\,(c_1, c_2, \cdots, c_n)$. Show that there is a unimodular matrix U such that $\mathbf{c}U = (g, 0, 0, \cdots, 0)$. (H)

***18.** Say that an integer n has the property P_k if n can be expressed as a sum of k positive squares. For any given m, prove that there exist infinitely many integers having all the properties $P_1, P_2, \cdots, P_m$.

***19.** Let $\mathbf{b} = (b_1, b_2, \cdots, b_n)$ be a row vector with integral coordinates, and $\text{g.c.d.}(b_1, b_2, \cdots, b_n) = 1$. Let $\Lambda = A\mathbb{Z}^n$ where A has integral elements. Let $\mathbf{a}_1, \mathbf{a}_2, \cdots, \mathbf{a}_n$ denote the columns of A, and put $g = \text{g.c.d.}(\mathbf{ba}_1, \mathbf{ba}_2, \cdots, \mathbf{ba}_n)$. Show that if $\mathbf{x} \in \Lambda$ then $g \mid \mathbf{bx}$, and that there is an $\mathbf{x} \in \Lambda$ such that $\mathbf{bx} = g$. Show that there is a basis $\mathbf{f}_1, \mathbf{f}_2, \cdots, \mathbf{f}_n$ of Λ such that $\mathbf{b}f_1 = g$, $\mathbf{bf}_i = 0$ for $i > 1$.

***20.** Let $\mathbf{b} = (b_1, b_2, \cdots, b_n)$ be a row vector with integral coordinates, and $\text{g.c.d.}(b_1, b_2, \cdots, b_n) = 1$. Let $\Lambda_1 = A\mathbb{Z}^n$, where A has integral elements. Let $\mathbf{a}_1, \mathbf{a}_2, \cdots, \mathbf{a}_n$ denote the columns of A, and put $g = \text{g.c.d.}(\mathbf{ba}_1, \mathbf{ba}_2, \cdots, \mathbf{ba}_n)$. Let m be a positive integer, and put $\Lambda_2 = \{\mathbf{x} \in \Lambda_1 \colon \mathbf{bx} \equiv 0 \pmod{m}\}$. Show that Λ_2 is a lattice, and that $d(\Lambda_2) = d(\Lambda_1)m/(g, m)$.

***21.** Suppose that Λ_2 is a sublattice of Λ_1. For $\mathbf{x}, \mathbf{y} \in \Lambda_1$ we say that $\mathbf{x} \equiv \mathbf{y} \pmod{\Lambda_2}$ if $\mathbf{x} - \mathbf{y} \in \Lambda_2$. Show that this defines an equivalence relation that partitions Λ_1 into precisely $d(\Lambda_2)/d(\Lambda_1)$ equivalence classes.

***22.** Let Λ be a set in $\mathbb{R}^n$ with the following properties: (i) Λ is an additive group; (ii) Λ is not contained in any proper subspace of $\mathbb{R}^n$; (iii) There is an $r > 0$ such that $\mathbf{x} \in \Lambda$, $|\mathbf{x}| < r$ implies that $\mathbf{x} = \mathbf{0}$. Show that Λ is a lattice.

***23.** Let $\mathscr{B}_n(r)$ denote a ball of radius r in $\mathbb{R}^n$. (*a*) Show that $v(\mathscr{B}_n(r)) = v_n r^n$ for some constant v_n. (*b*) For a set $\mathscr{S} \subseteq \mathbb{Z}^n$ let $|\partial\mathscr{S}|$ denote the surface area of $\mathscr{S}$ (i.e., the $(n-1)$-dimensional content of the boundary). Show that $|\partial\mathscr{B}_n(r)| = s_n r^{n-1}$ for some constant s_n. (*c*) Show that $\dfrac{d}{dr}v(\mathscr{B}_n(r)) = |\partial\mathscr{B}_n(r)|$. Deduce that $s_n = nv_n$. (*d*) Show that

$$\int_{\mathbb{R}^n} e^{-|\mathbf{x}|^2}\,d\mathbf{x} = \int_0^\infty |\partial\mathscr{B}_n(r)|e^{-r^2}\,dr.$$

(*e*) For $s > 0$ put $\Gamma(s) = \int_0^\infty x^{s-1}e^{-x}\,dx$. (This is Euler's integral for the gamma function.) Show that if $\alpha > -1/2$ then $\int_0^\infty r^\alpha e^{-r^2}\,dr = \frac{1}{2}\Gamma\left(\dfrac{\alpha+1}{2}\right)$. (*f*) Show that $\int_{\mathbb{R}^n} e^{-|\mathbf{x}|^2}\,d\mathbf{x} = \Gamma(1/2)^n$. (*g*) Deduce that $s_n = 2\Gamma(1/2)^n/\Gamma(n/2)$. ($h$) Use integration by parts to show that if $s > 0$ then $s\Gamma(s) = \Gamma(s+1)$. Show that $\Gamma(1) = 1$. Use induction to show that $\Gamma(n) = (n-1)!$. (i) From the known value $s_2 = 2\pi$, deduce that $\Gamma(1/2) = \sqrt{\pi}$, and by induction that

$$\Gamma\left(\frac{2m+1}{2}\right) = \sqrt{\pi}\,1 \cdot 3 \cdot \cdots \cdot (2m-1)/2^m$$

$$= \sqrt{\pi}\,(2m)!/(2^{2m}m!).$$

(j) Show that $s_{2m} = 2\pi^m/(m-1)!$, $v_{2m} = \pi^m/m!$ for $m = 1, 2, 3, \cdots$. (k) Show that $s_{2m+1} = 2^{2m+1}\pi^m m!/(2m)!$, $v_{2m+1} = 2^{2m+1}\pi^m m!/(2m+1)!$ for $m = 1, 2, 3, \cdots$.

NOTES ON CHAPTER 6

§6.2 A second proof of Hurwitz's theorem (Theorem 6.11) is given in the next chapter, using continued fractions (see Theorem 7.17).

§6.3 The proof of Theorem 6.16 follows that of A. E. Maier, "On the Irrationality of Certain Trigonometrical Numbers," *Amer. Math. Monthly* 72, (1962), 1012. Further results on the topic of this section can be found in the book by Niven listed in the General References.

§6.4 The geometry of numbers was initiated and named by Hermann Minkowski (1864–1909), who published a book on the subject in 1894. Minkowski's fruitful work was cut short by his untimely death. Theorems 6.20 and 6.21 give two formulations of Minkowski's first theorem concerning convex bodies. Blichfeldt's principle, which provides an elegant path to Minkowski's first theorem, was discovered by H. F. Blichfeldt in 1914. Detailed accounts of the subject are found in the books by Cassels and also in the book by Gruber and Lekkerkerker listed in the General References. One may also consult the interesting book of J. Hammer, *Unsolved Problems Concerning Lattice Points*, Pitman (London), 1977.

Theorem 6.25 is due to Dirichlet. By this theorem we see that there is a number b such that each of the numbers bx_i is near an integer. More generally, for given real numbers α_i one may ask whether there is an integer b such that each of the numbers $bx_i + \alpha_i$ is near an integer. The precise conditions that ensure the existence of such b were determined by Kronecker. For a simple proof of Kronecker's theorem, see Ka-Lam Kueh, "A note on Kronecker's approximation theorem," *Amer. Math. Monthly*, 93 (1986), 555–556.

The first known proof of Theorem 6.26 was given in 1770 by Lagrange, though it had been stated earlier without proof, and Fermat had once claimed to have a proof by descent. Our exposition follows that of H. Davenport, "The geometry of numbers," *Math. Gazette* 31 (1947),

206–210. In 1828, Jacobi showed that if n is a positive integer then the number of ordered quadruples (x_1, x_2, x_3, x_4) of integers for which $n = x_1^2 + x_2^2 + x_3^2 + x_4^2$ is 8 times the sum of those positive divisors d of n for which $4 \nmid d$. G. Rousseau, "On a construction for the representation of a positive integer as a sum of four squares," *L'Enseign. Math.* 33 (1987), 301–306, has formulated an efficient calculational procedure that provides an explicit representation of n as a sum of four squares. The method involves extending the continued fraction process to Gaussian integers.

The observation that if $n \equiv 7 \pmod 8$ then n is not the sum of three squares can be extended to show that if n is of the form $4^a(8k + 7)$ then n is not a sum of three squares. In 1798, Legendre outlined a proof that all other numbers are sums of three squares, and he supported his arguments with numerical evidence. Legendre later constructed a proof in the manner he had described, but in the meantime Gauss had proved a much more precise formula for the number of primitive representations of n as a sum of three squares, in 1801. From Gauss's formula it is at once evident that n is a sum of three squares if and only if n is not of the form $4^a(8k + 7)$. A short proof of Gauss's three squares theorem is given in the book of Serre, and other proofs are found in the book of Grosswald. Further proofs are discussed by C. Small, "Sums of three squares and levels of quadratic number fields," *Amer. Math. Monthly* 93 (1986), 276–279. Additional historical details are found in the book of Weil, as well as an elegant proof discovered in 1912 by L. Aubry that if a positive integer n is a sum of three rational squares then it is the sum of three integral squares.

In 1770, Edward Waring asserted without proof that every positive integer is a sum of nine cubes, is also a sum of 19 fourth powers, and so on. Thus *Waring's problem* was first interpreted as the question whether for each positive integer k there is an integer $s(k)$ such that every natural number is a sum of at most $s(k)$ positive kth powers. The answer is yes. This was established first for several special values of k, and in 1909 D. Hilbert solved the problem in general, using a family of complicated algebraic identities.

Once Hilbert had shown the existence of $s(k)$, attention then turned to the problem of estimating $s(k)$ and, if possible, of finding the least positive value of $s(k)$, traditionally denoted by $g(k)$. (For example, Theorem 6.26 and the remark following imply that $g(2) = 4$.) In the 1920s, G. H. Hardy and J. E. Littlewood developed an analytic method, sharpened later by I. M. Vinogradov, which gives asymptotic estimates for the number of representations. A simplified account of Hilbert's proof, and an elementary description of the analytic approach is found in the paper of W. J. Ellison, "Waring's problem," *Amer. Math. Monthly* (1971), 78, 10–36; see also C. Small, "Waring's problem," *Math. Mag.* 50 (1977),

12–16. The asymptotic analysis involves a number of technical problems, which are fully discussed by R. C. Vaughan in *The Hardy-Littlewood Method*, Cambridge Tracts 80, Cambridge University Press (Cambridge, UK), 1981.

Another fundamental number, $G(k)$, is defined to be the least positive integer such that every sufficiently large natural number is a sum of at most $G(k)$ positive kth powers. For example, although $g(3) = 9$, no integers other than 23 and 239 require as many as 9 cubes in their representations, and only a finite number of integers require 8 cubes, so that $G(3) \leqslant 7$. Details about the values of, or bounds on, $g(k)$ and $G(k)$ are given in Ribenboim (1989).

CHAPTER 7

Simple Continued Fractions

The following example illustrates the power of the theory of this chapter: the smallest solution of $x^2 - 61y^2 = 1$ in positive integers, which can be used to generate all solutions, has a value of x exceeding 10^9. This solution is easily calculated in Problem 10 of Section 7.8. Speaking more generally, continued fractions provide another representation of real numbers, offering insights that are not revealed by the decimal representation.

7.1 THE EUCLIDEAN ALGORITHM

Given any rational fraction u_0/u_1, in lowest terms so that $(u_0, u_1) = 1$ and $u_1 > 0$, we apply the Euclidean algorithm as formulated in Theorem 1.11 to get

$$\begin{aligned} u_0 &= u_1 a_0 + u_2, & 0 &< u_2 < u_1 \\ u_1 &= u_2 a_1 + u_3, & 0 &< u_3 < u_2 \\ u_2 &= u_3 a_2 + u_4, & 0 &< u_4 < u_3 \\ u_{j-1} &= u_j a_{j-1} + u_{j+1}, & 0 &< u_{j+1} < u_j \\ u_j &= u_{j+1} a_j. \end{aligned} \tag{7.1}$$

The notation has been altered from that of Theorem 1.11 by the replacement of b, c by u_0, u_1, of $r_1, r_2, \cdots, r_j$ by $u_2, u_3, \cdots, u_{j+1}$, and of $q_1, q_2, \cdots, q_{j+1}$ by $a_0, a_1, \cdots, a_j$. The form (7.1) is a little more suitable for our present purposes. If we write ξ_i in place of u_i/u_{i+1} for all values of i

in the range $0 \leqslant i \leqslant j$, then equations (7.1) become

$$\xi_i = a_i + \frac{1}{\xi_{i+1}}, \qquad 0 \leqslant i \leqslant j-1; \qquad \xi_j = a_j. \tag{7.2}$$

If we take the first two of these equations, those for which $i = 0$ and $i = 1$, and eliminate ξ_1, we get

$$\xi_0 = a_0 + \cfrac{1}{a_1 + \cfrac{1}{\xi_2}}.$$

In this result we replace ξ_2 by its value from (7.2), and then we continue with the replacement of $\xi_3, \xi_4, \cdots$, to get

$$\frac{u_0}{u_1} = \xi_0 = a_0 + \cfrac{1}{a_1 + \cfrac{\ddots}{+ \cfrac{1}{a_{j-1} + \cfrac{1}{a_j}}}}. \tag{7.3}$$

This is a *continued fraction expansion* of ξ_0, or of u_0/u_1. The integers a_i are called the *partial quotients* since they are the quotients in the repeated application of the division algorithm in equations (7.1). We presumed that the rational fraction u_0/u_1 had positive denominator u_1, but we cannot make a similar assumption about u_0. Hence a_0 may be positive, negative, or zero. However, since $0 < u_2 < u_1$, we note that the quotient a_1 is positive, and similarly the subsequent quotients $a_2, a_3, \cdots, a_j$ are positive integers. In case $j \geqslant 1$, that is if the set (7.1) contains more than one equation, then $a_j = u_j/u_{j+1}$ and $0 < u_{j+1} < u_j$ imply that $a_j > 1$.

We shall use the notation $\langle a_0, a_1, \cdots, a_j \rangle$ to designate the continued fraction in (7.3). In general, if $x_0, x_1, \cdots, x_j$ are any real numbers, all positive except perhaps x_0, we shall write

$$\langle x_0, x_1, \cdots, x_j \rangle = x_0 + \cfrac{1}{x_1 + \cfrac{\ddots}{+ \cfrac{1}{x_{j-1} + \cfrac{1}{x_j}}}}.$$

Such a finite continued fraction is said to be *simple* if all the x_i are integers. The following obvious formulas are often useful:

$$\langle x_0, x_1, \cdots, x_j \rangle = x_0 + \frac{1}{\langle x_1, \cdots, x_j \rangle}$$

$$= \left\langle x_0, x_1, \cdots, x_{j-2}, x_{j-1} + \frac{1}{x_j} \right\rangle.$$

The symbol $[x_0, x_1, \cdots, x_j]$ is often used to represent a continued fraction. We use the notation $\langle x_0, x_1, \cdots, x_j \rangle$ to avoid confusion with the least common multiple and the greatest integer.

PROBLEMS

1. Expand the rational fractions 17/3, 3/17, and 8/1 into finite simple continued fractions.
2. Prove that the set (7.1) consists of exactly one equation if and only if $u_1 = 1$. Under what circumstances is $a_0 = 0$?
3. Convert into rational numbers: $\langle 2, 1, 4 \rangle$; $\langle -3, 2, 12 \rangle$; $\langle 0, 1, 1, 100 \rangle$.
4. Given positive integers b, c, d with $c > d$, prove that $\langle a, c \rangle < \langle a, d \rangle$ but $\langle a, b, c \rangle > \langle a, b, d \rangle$ for any integer a.
5. Let $a_1, a_2, \cdots, a_n$ and c be positive real numbers. Prove that

$$\langle a_0, a_1, \cdots, a_n \rangle > \langle a_0, a_1, \cdots, a_n + c \rangle$$

holds if n is odd, but is false if n is even.

7.2 UNIQUENESS

In the last section we saw that such a fraction as 51/22 can be expanded into a simple continued fraction, $51/22 = \langle 2, 3, 7 \rangle$. It can be verified that 51/22 can also be expressed as $\langle 2, 3, 6, 1 \rangle$, but it turns out that these are the only two representations of 51/22. In general, we note that the simple continued fraction expansion (7.3) has an alternate form,

$$\frac{u_0}{u_1} = \langle a_0, a_1, \cdots, a_{j-1}, a_j \rangle = \langle a_0, a_1, \cdots, a_{j-2}, a_{j-1}, a_j - 1, 1 \rangle. \quad (7.4)$$

The following result establishes that these are the only two simple continued fraction expansions of a fixed rational number.

Theorem 7.1 *If* $\langle a_0, a_1, \cdots, a_j\rangle = \langle b_0, b_1, \cdots, b_n\rangle$ *where these finite continued fractions are simple, and if* $a_j > 1$ *and* $b_n > 1$, *then* $j = n$ *and* $a_i = b_i$ *for* $i = 0, 1, \cdots, n$.

Proof We write y_i for the continued fraction $\langle b_i, b_{i+1}, \cdots, b_n\rangle$ and observe that

$$y_i = \langle b_i, b_{i+1}, \cdots, b_n\rangle = b_i + \frac{1}{\langle b_{i+1}, b_{i+2}, \cdots, b_n\rangle} = b_i + \frac{1}{y_{i+1}}. \quad (7.5)$$

Thus we have $y_i > b_i$ and $y_i > 1$ for $i = 1, 2, \cdots, n-1$, and $y_n = b_n > 1$. Consequently $b_i = [y_i]$ for all values of i in the range $0 \leqslant i \leqslant n$. The hypothesis that the continued fractions are equal can be written in the form $y_0 = \xi_0$, where we are using the notation of equation (7.3). Now the definition of ξ_i as u_i/u_{i+1} implies that $\xi_{i+1} > 1$ for all values of $i \geqslant 0$, and so $a_i = [\xi_i]$ for $0 \leqslant i \leqslant j$ by equations (7.2). It follows from $y_0 = \xi_0$ that, taking integral parts, $b_0 = [y_0] = [\xi_0] = a_0$. By equations (7.2) and (7.5) we get

$$\frac{1}{\xi_1} = \xi_0 - a_0 = y_0 - b_0 = \frac{1}{y_1}, \quad \xi_1 = y_1, \quad a_1 = [\xi_1] = [y_1] = b_1.$$

This gives us the start of a proof by mathematical induction. We now establish that $\xi_i = y_i$ and $a_i = b_i$ imply that $\xi_{i+1} = y_{i+1}$ and $a_{i+1} = b_{i+1}$. To see this, we again use equations (7.2) and (7.5) to write

$$\frac{1}{\xi_{i+1}} = \xi_i - a_i = y_i - b_i = \frac{1}{y_{i+1}},$$

$$\xi_{i+1} = y_{i+1}, \quad a_{i+1} = [\xi_{i+1}] = [y_{i+1}] = b_{i+1}.$$

It must also follow that the continued fractions have the same length, that is, that $j = n$. For suppose that, say, $j < n$. From the preceding argument we have $\xi_j = y_j$, $a_j = b_j$. But $\xi_j = a_j$ by (7.2) and $y_j > b_j$ by (7.5), and so we have a contradiction. If we had assumed $j > n$, a symmetrical contradiction would have arisen, and thus j must equal n, and the theorem is proved.

Theorem 7.2 *Any finite simple continued fraction represents a rational number. Conversely, any rational number can be expressed as a finite simple continued fraction, and in exactly two ways.*

Proof The first assertion can be established by mathematical induction on the number of terms in the continued fraction, by use of the formula

$$\langle a_0, a_1, \cdots, a_j \rangle = a_0 + \frac{1}{\langle a_1, a_2, \cdots, a_j \rangle}.$$

The second assertion follows from the development of u_0/u_1 into a finite simple continued fraction in Section 7.1, together with equation (7.4) and Theorem 7.1.

PROBLEM

1. Let $a_0, a_1, \cdots, a_n$ and $b_0, b_1, \cdots, b_{n+1}$ be positive integers. State the conditions for

$$\langle a_0, a_1, \cdots, a_n \rangle < \langle b_0, b_1, \cdots, b_{n+1} \rangle.$$

7.3 INFINITE CONTINUED FRACTIONS

Let $a_0, a_1, a_2, \cdots$ be an infinite sequence of integers, all positive except perhaps a_0. We define two sequences of integers $\{h_n\}$ and $\{k_n\}$ inductively as follows:

$$\begin{aligned} h_{-2} &= 0, h_{-1} = 1, h_i = a_i h_{i-1} + h_{i-2} \quad \text{for } i \geqslant 0 \\ k_{-2} &= 1, k_{-1} = 0, k_i = a_i k_{i-1} + k_{i-2} \quad \text{for } i \geqslant 0. \end{aligned} \tag{7.6}$$

We note that $k_0 = 1$, $k_1 = a_1 k_0 \geqslant k_0$, $k_2 > k_1$, $k_3 > k_2$, etc., so that $1 = k_0 \leqslant k_1 < k_2 < k_3 < \cdots < k_n < \cdots$.

Theorem 7.3 *For any positive real number* x,

$$\langle a_0, a_1, \cdots, a_{n-1}, x \rangle = \frac{x h_{n-1} + h_{n-2}}{x k_{n-1} + k_{n-2}}.$$

Proof If $n = 0$, the result is to be interpreted as

$$x = \frac{x h_{-1} + h_{-2}}{x k_{-1} + k_{-2}}$$

which is true by equations (7.6). If $n = 1$, the result is

$$\langle a_0, x \rangle = \frac{xh_0 + h_{-1}}{xk_0 + k_{-1}}$$

which can be verified from (7.6) and the fact that $\langle a_0, x \rangle$ stands for $a_0 + 1/x$. We establish the theorem in general by induction. Assuming that the result holds for $\langle a_0, a_1, \cdots, a_{n-1}, x \rangle$, we see that

$$\langle a_0, a_1, \cdots, a_n, x \rangle = \left\langle a_0, a_1, \cdots, a_{n-1}, a_n + \frac{1}{x} \right\rangle$$

$$= \frac{(a_n + 1/x)h_{n-1} + h_{n-2}}{(a_n + 1/x)k_{n-1} + k_{n-2}}$$

$$= \frac{x(a_n h_{n-1} + h_{n-2}) + h_{n-1}}{x(a_n k_{n-1} + k_{n-2}) + k_{n-1}} = \frac{xh_n + h_{n-1}}{xk_n + k_{n-1}}.$$

Theorem 7.4 *If we define* $r_n = \langle a_0, a_1, \cdots, a_n \rangle$ *for all integers* $n \geqslant 0$, *then* $r_n = h_n/k_n$.

Proof We apply Theorem 7.3 with x replaced by a_n and then use equations (7.6) thus:

$$r_n = \langle a_0, a_1, \cdots, a_n \rangle = \frac{a_n h_{n-1} + h_{n-2}}{a_n k_{n-1} + k_{n-2}} = \frac{h_n}{k_n}.$$

Theorem 7.5 *The equations*

$$h_i k_{i-1} - h_{i-1} k_i = (-1)^{i-1} \quad \text{and} \quad r_i - r_{i-1} = \frac{(-1)^{i-1}}{k_i k_{i-1}}$$

hold for $i \geqslant 1$. *The identities*

$$h_i k_{i-2} - h_{i-2} k_i = (-1)^i a_i \quad \text{and} \quad r_i - r_{i-2} = \frac{(-1)^i a_i}{k_i k_{i-2}}$$

hold for $i > 1$. *The fraction* h_1/k_i *is reduced, that is* $(h_i, k_i) = 1$.

Proof The equations (7.6) imply that $h_{-1}k_{-2} - h_{-2}k_{-1} = 1$. Continuing the proof by induction, we assume that $h_{i-1}k_{i-2} - h_{i-2}k_{i-1} = (-1)^{i-2}$.

Again we use equations (7.6) to get $h_i k_{i-1} - h_{i-1} k_i = (a_i h_{i-1} + h_{i-2}) k_{i-1} - h_{i-1}(a_i k_{i-1} + k_{i-2}) = -(h_{i-1} k_{i-2} - h_{i-2} k_{i-1}) = (-1)^{i-1}$. This proves the first result stated in the theorem. We divide by $k_{i-1} k_i$ to get the second result, the formula for $r_i - r_{i-1}$. Furthermore, the fraction h_i/k_i is in lowest terms since any factor of h_i and k_i is also a factor of $(-1)^{i-1}$.

The other formulas can be derived in much the same way from (7.6), although we do not need induction in this case. First we observe that $h_0 k_{-2} - h_{-2} k_0 = a_0$, and that in general $h_i k_{i-2} - h_{i-2} k_i = (a_i h_{i-1} + h_{i-2}) k_{i-2} - h_{i-2}(a_i k_{i-1} + k_{i-2}) = a_i(h_{i-1} k_{i-2} - h_{i-2} k_{i-1}) = (-1)^i a_i$. The final identity can be obtained by dividing by $k_{i-2} k_i$.

Theorem 7.6 *The values r_n defined in Theorem 7.4 satisfy the infinite chain of inequalities $r_0 < r_2 < r_4 < r_6 < \cdots < r_7 < r_5 < r_3 < r_1$. Stated in words, the r_n with even subscripts form an increasing sequence, those with odd subscripts form a decreasing sequence, and every r_{2n} is less than every r_{2j-1}. Furthermore, $\lim_{n \to \infty} r_n$ exists, and for every $j \geq 0$, $r_{2j} < \lim_{n \to \infty} r_n < r_{2j+1}$.*

Proof The identities of Theorem 7.5 for $r_i - r_{i-1}$ and $r_i - r_{i-2}$ imply that $r_{2j} < r_{2j+2}$, $r_{2j-1} > r_{2j+1}$, and $r_{2j} < r_{2j-1}$ because the k_i are positive for $i \geq 0$ and the a_i are positive for $i \geq 1$. Thus we have $r_0 < r_2 < r_4 < \cdots$ and $r_1 > r_3 > r_5 > \cdots$. To prove that $r_{2n} < r_{2j-1}$, we put the previous results together in the form

$$r_{2n} < r_{2n+2j} < r_{2n+2j-1} \leq r_{2j-1}.$$

The sequence $r_0, r_2, r_4, \cdots$ is monotonically increasing and is bounded above by r_1, and so has a limit. Analogously, the sequence $r_1, r_3, r_5, \cdots$ is monotonically decreasing and is bounded below by r_0, and so has a limit. These two limits are equal because, by Theorem 7.5, the difference $r_i - r_{i-1}$ tends to zero as i tends to infinity, since the integers k_i are increasing with i. Another way of looking at this is to observe that $(r_0, r_1), (r_2, r_3), (r_4, r_5), \cdots$ is a chain of nested intervals defining a real number, namely $\lim_{n \to \infty} r_n$.

These theorems suggest the following definition.

Definition 7.1 *An infinite sequence $a_0, a_1, a_2, \cdots$ of integers, all positive except perhaps for a_0, determines an infinite simple continued fraction $\langle a_0, a_1, a_2, \cdots \rangle$. The value of $\langle a_0, a_1, a_2, \cdots \rangle$ is defined to be $\lim_{n \to \infty} \langle a_0, a_1, a_2, \cdots, a_n \rangle$.*

This limit, being the same as $\lim_{n \to \infty} r_n$, exists by Theorem 7.6. Another way of writing this limit is $\lim_{n \to \infty} h_n/k_n$. The rational number

$\langle a_0, a_1, \cdots, a_n \rangle = h_n/k_n = r_n$ is called the nth *convergent* to the infinite continued fraction. We say that the infinite continued fraction converges to the value $\lim_{n \to \infty} r_n$. In the case of a finite simple continued fraction $\langle a_0, a_1, \cdots, a_n \rangle$ we similarly call the number $\langle a_0, a_1, \cdots, a_m \rangle$ the mth *convergent* to $\langle a_0, a_1, \cdots, a_n \rangle$.

Theorem 7.7 *The value of any infinite simple continued fraction $\langle a_0, a_1, a_2, \cdots \rangle$ is irrational.*

Proof Writing θ for $\langle a_0, a_1, a_2, \cdots \rangle$ we observe by Theorem 7.6 that θ lies between r_n and r_{n+1}, so that $0 < |\theta - r_n| < |r_{n+1} - r_n|$. Multiplying by k_n, and making use of the result from Theorem 7.5 that $|r_{n+1} - r_n| = (k_n k_{n+1})^{-1}$, we have

$$0 < |k_n\theta - h_n| < \frac{1}{k_{n+1}}.$$

Now suppose that θ were rational, say $\theta = a/b$ with integers a and b, $b > 0$. Then the above inequality would become, upon multiplication by b,

$$0 < |k_n a - h_n b| < \frac{b}{k_{n+1}}.$$

The integers k_n increase with n, so we could choose n sufficiently large so that $b < k_{n+1}$. Then the integer $|k_n a - h_n b|$ would lie between 0 and 1, which is impossible.

Suppose we have two different infinite simple continued fractions, $\langle a_0, a_1, a_2, \cdots \rangle$ and $\langle b_0, b_1, b_2, \cdots \rangle$. Can these converge to the same value? The answer is no, and we establish this in the next two results.

Lemma 7.8 *Let $\theta = \langle a_0, a_1, a_2, \cdots \rangle$ be a simple continued fraction. Then $a_0 = [\theta]$. Furthermore if θ_1 denotes $\langle a_1, a_2, a_3, \cdots \rangle$ then $\theta = a_0 + 1/\theta_1$.*

Proof By Theorem 7.6 we see that $r_0 < \theta < r_1$, that is $a_0 < \theta < a_0 + 1/a_1$. Now $a_1 \geqslant 1$, so we have $a_0 < \theta < a_0 + 1$, and hence $a_0 = [\theta]$. Also

$$\theta = \lim_{n \to \infty} \langle a_0, a_1, \cdots, a_n \rangle = \lim_{n \to \infty} \left(a_0 + \frac{1}{\langle a_1, \cdots, a_n \rangle} \right)$$

$$= a_0 + \frac{1}{\lim_{n \to \infty} \langle a_1, \cdots, a_n \rangle} = a_0 + \frac{1}{\theta_1}.$$

Theorem 7.9 *Two distinct infinite simple continued fractions converge to different values.*

Proof Let us suppose that $\langle a_0, a_1, a_2, \cdots \rangle = \langle b_0, b_1, b_2, \cdots \rangle = \theta$. Then by Lemma 7.8, $[\theta] = a_0 = b_0$ and

$$\theta = a_0 + \frac{1}{\langle a_1, a_2, \cdots \rangle} = b_0 + \frac{1}{\langle b_1, b_2, \cdots \rangle}.$$

Hence $\langle a_1, a_2, \cdots \rangle = \langle b_1, b_2, \cdots \rangle$. Repetition of the argument gives $a_1 = b_1$, and so by mathematical induction $a_n = b_n$ for all n.

PROBLEMS

1. Evaluate the infinite continued fraction $\langle 1, 1, 1, 1, \cdots \rangle$. (H)
2. Evaluate the infinite continued fractions $\langle 2, 1, 1, 1, 1, \cdots \rangle$ and $\langle 2, 3, 1, 1, 1, 1, \cdots \rangle$. (H)
3. Evaluate the infinite continued fractions:
   (*a*) $\langle 2, 2, 2, 2, \cdots \rangle$; (*b*) $\langle 1, 2, 1, 2, 1, 2, \cdots \rangle$;
   (*c*) $\langle 2, 1, 2, 1, 2, 1, \cdots \rangle$; (*d*) $\langle 1, 3, 1, 2, 1, 2, 1, 2, \cdots \rangle$.
4. For $n \geqslant 1$, prove that $k_n/k_{n-1} = \langle a_n, a_{n-1}, \cdots, a_2, a_1 \rangle$. Find and prove a similar continued fraction expansion for h_n/h_{n-1}, assuming $a_0 \geqslant 0$.
5. Let u_0/u_1 be a rational number in lowest terms, and write $u_0/u_1 = \langle a_0, a_1, \cdots, a_n \rangle$. Show that if $0 \leqslant i < n$, then $|r_i - u_0/u_1| \leqslant 1/(k_i k_{i+1})$, with equality if and only if $i = n - 1$.
6. Let p be a prime number, $p \equiv 1 \pmod 4$, and suppose that $u^2 \equiv -1 \pmod p$. (A quick method for finding such a u is described in Section 2.9., in the remarks preceding Theorem 2.45.) Write $u/p = \langle a_0, a_1, \cdots, a_n \rangle$, and let i be the largest integer such that $k_i \leqslant \sqrt{p}$. Show that $|h_i/k_i - u/p| < 1/(k_i\sqrt{p})$, and hence that $|h_i p - uk_i| < \sqrt{p}$. Put $x = k_i$, $y = h_i p - uk_i$. Show that $0 < x^2 + y^2 < 2p$, and that $x^2 + y^2 \equiv 0 \pmod p$. Deduce that $x^2 + y^2 = p$. (This method was given in 1848 by Ch. Hermite. An even simpler procedure, which depends on the Euclidean algorithm, is discussed by S. Wagon, "The Euclidean algorithm strikes again," *Amer. Math. Monthly*, 97 (1990), 125–129.)

7.4 IRRATIONAL NUMBERS

We have shown that any infinite simple continued fraction represents an irrational number. Conversely, if we begin with an irrational number ξ, or ξ_0, we can expand it into an infinite simple continued fraction. To do this we define $a_0 = [\xi_0]$, $\xi_1 = 1/(\xi_0 - a_0)$, and next $a_1 = [\xi_1]$, $\xi_2 = 1/(\xi_1 - a_1)$, and so by an inductive definition

$$a_i = [\xi_i], \qquad \xi_{i+1} = \frac{1}{\xi_i - a_i}. \tag{7.7}$$

The a_i are integers by definition, and the ξ_i are all irrational since the irrationality of ξ_1 is implied by that of ξ_0, that of ξ_2 by that of ξ_1, and so on. Furthermore, $a_i \geqslant 1$ for $i \geqslant 1$ because $a_{i-1} = [\xi_{i-1}]$ and the fact that ξ_{i-1} is irrational implies that

$$a_{i-1} < \xi_{i-1} < 1 + a_{i-1}, \qquad 0 < \xi_{i-1} - a_{i-1} < 1,$$

$$\xi_i = \frac{1}{\xi_{i-1} - a_{i-1}} > 1, \qquad a_i = [\xi_i] \geqslant 1.$$

Next we use repeated application of (7.7) in the form $\xi_i = a_i + 1/\xi_{i+1}$ to get the chain

$$\xi = \xi_0 = a_0 + \frac{1}{\xi_1} = \langle a_0, \xi_1 \rangle$$

$$= \left\langle a_0, a_1 + \frac{1}{\xi_2} \right\rangle = \langle a_0, a_1, \xi_2 \rangle$$

$$= \left\langle a_0, a_1, \cdots, a_{m-2}, a_{m-1} + \frac{1}{\xi_m} \right\rangle = \langle a_0, a_1, \cdots, a_{m-1}, \xi_m \rangle.$$

This suggests, but does not establish, that ξ is the value of the infinite continued fraction $\langle a_0, a_1, a_2, \cdots \rangle$ determined by the integers a_i.

To prove this we use Theorem 7.3 to write

$$\xi = \langle a_0, a_1, \cdots, a_{n-1}, \xi_n \rangle = \frac{\xi_n h_{n-1} + h_{n-2}}{\xi_n k_{n-1} + k_{n-2}} \tag{7.8}$$

with the h_i and k_i defined as in (7.6). By Theorem 7.5 we get

$$\begin{aligned}\xi - r_{n-1} = \xi - \frac{h_{n-1}}{k_{n-1}} &= \frac{\xi_n h_{n-1} + h_{n-2}}{\xi_n k_{n-1} + k_{n-2}} - \frac{h_{n-1}}{k_{n-1}} \\ &= \frac{-(h_{n-1}k_{n-2} - h_{n-2}k_{n-1})}{k_{n-1}(\xi_n k_{n-1} + k_{n-2})} = \frac{(-1)^{n-1}}{k_{n-1}(\xi_n k_{n-1} + k_{n-2})}.\end{aligned} \tag{7.9}$$

This fraction tends to zero as n tends to infinity because the integers k_n are increasing with n, and ξ_n is positive. Hence $\xi - r_{n-1}$ tends to zero as n tends to infinity and then, by Definition 7.1,

$$\xi = \lim_{n\to\infty} r_n = \lim_{n\to\infty} \langle a_0, a_1, \cdots, a_n \rangle = \langle a_0, a_1, a_2, \cdots \rangle.$$

We summarize the results of the last two sections in the following theorem.

Theorem 7.10 *Any irrational number ξ is uniquely expressible, by the procedure that gave equations* (7.7), *as an infinite simple continued fraction $\langle a_0, a_1, a_2, \cdots \rangle$. Conversely, any such continued fraction determined by integers a_i that are positive for all $i > 0$ represents an irrational number, ξ. The finite simple continued fraction $\langle a_0, a_1, \cdots, a_n \rangle$ has the rational value $h_n/k_n = r_n$, and is called the* nth *convergent to ξ. Equations* (7.6) *relate the h_i and k_i to the a_i. For $n = 0, 2, 4, \cdots$ these convergents form a monotonically increasing sequence with ξ as a limit. Similarly, for $n = 1, 3, 5, \cdots$ the convergents form a monotonically decreasing sequence tending to ξ. The denominators k_n of the convergents are an increasing sequence of positive integers for $n > 0$. Finally, with ξ_i defined by* (7.7), *we have $\langle a_0, a_1, \cdots \rangle = \langle a_0, a_1, \cdots, a_{n-1}, \xi_n \rangle$ and $\xi_n = \langle a_n, a_{n+1}, a_{n+2}, \cdots \rangle$.*

Proof Only the last equation is new, and it becomes obvious if we apply to ξ_n the process described at the opening of this section.

Example 1 Expand $\sqrt{5}$ as an infinite simple continued fraction.

Solution We see that

$$\sqrt{5} = 2 + (\sqrt{5} - 2) = 2 + 1/(\sqrt{5} + 2)$$

and

$$\sqrt{5} + 2 = 4 + (\sqrt{5} - 2) = 4 + 1/(\sqrt{5} + 2).$$

In view of the repetition of $1/(\sqrt{5}+2)$, it follows that $\sqrt{5} = \langle 2, 4, 4, 4, \cdots \rangle$.

PROBLEMS

1. Expand each of the following as infinite simple continued fractions: $\sqrt{2}$, $\sqrt{2}-1$, $\sqrt{2}/2$, $\sqrt{3}$, $1/\sqrt{3}$.
2. Given that two irrational numbers have identical convergents $h_0/k_0, h_1/k_1, \cdots$, up to h_n/k_n, prove that their continued fraction expansions are identical up to a_n.
3. Let α, β, γ be irrational numbers satisfying $\alpha < \beta < \gamma$. If α and γ have identical convergents $h_0/k_0, h_1/k_1, \cdots$, up to h_n/k_n, prove that β also has these same convergents up to h_n/k_n.
4. Let ξ be an irrational number with continued fraction expansion $\langle a_0, a_1, a_2, a_3, \cdots \rangle$. Let $b_1, b_2, b_3, \cdots$ be any finite or infinite sequence of positive integers. Prove that
$$\lim_{n\to\infty} \langle a_0, a_1, a_2, \cdots, a_n, b_1, b_2, b_3, \cdots \rangle = \xi.$$
5. In the notation used in the text, prove that
$$\xi_n = \langle a_n, a_{n+1}, a_{n+2}, \cdots \rangle.$$

*6. Prove that for $n \geqslant 1$,
$$\xi - \frac{h_n}{k_n} = (-1)^n k_n^{-2}\{\xi_{n+1} + \langle 0, a_n, a_{n-1}, \cdots, a_2, a_1 \rangle\}^{-1}.$$

7. Prove that
$$k_n|k_{n-1}\xi - h_{n-1}| + k_{n-1}|k_n\xi - h_n| = 1.$$

7.5 APPROXIMATIONS TO IRRATIONAL NUMBERS

Continuing to use the notation of the preceding sections, we now show that the convergents h_n/k_n form a sequence of "best" rational approximations to the irrational number ξ.

Theorem 7.11 *We have for any $n \geqslant 0$,*
$$\left|\xi - \frac{h_n}{k_n}\right| < \frac{1}{k_n k_{n+1}} \quad \text{and} \quad |\xi k_n - h_n| < \frac{1}{k_{n+1}}.$$

Proof The second inequality follows from the first by multiplication by k_n. By (7.9) and (7.7) we see that

$$\left|\xi - \frac{h_n}{k_n}\right| = \frac{1}{k_n(\xi_{n+1}k_n + k_{n-1})} < \frac{1}{k_n(a_{n+1}k_n + k_{n-1})}.$$

Using (7.6), we replace $a_{n+1}k_n + k_{n-1}$ by k_{n+1} to obtain the first inequality.

Theorem 7.12 *The convergents h_n/k_n are successively closer to ξ, that is*

$$\left|\xi - \frac{h_n}{k_n}\right| < \left|\xi - \frac{h_{n-1}}{k_{n-1}}\right|.$$

In fact the stronger inequality $|\xi k_n - h_n| < |\xi k_{n-1} - h_{n-1}|$ holds.

Proof To see that the second inequality is stronger in that it implies the first, we use $k_{n-1} \leqslant k_n$ to write

$$\left|\xi - \frac{h_n}{k_n}\right| = \frac{1}{k_n}|\xi k_n - h_n| < \frac{1}{k_n}|\xi k_{n-1} - h_{n-1}|$$

$$\leqslant \frac{1}{k_{n-1}}|\xi k_{n-1} - h_{n-1}| = \left|\xi - \frac{h_{n-1}}{k_{n-1}}\right|.$$

Now to prove the stronger inequality we observe that $a_n + 1 > \xi_n$ by (7.7), and so by (7.6),

$$\xi_n k_{n-1} + k_{n-2} < (a_n + 1)k_{n-1} + k_{n-2}$$

$$= k_n + k_{n-1} \leqslant a_{n+1}k_n + k_{n-1} = k_{n+1}.$$

This inequality and (7.9) imply that

$$\left|\xi - \frac{h_{n-1}}{k_{n-1}}\right| = \frac{1}{k_{n-1}(\xi_n k_{n-1} + k_{n-2})} > \frac{1}{k_{n-1}k_{n+1}}.$$

We multiply by k_{n-1} and use Theorem 7.11 to get

$$|\xi k_{n-1} - h_{n-1}| > \frac{1}{k_{n+1}} > |\xi k_n - h_n|.$$

The convergent h_n/k_n is the best approximation to ξ of all the rational fractions with denominator k_n or less. The following theorem states this in a different way.

Theorem 7.13 *If a/b is a rational number with positive denominator such that $|\xi - a/b| < |\xi - h_n/k_n|$ for some $n \geqslant 1$, then $b > k_n$. In fact if $|\xi b - a| < |\xi k_n - h_n|$ for some $n \geqslant 0$, then $b \geqslant k_{n+1}$.*

Proof First we show that the second part of the theorem implies the first. Suppose that the first part is false so that there is an a/b with

$$\left|\xi - \frac{a}{b}\right| < \left|\xi - \frac{h_n}{k_n}\right| \quad \text{and} \quad b \leqslant k_n.$$

The product of these inequalities gives $|\xi b - a| < |\xi k_n - h_n|$. But the second part of the theorem says that this implies $b \geqslant k_{n+1}$, so we have a contradiction, since $k_n < k_{n+1}$ for $n \geqslant 1$.

To prove the second part of the theorem we proceed again by indirect argument, assuming that $|\xi b - a| < |\xi k_n - h_n|$ and $b < k_{n+1}$. Consider the linear equations in x and y,

$$xk_n + yk_{n+1} = b, \qquad xh_n + yh_{n+1} = a.$$

The determinant of coefficients is ± 1 by Theorem 7.5, and consequently these equations have an integral solution x, y. Moreover, neither x nor y is zero. For if $x = 0$ then $b = yk_{n+1}$, which implies that $y \neq 0$, in fact that $y > 0$ and $b \geqslant k_{n+1}$, in contradiction to $b < k_{n+1}$. If $y = 0$ then $a = xh_n$, $b = xk_n$, and

$$|\xi b - a| = |\xi xk_n - xh_n| = |x|\,|\xi k_n - h_n| \geqslant |k_n \xi - h_n|$$

since $|x| \geqslant 1$, and again we have a contradiction.

Next we prove that x and y have opposite signs. First, if $y < 0$, then $xk_n = b - yk_{n+1}$ shows that $x > 0$. Second, if $y > 0$, then $b < k_{n+1}$ implies that $b < yk_{n+1}$ and so xk_n is negative, whence $x < 0$. Now it follows from Theorem 7.10 that $\xi k_n - h_n$ and $\xi k_{n+1} - h_{n+1}$ have opposite signs, and hence $x(\xi k_n - h_n)$ and $y(\xi k_{n+1} - h_{n+1})$ have the same sign. From the equations defining x and y we get $\xi b - a = x(\xi k_n - h_n) + y(\xi k_{n+1} - h_{n+1})$. Since the two terms on the right have the same sign, the absolute value of the whole equals the sum of the separate absolute values. Thus

$$\begin{aligned}|\xi b - a| &= |x(\xi k_n - h_n) + y(\xi k_{n+1} - h_{n+1})| \\ &= |x(\xi k_n - h_n)| + |y(\xi k_{n+1} - h_{n+1})| \\ &> |x(\xi k_n - h_n)| = |x|\,|\xi k_n - h_n| \geqslant |\xi k_n - h_n|.\end{aligned}$$

This is a contradiction, and so the theorem is established.

Theorem 7.14 *Let ξ denote any irrational number. If there is a rational number a/b with $b \geqslant 1$ such that*

$$\left|\xi - \frac{a}{b}\right| < \frac{1}{2b^2}$$

then a/b equals one of the convergents of the simple continued fraction expansion of ξ.

Proof It suffices to prove the result in the case $(a, b) = 1$. Let the convergents of the simple continued fraction expansion of ξ be h_j/k_j, and suppose that a/b is not a convergent. The inequalities $k_n \leqslant b < k_{n+1}$ determine an integer n. For this n, the inequality $|\xi b - a| < |\xi k_n - h_n|$ is impossible because of Theorem 7.13.

Therefore we have

$$|\xi k_n - h_n| \leqslant |\xi b - a| < \frac{1}{2b},$$

$$\left|\xi - \frac{h_n}{k_n}\right| < \frac{1}{2bk_n}.$$

Using the facts that $a/b \neq h_n/k_n$ and that $bh_n - ak_n$ is an integer, we find that

$$\frac{1}{bk_n} \leqslant \frac{|bh_n - ak_n|}{bk_n} = \left|\frac{h_n}{k_n} - \frac{a}{b}\right| \leqslant \left|\xi - \frac{h_n}{k_n}\right| + \left|\xi - \frac{a}{b}\right| < \frac{1}{2bk_n} + \frac{1}{2b^2}.$$

This implies $b < k_n$ which is a contradiction.

Theorem 7.15 *The* nth *convergent of* $1/x$ *is the reciprocal of the* $(n-1)$*st convergent of* x *if* x *is any real number* > 1.

Proof We have $x = \langle a_0, a_1, \cdots \rangle$ and $1/x = \langle 0, a_0, a_1, \cdots \rangle$. If h_n/k_n and h'_n/k'_n are the convergents for x and $1/x$, respectively, then

$$\begin{array}{llll} h'_0 = 0, & h'_1 = 1, & h'_2 = a_1, & h'_n = a_{n-1}h'_{n-1} + h'_{n-2} \\ & k_0 = 1, & k_1 = a_1, & k_{n-1} = a_{n-1}k_{n-2} + k_{n-3} \\ k'_0 = 1, & k'_1 = a_0, & k'_2 = a_0a_1 + 1, & k'_n = a_{n-1}k'_{n-1} + k'_{n-2} \\ & h_0 = a_0, & h_1 = a_0a_1 + 1, & h_{n-1} = a_{n-1}h_{n-2} + h_{n-3}. \end{array}$$

The theorem now follows by mathematical induction.

PROBLEMS

1. Prove that the first assertion in Theorem 7.13 holds in case $n = 0$ if $k_1 > 1$.
2. Prove that the first assertion in Theorem 7.13 becomes false if $b > k_n$ is replaced by $b \geqslant k_{n+1}$. (H)
3. Say that a rational number a/b with $b > 0$ is a "good approximation" to the irrational number ξ if

$$|\xi b - a| = \min_{\substack{\text{all } x \\ 0 < y \leqslant b}} |\xi y - x|,$$

where, as indicated, the minimum on the right is to be taken over all pairs of integers x, y satisfying $0 < y \leqslant b$. Prove that every convergent h_n/k_n to ξ with $n > 0$ is a "good approximation".
4. Prove that every "good approximation" to ξ is a convergent.

*5. (*a*) Prove that if r/s lies between a/b and c/d, where the denominators of these rational fractions are positive, and if $ad - bc = \pm 1$, then $s > b$ and $s > d$.
(*b*) Let ξ be an irrational with convergents $\{h_n/k_n\}$. Prove that the sequence

$$\frac{h_{n-1}}{k_{n-1}}, \frac{h_{n-1} + h_n}{k_{n-1} + k_n}, \frac{h_{n-1} + 2h_n}{k_{n-1} + 2k_n}, \cdots, \frac{h_{n-1} + a_{n+1}h_n}{k_{n-1} + a_{n+1}k_n} = \frac{h_{n+1}}{k_{n+1}}$$

is increasing if n is odd, decreasing if n is even. If a/b and c/d denote any consecutive pair of this sequence, prove that $ad - bc = \pm 1$. The terms of this sequence, except the first and last, are called the *secondary convergents*; here n runs through all values $1, 2, \cdots$.
(*c*) Say that a rational number a/b is a "fair approximation" to ξ if $|\xi - a/b| = \min|\xi - x/y|$, the minimum being taken over all integers

x and y with $0 < y \leqslant b$. Prove that every good approximation is a fair approximation. Prove that every fair approximation is either a convergent or a secondary convergent to ξ.
(d) Prove that not every secondary convergent is a "fair approximation". *Suggestion:* Consider $\xi = \sqrt{2}$.
(e) Say that an infinite sequence of rational numbers, $r_1, r_2, r_3, \cdots$ with limit ξ is an "approximating sequence" to an irrational number ξ if $|\xi - r_{j+1}| < |\xi - r_j|$, $j = 1, 2, 3, \cdots$, and if the positive denominators of the r_j are increasing with j. Prove that the "fair approximations" to ξ form an "approximating sequence."
(f) Let S_{n-1} denote the finite sequence of (b) with the first term deleted, so that S_{n-1} has a_{n+1} terms, the last term being h_{n+1}/k_{n+1}. Prove that the infinite sequence of rational numbers obtained by first taking the terms of S_0 in order, then the terms of S_2, then S_4, then $S_6, \cdots$, is also an "approximating sequence" to ξ. Prove also that this sequence is maximal in the sense that if any other rational number $< \xi$ is introduced into the sequence as a new member, we no longer have an approximating sequence.
(g) Establish analogous properties for the sequence obtained by taking the terms of $S_{-1}, S_1, S_3, S_5, \cdots$.

6. Let ξ be irrational, $\xi = \langle a_0, a_1, a_2, \cdots \rangle$. Verify that

$$-\xi = \langle -a_0 - 1, 1, a_1 - 1, a_2, a_3, \cdots \rangle \text{ if } a_1 > 1.$$

and $-\xi = \langle -a_0 - 1, a_2 + 1, a_3, a_4, \cdots \rangle$ if $a_1 = 1$.

7.6 BEST POSSIBLE APPROXIMATIONS

Theorem 7.11 provides another method of proving Theorem 6.9. For in the statement of Theorem 7.11 we can replace k_{n+1} by the smaller integer k_n to get the weaker, but still correct, inequality

$$\left| \xi - \frac{h_n}{k_n} \right| < \frac{1}{k_n^2}.$$

Moreover the process described in Section 7.4 enables us to determine for any give irrational ξ as many convergents h_n/k_n as we please. We can also use continued fractions to get different proofs of Theorems 6.11 and 6.12. These results are repeated here because of their considerable importance in the theory and to reveal more about continued fractions.

Lemma 7.16 *If x is real, $x > 1$, and $x + x^{-1} < \sqrt{5}$, then $x < \frac{1}{2}(\sqrt{5} + 1)$ and $x^{-1} > \frac{1}{2}(\sqrt{5} - 1)$.*

Proof For real $x \geqslant 1$ we note that $x + x^{-1}$ increases with x, and $x + x^{-1} = \sqrt{5}$ if $x = \frac{1}{2}(\sqrt{5} + 1)$.

Theorem 7.17 *Hurwitz. Given any irrational number ξ, there exist infinitely many rational numbers h/k such that*

$$\left|\xi - \frac{h}{k}\right| < \frac{1}{\sqrt{5}\,k^2}. \tag{7.10}$$

Proof We will establish that, of every three consecutive convergents of the simple continued fraction expansion of ξ, at least one satisfies the inequality.

Let q_n denote k_n/k_{n-1}. We first prove that

$$q_j + q_j^{-1} < \sqrt{5} \tag{7.11}$$

if (7.10) is false for both $h/k = h_{j-1}/k_{j-1}$ and $h/k = h_j/k_j$. Suppose (7.10) is false for these two values of h/k. We have

$$\left|\xi - \frac{h_{j-1}}{k_{j-1}}\right| + \left|\xi - \frac{h_j}{k_j}\right| \geqslant \frac{1}{\sqrt{5}\,k_{j-1}^2} + \frac{1}{\sqrt{5}\,k_j^2}.$$

But ξ lies between h_{j-1}/k_{j-1} and h_j/k_j and hence we find, using Theorem 7.5, that

$$\left|\xi - \frac{h_{j-1}}{k_{j-1}}\right| + \left|\xi - \frac{h_j}{k_j}\right| = \left|\frac{h_{j-1}}{k_{j-1}} - \frac{h_j}{k_j}\right| = \frac{1}{k_{j-1}k_j}.$$

Combining these results we get

$$\frac{k_j}{k_{j-1}} + \frac{k_{j-1}}{k_j} \leqslant \sqrt{5}.$$

Since the left side is rational we actually have a strict inequality, and (7.11) follows.

Now suppose (7.10) is false for $h/k = h_i/k_i$, $i = n-1, n, n+1$. We then have (7.11) for both $j = n$ and $j = n+1$. By Lemma 7.16 we see that $q_n^{-1} > \frac{1}{2}(\sqrt{5} - 1)$ and $q_{n+1} < \frac{1}{2}(\sqrt{5} + 1)$, and, by (7.6) we find $q_{n+1} =$

$a_{n+1} + q_n^{-1}$. This gives us

$$\tfrac{1}{2}(\sqrt{5} + 1) > q_{n+1} = a_{n+1} + q_n^{-1} > a_{n+1} + \tfrac{1}{2}(\sqrt{5} - 1)$$

$$\geqslant 1 + \tfrac{1}{2}(\sqrt{5} - 1) = \tfrac{1}{2}(\sqrt{5} + 1)$$

and this is a contradiction.

Theorem 7.18 *The constant $\sqrt{5}$ in the preceding theorem is best possible. In other words Theorem 7.17 does not hold if $\sqrt{5}$ is replaced by any larger value.*

Proof It suffices to exhibit an irrational number ξ for which $\sqrt{5}$ is the largest possible constant. Consider the irrational ξ whose continued fraction expansion is $\langle 1, 1, 1, \cdots \rangle$. We see that

$$\xi = 1 + \frac{1}{\langle 1, 1, \cdots \rangle} = 1 + \frac{1}{\xi}, \qquad \xi^2 = \xi + 1, \qquad \xi = \frac{1}{2}(\sqrt{5} + 1).$$

Using (7.7) we can prove by induction that $\xi_i = (\sqrt{5} + 1)/2$ for all $i \geqslant 0$, for if $\xi_i = (\sqrt{5} + 1)/2$ then

$$\xi_{i+1} = (\xi_i - a_i)^{-1} = \left(\tfrac{1}{2}(\sqrt{5} + 1) - 1\right)^{-1} = \tfrac{1}{2}(\sqrt{5} + 1).$$

A simple calculation yields $h_0 = k_0 = k_1 = 1$, $h_1 = k_2 = 2$. Equations (7.6) become $h_i = h_{i-1} + h_{i-2}$, $k_i = k_{i-1} + k_{i-2}$, and so by mathematical induction $k_n = h_{n-1}$ for $n \geqslant 1$. Hence we have

$$\lim_{n\to\infty} \frac{k_{n-1}}{k_n} = \lim_{n\to\infty} \frac{k_{n-1}}{h_{n-1}} = \frac{1}{\xi} = \frac{\sqrt{5} - 1}{2}$$

$$\lim_{n\to\infty} \left(\xi_{n+1} + \frac{k_{n-1}}{k_n} \right) = \frac{\sqrt{5} + 1}{2} + \frac{\sqrt{5} - 1}{2} = \sqrt{5}.$$

If c is any constant exceeding $\sqrt{5}$, then

$$\xi_{n+1} + \frac{k_{n-1}}{k_n} > c$$

holds for only a finite number of values of n. Thus, by (7.9),

$$\left|\xi - \frac{h_n}{k_n}\right| = \frac{1}{k_n^2(\xi_{n+1} + k_{n-1}/k_n)} < \frac{1}{ck_n^2}$$

holds for only a finite number of values of n. Thus there are only a finite number of rational numbers h/k satisfying $|\xi - h/k| < 1/(ck^2)$, because any such h/k is one of the convergents to ξ by Theorem 7.14.

PROBLEMS

1. Find two rational numbers a/b satisfying

$$\left|\sqrt{2} - \frac{a}{b}\right| < \frac{1}{\sqrt{5}\,b^2}.$$

2. Find two rational numbers a/b satisfying

$$\left|\pi - \frac{a}{b}\right| < \frac{1}{\sqrt{5}\,b^2}.$$

3. Prove that the following is false for any constant $c > 2$: Given any irrational number ξ, there exist infinitely many rational numbers h/k such that

$$\left|\xi - \frac{h}{k}\right| < \frac{1}{k^c}.$$

***4.** Given any constant c, prove that there exists an irrational number ξ and infinitely many rational numbers h/k such that

$$\left|\xi - \frac{h}{k}\right| < \frac{1}{k^c}.$$

***5.** Prove that of every two consecutive convergents h_n/k_n to ξ with $n \geqslant 0$, at least one satisfies

$$\left|\xi - \frac{h}{k}\right| < \frac{1}{2k^2}.$$

7.7 PERIODIC CONTINUED FRACTIONS

An infinite simple continued fraction $\langle a_0, a_1, a_2, \cdots \rangle$ is said to be *periodic* if there is an integer n such that $a_r = a_{n+r}$ for all sufficiently large integers r. Thus a periodic continued fraction can be written in the form

$$\begin{aligned} &\langle b_0, b_1, b_2, \cdots, b_j, a_0, a_1, \cdots, a_{n-1}, a_0, a_1, \cdots, a_{n-1}, \cdots \rangle \\ &= \langle b_0, b_1, b_2, \cdots, b_j, \overline{a_0, a_1, \cdots, a_{n-1}} \rangle \end{aligned} \tag{7.12}$$

where the bar over the $a_0, a_1, \cdots, a_{n-1}$ indicates that this block of integers is repeated indefinitely. For example $\langle \overline{2,3} \rangle$ denotes $\langle 2,3,2,3,2,3,\cdots \rangle$ and its value is easily computed. Writing θ for $\langle \overline{2,3} \rangle$ we have

$$\theta = 2 + \cfrac{1}{3 + \cfrac{1}{\theta}}.$$

This is a quadratic equation in θ, and we discard the negative root to get the value $\theta = (3 + \sqrt{15})/3$. As a second example consider $\langle 4, 1, \overline{2,3} \rangle$. Calling this ξ, we have $\xi = \langle 4, 1, \theta \rangle$, with θ as above, and so

$$\xi = 4 + (1 + \theta^{-1})^{-1} = 4 + \frac{\theta}{\theta + 1} = \frac{29 + \sqrt{15}}{7}.$$

These two examples illustrate the following result.

Theorem 7.19 *Any periodic simple continued fraction is a quadratic irrational number, and conversely.*

Proof Let us write ξ for the periodic continued fraction of (7.12) and θ for its purely periodic part,

$$\theta = \langle \overline{a_0, a_1, \cdots, a_{n-1}} \rangle = \langle a_0, a_1, \cdots, a_{n-1}, \theta \rangle.$$

Then equation (7.8) gives

$$\theta = \frac{\theta h_{n-1} + h_{n-2}}{\theta k_{n-1} + k_{n-2}}$$

and this is a quadratic equation in θ. Hence θ is either a quadratic irrational number or a rational number, but the latter is ruled out by Theorem 7.7. Now ξ can be written in terms of θ,

$$\xi = \langle b_0, b_1, \cdots, b_j, \theta \rangle = \frac{\theta m + m'}{\theta q + q'}$$

where m'/q' and m/q are the last two convergents to $\langle b_0, b_1, \cdots, b_j \rangle$. But θ is of the form $(a + \sqrt{b})/c$, and hence ξ is of similar form because, as with θ, we can rule out the possibility that ξ is rational.

To prove the converse, let us begin with any quadratic irrational ξ, or ξ_0, of the form $\xi = \xi_0 = (a + \sqrt{b})/c$, with integers $a, b, c, b > 0$, $c \neq 0$. The integer b is not a perfect square since ξ is irrational. We multiply numerator and denominator by $|c|$ to get

$$\xi_0 = \frac{ac + \sqrt{bc^2}}{c^2} \quad \text{or} \quad \xi_0 = \frac{-ac + \sqrt{bc^2}}{-c^2}$$

according as c is positive or negative. Thus we can write ξ in the form

$$\xi_0 = \frac{m_0 + \sqrt{d}}{q_0}$$

where $q_0 | (d - m_0^2)$, d, m_0, and q_0 are integers, $q_0 \neq 0$, d not a perfect square. By writing ξ_0 in this form we can get a simple formulation of its continued fraction expansion $\langle a_0, a_1, a_2, \cdots \rangle$. We shall prove that the equations

$$a_i = [\xi_i], \qquad \xi_i = \frac{m_i + \sqrt{d}}{q_i}$$

$$m_{i+1} = a_i q_i - m_i, \qquad q_{i+1} = \frac{d - m_{i+1}^2}{q_i} \tag{7.13}$$

define infinite sequences of integers m_i, q_i, a_i, and irrationals ξ_i in such a way that equations (7.7) hold, and hence we will have the continued fraction expansion of ξ_0.

In the first place, we start with ξ_0, m_0, q_0 as determined above, and we let $a_0 = [\xi_0]$. If ξ_i, m_i, q_i, a_i are known, then we take $m_{i+1} = a_i q_i - m_i$, $q_{i+1} = (d - m_{i+1}^2)/q_i$, $\xi_{i+1} = (m_{i+1} + \sqrt{d})/q_{i+1}$, $a_{i+1} = [\xi_{i+1}]$. That is, (7.13) actually does determine sequences ξ_i, m_i, q_i, a_i that are at least real.

Now we use mathematical induction to prove that the m_i and q_i are integers such that $q_i \neq 0$ and $q_i | (d - m_i^2)$. This holds for $i = 0$. If it is true at the ith stage, we observe that $m_{i+1} = a_i q_i - m_i$ is an integer. Then the equation

$$q_{i+1} = \frac{d - m_{i+1}^2}{q_i} = \frac{d - m_i^2}{q_i} + 2a_i m_i - a_i^2 q_i$$

establishes that q_{i+1} is an integer. Moreover, q_{i+1} cannot be zero, since if it were, we would have $d = m_{i+1}^2$, whereas d is not a perfect square. Finally, we have $q_i = (d - m_{i+1}^2)/q_{i+1}$, so that $q_{i+1} | (d - m_{i+1}^2)$.

Next we can verify that

$$\xi_i - a_i = \frac{-a_i q_i + m_i + \sqrt{d}}{q_i} = \frac{\sqrt{d} - m_{i+1}}{q_i} = \frac{d - m_{i+1}^2}{q_i(\sqrt{d} + m_{i+1})}$$

$$= \frac{q_{i+1}}{\sqrt{d} + m_{i+1}} = \frac{1}{\xi_{i+1}}$$

which verifies (7.7) and so we have proved that $\xi_0 = \langle a_0, a_1, a_2, \cdots \rangle$, with the a_i defined by (7.13).

By ξ_i' we denote the *conjugate* of ξ_i, that is, $\xi_i' = (m_i - \sqrt{d})/q_i$. Since the conjugate of a quotient equals the quotient of the conjugates, we get the equation

$$\xi_0' = \frac{\xi_n' h_{n-1} + h_{n-2}}{\xi_n' k_{n-1} + k_{n-2}}$$

by taking conjugates in (7.8). Solving for ξ_n' we have

$$\xi_n' = -\frac{k_{n-2}}{k_{n-1}}\left(\frac{\xi_0' - h_{n-2}/k_{n-2}}{\xi_0' - h_{n-1}/k_{n-1}}\right).$$

As n tends to infinity, both h_{n-1}/k_{n-1} and h_{n-2}/k_{n-2} tend to ξ_0, which is different from ξ_0', and hence the fraction in parentheses tends to 1. Thus for sufficiently large n, say $n > N$ where N is fixed, the fraction in parentheses is positive, and ξ_n' is negative. But ξ_n is positive for $n \geqslant 1$ and hence $\xi_n - \xi_n' > 0$ and $n > N$. Applying (7.13) we see that this gives $2\sqrt{d}/q_n > 0$ and hence $q_n > 0$ for $n > N$.

It also follows from (7.13) that

$$q_n q_{n+1} = d - m_{n+1}^2 \leqslant d, \qquad q_n \leqslant q_n q_{n+1} \leqslant d$$

$$m_{n+1}^2 < m_{n+1}^2 + q_n q_{n+1} = d, \qquad |m_{n+1}| < \sqrt{d}$$

for $n > N$. Since d is a fixed positive integer we conclude that q_n and m_{n+1} can assume only a fixed number of possible values for $n > N$. Hence the ordered pairs (m_n, q_n) can assume only a fixed number of possible pair values for $n > N$, and so there are distinct integers j and k such that $m_j = m_k$ and $q_j = q_k$. We can suppose we have chosen j and k so that

$j < k$. By (7.13) this implies that $\xi_j = \xi_k$ and hence that

$$\xi_0 = \langle a_0, a_1, \cdots, a_{j-1}, \overline{a_j, a_{j+1}, \cdots, a_{k-1}} \rangle.$$

The proof of Theorem 7.19 is now complete.

Next we determine the subclass of real quadratic irrationals that have *purely periodic* continued fraction expansions, that is, expressions of the form $\langle \overline{a_0, a_1, \cdots, a_n} \rangle$.

Theorem 7.20 *The continued fraction expansion of the real quadratic irrational number ξ is purely periodic if and only if $\xi > 1$ and $-1 < \xi' < 0$, where ξ' denotes the conjugate of ξ.*

Proof First we assume that $\xi > 1$ and $-1 < \xi' < 0$. As usual we write ξ_0 for ξ and take conjugates in (7.7) to obtain

$$\frac{1}{\xi'_{i+1}} = \xi'_i - a_i. \tag{7.14}$$

Now $a_i \geq 1$ for all i, even for $i = 0$, since $\xi_0 > 1$. Hence if $\xi'_i < 0$, then $1/\xi'_{i+1} < -1$, and we have $-1 < \xi'_{i+1} < 0$. Since $-1 < \xi'_0 < 0$ we see, by mathematical induction, that $-1 < \xi'_i < 0$ holds for all $i \geq 0$. Then, since $\xi'_i = a_i + 1/\xi'_{i+1}$ by (7.14), we have

$$0 < -\frac{1}{\xi'_{i+1}} - a_i < 1, \qquad a_i = \left[-\frac{1}{\xi'_{i+1}} \right].$$

Now ξ is a quadratic irrational, so $\xi_j = \xi_k$ for some integers j and k with $0 < j < k$. Then we have $\xi'_j = \xi'_k$ and

$$a_{j-1} = \left[-\frac{1}{\xi'_j} \right] = \left[-\frac{1}{\xi'_k} \right] = a_{k-1}$$

$$\xi_{j-1} = a_{j-1} + \frac{1}{\xi_j} = a_{k-1} + \frac{1}{\xi_k} = \xi_{k-1}.$$

Thus $\xi_j = \xi_k$ implies $\xi_{j-1} = \xi_{k-1}$. A j-fold iteration of this implication gives us $\xi_0 = \xi_{k-j}$, and we have

$$\xi = \xi_0 = \langle \overline{a_0, a_1, \cdots, a_{k-j-1}} \rangle.$$

To prove the converse, let us assume that ξ is purely periodic, say $\xi = \langle \overline{a_0, a_1, \cdots, a_{n-1}} \rangle$, where $a_0, a_1, \cdots, a_{n-1}$ are positive integers. Then $\xi > a_0 \geqslant 1$. Also, by (7.8) we have

$$\xi = \langle a_0, a_1, \cdots, a_{n-1}, \xi \rangle = \frac{\xi h_{n-1} + h_{n-2}}{\xi k_{n-1} + k_{n-2}}.$$

Thus ξ satisfies the equation

$$f(x) = x^2 k_{n-1} + x(k_{n-2} - h_{n-1}) - h_{n-2} = 0.$$

This quadratic equation has two roots, ξ and its conjugate ξ'. Since $\xi > 1$, we need only prove that $f(x)$ has a root between -1 and 0 in order to establish that $-1 < \xi' < 0$. We shall do this by showing that $f(-1)$ and $f(0)$ have opposite signs. First we observe that $f(0) = -h_{n-2} < 0$ by (7.6) since $a_i > 0$ for $i \geqslant 0$. Next we see that for $n \geqslant 1$

$$\begin{aligned} f(-1) &= k_{n-1} - k_{n-2} + h_{n-1} - h_{n-2} \\ &= (k_{n-2} + h_{n-2})(a_{n-1} - 1) + k_{n-3} + h_{n-3} \\ &\geqslant k_{n-3} + h_{n-3} > 0. \end{aligned}$$

We now turn to the continued fraction expansion of $\sqrt{d}$ for a positive integer d not a perfect square. We get at this by considering the closely related irrational number $\sqrt{d} + [\sqrt{d}\,]$. This number satisfies the conditions of Theorem 7.20, and so its continued fraction is purely periodic,

$$\sqrt{d} + \left[\sqrt{d}\right] = \langle \overline{a_0, a_1, \cdots, a_{r-1}} \rangle = \langle a_0, \overline{a_1, \cdots, a_{r-1}, a_0} \rangle. \quad (7.15)$$

We can suppose that we have chosen r to be the smallest integer for which $\sqrt{d} + [\sqrt{d}\,]$ has an expansion of the form (7.15). Now we note that $\xi_i = \langle a_i, a_{i+1}, \cdots \rangle$ is purely periodic for all values of i, and that $\xi_0 = \xi_r = \xi_{2r} = \cdots$. Furthermore, $\xi_1, \xi_2, \cdots, \xi_{r-1}$ are all different from ξ_0, since otherwise there would be a shorter period. Thus $\xi_i = \xi_0$ if and only if i is of the form mr.

Now we can start with $\xi_0 = \sqrt{d} + [\sqrt{d}\,]$, $q_0 = 1$, $m_0 = [\sqrt{d}\,]$ in (7.13) because $1 \mid (d - [\sqrt{d}\,]^2)$. Then, for all $j \geqslant 0$,

$$\frac{m_{jr} + \sqrt{d}}{q_{jr}} = \xi_{jr} = \xi_0 = \frac{m_0 + \sqrt{d}}{q_0} = \left[\sqrt{d}\right] + \sqrt{d}$$

$$m_{jr} - q_{jr}\left[\sqrt{d}\right] = (q_{jr} - 1)\sqrt{d} \quad (7.16)$$

and hence $q_{jr} = 1$ since the left side is rational and $\sqrt{d}$ is irrational. Moreover $q_i = 1$ for no other values of the subscript i. For $q_i = 1$ implies $\xi_i = m_i + \sqrt{d}$, but ξ_i has a purely periodic expansion so that, by Theorem 7.20 we have $-1 < m_i - \sqrt{d} < 0$, $\sqrt{d} - 1 < m_i < \sqrt{d}$, and hence $m_i = [\sqrt{d}]$. Thus $\xi_i = \xi_0$ and i is a multiple of r.

We also establish that $q_i = -1$ does not hold for any i. For $q_i = -1$ implies $\xi_i = -m_i - \sqrt{d}$ by (7.13), and by Theorem 7.20 we would have $-m_i - \sqrt{d} > 1$ and $-1 < -m_i + \sqrt{d} < 0$. But this implies $\sqrt{d} < m_i < -\sqrt{d} - 1$, which is impossible.

Noting that $a_0 = [\sqrt{d} + [\sqrt{d}]] = 2[\sqrt{d}]$, we can now turn to the case $\xi = \sqrt{d}$. Using (7.15) we have

$$\begin{aligned}\sqrt{d} &= -\left[\sqrt{d}\right] + \left(\sqrt{d} + \left[\sqrt{d}\right]\right)\\ &= -\left[\sqrt{d}\right] + \left\langle 2\left[\sqrt{d}\right], \overline{a_1, a_2, \cdots, a_{r-1}, a_0}\right\rangle\\ &= \left\langle \left[\sqrt{d}\right], \overline{a_1, a_2, \cdots, a_{r-1}, a_0}\right\rangle\end{aligned}$$

with $a_0 = 2[\sqrt{d}]$.

When we apply (7.13) to $\sqrt{d} + [\sqrt{d}]$, $q_0 = 1$, $m_0 = [\sqrt{d}]$ we have $a_0 = 2[\sqrt{d}]$, $m_1 = [\sqrt{d}]$, $q_1 = d - [\sqrt{d}]^2$. But we can also apply (7.13) to $\sqrt{d}$ with $q_0 = 1$, $m_0 = 0$, and we find $a_0 = [\sqrt{d}]$, $m_1 = [\sqrt{d}]$, $q_1 = d - [\sqrt{d}]^2$. The value of a_0 is different, but the values of m_1, and of q_1, are the same in both cases. Since $\xi_i = (m_i + \sqrt{d})/q_i$ we see that further application of (7.13) yields the same values for the a_i, for the m_i, and for the q_i, in both cases. In other words, the expansions of $\sqrt{d} + [\sqrt{d}]$ and $\sqrt{d}$ differ only in the values of a_0 and m_0. Stating our results explicitly for the case $\sqrt{d}$ we have the following theorem.

Theorem 7.21 *If the positive integer d is not a perfect square, the simple continued fraction expansion of $\sqrt{d}$ has the form*

$$\sqrt{d} = \langle a_0, \overline{a_1, a_2, \cdots, a_{r-1}, 2a_0}\rangle$$

with $a_0 = [\sqrt{d}]$. Furthermore with $\xi_0 = \sqrt{d}$, $q_0 = 1$, $m_0 = 0$, in equations (7.13), we have $q_i = 1$ if and only if $r \mid i$, and $q_i = -1$ holds for no subscript i. Here r denotes the length of the shortest period in the expansion of $\sqrt{d}$.

Example 2 Find the irrational number having continued fraction expansion $\langle 8, \overline{1, 16}\rangle$.

Solution Write this as $8 + x^{-1}$, so that $x = \langle \overline{1, 16} \rangle$. Observing that $x = \langle 1, 16, \overline{1, 16} \rangle = \langle 1, 16, x \rangle$, we get the equation $x = 1 + (16 + x^{-1})^{-1}$, which is equivalent to the quadratic equation

$$x^{-2} + 16x^{-1} - 16 = 0.$$

Solving this for x^{-1} and discarding the negative solution, we get $x^{-1} = -8 + \sqrt{80}$. Hence the answer is $\sqrt{80}$.

PROBLEMS

1. For what positive integers c does the quadratic irrational $([\sqrt{d}] + \sqrt{d})/c$ have a purely periodic expansion?
2. Find the irrational number having continued fraction expansion $\langle 9, \overline{9, 18} \rangle$.
3. Expand $\sqrt{15}$ into an infinite simple continued fraction.

7.8 PELL'S EQUATION

The equation $x^2 - dy^2 = N$, with given integers d and N and unknowns x and y, is usually called *Pell's equation*. If d is negative, it can have only a finite number of solutions. If d is a perfect square, say $d = a^2$, the equation reduces to $(x - ay)(x + ay) = N$ and again there is only a finite number of solutions. The most interesting case of the equation arises when d is a positive integer not a perfect square. For this case, simple continued fractions are very useful.

Although John Pell contributed very little to the analysis of the equation, it bears his name because of a mistake by Euler. Lagrange was the first to prove that $x^2 - dy^2 = 1$ has infinitely many solutions in integers if d is a fixed positive integer, not a perfect square. As we shall see in Section 9.6, the solutions of this equation are very significant in the theory of quadratic fields. Let us now turn to a method of solution.

We expand $\sqrt{d}$ into a continued fraction as in Theorem 7.21, with convergents h_n/k_n, and with q_n defined by equations (7.13) with $\xi_0 = \sqrt{d}$, $q_0 = 1$, $m_0 = 0$.

Theorem 7.22 *If d is a positive integer not a perfect square, then $h_n^2 - dk_n^2 = (-1)^{n-1} q_{n+1}$ for all integers $n \geqslant -1$.*

Proof From equations (7.8) and (7.13), we have

$$\sqrt{d} = \xi_0 = \frac{\xi_{n+1}h_n + h_{n-1}}{\xi_{n+1}k_n + k_{n-1}} = \frac{(m_{n+1} + \sqrt{d})h_n + q_{n+1}h_{n-1}}{(m_{n+1} + \sqrt{d})k_n + q_{n+1}k_{n-1}}.$$

We simplify this equation and separate it into a rational and a purely irrational part much as we did in (7.16). Each part must be zero so we get two equations, and we can eliminate m_{n+1} from them. The final result is

$$h_n^2 - dk_n^2 = (h_n k_{n-1} - h_{n-1}k_n)q_{n+1} = (-1)^{n-1}q_{n+1}$$

where we used Theorem 7.5 in the last step.

Corollary 7.23 *Taking r as the length of the period of the expansion of* $\sqrt{d}$, *as in Theorem* 7.21, *we have for* $n \geqslant 0$,

$$h_{nr-1}^2 - dk_{nr-1}^2 = (-1)^{nr}q_{nr} = (-1)^{nr}.$$

With n even, this gives infinitely many solutions of $x^2 - dy^2 = 1$ in integers, provided d is positive and not a perfect square.

It can be seen that Theorem 7.22 gives us solutions of Pell's equation for certain values of N. In particular, Corollary 7.23 gives infinitely many solutions of $x^2 - dy^2 = 1$ by the use of even values nr. Of course if r is even, all values of nr are even. If r is odd, Corollary 7.23 gives infinitely many solutions of $x^2 - dy^2 = -1$ by the use of odd integers $n \geqslant 1$. The next theorem shows that every solution of $x^2 - dy^2 = \pm 1$ can be obtained from the continued fraction expansion of $\sqrt{d}$. But first we make this simple observation: Apart from such trivial solutions as $x = \pm 1$, $y = 0$ of $x^2 - dy^2 = 1$, all solutions of $x^2 - dy^2 = N$ fall into sets of four by all combinations of signs $\pm x$, $\pm y$. Hence it is sufficient to discuss the positive solutions $x > 0$, $y > 0$.

Theorem 7.24 *Let d be a positive integer not a perfect square, and let the convergents to the continued fraction expansion of* $\sqrt{d}$ *be* h_n/k_n. *Let the integer N satisfy* $|N| < \sqrt{d}$. *Then any positive solution* $x = s$, $y = t$ *of* $x^2 - dy^2 = N$ *with* $(s, t) = 1$ *satisfies* $s = h_n$, $t = k_n$ *for some positive integer n.*

Proof Let E and M be positive integers such that $(E, M) = 1$ and $E^2 - \rho M^2 = \sigma$, where $\sqrt{\rho}$ is irrational and $0 < \sigma < \sqrt{\rho}$. Here ρ and σ

are real numbers, not necessarily integers. Then

$$\frac{E}{M} - \sqrt{\rho} = \frac{\sigma}{M\left(E + M\sqrt{\rho}\right)}$$

and hence

$$0 < \frac{E}{M} - \sqrt{\rho} < \frac{\sqrt{\rho}}{M\left(E + M\sqrt{\rho}\right)} = \frac{1}{M^2\left(E/\left(M\sqrt{\rho}\right) + 1\right)}.$$

Also $0 < E/M - \sqrt{\rho}$ implies $E/(M\sqrt{\rho}) > 1$, and therefore

$$\left|\frac{E}{M} - \sqrt{\rho}\right| < \frac{1}{2M^2}.$$

By Theorem 7.14, E/M is a convergent in the continued fraction expansion of $\sqrt{\rho}$.

If $N > 0$, we take $\sigma = N$, $\rho = d$, $E = s$, $M = t$, and the theorem holds in this case.

If $N < 0$, then $t^2 - (1/d)s^2 = -N/d$, and we take $\sigma = -N/d$, $\rho = 1/d$, $E = t$, $M = s$. We find that t/s is a convergent in the expansion of $1/\sqrt{d}$. Then Theorem 7.15 shows that s/t is a convergent in the expansion of $\sqrt{d}$.

Theorem 7.25 *All positive solutions of $x^2 - dy^2 = \pm 1$ are to be found among $x = h_n$, $y = k_n$, where h_n/k_n are the convergents of the expansion of $\sqrt{d}$. If r is the period of the expansion of $\sqrt{d}$, as in Theorem 7.21, and if r is even, then $x^2 - dy^2 = -1$ has no solution, and all positive solutions of $x^2 - dy^2 = 1$ are given by $x = h_{nr-1}$, $y = k_{nr-1}$ for $n = 1, 2, 3, \cdots$. On the other hand, if r is odd, then $x = h_{nr-1}$, $y = k_{nr-1}$ give all positive solutions of $x^2 - dy^2 = -1$ by use of $n = 1, 3, 5, \cdots$, and all positive solutions of $x^2 - dy^2 = 1$ by use of $n = 2, 4, 6, \cdots$.*

Proof This result is a corollary of Theorems 7.21, 7.22, and 7.24.

The sequences of pairs $(h_0, k_0), (h_1, k_1), \cdots$ will include all positive solutions of $x^2 - dy^2 = 1$. Furthermore, $a_0 = [\sqrt{d}] > 0$, so the sequence $h_0, h_1, h_2, \cdots$ is strictly increasing. If we let x_1, y_1 denote the first solution that appears, then for every other solution x, y we shall have $x > x_1$, and hence $y > y_1$ also. Having found this least positive solution by means of continued fractions, we can find all the remaining positive solutions by a simpler method, as follows

Theorem 7.26 *If x_1, y_1 is the least positive solution of $x^2 - dy^2 = 1$, d being a positive integer not a perfect square, then all positive solutions are given by x_n, y_n for $n = 1, 2, 3, \cdots$ where x_n and y_n are the integers defined by $x_n + y_n\sqrt{d} = (x_1 + y_1\sqrt{d})^n$.*

The values of x_n and y_n are determined by expanding the power and equating the rational parts, and the purely irrational parts. For example, $x_3 + y_3\sqrt{d} = (x_1 + y_1\sqrt{d})^3$ so that $x_3 = x_1^3 + 3x_1y_1^2d$ and $y_3 = 3x_1^2y_1 + y_1^3d$.

Proof First we establish that x_n, y_n is a solution. We have $x_n - y_n\sqrt{d} = (x_1 - y_1\sqrt{d})^n$, since the conjugate of a product is the product of the conjugates. Hence we can write

$$x_n^2 - y_n^2d = \left(x_n - y_n\sqrt{d}\right)\left(x_n + y_n\sqrt{d}\right)$$

$$= \left(x_1 - y_1\sqrt{d}\right)^n\left(x_1 + y_1\sqrt{d}\right)^n = \left(x_1^2 - y_1^2d\right)^n = 1.$$

Next we show that every positive solution is obtained in this way. Suppose there is a positive solution s, t that is not in the collection $\{x_n, y_n\}$. Since both $x_1 + y_1\sqrt{d}$ and $s + t\sqrt{d}$ are greater than 1, there must be some integer m such that $(x_1 + y_1\sqrt{d})^m \leqslant s + t\sqrt{d} < (x_1 + y_1\sqrt{d})^{m+1}$. We cannot have $(x_1 + y_1\sqrt{d})^m = s + t\sqrt{d}$, for this would imply $x_m + y_m\sqrt{d} = s + t\sqrt{d}$, and hence $s = x_m$, $t = y_m$. Now

$$\left(x_1 - y_1\sqrt{d}\right)^m = \left(x_1 + y_1\sqrt{d}\right)^{-m},$$

and we can multiply this inequality by $(x_1 - y_1\sqrt{d})^m$ to obtain

$$1 < \left(s + t\sqrt{d}\right)\left(x_1 - y_1\sqrt{d}\right)^m < x_1 + y_1\sqrt{d}.$$

Defining integers a and b by $a + b\sqrt{d} = (s + t\sqrt{d})(x_1 - y_1\sqrt{d})^m$ we have

$$a^2 - b^2d = (s^2 - t^2d)\left(x_1^2 - y_1^2d\right)^m = 1$$

so a, b is a solution of $x^2 - dy^2 = 1$ such that $1 < a + b\sqrt{d} < x_1 + y_1\sqrt{d}$. But then $0 < (a + b\sqrt{d})^{-1}$, and hence $0 < a - b\sqrt{d} < 1$. Now we have

$$a = \tfrac{1}{2}\left(a + b\sqrt{d}\right) + \tfrac{1}{2}\left(a - b\sqrt{d}\right) > \tfrac{1}{2} + 0 > 0,$$

$$b\sqrt{d} = \tfrac{1}{2}\left(a + b\sqrt{d}\right) - \tfrac{1}{2}\left(a - b\sqrt{d}\right) > \tfrac{1}{2} - \tfrac{1}{2} = 0,$$

so a, b is a positive solution. Therefore $a > x_1, b > y_1$, but this contradicts $a + b\sqrt{d} < x_1 + y_1\sqrt{d}$, and hence our supposition was false. All positive solutions are given by x_n, y_n, $n = 1, 2, 3, \cdots$.

It may be noted that the definition of x_n, y_n can be extended to zero and negative n. They then give nonpositive solutions.

In case $N \neq \pm 1$, certain results can be proved about $x^2 - dy^2 = N$, but they are not as complete as what we have shown to be true in the case $N = 1$. For example, if x_1, y_1 is the smallest positive solution of $x^2 - dy^2 = 1$, and if $r_0^2 - ds_0^2 = N$, then integers r_n, s_n can be defined by $r_n + s_n\sqrt{d} = (r_0 + s_0\sqrt{d})(x_1 + y_1\sqrt{d})^n$, and it is easy to show that r_n, s_n are solutions of $x^2 - dy^2 = N$. However, there is no assurance that all positive solutions can be obtained in this way starting from a fixed r_0, s_0.

Numerical Examples Although Theorem 7.25 gives an assured procedure for solving $x^2 - dy^2 = \pm 1$, it may be noted that the equation can be solved by inspection for many small values of d. For example, it is obvious that the least positive solution of $x^2 - 82y^2 = -1$ is $x = 9$, $y = 1$. How can we get the least positive solution of $x^2 - 82y^2 = 1$? Looking ahead to Problem 1 at the end of this section, we see that it can be found by equating the rational and irrational parts of

$$x + y\sqrt{82} = (9 + \sqrt{82})^2$$

giving the least positive solution $x = 163$, $y = 18$.

For certain values of d, it is possible to see that $x^2 - dy^2 = -1$ has no solution in integers. In fact, this is established in Problem 3 of this section for all $d \equiv 3 \pmod 4$. Thus, for example, $x^2 - 7y^2 = -1$ has no solution. The least positive solution of $x^2 - 7y^2 = 1$ is seen by inspection to be $x = 8$, $y = 3$. Then, according to Problem 1, all solutions of $x^2 - 7y^2 = 1$ in positive integers can be obtained by equating the rational and irrational parts of

$$x_n + y_n\sqrt{7} = (8 + 3\sqrt{7})^n$$

for $n = 1, 2, 3, \cdots$.

As another example, consider the equation $x^2 - 30y^2 = 1$, with the rather obvious least positive solution $x = 11$, $y = 2$. Now by Theorem 7.25, or Problem 1, if there are any solutions of $x^2 - 30y^2 = -1$, there must be a least positive solution satisfying $x < 11$, $y < 2$. But $y = 1$ gives no solution, and hence we conclude that $x^2 - 30y^2 = -1$ has no solutions.

All the preceding examples depend on observing some solution by inspection. Now we turn to a case where inspection yields nothing, except perhaps to persons who are very facile with calculations.

Example 3 Find the least positive solution of $x^2 - 73y^2 = -1$ (if it exists) and of $x^2 - 73y^2 = 1$, given that $\sqrt{73} = \langle 8, \overline{1, 1, 5, 5, 1, 1, 16} \rangle$.

Solution Since the period of this continued fraction expansion is 7, an odd number, we know from Theorem 7.25 that $x^2 - 73y^2 = -1$ has solutions. Moreover, the least positive solution is $x = h_6$, $y = k_6$ from the convergent h_6/k_6. Using Equations (7.6), we see that the convergents are, starting with h_0/k_0,

$$8/1, 9/1, 17/2, 94/11, 487/57, 561/68, 1068/125.$$

Hence, $x = 1068$, $y = 125$ gives the least positive solution of $x^2 - 73y^2 = -1$. To get the least positive solution of $x^2 - 73y^2 = 1$, we use Problem 1 below and so calculate x and y y equating the rational and irrational parts of

$$x + y\sqrt{73} = \left(1068 + 125\sqrt{73}\right)^2.$$

The answer is $x = 2{,}281{,}249$, $y = 267{,}000$.

This easy solution of the problem depends on knowing the continued fraction expansion of $\sqrt{73}$. Although this expansion can be calculated by formula (7.13), we give a variation of this in Section 7.9 that makes the work easier, using $\sqrt{73}$ as an actual example.

PROBLEMS

The symbol d denotes a positive integer, not a perfect square.

***1.** Assuming that $x^2 - dy^2 = -1$ is solvable, let x_1, y_1 be the smallest positive solution. Prove that x_2, y_2 defined by $x_2 = y_2\sqrt{d} = (x_1 + y_1\sqrt{d})^2$ is the smallest positive solution of $x^2 - dy^2 = 1$. Also prove that all solutions of $x^2 - dy^2 = -1$ are given by x_n, y_n, where $x_n + y_n\sqrt{d} = (x_1 + y_1\sqrt{d})^n$, with $n = 1, 3, 5, 7, \cdots$, and that all solutions of $x^2 - dy^2 = 1$ are given by x_n, y_n with $n = 2, 4, 6, 8, \cdots$.

2. Suppose that N is a nonzero integer. Prove that if $x^2 - dy^2 = N$ has one solution, then it has infinitely many. (H)

3. Prove that $x^2 - dy^2 = -1$ has no solution if $d \equiv 3 \pmod 4$.

4. Let d be a positive integer, not a perfect square. If k is any positive integer, prove that there are infinitely many solutions in integers of $x^2 - dy^2 = 1$ with $k \mid y$.

***5.** Prove that the sum of the first n natural numbers is a perfect square for infinitely many values of n.

6. Prove that $n^2 + (n+1)^2$ is a perfect square for infinitely many values of n. (H)

7. Observe that $x^2 - 80y^2 = 1$ has a solution in positive integers by inspection. Hence, prove that $x^2 - 80y^2 = -1$ has no solution in integers. Generalize the argument to prove that for any integer k, $x^2 - (k^2 - 1)y^2 = -1$ has no solutions in integers.

8. Given $\sqrt{18} = \langle 4, \overline{4, 8} \rangle$, find the least positive solution of $x^2 - 18y^2 = -1$ (if any) and of $x^2 - 18y^2 = 1$.

9. *Calculator problem*. Find the least positive solution of $x^2 - 29y^2 = -1$ (if any) and of $x^2 - 29y^2 = 1$, given $\sqrt{29} = \langle 5, \overline{2, 1, 1, 2, 10} \rangle$.

10. *Calculator problem*. Find the least positive solution of $x^2 - 61y^2 = -1$ and also of $x^2 - 61y^2 = 1$, given

$$\sqrt{61} = \langle 7, \overline{1, 4, 3, 1, 2, 2, 1, 3, 4, 1, 14} \rangle.$$

(One value of x in the answer exceeds 10^9, so the calculation is sizable. The procedure in Example 3 for $x^2 - 73y^2 = \pm 1$ in the text can serve as a model. A calculator with an eight-digit display is adequate, because, for example, to square 1234567 we can use $(a + b)^2 = a^2 + 2ab + b^2$, with $a = 1234000$ and $b = 567$.)

11. Show that if d is divisible by a prime number p, $p \equiv 3 \pmod 4$, then the equation $x^2 - dy^2 = -1$ has no solution in integers.

12. Suppose that $p \equiv 1 \pmod 4$. Show that if $x^2 - py^2 = 1$ then x is odd and y is even. Suppose that $x_0^2 - py_0^2 = 1$ with $y_0 > 0$, y_0 minimal. Show that g.c.d.$(x_0 + 1, x_0 - 1) = 2$. Deduce that one of two cases arises: Case 1. $x_0 - 1 = 2pu^2$, $x_0 + 1 = 2v^2$. Case 2. $x_0 - 1 = 2u^2$, $x_0 + 1 = 2pv^2$. Show that in Case 1, $v^2 - pu^2 = 1$ with $|u| < y_0$, a contradiction to the minimality of y_0. Show that in Case 2, $u^2 - pv^2 = -1$. Conclude that if $p \equiv 1 \pmod 4$ then the equation $x^2 - py^2 = -1$ has an integral solution.

13. Show that the solution of $x^2 - 34y^2 = 1$ with $y > 0$, y minimal, is $(\pm 35, 6)$. By examining $y = 1, 2, 3, 4, 5$, deduce that the equation $x^2 - 34y^2 = -1$ has no integral solution. Observe that this latter equation has the rational solutions $(5/3, 1/3)$, $(3/5, 1/5)$. Using the first of these rational solutions, show that the congruence $x^2 - 34y^2 \equiv -1 \pmod m$ has a solution provided that $3 \nmid m$. Similarly, use the second rational solution to show that the congruence has a solution if $5 \nmid m$. Use the Chinese Remainder Theorem to show that the congruence has a solution for all positive m.

7.9 NUMERICAL COMPUTATION

The numerical computations involved in finding a simple continued fraction can be rather lengthy. In general the algorithm (7.7) must be used. However, if ξ_0 is a quadratic irrational the work can be simplified. It is probably best to use (7.13) in a slightly altered form. From (7.13) we have

$$q_{i+1} = \frac{d - m_{i+1}^2}{q_i} = \frac{d - (a_i q_i - m_i)^2}{q_i} = \frac{d - m_i^2}{q_i} - a_i^2 q_i + 2a_i m_i$$

$$= q_{i-1} - a_i(a_i q_i - m_i) + a_i m_i = q_{i-1} + a_i(m_i - m_{i+1}).$$

Starting with $\xi_0 = (m_0 + \sqrt{d})/q_0$, $q_0 | (d - m_0^2)$, we obtain, in turn,

$$a_0 = \left[\frac{m_0 + \sqrt{d}}{q_0}\right], \qquad m_1 = a_0 q_0 - m_0, \qquad q_1 = \frac{d - m_1^2}{q_0}$$

$$a_1 = \left[\frac{m_1 + \sqrt{d}}{q_1}\right], \qquad m_2 = a_1 q_1 - m_1, \qquad q_2 = q_0 + a_1(m_1 - m_2)$$

$$\cdots\cdots\cdots\cdots\cdots$$

$$a_{i-1} = \left[\frac{m_{i-1} + \sqrt{d}}{q_{i-1}}\right], \qquad m_i = a_{i-1} q_{i-1} - m_{i-1},$$

$$q_i = q_{i-2} + a_{i-1}(m_{i-1} - m_i), \quad i \geqslant 1.$$

The formula $q_i q_{i+1} = d - m_{i+1}^2$ serves as a good check. Even for large numbers, this procedure is fairly simple to carry out.

In order to calculate the continued fraction expansion of $\sqrt{d}$ by this method, d being a positive integer and not a perfect square, we see that $m_0 = 0$ and $q_0 = 1$ in such a case. For $\sqrt{73}$ for example, we see that the sequence of calculations begins as follows.

$$m_0 = 0,\ q_0 = 1,\ a_0 = 8,\ m_1 = 8,\ q_1 = 9,\ a_1 = 1,$$

$$m_2 = 1,\ q_2 = 8,\ a_2 = 1,\ m_3 = 7,\ q_3 = 3,\ a_4 = 5,\ \cdots .$$

PROBLEM

1. Continue the calculation started above for $\sqrt{73}$, and verify the continued fraction expansion given in Example 3 in the preceding section.

NOTES ON CHAPTER 7

A completely different approach to continued fractions, specifically with the continued fractions arising naturally out of the approximations rather than the other way about, can be found (for example) in Chapter 1 of Cassels (1957), listed in the General References. The reader interested in statistical questions concerning the usual size of the partial quotients a_i and the expected rate of growth of the denominators k_i should consult the beautiful little book by Khinchin.

CHAPTER 8

Primes and Multiplicative Number Theory

In this chapter we study the asymptotics connected with the multiplicative structure of the integers. The estimates we derive concern prime numbers or the size of multiplicative functions. From such estimates we gain insights concerning the number and size of the prime factors of a typical integer.

8.1 ELEMENTARY PRIME NUMBER ESTIMATES

Let $\pi(x)$ denote the number of primes not exceeding the real number x. In our remarks at the end of Section 1.2 we mentioned the Prime Number Theorem, which asserts that

$$\pi(x) \sim \frac{x}{\log x} \tag{8.1}$$

as $x \to \infty$. This was first proved in 1896, independently by J. Hadamard and Ch. de la Vallée Poussin. We do not prove this, but instead establish a weaker estimate, namely that there exist positive real numbers a and b such that

$$a\frac{x}{\log x} < \pi(x) < b\frac{x}{\log x} \tag{8.2}$$

for all large x. Estimates of this kind were first proved by P. L. Chebyshev in 1852, and we follow his method quite closely. Chebyshev observed that it is fruitful to begin by counting all prime powers $p^k \leqslant x$, each with weight $\log p$, and then derive a corresponding estimate for $\pi(x)$ as a consequence.

Definition 8.1 *The von Mangoldt function* $\Lambda(n)$ *is the arithmetic function* $\Lambda(n) = \log p$ *if* $n = p^k$, $\Lambda(n) = 0$ *otherwise. We let* $\psi(x) = \sum_{1 \leqslant n \leqslant x} \Lambda(n)$, $\vartheta(x) = \sum_{p \leqslant x} \log p$, *and* $\pi(x) = \sum_{p \leqslant x} 1$.

The motivation for considering $\Lambda(n)$ lies in the following observation.

Theorem 8.1 *For every positive integer* n, $\sum_{d|n} \Lambda(d) = \log n$.

Proof Write n as a product of primes in the canonical manner, $n = \prod_p p^a$, where $a = a(p, n)$. Taking logarithms, we find that $\log n = \sum_p a \log p$. But $p^a \| n$, and hence $p^k | n$ if and only if k is one of the numbers $1, 2, \cdots, a$. Thus the sum over p is

$$\sum_{\substack{p,k \\ p^k | n}} \log p = \sum_{d|n} \Lambda(d).$$

Since the function $\log n$ increases very smoothly, we can estimate the sum of $\log n$ quite accurately.

Lemma 8.2 *Let* $T(x) = \sum_{1 \leqslant n \leqslant x} \log n$. *Then for every real number* $x \geqslant 1$ *there is a real number* θ, $|\theta| \leqslant 1$, *such that*

$$T(x) = x \log x - x + \theta \log ex.$$

Proof Let $N = [x]$. We first derive a lower bound for $T(x)$. Since the function $\log u$ is increasing, it follows that $\int_{n-1}^{n} \log u \, du \leqslant \log n$. As $\log 1 = 0$, we deduce that

$$T(x) = \sum_{n=2}^{N} \log n \geqslant \sum_{n=2}^{N} \int_{n-1}^{n} \log u \, du = \int_1^N \log u \, du$$

$$= \int_1^x \log u \, du - \int_N^x \log u \, du.$$

Here the first integral is $[u \log u - u|_1^x = x \log x - x + 1 \geqslant x \log x - x$, and the second integral is $\leqslant \log x$. Hence

$$T(x) \geqslant x \log x - x - \log x.$$

To derive a similar upper bound for $T(x)$ we observe that $\int_n^{n+1} \log u \, du \geqslant$

log n, so that

$$T(x) = \log N + \sum_{n=1}^{N-1} \log n \leqslant \log x + \sum_{n=1}^{N-1} \int_n^{n+1} \log u \, du$$

$$= \log x + \int_1^N \log u \, du \leqslant \log x + \int_1^x \log u \, du,$$

and hence $T(x) \leqslant x \log x - x + 1 + \log x$. The stated estimate follows on combining these two bounds.

By applying the Möbius inversion formula (Theorem 4.8) to the formula of Theorem 8.1, we see that

$$\Lambda(n) = \sum_{d|n} \mu(d) \log n/d.$$

On summing both sides over $n \leqslant x$ we find that

$$\psi(x) = \sum_{n \leqslant x} \sum_{d|n} \mu(d) \log n/d.$$

Writing $n = dm$, the iterated sum may be expressed as a double sum over pairs d, m of positive integers for which $dm \leqslant x$. This may be expressed as an iterated sum, by summing first over $d \leqslant x$, and then over $m \leqslant x/d$. Thus the sum above is

$$= \sum_{d \leqslant x} \mu(d) \sum_{m \leqslant x/d} \log m = \sum_{d \leqslant x} \mu(d) T(x/d).$$

Here we have expressed $\psi(x)$ in terms of the sum $T(y)$ whose asymptotic size we know quite accurately, but new problems arise when we insert the approximation provided by Lemma 8.2 into this relation. Not only do we not know how to estimate the main terms, which contain sums involving the Möbius function, but the error term also makes a large contribution. The large values of d are especially troublesome in this regard. On the other hand, if we are given a sequence of real numbers $\nu(d)$ with $\nu(d) = 0$ for $d > D$, then we could use Lemma 8.2 to estimate the sum

$$\sum_{d \leqslant D} \nu(d) T(x/d). \tag{8.3}$$

Indeed, by Lemma 8.2 this is

$$= x(\log x - 1)\Big(\sum_{d \leqslant D} \nu(d)/d\Big) - x\Big(\sum_{d \leqslant x} \nu(d)(\log d)/d\Big)$$

$$+ \theta\Big(\sum_{d \leqslant D} |\nu(d)|\Big) \log ex \tag{8.4}$$

for $x \geqslant D$, where θ satisfies $|\theta| \leqslant 1$. In order to eliminate the first main term we restrict ourselves to choices of the $\nu(d)$ for which

$$\sum_{d \leqslant D} \nu(d)/d = 0. \tag{8.5}$$

Since $T(y)$ is a sum of $\log n$, which in turn is a sum of $\Lambda(r)$, we may write the expression (8.3) in the form $\sum_{r \leqslant x} N(r, x)\Lambda(r)$. Our strategy is to choose the $\nu(d)$ in such a way that these coefficients $N(r, x)$ are near 1 throughout most of the range, so that this sum is near $\psi(x)$. It is to be expected that the numbers $\nu(d)$ will bear some resemblance to $\mu(d)$. To find a formula for the $N(r, x)$ we note that the expression (8.3) is

$$= \sum_{d=1}^{D} \nu(d) \sum_{n \leqslant x/d} \log n = \sum_{\substack{d, n \\ dn \leqslant x}} \nu(d) \log n.$$

Using Theorem 8.1, we write $\log n = \sum_{r|n} \Lambda(r)$, and choose k so that $rk = n$. Thus the above is

$$= \sum_{\substack{d, n \\ dn \leqslant x}} \nu(d) \sum_{r|n} \Lambda(r) = \sum_{\substack{d, r, k \\ drk \leqslant x}} \nu(d)\Lambda(r).$$

We write this triple sum as an iterated sum, summing first over $r \leqslant x$, then over $d \leqslant x/r$, and finally over $k \leqslant x/(rd)$. Thus the above is

$$= \sum_{r \leqslant x} \Lambda(r) \sum_{d \leqslant x/r} \nu(d) \sum_{k \leqslant x/(rd)} 1 = \sum_{r \leqslant x} \Lambda(r) \sum_{d \leqslant x/r} \nu(d)[x/(rd)],$$

and hence the expression (8.3) equals $\sum_{r \leqslant x} \Lambda(r)N(x/r)$ where

$$N(y) = \sum_{d \leqslant D} \nu(d)[y/d]. \tag{8.6}$$

To summarize our argument thus far, we have shown that if the numbers

$\nu(d)$ satisfy (8.5) then

$$\sum_{r\leqslant x}\Lambda(r)N(x/r) = -x\Big(\sum_{d\leqslant D}\nu(d)(\log d)/d\Big) + \theta\Big(\sum_{d\leqslant D}|\nu(d)|\Big)\log ex \tag{8.7}$$

for $x \geqslant D$, where $N(y)$ is given by (8.6) and $|\theta| \leqslant 1$. Writing $[y/d] = y/d - \{y/d\}$ in (8.6), and appealing to (8.5), we see that

$$N(y) = -\sum_{d\leqslant D}\nu(d)\{y/d\}.$$

Since the function $\{y/d\}$ has period d, it follows that $N(y)$ has period q where q is the least common multiple of those numbers d for which $\nu(d) \neq 0$. (This number q is not necessarily the least period of $N(y)$.) By selecting suitable values for the numbers $\nu(d)$ we can derive upper and lower bounds for $\psi(x)$.

Theorem 8.3 *Put* $a_0 = \frac{1}{3}\log 2 + \frac{1}{2}\log 3 = 0.780355\cdots$, $b_0 = \frac{3}{2}a_0 = \frac{1}{2}\log 2 + \frac{3}{4}\log 3 = 1.170533\cdots$. *If* $a < a_0$ *and* $b > b_0$, *then there is a number* x_0 *(depending on* a *and* b*) such that* $ax < \psi(x) < bx$ *whenever* $x > x_0$.

Proof We take $\nu(1) = 1$, $\nu(2) = -1$, $\nu(3) = -2$, $\nu(6) = 1$, $\nu(d) = 0$ otherwise, and verify that (8.5) holds. Moreover, $N(y)$ has period 6, and from (8.6) we see that $N(y) = 0$ for $0 \leqslant y < 1$, $N(y) = 1$ for $1 \leqslant y < 3$, $N(y) = 0$ for $3 \leqslant y < 5$, $N(y) = 1$ for $5 \leqslant y < 6$. Since $N(y) \leqslant 1$ for all y, it follows that the left side of (8.7) does not exceed $\psi(x)$. Hence (8.7) gives the lower bound

$$\psi(x) \geqslant a_0 x - 5\log ex \tag{8.8}$$

for $x \geqslant 6$. Thus $\psi(x) \geqslant ax$ for all sufficiently large x if $a < a_0$.

To derive an upper bound for $\psi(x)$ we note that $N(y) \geqslant 0$ for all y and that $N(y) = 1$ for $1 \leqslant y < 3$. Hence the left side of (8.7) is $\geqslant \sum_{x/3<n\leqslant x}\Lambda(n) = \psi(x) - \psi(x/3)$. That is,

$$\psi(x) - \psi(x/3) \leqslant a_0 x + 5\log ex$$

for $x \geqslant 6$. By direct calculation we verify that this also holds for $1 \leqslant x \leqslant 6$. Let 3^K be the largest power of 3 not exceeding x. Replacing x by $x/3^k$ and summing over $k = 0, 1, \cdots, K$, we see that

$$\psi(x) = \sum_{k=0}^{K}\psi(x/3^k) - \psi(x/3^{k+1}) \leqslant \sum_{k=0}^{K}\big(a_0x/3^k + 5\log ex\big).$$

As $\sum_0^\infty 1/3^k = (1 - 1/3)^{-1} = 3/2$ and $K = [\log x/\log 3] < \log x$, we conclude that

$$\psi(x) < b_0 x + 5(\log ex)^2 \tag{8.9}$$

for $x \geqslant 1$. Thus if $b > b_0$, then $\psi(x) < bx$ for all sufficiently large x, and the proof is complete.

Having determined the order of magnitude of $\psi(x)$, we now relate $\psi(x)$ to $\vartheta(x)$, and then $\vartheta(x)$ to $\pi(x)$, to establish (8.2). Thus far we have kept close track of the constants that arise in the secondary terms. To focus attention on the salient features of our estimates, and to free ourselves of the need to calculate all constants, we use the "big-O" notation. We let $O(g(x))$ denote a function $f(x)$ with the property that there is an absolute constant C for which $|f(x)| \leqslant Cg(x)$ uniformly in x, and we say, "$f(x)$ is of order $g(x)$," or, "$f(x)$ is big-oh of $g(x)$." For example, since $[x] = x - \{x\}$ and $\{x\}$ is a bounded function, we may write $[x] = x + O(1)$.

Theorem 8.4 *For $x \geqslant 1$, $\vartheta(x) = \psi(x) + O(x^{1/2})$.*

Proof From the definitions of $\psi(x)$ and $\vartheta(x)$ we see that $\vartheta(x) \leqslant \psi(x)$ for all x. To derive an upper bound for the difference $\psi(x) - \vartheta(x)$ we note that

$$\psi(x) = \sum_{n \leqslant x} \Lambda(n) = \sum_{p^k \leqslant x} \log p = \sum_k \sum_{p \leqslant x^{1/k}} \log p = \sum_k \vartheta(x^{1/k}).$$

Put $K = [\log x/\log 2]$. If $k > K$ then $x^{1/k} < 2$, and hence $\vartheta(x^{1/k}) = 0$. Thus we may confine our attention to those k for which $k \leqslant K$. Subtracting $\vartheta(x)$ from both sides, we see that

$$\psi(x) - \vartheta(x) = \sum_{2 \leqslant k \leqslant K} \vartheta(x^{1/k}) \leqslant \sum_{2 \leqslant k \leqslant K} \psi(x^{1/k}) = \sum_{2 \leqslant k \leqslant K} O(x^{1/k})$$

by Theorem 8.3. The implicit constant does not depend on k, and the terms are decreasing, so the above is

$$= O(x^{1/2} + Kx^{1/3}) = O(x^{1/2} + x^{1/3}\log x) = O(x^{1/2}).$$

This gives the stated estimate.

Theorem 8.5 *For* $x \geqslant 2$, $\pi(x) = \dfrac{\vartheta(x)}{\log x} + O(x/(\log x)^2)$.

From this we see that the Prime Number Theorem (8.1) is equivalent to the assertion that

$$\vartheta(x) \sim x \tag{8.10}$$

as $x \to \infty$. By Theorem 8.4 this is in turn equivalent to the assertion that

$$\psi(x) \sim x \tag{8.11}$$

as $x \to \infty$.

Proof We first show that if $x \geqslant 2$ then

$$\pi(x) = \frac{\vartheta(x)}{\log x} + \int_2^x \vartheta(u) u^{-1} (\log u)^{-2}\, du. \tag{8.12}$$

To evaluate the integral we write $\vartheta(u)$ as a sum over prime numbers and interchange the order of summation and integration. Thus the integral is

$$\begin{aligned}\int_2^x \Big(\sum_{p \leqslant u} \log p\Big) u^{-1} (\log u)^{-2}\, du &= \sum_{p \leqslant x} (\log p) \int_p^x u^{-1} (\log u)^{-2}\, du \\ &= \sum_{p \leqslant x} (\log p) \left(\frac{1}{\log p} - \frac{1}{\log x} \right),\end{aligned}$$

which gives (8.12). Since $0 \leqslant \vartheta(x) \leqslant \psi(x)$, it follows from Theorem 8.3 that $\vartheta(x) = O(x)$. Hence the integral in (8.12) is $O(\int_2^x (\log u)^{-2}\, du)$. We consider $2 \leqslant u \leqslant \sqrt{x}$ and $\sqrt{x} \leqslant u \leqslant x$ separately. In the first range the integrand is uniformly bounded, and thus the first portion contributes an amount that is $O(\sqrt{x})$. In the second range, the integrand is uniformly $\leqslant 4/(\log x)^2$, and hence the integral over the second range is $O(x/(\log x)^2)$. This completes the proof, but we remark that more precise estimates of this integral can be derived by integrating by parts.

Corollary 8.6 *Let* a_0 *and* b_0 *be as in Theorem* 8.3. *If* $a < a_0$ *and* $b > b_0$, *then the inequalities* (8.2) *hold for all large* x.

Proof We appeal successively to Theorems 8.5, 8.4, and 8.3 to see that

$$\pi(x) = \frac{\vartheta(x)}{\log x} + O\big(x/(\log x)^2\big) = \frac{\psi(x)}{\log x} + O\big(x/(\log x)^2\big)$$

$$\leqslant b_0 \frac{x}{\log x} + O\big(x/(\log x)^2\big). \tag{8.13}$$

This gives the upper bound in (8.2) for all large, x, if $b > b_0$. Similarly,

$$\pi(x) \geqslant a_0 \frac{x}{\log x} + O\big(x/(\log x)^2\big), \tag{8.14}$$

which gives the lower bound of (8.2) for all large x.

Let c be an absolute constant, $c > 1$. Then $\log cx = \log x + O(1)$, and hence $1/\log cx = 1/\log x + O(1/(\log x)^2)$ for $x \geqslant 2$. Thus if we apply (8.14) with x replaced by cx, and combine this with (8.13), we find that

$$\pi(cx) - \pi(x) \geqslant (ca_0 - b_0)\frac{x}{\log x} + O\big(x/(\log x)^2\big).$$

From Theorem 8.3 we recall that $b_0/a_0 = 3/2$. If $c < 3/2$ then the inequality above is useless, for then the right side is negative while the left side is trivially non-negative. On the other hand, if $c > 3/2$ then the right side is positive, and we deduce that the interval $(x, cx]$ contains at least one prime number, provided that $x > x_0(c)$. After determining an admissible value for $x_0(c)$, one may examine smaller x directly, and thus determine the least acceptable value of $x_0(c)$. We perform this calculation when $c = 2$.

Theorem 8.7 *Bertrand's Postulate. If x is a real number, $x > 1$, then there exists at least one prime number in the open interval $(x, 2x)$.*

Proof Suppose that the interval $(x, 2x)$ contains no prime number. If p is prime then there is at most one value of k for which $p^k \in (x, 2x)$, since $p^{k+1}/p^k = p \geqslant 2$. Furthermore, $k > 1$, since the interval contains no primes. Hence

$$\psi(2x) - \psi(x) = \sum_{x < p^k \leqslant 2x} \log p \leqslant \psi(\sqrt{2x}) + \log 2x.$$

Here the last term on the right is required because $2x$ may be a prime number. We use (8.8) to provide a lower bound for $\psi(2x)$, and use (8.9) to provide upper bounds for $\psi(x)$ and $\psi(\sqrt{2x})$. Thus we find that

$$(2a_0 - b_0)x - 5\log 2ex - 5(\log ex)^2$$

$$\leqslant b_0\sqrt{2x} + 5\left(\log e\sqrt{2x}\right)^2 + \log 2x. \tag{8.15}$$

Here the left side is comparable to x as $x \to \infty$, while the right side is comparable to $\sqrt{x}$. Hence the set of x for which this holds is bounded. In fact, we show that if (8.15) holds then $x < 1600$. That is, if $x \geqslant 1600$ then

$$2a_0 - b_0 \geqslant 5(\log 2ex)/x + 5(\log ex)^2/x$$

$$+ 5\left(\log e\sqrt{2x}\right)^2/x + (\log 2x)/x + b_0\sqrt{2}/\sqrt{x}. \tag{8.16}$$

To this end let $f(x)$ be a function of the form $f(x) = (\log ax^b)^c/x$ where a, b, c are positive real constants. Then $\log f(x) = c \log\log ax^b - \log x$, and by differentiating it follows that

$$\frac{f'(x)}{f(x)} = \left(bc/(\log ax^b) - 1\right)/x.$$

Thus if $ax^b > e^{bc}$, then $f(x) > 0$ and the above expression is negative, so that $f'(x) < 0$. In other words, $f(x)$ is decreasing in the interval $[x_0, \infty)$ where $x_0 = e^c/a^{1/b}$. Thus in particular the first term on the right side of (8.16) is decreasing for $x \geqslant x_1 = 1/2$, the second is decreasing for $x \geqslant x_2 = e$, the third is decreasing for $x \geqslant x_3 = 1/2$, and the fourth is decreasing for $x \geqslant x_4 = e/2$. Since the last term on the right side of (8.16) is decreasing for all positive values of x, we conclude that the right side is decreasing for $x \geqslant x_2 = 2.71828\cdots$. By direct calculation we discover that the right side of (8.16) is less than $3/8$ when $x = 1600$, while the left side is $> 3/8$. Since the right side is decreasing, it follows that (8.16) holds for all $x \geqslant 1600$.

We have shown that Bertrand's postulate is true for $x \geqslant 1600$. To verify it for $1 < x < 1600$ we note that the following thirteen numbers are prime: 2, 3, 5, 7, 13, 23, 43, 83, 163, 317, 631, 1259, 2503. As each term of this sequence is less than twice the preceding member, Bertrand's postulate is valid for $1 < x < 2503$, and the proof is complete.

We have determined the order of magnitude of $\pi(x)$, but not the stronger asymptotic relation (8.1). We now consider sums involving primes whose asymptotic size we can determine more precisely.

Theorem 8.8 *Suppose that* $x \geqslant 2$. *Then*

(a) $\sum_{n \leqslant x} \Lambda(n)/n = \log x + O(1)$;

(b) $\sum_{p \leqslant x} (\log p)/p = \log x + O(1)$;

(c) $\int_1^x \psi(u)/u^2\, du = \log x + O(1)$;

(d) *for a suitable constant* b,

$$\sum_{p \leqslant x} 1/p = \log\log x + b + O(1/\log x);$$

(e) *for a suitable constant* $c > 0$,

$$\prod_{p \leqslant x} (1 - 1/p) = \frac{c}{\log x}(1 + O(1/\log x)).$$

Let γ denote Euler's constant (i.e., the constant in Lemma 8.27). It may be shown that the constant c above is $e^{-\gamma}$. A proof of this is outlined in Problem 27 at the end of Section 8.3.

Proof (*a*) Let $T(x)$ be as in Lemma 8.2. Then by Theorem 8.1, $T(x) = \sum_{n \leqslant x}\sum_{d|n}\Lambda(d)$. Writing $n = md$, we see that

$$T(x) = \sum_{\substack{m,d \\ md \leqslant x}} \Lambda(d) = \sum_{d \leqslant x} \Lambda(d) \sum_{m \leqslant x/d} 1 = \sum_{d \leqslant x} \Lambda(d)[x/d].$$

Since $[x] = x + O(1)$, the sum on the right is

$$x \sum_{d \leqslant x} \Lambda(d)/d + O\Big(\sum_{d \leqslant x} \Lambda(d)\Big).$$

The sum in the error term is $\psi(x)$, which is $O(x)$ by Theorem 8.3. Since Lemma 8.2 gives $T(x) = x\log x + O(x)$, the assertion (*a*) follows by dividing through by x.

(*b*) The sum in (*b*) is smaller than the sum in (*a*) by the amount

$$\sum_{\substack{p^k \leqslant x \\ k > 1}} \log p/p^k < \sum_p \log p \sum_{k=2}^{\infty} p^{-k} = \sum_p \frac{\log p}{p(p-1)}.$$

This latter series converges, since it is a subseries of the convergent series

$\sum_{n=2}^{\infty}\dfrac{\log n}{n(n-1)}$. Thus the difference between the sum (a) and the sum (b) is uniformly bounded, and hence the assertion (b) follows from (a).

(c) By definition, $\psi(x) = \sum_{n \leqslant x}\Lambda(n)$. On inverting the order of summation and integration, we find that

$$\begin{aligned}\int_1^x \psi(u)/u^2\,du &= \int_1^x \sum_{n\leqslant u}\Lambda(n)u^{-2}\,du\\ &= \sum_{n\leqslant x}\Lambda(n)\int_n^x u^{-2}\,du = \sum_{n\leqslant x}\Lambda(n)\left(\frac{1}{n}-\frac{1}{x}\right)\\ &= \left(\sum_{n\leqslant x}\Lambda(n)/n\right) - \psi(x)/x.\end{aligned}$$

By Theorem 8.3, $\psi(x)/x = O(1)$. Thus the result follows from (a).

(d) Let $L(x)$ denote the sum in (b). Then

$$\int_2^x \frac{L(u)}{u(\log u)^2}\,du = \int_2^x \sum_{p\leqslant u}\frac{\log p}{p}\,\frac{1}{u(\log u)^2}\,du.$$

On inverting the order of summation and integration, we find that this is

$$= \sum_{p\leqslant x}\frac{\log p}{p}\int_p^x \frac{1}{u(\log u)^2}\,du = \sum_{p\leqslant x}\frac{\log p}{p}\left(\frac{1}{\log p}-\frac{1}{\log x}\right).$$

This is the sum in question minus $L(x)/\log x$. That is,

$$\sum_{p\leqslant x} 1/p = \frac{L(x)}{\log x} + \int_2^x \frac{L(u)}{u(\log u)^2}\,du.$$

Now let $E(x)$ denote the error term in (b), so that (b) takes the form $L(x) = \log x + E(x)$, where $E(x)$ is uniformly bounded. Then the right side above is

$$= 1 + \log\log x - \log\log 2 + \frac{E(x)}{\log x} + \int_2^x \frac{E(u)}{u(\log u)^2}\,du.$$

We set

$$b = 1 - \log\log 2 + \int_2^{\infty}\frac{E(u)}{u(\log u)^2}\,du,$$

so that the sum in question is

$$\log\log x + b + \frac{E(x)}{\log x} - \int_x^\infty \frac{E(u)}{u(\log u)^2}\,du.$$

Since $E(u)$ is uniformly bounded, these last two terms are $O(1/\log x)$, and we have (d).

(e) Let $u(\delta) = \log(1-\delta) + \delta$. Then $u(\delta) = O(\delta^2)$ uniformly for $|\delta| \leqslant 1/2$, so that

$$\sum_{p\leqslant x} \log(1 - 1/p) = -\sum_{p\leqslant x} 1/p + \sum_p u(1/p) - \sum_{p>x} u(1/p).$$

Here the second sum on the right is absolutely convergent and thus denotes an absolute constant, say b', while the third sum on the right is $O(\Sigma_{p>x} 1/p^2) = O(\Sigma_{n>x} 1/n^2) = O(1/x)$. Thus by (d) it follows that the right side above is $= -\log\log x - b + b' + O(1/\log x)$. On exponentiating, we find that

$$\prod_{p\leqslant x} (1 - 1/p) = \frac{c}{\log x} \exp(O(1/\log x))$$

where $c = \exp(-b + b') > 0$. Since $\exp(\delta) = 1 + O(\delta)$ uniformly for $|\delta| \leqslant 2$, we obtain (e), and the proof is complete.

Corollary 8.9 $\limsup_{x\to\infty} \dfrac{\pi(x)}{x/\log x} \geqslant 1$ *and* $\liminf_{x\to\infty} \dfrac{\pi(x)}{x/\log x} \leqslant 1.$

From this we see that if $\dfrac{\pi(x)}{x/\log x}$ has a limit as $x \to \infty$, then its value must be 1.

Proof We treat the lim sup; the proof for the lim inf is similar. From Theorems 8.5 and 8.4 it is evident that

$$\limsup_{x\to\infty} \frac{\pi(x)}{x/\log x} = \limsup_{x\to\infty} \frac{\vartheta(x)}{x} = \limsup_{x\to\infty} \frac{\psi(x)}{x}. \tag{8.17}$$

If this last lim sup were less than 1 then there would be an $\varepsilon > 0$ such that $\psi(x) < (1-\varepsilon)x$ for all $x > x_0$, and then it would follow that the integral in Theorem 8.8(c) is $\leqslant (1-\varepsilon)\log x + O(1)$. Since this contradicts the estimate of Theorem 8.8(c), it follows that this lim sup is $\geqslant 1$.

In Theorem 1.18 we established that there exist arbitrarily long intervals containing no prime number. We are now in a position to put this in the following more quantitative form.

Corollary 8.10 *Let p' denote the least prime exceeding p. Then*

$$\limsup_{p\to\infty} \frac{p'-p}{\log p} \geqslant 1 \text{ and } \liminf_{p\to\infty} \frac{p'-p}{\log p} \leqslant 1.$$

Proof Suppose that $0 < x_1 < x_2$, and let p_1 denote the least prime exceeding x_1, p_2 the least prime exceeding x_2. We compare the telescoping sum

$$\sum_{x_1 < p \leqslant x_2} (p'-p) = p_2 - p_1 \tag{8.18}$$

with the sum

$$\sum_{x_1 < p \leqslant x_2} \log p = \vartheta(x_2) - \vartheta(x_1). \tag{8.19}$$

Suppose that c is a number such that $p' - p \leqslant c \log p$ for all primes p in the interval $(x_1, x_2]$. Then

$$p_2 - p_1 \leqslant c(\vartheta(x_2) - \vartheta(x_1)). \tag{8.20}$$

By Corollary 8.9 and (8.17), there exist arbitrarily large numbers x_2 for which $\vartheta(x_2) < (1+\varepsilon)x_2$. For such x_2 the right side of (8.20) is $< c(1+\varepsilon)x_2$. By Bertrand's postulate $p_1 < 2x_1$, so the left side of (8.20) is $> x_2 - 2x_1$. Thus if $x_2 > x_1/\varepsilon$, then $c > 1 - 3\varepsilon$. That is, there exist arbitrarily large primes p such that $p' - p > (1-3\varepsilon)\log p$. Since ε is arbitrarily small, this gives the stated lower bound for the lim sup.

Suppose now that c is a number such that $p' - p \geqslant c \log p$ for all primes p in the interval $(x_1, x_2]$. Then

$$p_2 - p_1 \geqslant c(\vartheta(x_2) - \vartheta(x_1)). \tag{8.21}$$

By Corollary 8.9 and (8.17), there exist arbitrarily large numbers x_0 for which $\vartheta(x_0) > (1-\varepsilon)x_0$. For such an x_0 let p_0 be the largest prime not exceeding x_0, and take $x_2 = p_0 - 1$. Then $\vartheta(x_2) = \vartheta(x_0) - \log p_0 > (1-2\varepsilon)x_0 > (1-2\varepsilon)x_2$. We suppose also that $x_2 > x_1/\varepsilon$, so that $\vartheta(x_1) < 2x_1 < 2\varepsilon x_2$. Hence the right side of (8.21) is $> c(1-4\varepsilon)x_2$. Since $p_2 = x_2 + 1$, the left side of (8.21) is $< x_2$. It follows that $c < (1-4\varepsilon)^{-1}$. That is, there exist arbitrarily large primes p such that $p' - p <$

$(1-4\varepsilon)^{-1}\log p$. Since ε is arbitrarily small, this gives the stated upper bound for the lim inf.

PROBLEMS

1. Show that $\Sigma_{d\leqslant x}\mu(d)[x/d]=1$ for all real numbers $x\geqslant 1$. Deduce that $|\Sigma_{d\leqslant x}\mu(d)/d|\leqslant 1$ for all real $x\geqslant 1$.
2. Show that $\Lambda(n)=-\Sigma_{d|n}\mu(d)\log d$ for every positive integer n.
3. For $1\leqslant d\leqslant D$ let $\nu(d)$ be real numbers satisfying (8.5), and let q denote the least common multiple of those d for which $\nu(d)\neq 0$. Show that if y is not an integer then $N(y)+N(q-y)=-\Sigma_{d\leqslant D}\nu(d)$ where $N(y)$ is given by (8.6). (H)
4. Show that $2^x<\prod_{p\leqslant x}p<(13/4)^x$ for all sufficiently large x. (H)
5. Let d_n denote the least common multiple of the integers $1,2,\cdots,n$. Show that $d_n=e^{\psi(n)}$. Show also that $2^n<d_n<(13/4)^n$ for all sufficiently large integers n.
6. Show that if n is a positive integer then $T(n)=\log n!$. Show that $2^{2n}/(2n)\leqslant\binom{2n}{n}\leqslant 2^{2n}$. Deduce that $(2\log 2)n-\log 2n\leqslant T(2n)-2T(n)\leqslant(2\log 2)n$. (H)
7. Set $\nu(1)=1$, $\nu(2)=-2$, $\nu(d)=0$ for $d>2$. Show that (8.5) is satisfied. Show that $N(y)$ defined by (8.6) has period 2, and that $N(y)=0$ for $0\leqslant y<1$, $N(y)=1$ for $1\leqslant y<2$. Use de Polignac's formula (Theorem 4.2) to determine the canonical factorization of $\binom{2n}{n}$ into primes. Show that this factorization is equivalent to the identity $T(2n)-2T(n)=\Sigma_{r\leqslant 2n}\Lambda(r)N(2n/r)$. Explain why $\psi(2n)-\psi(n)\leqslant T(2n)-2T(n)\leqslant\psi(2n)$, and derive a weaker form of Theorem 8.3 with a_0 replaced by $\log 2=0.6931\cdots$ and b_0 replaced by $2\log 2=1.3863\cdots$.
8. Prove that $n!=m^k$ is impossible in integers $k>1$, $m>1$, $n>1$. (H)
9. Let k and r be integers, $k>1$, $r>1$. Show that there is a prime number whose representation in base r has exactly k digits.
10. For this problem include 1 as a prime. Prove that every positive integer can be written as a sum of one or more distinct primes.
11. Show that $\Sigma_{n\leqslant x}\psi(x/n)=T(x)$ for $x\geqslant 1$, where $T(x)$ is defined as in Lemma 8.2.
12. Let $P(x)$ be a polynomial with integral coefficients and degree not exceeding n, and put $I(P)=\int_0^1 P(x)\,dx$. Show that $I(P)d_{n+1}$ is an

integer, where d_n is defined as in Problem 5. Show that there is such a polynomial $P(x)$ for which $I(P)d_{n+1} = 1$.

13. Put $Q(x) = x^2(1-x)^2(2x-1)$. Show that $\max_{0\leqslant x\leqslant 1} |Q(x)| = 5^{-5/2}$. In the notation of the preceding problem, show that if $P(x) = Q(x)^{2k}$ then $0 < I(P) < 5^{-5k}$. Deduce that $d_{10k+1} \geqslant 5^{5k}$, and hence that $\psi(10k+1) \geqslant c10k$ where $c = (\log 5)/2$.

14. Chebyshev took $\nu(1) = 1$, $\nu(2) = -1$, $\nu(3) = -1$, $\nu(5) = -1$, $\nu(30) = 1$, $\nu(d) = 0$ otherwise. Show that these $\nu(d)$ satisfy (8.5). Let $N(y)$ be given by (8.6). Show that $N(y)$ has period 30, that $N(y)$ takes only the values 0, 1, and that $N(y) = 1$ for $1 \leqslant y < 6$. Use this to derive a version of Theorem 8.3 with a_0 replaced by the larger constant $a_1 = (7/15)\log 2 + (3/10)\log 3 + (1/6)\log 5 = 0.9212\cdots$, and with b_0 replaced by the smaller constant $b_1 = 6a_1/5 = 1.1056\cdots$. Deduce that the interval (x, cx) contains a prime number for all large x provided that $c > 6/5$.

15. Let c be an absolute constant, $c > 1$. Show that for $x \geqslant 2$,

$$\int_2^x (\psi(cu) - \psi(u))u^{-2}\,du = (c-1)\log x + O(1).$$

16. Show that for $x \geqslant 2$,

$$\vartheta(x) = \pi(x)(\log x) - \int_2^x \pi(u)/u\,du.$$

***17.** Show that $\prod_{p\leqslant x} p < 4^x$ for all real numbers $x \geqslant 2$.

***18.** Suppose that the Prime Number Theorem (8.1) holds. Deduce that if $c > 1$ then there is a number $x_0(c)$ such that the interval (x, cx) contains a prime number for all $x > x_0(c)$.

***19.** Let p_n denote the nth prime. Show that $\limsup_{n\to\infty} p_n/(n\log n) \geqslant 1$, and that $\liminf_{n\to\infty} p_n/(n\log n) \leqslant 1$.

***20.** Let p_n denote the nth prime number. Show that the Prime Number Theorem (8.1) is equivalent to the assertion that $p_n \sim n\log n$ as $n \to \infty$.

8.2 DIRICHLET SERIES

A *Dirichlet series* is any series of the form $\sum_{n=1}^{\infty} a_n/n^s$. Here s is a real number, so that the series defines a function $A(s)$ of the real variable s, provided that the series converges. The *Riemann zeta function* is an important example of a Dirichlet series. For $s > 1$ it is defined to be

$$\zeta(s) = \sum_{n=1}^{\infty} 1/n^s. \tag{8.22}$$

Here the summands are monotonically decreasing, so we may use the integral test to determine when this series converges. Since $\int_1^\infty 1/u^s\, du < \infty$ if and only if $s > 1$, we see that this series is absolutely convergent for $s > 1$, but divergent for $s \leqslant 1$.

In this section we establish the basic analytic properties of Dirichlet series in a manner analogous to the basic theory of power series. However, our main object is to discover useful relationships among arithmetic functions by manipulating the Dirichlet series they generate.

Questions of convergence of Dirichlet series can be subtle, but for our present purposes it is enough to consider absolutely convergent Dirichlet series. We have already shown that the Dirichlet series (8.22) is absolutely convergent if and only if $s \in (1, \infty)$. This behavior is typical of more general Dirichlet series. Suppose that α is a real number such that

$$\sum_{n=1}^{\infty} |a_n| n^{-\alpha} < \infty. \tag{8.23}$$

Since n^{-s} is a monotonically decreasing function of s, it follows by the comparison test that the series $\sum_{n=1}^{\infty} a_n n^{-s}$ is uniformly and absolutely convergent for $\alpha \leqslant s < \infty$. We let σ_a denote the infimum of those real numbers α for which (8.23) holds. This number is called the *abscissa of absolute convergence* of the series $\sum a_n n^{-s}$, since the series is absolutely convergent for every $s > \sigma_a$, but not for any $s < \sigma_a$. It may happen that $\sigma_a = -\infty$, in which case the series is absolutely convergent for all real s, or it may happen that $\sigma_a = +\infty$, which is to say the series is absolutely convergent for no real number s. Examples of these two extreme cases are found in Problems 2 and 3 at the end of this section. We have established the following theorem.

Theorem 8.11 *For each Dirichlet series $A(s) = \sum_{n=1}^{\infty} a_n/n^s$ there exists a unique real number σ_a such that the series $A(s)$ is absolutely convergent for $s > \sigma_a$, but is not absolutely convergent for $s < \sigma_a$. If $c > \sigma_a$, then the series $A(s)$ is uniformly convergent for s in the interval $[c, +\infty)$.*

Corollary 8.12 *Let σ_a be the abscissa of absolute convergence of the Dirichlet series $A(s) = \sum_{n=1}^{\infty} a_n/n^s$. Then $A(s)$ is a continuous function on the open interval $(\sigma_a, +\infty)$.*

Proof Each term a_n/n^s is a continuous function of s. Take $c > \sigma_a$. On the interval $[c, +\infty)$ the series $A(s)$ is a uniformly convergent series of continuous functions, and therefore $A(s)$ is continuous on this interval. Since c may be arbitrarily close to σ_a, we conclude that $A(s)$ is continuous on the open interval $(\sigma_a, +\infty)$.

We now show that the abscissa of absolute convergence is related to the average size of the numbers $|a_n|$.

Theorem 8.13 *Let σ_a be the abscissa of absolute convergence of the Dirichlet series $A(s) = \sum_{n=1}^{\infty} a_n/n^s$. If c is a non-negative real number such that*

$$\sum_{n \leqslant x} |a_n| = O(x^c) \tag{8.24}$$

as $x \to \infty$, then $\sigma_a \leqslant c$. Conversely, if $c > \max(0, \sigma_a)$ then (8.24) holds as $x \to \infty$.

Proof Suppose that (8.24) holds and that $\varepsilon > 0$. Then

$$\sum_{N<n\leqslant 2N} |a_n| n^{-c-\varepsilon} \leqslant N^{-c-\varepsilon} \sum_{N<n\leqslant 2N} |a_n| = O\big(N^{-c-\varepsilon}(2N)^c\big) = O(N^{-\varepsilon}).$$

We take $N = 2^k$ and sum over k. Since $\sum_k 2^{-k\varepsilon} < \infty$, it follows that $\sum_n |a_n| n^{-c-\varepsilon} < \infty$. Since ε may be arbitrarily small, we conclude that $\sigma_a \leqslant c$.

Conversely, if $c \geqslant 0$ then $(x/n)^c \geqslant 1$ for all $n \leqslant x$, and consequently

$$\sum_{n\leqslant x} |a_n| \leqslant \sum_{n\leqslant x} |a_n| (x/n)^c \leqslant x^c \sum_{n=1}^{\infty} |a_n| n^{-c}.$$

If in addition $c > \sigma_a$ then the series on the right converges, and we have (8.24).

Let $A(s) = \sum a_m/m^s$ and $B(s) = \sum b_n/n^s$ be two Dirichlet series. Here m and n run from 1 to ∞. For brevity we sometimes omit the limits of summation, when they may be inferred from the context. We now consider the product function $A(s)B(s)$. Ignoring questions of convergence for the moment, we see that this product is a double series in which the general term is

$$\frac{a_m}{m^s} \cdot \frac{b_n}{n^s} = \frac{a_m b_n}{(mn)^s}. \tag{8.25}$$

Here the base of the exponential is the product mn, so it is natural to group together those terms for which mn has a given value, say $mn = r$. With this in mind, we set

$$c_r = \sum_{d|r} a_d b_{r/d}. \tag{8.26}$$

This new sequence $\{c_r\}$ is called the *Dirichlet convolution* of the two sequences $\{a_m\}$ and $\{b_n\}$. We express this in symbols by writing $c = a * b$. It is reasonable to expect that $A(s)B(s) = C(s)$, where $C(s)$ is the Dirichlet series $C(s) = \Sigma c_r/r^s$. We now show that this is indeed the case if the two given series are absolutely convergent.

Theorem 8.14 *Suppose that s is a real number for which the Dirichlet series* $\Sigma a_m/m^s$ *and* $\Sigma b_n/n^s$ *are both absolutely convergent. Let the numbers* c_r *be defined by* (8.26). *Then the Dirichlet series* $C(s) = \Sigma c_r/r^s$ *is absolutely convergent, and* $C(s) = A(s)B(s)$.

Here we encounter a special case of the general principle that an absolutely convergent series may be arbitrarily rearranged without disturbing the absolute convergence or altering the value of the sum. Rather than appeal to the general principle, we give a self-contained proof that applies to the present situation.

Proof For positive real numbers R let $S_c(R) = \Sigma_{1 \leqslant r \leqslant R} c_r/r^s$, and similarly let $S_a(M)$ and $S_b(N)$ denote partial sums of $A(s)$ and of $B(s)$. We show first that $S_c(R)$ tends to $A(s)B(s)$ as $R \to \infty$. In (8.26) we replace d by m and r/d by n, and thus find that $S_c(R)$ may be written in the form

$$S_c(R) = \sum_{mn \leqslant R} (a_m/m^s)(b_n/n^s). \tag{8.27}$$

Here the sum is over those pairs m, n of positive integers for which $mn \leqslant R$. Let T_1 be the sum formed by restricting m and n by the conditions $1 \leqslant m \leqslant \sqrt{R}$, $1 \leqslant n \leqslant \sqrt{R}$; let T_2 be the sum over those m, n for which $1 \leqslant m \leqslant \sqrt{R} < n \leqslant x/m$; and let T_3 be the sum over those m, n for which $1 \leqslant n \leqslant \sqrt{R} < m \leqslant x/n$, so that $S_c(R) = T_1 + T_2 + T_3$. We note that $T_1 = S_a(\sqrt{R})S_b(\sqrt{R})$, which tends to $A(s)B(s)$ as $R \to \infty$. On the other hand,

$$|T_2| \leqslant \sum_{1 \leqslant m \leqslant \sqrt{R}} |a_m| m^{-s} \sum_{\sqrt{R} < n \leqslant x/m} |b_n| n^{-s}.$$

We drop the condition $n \leqslant x/m$ in the inner sum and, having done this, drop the condition $m \leqslant \sqrt{R}$ in the outer sum. Thus we see that

$$|T_2| \leqslant \left(\sum_{m=1}^{\infty} |a_m| m^{-s}\right)\left(\sum_{n>\sqrt{R}} |b_n| n^{-s}\right).$$

Here the first factor is finite by hypothesis, and the second factor is the tail of an absolutely convergent series, which therefore tends to 0 as $R \to \infty$. Similarly

$$|T_3| \leqslant \left(\sum_{n=1}^{\infty} |b_n| n^{-s}\right)\left(\sum_{m>\sqrt{R}} |a_m| m^{-s}\right),$$

which tends to 0 as $R \to \infty$.

We have shown that the series $C(s)$ is convergent and that $C(s) = A(s)B(s)$. To complete the proof we must establish that the series $C(s)$ is absolutely convergent. To this end we apply the triangle inequality in (8.26) to see that

$$|c_r| \leqslant \sum_{d|r} |a_d|\, |b_{r/d}|.$$

Let C_r denote the sum on the right. We now apply the result that we have already demonstrated, with a_m replaced by $|a_m|$ and b_n replaced by $|b_n|$. This allows us to deduce that the series $\sum C_r/r^s$ is convergent. Since $|c_r| \leqslant C_r$ for all r, it follows by the comparison test that the series $C(s)$ is absolutely convergent. This completes the proof.

In (8.22) we defined $\zeta(s)$ as a sum of positive numbers. Thus it is obvious that $\zeta(s) > 0$ for all $s > 1$. We now express $1/\zeta(s)$ as a Dirichlet series.

Theorem 8.15 *If $s > 1$ then $1/\zeta(s) = \sum_{m=1}^{\infty} \mu(m)/m^s$.*

Proof We apply Theorem 8.14 with $a_m = \mu(m)$ and $b_n = 1$ for all n. To show that the series $A(s)$ is absolutely convergent, we note that $|\mu(m)| \leqslant 1$ for all m, so that by the comparison test

$$\sum |\mu(m)|/m^s \leqslant \zeta(s) < \infty$$

for $s > 1$. On comparing Theorem 4.7 with (8.26), we deduce that $c_1 = 1$, and that $c_r = 0$ for all $r > 1$. Thus $A(s)\zeta(s) = \sum c_r/r^s = 1$ for all $s > 1$.

For ease of reference we now state without proof a basic tool from the theory of series. For most of a century this was known as the *Weierstrass M-test*, though today it is more frequently called the *principle of dominated convergence*.

Lemma 8.16 *Let a be a real number, and for each positive integer n suppose that $M_n(x)$ is a function defined on the interval $[a,\infty)$. Let $M_1, M_2, \cdots$ be non-negative real numbers. If*

(i) $|M_n(x)| \leqslant M_n$ *for all real* $x \geqslant a$ *and all* $n = 1, 2, \cdots$,
(ii) $\lim_{x\to\infty} M_n(x)$ *exists for each* $n = 1, 2, \cdots$, *and*
(iii) $\sum_{n=1}^{\infty} M_n$ *converges,*

then $\lim_{x\to\infty} \sum_{n=1}^{\infty} M_n(x) = \sum_{n=1}^{\infty} \lim_{x\to\infty} M_n(x)$.

Theorem 8.17 *If $A(s) = \sum a_n/n^s$ is a Dirichlet series with abscissa of absolute convergence σ_a, $\sigma_a < \infty$, and if $A(s) = 0$ for all large s, then $a_n = 0$ for all n.*

More generally, if $B(s) = \sum b_n/n^s$ and $C(s) = \sum c_n/n^s$ are two Dirichlet series that are absolutely convergent for all large s, say for $s \geqslant \sigma_1$ and for $s \geqslant \sigma_2$, respectively, then the Dirichlet series with coefficients $a_n = b_n - c_n$ is absolutely convergent for $s \geqslant \sigma$ where $\sigma = \max(\sigma_1, \sigma_2)$. Thus Theorem 8.17 assures us that an expansion of a function as a Dirichlet series is *unique*. This is analogous to the corresponding uniqueness theorem for power series. The *existence* of a Dirichlet series expansion is quite a different matter. Here the theory of Dirichlet series departs from that of power series. While a power series expansion exists for any function of a wide class known as *analytic functions*, those functions expressible by Dirichlet series form a comparatively narrow subclass of analytic functions. Nevertheless, Dirichlet series are of great value in studying arithmetic functions.

Proof Suppose that $a_n = 0$ for $n < N$, and that c is a real number such that $\sum |a_n| n^{-c} < \infty$. We apply Lemma 8.16 with $M_n(s) = a_n(N/n)^s$ and $M_n = |a_n|(N/n)^c$. We note that $\lim_{s\to\infty} a_n(N/n)^s = a_N$ for $n = N$, and that this limit is 0 for $n > N$. Hence by Lemma 8.16,

$$\lim_{s\to+\infty} A(s)N^s = \lim_{s\to+\infty} \sum_{n=N}^{\infty} a_n(N/n)^s = \sum_{n=N}^{\infty} \lim_{s\to+\infty} a_n(N/n)^s = a_N.$$

Since $A(s) = 0$ for all large s by hypothesis, it follows that the limit on the left is 0, and hence that $a_N = 0$. Hence $a_n = 0$ for all n, and the theorem is proved.

Suppose that $a_m = 1$ for all m and that $b_n = 1$ for all n in (8.26). Then $c_r = d(r)$, $A(s) = B(s) = \zeta(s)$, and by Theorem 8.14 it follows that

$$\sum_{r=1}^{\infty} d(r)/r^s = \zeta(s)^2 \tag{8.28}$$

for $s > 1$. Recalling (4.1), we take $a_m = \mu(m)$ and $b_n = n$ in Theorem 8.14 to see that

$$\sum_{r=1}^{\infty} \phi(r)/r^s = \frac{\zeta(s-1)}{\zeta(s)} \tag{8.29}$$

for $s > 2$. Similarly, we find that

$$\sum_{r=1}^{\infty} \sigma(r)/r^s = \zeta(s-1)\zeta(s) \tag{8.30}$$

for $s > 2$. On combining these three identities we see that

$$\left(\sum_{m=1}^{\infty} \phi(m)/m^s\right)\left(\sum_{n=1}^{\infty} d(n)/n^s\right) = \frac{\zeta(s-1)}{\zeta(s)} \cdot \zeta(s)^2$$

$$= \zeta(s-1)\zeta(s) = \sum_{r=1}^{\infty} \sigma(r)/r^s. \tag{8.31}$$

By a further application of Theorem 8.14 we see that the product of the two Dirichlet series on the left may be expressed as a Dirichlet series $C(s) = \Sigma c_r r^s$, which is absolutely convergent for $s > 2$ and has coefficients $c_r = \Sigma_{d|r}\phi(d)d(r/d)$. Then by Theorem 8.17 we deduce that

$$\sum_{d|r} \phi(d)\, d(r/d) = \sigma(r) \tag{8.32}$$

for all positive integers r. This identity may be proved by elementary reasoning, but the analytic approach offers new insights. For example, the hypothesis in the Möbius inversion formula (Theorem 4.8) amounts to the identity

$$\zeta(s)\sum f(n)/n^s = \sum F(n)/n^s, \tag{8.33}$$

while the conclusion in Theorem 4.8 similarly asserts that

$$\sum f(n)/n^s = \frac{1}{\zeta(s)} \sum F(n)/n^s. \tag{8.34}$$

In Theorem 4.9 it is shown that the second of these identities implies the first. Thus we have new proofs of Theorem 4.8 and Theorem 4.9, but only for functions f and F, which generate Dirichlet series whose abscissae of absolute convergence are less than infinity. To remove this restriction one could truncate the series. That is, let N be a large integer, and put $f_1(n) = f(n)$ for $n \leqslant N$, $f_1(n) = 0$ for $n > N$. If we replace f by f_1 in (8.33) then we obtain a new arithemetic function on the right, say F_1. Clearly $F_1(n) = F(n)$ for all $n \leqslant N$. All three series in (8.33) are absolutely convergent for $s > 1$. Thus we have (8.34) with f and F replaced by f_1 and F_1, and by comparing the coefficients on the two sides we deduce that $f_1(n) = \sum_{d|n}\mu(d)F_1(n/d)$ for all positive integers n. This gives the conclusion in Theorem 4.8 for all $n \leqslant N$. Since N may be taken arbitrarily large, we now have the conclusion without restriction. This truncation device can be used similarly to derive Theorem 4.9 analytically, without restriction on the sizes of the functions f and F. When employed in this way, the analytic approach not only yields short proofs of elementary identities but also helps one to discover useful relationships. The analytic method becomes more profound in more advanced work, as the asymptotic properties of an arithmetic function are related to the analytic properties of the associated Dirichlet series. In particular, the Prime Number Theorem may be derived from the deeper analytic properties of the Riemann zeta function.

The coefficients of a Dirichlet series need not be multiplicative, but in case the coefficients are multiplicative we may express the Dirichlet series as a product.

Theorem 8.18 *The Euler product formula. Suppose that $f(n)$ is a multiplicative function, and put $F(s) = \sum_{n=1}^{\infty} f(n)/n^s$. If s is a real number for which the series $F(s)$ is absolutely convergent, then*

$$F(s) = \prod_p \left(1 + f(p)/p^s + f(p^2)/p^{2s} + f(p^3)/p^{3s} + \cdots\right).$$

In case $f(n) = 1$ for all n, this is the Euler product for the Riemann zeta function,

$$\sum_{n=1}^{\infty} 1/n^s = \prod_p \left(1 + 1/p^s + 1/p^{2s} + 1/p^{3s} + \cdots\right), \qquad (8.35)$$

which is valid for $s > 1$. Ignoring questions of convergence for the moment, we observe that when the product on the right is expanded we obtain a sum of terms of the form $1/(p_1^{\alpha_1}p_2^{\alpha_2}\cdots p_r^{\alpha_r})^s$, where $p_1, p_2, \cdots, p_r$ are distinct primes. That is, the right side, when expanded, gives a sum

$\Sigma r(n)/n^s$ where $r(n)$ is the number of ways of expressing n as a product of prime powers. Since the Dirichlet series coefficients of a function are unique, the identity (8.35) asserts that $r(n) = 1$ for all n. That is, each positive integer is a product of prime powers in precisely one way. In this sense the important identity (8.35) constitutes an analytic formulation of the fundamental theorem of arithmetic.

Proof By the comparison test we see that

$$1 + |f(p)|/p^s + |f(p^2)|/p^{2s} + \cdots \leqslant \sum_{n=1}^{\infty} |f(n)|/n^s < \infty \quad (8.36)$$

for any prime number p. Thus by Theorem 8.14 we find that

$$(1 + f(2)/2^s + f(4)/4^s + \cdots)(1 + f(3)/3^s + f(9)/9^s + \cdots)$$
$$= \sum_{n \in \mathcal{N}} f(n)/n^s$$

where $\mathcal{N} = \{1, 2, 3, 4, 6, 8, 9, 12, \cdots\}$ is the set of all positive integers of the form $2^\alpha 3^\beta$. Here we have used the fact that $f(2^\alpha)f(3^\beta) = f(2^\alpha 3^\beta)$. More generally, let y be a positive real number, put $f_y(n) = f(n)$ if n is composed entirely of primes $p \leqslant y$, and put $f_y(n) = 0$ otherwise. Then by repeated applications of Theorem 8.14 we deduce that

$$\prod_{p \leqslant y} (1 + f(p)/p^s + f(p^2)/p^{2s} + f(p^3)/p^{3s} + \cdots) = \sum_{n=1}^{\infty} f_y(n)/n^s.$$

Here the sum on the right is a subsum of the series $F(s)$, and it remains to show that this series tends to $F(s)$ as $y \to \infty$. As y increases, the sum includes more of the terms in the series $F(s)$, so it is to be expected that the series would tend to $F(s)$. To construct a rigorous proof that this is the case, we apply the principle of dominated convergence (Lemma 8.16) with $M_n = |f(n)|/n^s$ and $M_n(y) = f_y(n)/n^s$. Since $\lim_{y \to \infty} f_y(n) = f(n)$ for each fixed n, we see by lemma 8.16 that

$$\lim_{y \to \infty} \sum_{n=1}^{\infty} f_y(n)/n^s = \sum_{n=1}^{\infty} \lim_{y \to \infty} f_y(n)/n^s = F(s),$$

and the proof is complete.

Corollary 8.19 *Suppose that $f(n)$ is a totally multiplicative function, and put $F(s) = \sum_{n=1}^{\infty} f(n)/n^s$. If s is a real number for which the series $F(s)$ is* *absolutely convergent, then*

$$F(s) = \prod_p (1 - f(p)/p^s)^{-1}.$$

In particular,

$$\zeta(s) = \prod_p (1 - 1/p^s)^{-1} \tag{8.37}$$

for $s > 1$.

Proof Since $f(n)$ is totally multiplicative, the series on the left in (8.36) is a geometric series, and we deduce from its convergence that $|f(p)| < 1$ for all primes p, and that this series converges to $1/(1 - f(p)/p^s)$. Inserting this in Theorem 8.18, we obtain the stated result.

We have noted that a multiplicative function is determined by its values on the prime powers. Since the Euler product involves only these values, this formula provides a quick means of spotting relationships between various Dirichlet series. For example, we consider the case $f(n) = \mu(n)$. Since $\mu(p) = -1$ for all primes p and $\mu(p^\alpha) = 0$ whenever $\alpha > 1$, we see at once that the product in Theorem 8.18 is $\Pi_p(1 - 1/p^s)$ for $s > 1$. On comparing this with (8.37), we obtain a second proof of Theorem 8.15. The identities (8.28), (8.29), and (8.30) can similarly be derived by comparing Euler products. We consider one more example of this technique.

Corollary 8.20 *For $s > 1$,*

$$\sum_{n=1}^{\infty} |\mu(n)|/n^s = \frac{\zeta(s)}{\zeta(2s)}. \tag{8.38}$$

Here the coefficient is 1 if n is square-free, and 0 otherwise.

Proof The function $f(n) = |\mu(n)|$ is multiplicative. Moreover, $f(p) = 1$ for all p and $f(p^\alpha) = 0$ when $\alpha > 1$. Thus when $s > 1$, the product in Theorem 8.18 is $\Pi_p(1 + 1/p^s)$. Using the identity $1 + z = (1 - z^2)/(1 - z)$, we deduce that this product is $\Pi_p(1 - 1/p^{2s})/(1 - 1/p^s)$. By the Euler product formula (8.37) for the zeta function we see that this product is $\zeta(s)/\zeta(2s)$.

To enlarge our repertoire of useful Dirichlet series, we show that a Dirichlet series may be differentiated term-by-term.

Theorem 8.21 *Let σ_a be the abscissa of absolute convergence of the Dirichlet series $A(s) = \sum_{n=1}^{\infty} a_n/n^s$. Then σ_a is also the abscissa of absolute convergence of the series $B(s) = -\sum_{n=1}^{\infty} a_n(\log n)/n^s$, and $\frac{d}{ds}A(s) = B(s)$ for $s > \sigma_a$.*

Proof Let σ_a' denote the abscissa of absolute convergence of $B(s)$. Since $\log n > 1$ for all $n > 2$, by the comparison test

$$\sum_{n=3}^{\infty} |a_n| n^{-s} \leqslant \sum_{n=3}^{\infty} |a_n| (\log n) n^{-s}.$$

Hence $A(s)$ is absolutely convergent whenever $B(s)$ is absolutely convergent, so that $\sigma_a \leqslant \sigma_a'$. To establish an inequality in the reverse direction we note that if $\varepsilon > 0$ is given then there is a number $N = N(\varepsilon)$ such that $\log n < n^{\varepsilon}$ for all $n \geqslant N$. Thus by the comparison test

$$\sum_{n=N}^{\infty} |a_n| (\log n) n^{-s} \leqslant \sum_{n=N}^{\infty} |a_n| n^{-s+\varepsilon}.$$

Hence $B(s)$ is absolutely convergent whenever $A(s - \varepsilon)$ is absolutely convergent, so that $\sigma_a' \leqslant \sigma_a + \varepsilon$. Since ε may be arbitrarily small, it follows that $\sigma_a' \leqslant \sigma_a$. Combining these two inequalities, we conclude that $\sigma_a' = \sigma_a$.

To prove the second assertion we suppose that s is fixed, $s > \sigma_a$, and we choose c so that $\sigma_a < c < s$. If $|h| \leqslant s - c$, then $(A(s + h) - A(s))/h = \sum_{n=1}^{\infty} M_n(h)$, where $M_n(h) = a_n n^{-s}(n^{-h} - 1)/h$. We note that $\lim_{h \to 0} M_n(h) = -a_n(\log n)n^{-s}$. Thus to complete the proof we have only to confirm that

$$\lim_{h \to 0} \sum_{n=1}^{\infty} M_n(h) = \sum_{n=1}^{\infty} \lim_{h \to 0} M_n(h). \tag{8.39}$$

To this end we appeal to the principle of dominated convergence (Lemma 8.16) with $\lim_{x \to \infty}$ replaced by $\lim_{h \to 0}$. We take $M_n = |a_n|(\log n)n^{-c}$ and note that by the mean value theorem of differential calculus there is a ξ between h and 0 such that $M_n(h) = -a_n(\log n)n^{-s-\xi}$. Thus $|M_n(h)| \leqslant M_n$ uniformly for $|h| \leqslant s - c$. As $\sum_n M_n < \infty$, the principle of dominated convergence applies, so we have (8.39), and the proof is complete.

Corollary 8.22 *The following Dirichlet series have abscissa of absolute convergence* 1, *and for* $s > 1$ *converge to the indicated values*:

$$-\zeta'(s) = \sum_{n=1}^{\infty} (\log n) n^{-s}, \tag{8.40}$$

$$\log \zeta(s) = \sum_{n=1}^{\infty} \frac{\Lambda(n)}{\log n} n^{-s}, \tag{8.41}$$

$$-\frac{\zeta'(s)}{\zeta(s)} = \sum_{n=1}^{\infty} \Lambda(n) n^{-s}. \tag{8.42}$$

Proof The first identity follows by applying Theorem 8.21 to (8.22). To derive the second formula we take logarithms of both sides of the Euler product identity (8.37). Thus we find that

$$\log \zeta(s) = \sum_{p} \log (1 - p^{-s})^{-1}.$$

Using the familiar power series expansion $\log(1 - z)^{-1} = \sum_{k=1}^{\infty} z^k / k$, which is valid for $|z| < 1$, we deduce that the above is

$$= \sum_{p} \sum_{k=1}^{\infty} \frac{1}{k} p^{-ks}.$$

This double series of positive numbers may be rearranged to put the numbers p^k in increasing order, without affecting either convergence or its value. Thus we see that we have a Dirichlet series whose coefficients may be written in the form $\Lambda(n)/\log n$. Since these coefficients are all $\leqslant 1$, by comparison with the Dirichlet series (8.22) for the zeta function we deduce that this series is absolutely convergent for $s > 1$. On the other hand, in Theorem 1.19 and again in Theorem 8.8(d) we have seen that the series $\sum 1/p$ diverges. Thus by the comparison test the series (8.41) diverges when $s = 1$. The third identity (8.42) follows immediately from (8.41) by Theorem 8.21, so the proof is complete.

In view of (8.40) and (8.42), the identity of Theorem 8.1 may be expressed analytically as $-\dfrac{\zeta'(s)}{\zeta(s)} \cdot \zeta(s) = -\zeta'(s)$.

Our main interest in this section has been to show how Dirichlet series may be used to discover identities among arithmetic functions, and especially how Euler products may be used to establish identities involving multiplicative functions. In more advanced work, the analytic properties of a Dirichlet series $\Sigma a_n n^{-s}$ are used to derive asymptotic estimates for the coefficient sum $\Sigma_{n \leqslant x} a_n$. As a first step in this direction, we establish some very simple asymptotic estimates.

Theorem 8.23 *The estimates*

$$\zeta(s) = \frac{1}{s-1} + O(1), \tag{8.43}$$

$$\zeta'(s) = -\frac{1}{(s-1)^2} + O(1), \tag{8.44}$$

$$\frac{\zeta'(s)}{\zeta(s)} = -\frac{1}{s-1} + O(1) \tag{8.45}$$

hold uniformly for $s > 1$.

From the first of these estimates we see that $\log \zeta(s) \to \infty$ as $s \to 1^+$. Then by using (8.41) one may deduce that $\Sigma 1/p = \infty$. In this case we have already proved more by elementary means, but in general one may use information concerning the asymptotic size of a Dirichlet series to give corresponding information regarding its coefficients.

Proof Let s be a positive real number. Then u^{-s} is a decreasing function of u, so that $(n+1)^{-s} \leqslant \int_n^{n+1} u^{-s}\,du \leqslant n^{-s}$ for $n = 1, 2, \cdots$. On summing over n we find that $\zeta(s) - 1 \leqslant \int_1^\infty u^{-s}\,du \leqslant \zeta(s)$ for $s > 1$. Here the integral is $1/(s-1)$, so it follows that

$$1/(s-1) \leqslant \zeta(s) \leqslant 1 + 1/(s-1) \tag{8.46}$$

uniformly for $s > 1$. Thus we have (8.43).

If $s > 1$, then $(\log u)u^{-s}$ is a decreasing function of u for $u \geqslant e$, so that $(\log(n+1))(n+1)^{-s} \leqslant \int_n^{n+1} (\log u)u^{-s}\,du \leqslant (\log n)n^{-s}$ for $n = 3, 4, \cdots$. On summing over n we find that

$$-\zeta'(s) - (\log 2)2^{-s} - (\log 3)3^{-s}$$

$$\leqslant \int_3^\infty (\log u)u^{-s}\,du \leqslant -\zeta'(s) - (\log 2)2^{-s}.$$

Since $\int_1^3 (\log u)u^{-s}\,du = O(1)$, we deduce that

$$-\zeta'(s) = \int_1^\infty (\log u)u^{-s}\,du + O(1)$$

uniformly for $s > 1$. By integrating by parts we find that this integral is $1/(s-1)^2$, which gives (8.44).

To derive (8.45) we note that (8.46) implies that $(s-1)/s \leqslant 1/\zeta(s) \leqslant s-1$ for $s > 1$. As $1/s = 1-(s-1)/s \geqslant 1-(s-1)$ for $s \geqslant 1$, it follows that $1/\zeta(s) = s-1+O((s-1)^2)$ for $s > 1$. By multiplying this estimate by the estimate of (8.44) we obtain (8.45) for $1 < s \leqslant 2$. Since (8.45) is obvious for $s > 2$, the proof is complete.

In this section we have found that it is fruitful to use Dirichlet series in the investigation of arithmetic functions, particularly multiplicative functions. One might try using some other kind of generating function, but our experience is that Dirichlet series offer the best approach in dealing with multiplicative questions. The explanation for this seems to lie in the simple identity (8.25), from which we saw that the coefficient of the product of two Dirichlet series is formed by collecting those terms for which the *product* mn is constant. In contrast, when the product of two power series is formed, one forms the new coefficients by grouping those terms for which the sum $m+n$ is constant. Thus power series are used to investigate additive questions. For example, in 1938, I. M. Vinogradov proved that there is an n_0 such that every odd integer $n > n_0$ can be written as a sum of three primes. His proof built on earlier work of G. H. Hardy and J. E. Littlewood that involved an analysis of the asymptotic properties of the power series $P(z) = \Sigma_p z^p$. In Chapter 10 we use power series to investigate the partition function $p(n)$, which is an arithmetic function arising from an additive problem (see Definition 10.1).

PROBLEMS

1. Show that if $f(n)$ and $g(n)$ are multiplicative functions then the Dirichlet convolution $f * g(n)$ is also a multiplicative function. If f and g are totally multiplicative, does it follow that $f * g$ is totally multiplicative?
2. Show that the Dirichlet series $\Sigma\, 2^n/n^s$ diverges for all s, so that $\sigma_a = +\infty$ for this series.
3. Show that the Dirichlet series $\Sigma 1/(2^n n^s)$ converges for all s, so that $\sigma_a = -\infty$ for this series.
4. Let $A(s) = \Sigma(-1)^{n-1}/n^s$. Prove that for this series $\sigma_a = 1$. Prove that this series is conditionally convergent for $0 < s \leqslant 1$, and divergent for $s \leqslant 0$. Prove that $A(s) = (1-2^{1-s})\zeta(s)$ for $s > 1$. (H)
5. Show that $1/\zeta'(s)$ cannot be expressed as a Dirichlet series. (H)
6. Let k be a given real number, and put $\sigma_k(n) = \Sigma_{d|n} d^k$. Show that $\Sigma\sigma_k(n)/n^s = \zeta(s)\zeta(s-k)$ for $s > \max(1, 1+k)$.

7. Use Dirichlet series to prove that $\sum_{d|n}\mu(d)\, d(n/d) = 1$ for all positive integers n.

8. Use Dirichlet series to prove that $\sum_{d|n}\mu(d)\sigma(n/d) = n$ for all positive integers n.

9. Use Dirichlet series to prove that $\sum_{d|n}\sigma(d) = n\sum_{d|n} d(d)/d$ for all positive integers n.

10. Use Euler products to give an analytic proof of the identity (4.1) at the end of Section 4.3.

11. Let $\lambda(n) = (-1)^{\Omega(n)}$ be Liouville's lambda function. Use Euler products to show that $\sum\lambda(n)/n^s = \zeta(2s)/\zeta(s)$ for $s > 1$. Let $f(n) = \sum_{d|n}|\mu(d)|\lambda(n/d)$. Use Dirichlet series to show that $f(1) = 1$ and that $f(n) = 0$ for all $n > 1$.

12. Use Euler products to show that $\sum 2^{\omega(n)}/n^s = \zeta(s)^2/\zeta(2s)$ for $s > 1$. Use Dirichlet series to show that $\sum_{d|n}\lambda(d)2^{\omega(n/d)} = 1$ for all positive integers n.

13. Let k be a given positive integer. Show that

$$\sum_{\substack{n=1\\(n,k)=1}}^{\infty} 1/n^s = \zeta(s)\prod_{p|k}(1 - 1/p^s) = \zeta(s)\sum_{d|k}\mu(d)/d^s \textit{ for } s > 1.$$

14. Let k be an integer > 1. We say that a positive integer n is kth *power free* if 1 is the largest kth power that divides n. Let $f(n) = 1$ if n is kth power free, and put $f(n) = 0$ otherwise. Prove that $\sum f(n)/n^s = \zeta(s)/\zeta(ks)$ for $s > 1$.

15. Let $d_k(n)$ denote the number of ordered k-tuples $(d_1, d_2, \cdots, d_k)$ of positive integers such that $d_1 d_2 \cdots d_k = n$. Show that $d(n) = d_2(n)$. Show that $\sum_{n=1}^{\infty} d_k(n)n^{-s} = \zeta(s)^k$ for $s > 1$ and $k = 1, 2, \cdots$.

16. Let $\mathcal{N}$ be the set of those positive integers n such that $3 \nmid d(n)$. Show that $\sum_{n\in\mathcal{N}} n^{-s} = \zeta(s)/(\zeta(2s)\zeta(3s))$ for $s > 1$.

***17.** Let $\mathcal{N}$ denote the set of those positive integers n whose base 10 representation does not contain the digit 9. Find the abscissa of absolute convergence of the Dirichlet series $\sum_{n\in\mathcal{N}} n^{-s}$.

***18.** Prove that $\sum d(n)^2/n^s = \zeta(s)^4/\zeta(2s)$ for $s > 1$.

***19.** Prove that $\sum_{d|n}\mu(d)\, d(n/d)^2 = \sum_{d|n}\mu(d)^2\, d(n/d)$ for all positive integers n.

***20.** Show that $\mu(n)^2 * d(n) = \mu(n) * d(n)^2$ for all positive integers n.

***21.** Suppose that the Dirichlet series $A(s) = \sum a_n/n^s$ is convergent when $s = s_0$. Prove that $A(s)$ is absolutely convergent when $s > s_0 + 1$.

***22.** Let $f(n)$ be an arithmetic function such that for any $\varepsilon > 0$ there exists an $n_0(\varepsilon)$ with the property that $n^{-\varepsilon} < |f(n)| < n^{\varepsilon}$ for all

$n > n_0(\varepsilon)$. (That is, $\lim_{n\to\infty} \log |f(n)| / \log n = 0$.) Show that the two Dirichlet series $\Sigma a_n/n^s$ and $\Sigma f(n)a_n/n^s$ have the same abscissa of absolute convergence.

***23.** Show that if $s > 1$ then $\zeta(s) = s\int_1^\infty [u]u^{-s-1}\,du$. (H)

***24.** Show that if $s > 1$ then $\zeta(s) = \dfrac{s}{s-1} - s\int_1^\infty \{u\}u^{-s-1}\,du$. Show that this latter integral is absolutely convergent for all $s > 0$. Let $\zeta(s)$ be defined for $0 < s < 1$ by this formula. Show that if $0 < s < 1$ then $\zeta(s) = -s\int_0^\infty \{u\}u^{-s-1}\,du$. Conclude that $\zeta(s) < 0$ for $0 < s < 1$ if the zeta function is defined in this way in this interval.

***25.** Use Dirichlet series to show that

$$\Lambda(n)\log n + \sum_{d|n}\Lambda(d)\Lambda(n/d) = \sum_{d|n}\mu(d)(\log n/d)^2$$

for all positive integers n.

8.3 ESTIMATES OF ARITHMETIC FUNCTIONS

In this section we investigate the size of some important arithmetic functions, both on average and in the extreme.

Suppose we wish to determine the asymptotic mean value of the arithmetic function $F(n)$. By the Möbius inversion formula (Theorems 4.7 and 4.8) we know that there is a unique function $f(n)$ such that

$$F(n) = \sum_{d|n} f(d). \tag{8.47}$$

If $f(n)$ is small on average then we can obtain a useful estimate of the average of $F(n)$ by writing

$$\sum_{n\le x} F(n) = \sum_{n\le x}\sum_{d|n} f(d) = \sum_{\substack{d,m\\ dm\le x}} f(d) = \sum_{d\le x} f(d)[x/d].$$

Since $[y] = y + O(1)$, this is

$$= x\sum_{d\le x} f(d)/d + O\Big(\sum_{d\le x}|f(d)|\Big). \tag{8.48}$$

If the first sum is a partial sum of a convergent series and the second sum is small compared with x, then this simple argument reveals that $F(n)$ has the asymptotic mean value $\Sigma_1^\infty f(d)/d$. We consider several applications that fit this description.

Theorem 8.24 *For $x \geqslant 2$,*

$$\sum_{n \leqslant x} \phi(n)/n = \frac{6}{\pi^2}x + O(\log x).$$

Proof Taking $F(n) = \phi(n)/n$, by (4.1) we see that (8.47) holds with $f(d) = \mu(d)/d$. Thus the first sum in (8.48) is $1/\zeta(2) - \Sigma_{d>x}\mu(d)/d^2$. This latter sum has absolute value less than

$$\sum_{d>x} 1/d^2 < \int_{x-1}^{\infty} 1/u^2\, du = 1/(x-1) = O(1/x). \tag{8.49}$$

When inserted in (8.48), this error term contributes an amount that is $O(1)$. In Appendix A.3 it is shown that $\zeta(2) = \pi^2/6$, which gives the constant in the main term. The second sum in (8.48) is $O(\Sigma_{d \leqslant x} 1/d) = O(\int_1^x 1/u\, du) = O(\log x)$, so the proof is complete.

In our next application of (8.48), we encounter a situation in which $f(d)$ is usually 0, but occasionally takes large values.

Theorem 8.25 *Let $Q(x)$ denote the number of square-free integers not exceeding x, that is, $Q(x) = \Sigma_{n \leqslant x}|\mu(n)|$. Then $Q(x) = \frac{6}{\pi^2}x + O(\sqrt{x})$.*

Proof From Corollary 8.20 we find that $|\mu(n)| = \Sigma_{d^2|n}\mu(d)$. This may be proved by elementary reasoning by observing that any positive integer n is uniquely of the form $n = rs^2$ where r is square free. Thus n is square free if and only if $s = 1$, and hence $|\mu(n)| = \Sigma_{d|s}\mu(d)$. Since $d|s$ if and only if $d^2|n$, this gives the stated identity. This identity is of the form (8.47) where $f(d) = \mu(k)$ if $d = k^2$, $f(d) = 0$ otherwise. Thus (8.48) gives

$$Q(x) = x \sum_{k \leqslant \sqrt{x}} \mu(k)/k^2 + O\Bigg(\sum_{k \leqslant \sqrt{x}} |\mu(k)|\Bigg).$$

By (8.49) the first sum is $1/\zeta(2) + O(1/\sqrt{x})$, and the second sum is $\leqslant \sqrt{x}$. To complete the proof it suffices to quote the value $\zeta(2) = \pi^2/6$ from Appendix A.3.

In most applications of (8.48), the functions $f(n)$, $F(n)$ are multiplicative, though this is not required. For example, in Section 4.2 we defined

$\omega(n)$ to be the number of distinct prime factors of n, $\omega(n) = \Sigma_{p|n}1$. This is of the form (8.47) with $f(d) = 1$ if d is prime, $f(d) = 0$ otherwise. Here $f(d)$ is not multiplicative, but (8.48) is still useful.

Theorem 8.26 *For* $x \geqslant 5$,

$$\sum_{n \leqslant x} \omega(n) = x \log\log x + bx + O(x/\log x)$$

where b *is the constant in Theorem* 8.8(*d*).

Since $\Sigma_{5<n\leqslant x} \log\log n = x \log\log x + O(x/\log x)$, we say that $\omega(n)$ has average value $\log\log n$. In particular, it follows that $\limsup_{n\to\infty} \omega(n)/\log\log n \geqslant 1$, and that $\liminf_{n\to\infty} \omega(n)/\log\log n \leqslant 1$.

Proof The estimate (8.48) gives

$$\sum_{n \leqslant x} \omega(n) = x \sum_{p \leqslant x} 1/p + O\Big(\sum_{p \leqslant x} 1\Big).$$

By Theorem 8.8(*d*) the first sum on the right is $= \log\log x + b + O(1/\log x)$. The second sum on the right is $\pi(x)$, which is $O(x/\log x)$ by Chebyshev's estimate (Corollary 8.6).

We may also estimate the mean of the divisor function $d(n) = \Sigma_{d|n}1$ using (8.48). Taking $f(d) = 1$ for all d, we find that

$$\sum_{n \leqslant x} d(n) = x \sum_{d \leqslant x} 1/d + O\Big(\sum_{d \leqslant x} 1\Big).$$

The first sum on the right may be approximated by an integral, which gives the approximation

$$\sum_{d \leqslant x} 1/d = \log x + O(1), \tag{8.50}$$

and hence we see that $\Sigma_{n\leqslant x} d(n) = x \log x + O(x)$. The leading term here is the same as in Lemma 8.2, so we say that the average size of $d(n)$ is $\log n$. In this case the function $f(d)$ is not very small, and the main term in (8.48) is only slightly larger than the error term.

By exercising greater care we shall establish a more precise estimate for the sum of the divisor function, but first we must refine the estimate

(8.50). To this end, for $n \geqslant 2$ let

$$\delta_n = \int_{n-1}^{n} 1/u\,du - 1/n. \tag{8.51}$$

Since the function $1/u$ is decreasing, the integral is less than $1/(n-1)$ but greater than $1/n$, so that $0 < \delta_n < 1/(n(n-1))$. We note that

$$\left(\sum_{n=1}^{N} 1/n\right) - \log N = 1 - \sum_{n=2}^{N} \delta_n.$$

Since the δ_n are positive, the right side is clearly a decreasing function of the integral variable N. Moreover, since $\delta_n = O(1/n^2)$, the right side converges to a finite limit

$$\gamma = 1 - \sum_{n=2}^{\infty} \delta_n$$

as $N \to \infty$. This number $\gamma = 0.57721\cdots$ is called *Euler's constant*. (It is conjectured that γ is irrational, but this has not yet been proved.) Substituting γ in the former expression, we see that

$$\sum_{n=1}^{N} 1/n = \log N + \gamma + \sum_{n=N+1}^{\infty} \delta_n$$

for all positive integers N. Here the sum on the right is

$$< \sum_{n=N+1}^{\infty} \frac{1}{n(n-1)} = \sum_{n=N+1}^{\infty} \left(\frac{1}{n-1} - \frac{1}{n}\right) = 1/N,$$

so we conclude that

$$\log N + \gamma < \sum_{n=1}^{N} 1/n < \log N + \gamma + 1/N \tag{8.52}$$

for all positive integers N. For our present purposes it is convenient to replace the upper limit N of summation by a real number x.

Lemma 8.27 *The estimate $\sum_{n \leqslant x} 1/n = \log x + \gamma + O(1/x)$ holds uniformly for $x \geqslant 1$.*

Proof We apply (8.52) with $N = [x]$. Since $\log N \leqslant \log x < \log(N+1) < \log N + 1/N$, we have the stated estimate.

Theorem 8.28 *For* $x \geqslant 2$,

$$\sum_{n \leqslant x} d(n) = x \log x + (2\gamma - 1)x + O(\sqrt{x}).$$

In Problems 29–37 at the end of this section we sketch a method of I. M. Vinogradov which shows that the above holds with the error term replaced by $O(x^{1/3}(\log x)^2)$.

Proof We write $d(n) = \Sigma_{d|n} 1$, and choose k so that $dk = n$. Thus the left side above may be written as a double sum

$$\sum_{\substack{d,k \\ dk \leqslant x}} 1.$$

This counts lattice points under the hyperbola $uv = x$ in the first quadrant of the u–v plane. We consider first those pairs d, k for which $d \leqslant \sqrt{x}$. Summing first over d and then over k, we see that such terms contribute an amount

$$\sum_{d \leqslant \sqrt{x}} \sum_{k \leqslant x/d} 1 = \sum_{d \leqslant \sqrt{x}} [x/d].$$

By symmetry, the terms for which $k \leqslant \sqrt{x}$ contribute the same amount. The terms for which both $d \leqslant \sqrt{x}$ and $k \leqslant \sqrt{x}$ have been counted twice, so their contribution, $[\sqrt{x}]^2$, must be subtracted. Thus we see that

$$\sum_{n \leqslant x} d(n) = 2 \sum_{d \leqslant \sqrt{x}} [x/d] - [\sqrt{x}]^2.$$

We replace $[x/d]$ by $x/d + O(1)$ and note that the sum of these error terms is $O(\sqrt{x})$. Similarly $[\sqrt{x}]^2 = (\sqrt{x} + O(1))^2 = x + O(\sqrt{x})$, so the sum above is

$$-x + O(\sqrt{x}) + 2x \sum_{d \leqslant \sqrt{x}} 1/d,$$

and the stated estimate now follows from Lemma 8.27.

Theorem 8.29 *Let q be a positive integer. The number of integers n such that $(n, q) = 1$ and $M + 1 \leqslant n \leqslant M + N$ is $(\phi(q)/q)N + O(2^{\omega(q)})$, uniformly for all integers M and all positive integers N.*

Proof Let $F(n) = 1$ if $(n, q) = 1$, $F(n) = 0$ otherwise. Then we have (8.47) with $f(d) = \mu(d)$ for $d|q$, $f(d) = 0$ otherwise. Thus the number in question is

$$\sum_{n=M+1}^{M+N} F(n) = \sum_{n=M+1}^{M+N} \sum_{\substack{d|n \\ d|q}} \mu(d) = \sum_{d|q} \mu(d) \sum_{\substack{n=M+1 \\ d|n}}^{M+N} 1.$$

The inner sum on the right is $[(M+N)/d] - [M/d]$, which is $(M+N)/d + O(1) - M/d + O(1) = N/d + O(1)$. Inserting this in the above, we obtain the main term $N\Sigma_{d|q}\mu(d)/d = N\phi(q)/q$, and the error term $O(\Sigma_{d|q}|\mu(d)|)$. This last sum is the number of square-free divisors of q, which is $2^{\omega(q)}$. This gives the stated result.

From Theorem 8.29 we see that any interval longer than $c2^{\omega(q)}q/\phi(q)$ must contain a number relatively prime to q. To put this in a more useful form, we must determine how large $2^{\omega(q)}$ is, in terms of more familiar functions of q.

Theorem 8.30 *For every $\varepsilon > 0$ there is an $n_0(\varepsilon)$ such that if $n > n_0(\varepsilon)$ then $\omega(n) < (1+\varepsilon)(\log n)/\log\log n$. This is best possible in the sense that there exist infinitely many n such that $\omega(n) > (1-\varepsilon)(\log n)/\log\log n$.*

At the opposite extreme, we observe that $\omega(n) \geqslant 1$ for all $n > 1$, and that $\omega(n) = 1$ for infinitely many n (the prime powers).

From the upper bound of Theorem 8.30 we see that

$$2^{\omega(q)} < q^{\frac{(1+\varepsilon)\log 2}{\log\log q}}$$

for all large integers q. Since the exponent on the right tends to 0 as q tends to infinity, it follows in particular that for every $\delta > 0$ there is a $q_0(\delta)$ such that every interval $(x, x + q^\delta)$ contains a reduced residue (mod q), provided that $q > q_0(\delta)$.

Proof We establish the upper bound for $\omega(n)$ first. Let ε be given, $\varepsilon > 0$, and put $f(u) = (1+\varepsilon)(\log u)/\log\log u$. By a simple application of differential calculus we see that $f(u)$ is increasing for $u > e^e = 15.154\cdots$. We call an integer $r \geqslant 16$ record-breaking if $\omega(r) > \omega(n)$ for all positive integers $n < r$. Let $\mathscr{R}$ be the set of all such record-breaking numbers r. We first prove the desired inequality for $r \in \mathscr{R}$. We note that $\omega(n) = k$ if and only if n is divisible by precisely k different primes. The least such n

is simply the product of the first k primes. That is, a number r is record-breaking if and only if r is of the form $r = \prod_{p \leqslant y} p$ for some suitable real number $y \geqslant 5$. But then $\log r = \vartheta(y)$ and $\omega(r) = \pi(y)$, and hence by Theorem 8.5 we have

$$\omega(r) = \frac{\log r}{\log y} + O\big(y/(\log y)^2\big).$$

Since $ay < \vartheta(y) < by$, by taking logarithms we find that $\log y = \log \vartheta(y) + O(1)$. That is, $\log y = \log\log r + O(1)$. Thus the above gives

$$\omega(r) = \frac{\log r}{\log\log r} + O\big((\log r)/(\log\log r)^2\big). \tag{8.53}$$

From this it follows that there is an $r_0(\varepsilon) \in \mathscr{R}$ such that $\omega(r) \leqslant f(r)$ whenever $r \in \mathscr{R}$, $r \geqslant r_0(\varepsilon)$. Now suppose that $n \geqslant r_0(\varepsilon)$, and let r be the largest member of $\mathscr{R}$ not exceeding n. Then $r_0(\varepsilon) \leqslant r \leqslant n$ and $\omega(n) \leqslant \omega(r) \leqslant f(r)$. Since f is increasing, it follows that $f(r) \leqslant f(n)$, so that $\omega(n) \leqslant f(n)$. Thus we have the stated upper bound.

From (8.53) we see that $\omega(r) > (1 - \varepsilon)(\log r)/(\log\log r)$ for all sufficiently large $r \in \mathscr{R}$. This suffices to give the lower bound, since the set $\mathscr{R}$ is infinite.

Since $2^{\omega(n)}$ is the number of square-free divisors of n, it follows that $2^{\omega(n)}$ does not exceed the total number $d(n)$ of divisors of n. This inequality $2^{\omega(n)} \leqslant d(n)$ is also evident from the formula $d(n) = \prod_p(\alpha + 1)$ given in Theorem 4.3. Here $\alpha = \alpha(p, n)$, and $n = \prod p^\alpha$ is the canonical factorization of n.

The simplest upper bound for $d(n)$ is obtained by observing that if $d \mid n$ then n/d also divides n, and of this pair of divisors at least one is $\leqslant \sqrt{n}$, so that $d(n) < 2\sqrt{n}$. We now show that for any given $\delta > 0$ we can determine the maximum of $d(n)/n^\delta$. Let $f_p(\alpha) = (\alpha + 1)/p^{\delta\alpha}$, so that $d(n)/n^\delta = \prod_p f_p(\alpha)$. We now let α be an integral variable, and for each prime p we find the α for which $f_p(\alpha)$ is maximal. We note that $f_p(\alpha) \geqslant f_p(\alpha - 1)$ if and only if $(\alpha + 1)/p^{\delta\alpha} \geqslant \alpha/p^{\delta(\alpha-1)}$, which is equivalent to the inequality $1 + 1/\alpha \geqslant p^\delta$, which in turn is equivalent to $\alpha \leqslant 1/(p^\delta - 1)$. Similarly we find that $f_p(\alpha) \geqslant f_p(\alpha + 1)$ if and only if $\alpha \geqslant 1/(p^\delta - 1) - 1$. Thus $f_p(\alpha)$ is maximal if and only if α lies in the interval $\mathscr{I} = [1/(p^\delta - 1) - 1,\ 1/(p^\delta - 1)]$. Take $\alpha_0(p) = [1/(p^\delta - 1)]$. If $1/(p^\delta - 1)$ is not an integer then $\alpha_0(p)$ is the unique integer in $\mathscr{I}$, and hence $f_p(0) < f_p(1) < \cdots < f_p(\alpha_0) > f_p(\alpha_0 + 1) > \cdots$. On the other hand, if $1/(p^\delta - 1)$ is an integer then $f_p(0) < f_p(1) < \cdots < f_p(\alpha_0 - 1)$

$= f_p(\alpha_0) > f_p(\alpha_0 + 1) > \cdots$. Thus in either case $f_p(\alpha)$ takes its maximum value when $\alpha = \alpha_0$. We also observe that if $p > 2^{1/\delta}$ then $\alpha_0(p) = 0$ and $f_p(\alpha_0) = 1$. Thus we have shown that for any $\delta > 0$ the inequality $d(n) \leqslant C_\delta n^\delta$ holds for all positive integers n, where

$$C_\delta = \prod_{p \leqslant 2^{1/\delta}} f_p(\alpha_0). \tag{8.54}$$

For example, if $\delta = 1/2$ we find that $\alpha_0(2) = 2$, so that $f_2(\alpha_0) = 3/2$, that $\alpha_0(3) = 1$, so that $f_3(\alpha_0) = 2/\sqrt{3}$, and that $\alpha_0(p) = 0$ for all $p > 3$. Hence $C_{1/2} = \sqrt{3}$, and we deduce that $d(n) \leqslant \sqrt{3n}$ for all positive integers n, with equality if and only if $n = 12$. By estimating the size of C_δ when δ is small, we obtain the following more general bound.

Theorem 8.31 *For every $\varepsilon > 0$ there is an $n_0(\varepsilon)$ such that if $n > n_0(\varepsilon)$ then $d(n) < n^{(1+\varepsilon)(\log 2)/\log\log n}$.*

Since $d(n) \geqslant 2^{\omega(n)}$ for all n, from Theorem 8.30 we know that for any $\varepsilon > 0$ there exist infinitely many n for which $d(n) > n^{(1-\varepsilon)(\log 2)/\log\log n}$.

Proof We take $\delta = (1 + \varepsilon/2)(\log 2)/\log\log n$, and show that $C_\delta < n^{(\varepsilon/2)(\log 2)/\log\log n}$ for all sufficiently large n. For this purpose it is enough to construct a crude bound for C_δ. We note that $f_p(\alpha) \leqslant \alpha + 1$. Since $p^\delta > 1 + \delta \log p$, it follows that $1/(p^\delta - 1) < 1/(\delta \log p) \leqslant 1/(\delta \log 2)$. Thus $f_p(\alpha_0) \leqslant 1 + 1/(\delta \log 2)$. Since we may assume that $\delta \leqslant 1$, we conclude that $f_p(\alpha_0) < e/\delta$. Since $\pi(2^{1/\delta}) \leqslant 2^{1/\delta}$, it follows that $C_\delta < (e/\delta)^{2^{1/\delta}}$. Expressed as a function of n, we note that $2^{1/\delta} = (\log n)^{1/(1+\varepsilon/2)} = (\log n)(\log n)^{-\varepsilon/(2+\varepsilon)}$. Consequently, $C_\delta < n^\eta$ where

$$\eta = (\log n)^{-\varepsilon/(2+\varepsilon)} \log(4 \log\log n).$$

Since $\eta(\log\log n)$ tends to 0 as $n \to \infty$, it follows that $\eta < (\varepsilon/2)(\log 2)/\log\log n$ for all large n, and the proof is complete.

From Theorem 8.26 we see that the average size of $\omega(n)$ is asymptotically $\log\log n$. We now show that $\omega(n)$ is quite near $\log\log n$ for most n.

Theorem 8.32 *For $n > 5$, $\sum_{1<n\leqslant x}(\omega(n) - \log\log n)^2 = O(x \log\log x)$.*

Proof We shall prove the following three estimates:

$$\sum_{1<n\leqslant x} \omega(n)^2 \leqslant x(\log\log x)^2 + O(x \log\log x), \tag{8.55}$$

$$-2 \sum_{1<n\leqslant x} \omega(n)(\log\log n) = -2x(\log\log x)^2 + O(x \log\log x), \tag{8.56}$$

$$\sum_{1<n\leqslant x} (\log\log n)^2 = x(\log\log x)^2 + O(x \log\log x). \tag{8.57}$$

The stated result then follows by adding these three quantities. With a little more work we could show that (8.55) holds with the inequality replaced by equality (see Problem 23 at the end of this section), but the weaker estimate (8.55) is sufficient for our purposes.

Letting p and q denote primes, we see from the definition of $\omega(n)$ that the left side of (8.55) is

$$\sum_{n\leqslant x}\sum_{p\mid n}\sum_{q\mid n}1=\sum_{p}\sum_{q}\sum_{\substack{n\leqslant x\\ p\mid n,\, q\mid n}}1.$$

If $p = q$, then the inner sum on the right is $[x/p]$, while if $p \neq q$, then this sum is $[x/pq]$. Thus the above is

$$\sum_{p\leqslant x}[x/p]+\sum_{p\neq q,\, pq\leqslant x}[x/pq].$$

Since $[u] \leqslant u$ for all real u, we obtain a larger quantity by dropping the square brackets. We also drop the condition $p \neq q$, and sum over all pairs p, q of primes for which $p \leqslant x$, $q \leqslant x$. Thus the above is

$$\leqslant \sum_{p\leqslant x}x/p+\sum_{p\leqslant x,\, q\leqslant x}x/pq=x\Big(\sum_{p\leqslant x}1/p\Big)+x\Big(\sum_{p\leqslant x}1/p\Big)^2,$$

and (8.55) follows by appealing to Theorem 8.26.

To prove (8.56) we write the sum on the left as

$$(\log\log x)\sum_{1<n\leqslant x}\omega(n)-\sum_{1<n\leqslant x}\omega(n)\log\frac{\log x}{\log n}.$$

By Theorem 8.26, the first term above is $x(\log\log x)^2 + O(x\log\log x)$. To estimate the second sum we consider separately $1 < n \leqslant \sqrt{x}$ and $\sqrt{x} < n \leqslant x$. In the first interval the logarithmic factor is $O(\log\log x)$, so by Theorem 8.26 the first interval contributes an amount that is $O(\sqrt{x}(\log\log x)^2)$. In the second interval the logarithmic factor is $O(1)$, so by Theorem 8.26 the second interval contributes an amount that is $O(x\log\log x)$. On combining these estimates we obtain (8.56).

To prove (8.57) we note that the summand is increasing for $n \geqslant 3$, so that the sum is $= \int_3^x(\log\log u)^2\,du + O((\log\log x)^2)$. By integrating by parts, we see that this integral is

$$x(\log\log x)^2-3(\log\log 3)^2-2\int_3^x\frac{\log\log u}{\log u}\,du.$$

Since the integrand here is bounded, the integral from 3 to $\sqrt{x}$ is $O(\sqrt{x})$.

For $\sqrt{x} \leqslant u \leqslant x$ the integrand is $O((\log\log x)/\log x)$, so the integral over this second range is $O(x(\log\log x)/\log x)$. On combining these estimates we obtain (8.57), and the proof is complete.

Corollary 8.33 *The inequality*

$$|\omega(n) - \log\log n| < (\log\log n)^{3/4} \tag{8.58}$$

holds for all n, $1 < n \leqslant x$, with the exception of at most $O(x/(\log\log x)^{1/2})$ integers n.

Proof We may ignore the $n \leqslant \sqrt{x}$, since there are at most $\sqrt{x}$ such n. Suppose that $\sqrt{x} < n \leqslant x$ and that (8.58) fails. Then n contributes at least $(\log\log n)^{3/2}$ to the sum in Theorem 8.32. Since $(\log\log n)^{3/2} \geqslant \frac{1}{2}(\log\log x)^{3/2}$ for $\sqrt{x} < n \leqslant x$, it follows from Theorem 8.32 that there can be at most $O(x/(\log\log x)^{1/2})$ such n.

By the same method that we used to prove Theorem 8.32 we can also show that Theorem 8.32 holds with $\omega(n)$ replaced by $\Omega(n)$ (or see Problem 24 below). Here $\Omega(n)$ denotes the total number of prime factors of n, counting multiplicity. That is, if $n = \prod p^{\alpha}$, then $\Omega(n) = \sum_p \alpha$. Since $2^{\omega(n)} \leqslant d(n) \leqslant 2^{\Omega(n)}$ for all n, by arguing as in the proof of Theorem 8.33 we find that $d(n)$ lies between $(\log n)^{(1-\varepsilon)\log 2}$ and $(\log n)^{(1+\varepsilon)\log 2}$ for most integers n. Since $\log 2 = 0.693\cdots$, the normal size of $d(n)$ is smaller than the average size, $\log n$, which we estimated in Theorem 8.28. By using more advanced techniques it may be shown that the larger average reflects a relatively sparse sequence of n for which $d(n)$ is disproportionately large. That is, there are roughly $x/(\log x)^{2\log 2}$ integers $n \leqslant x$ for which $d(n)$ is roughly $(\log n)^{1+2\log 2}$.

PROBLEMS

1. Show that $\sum_{n \leqslant N} \phi(n)[N/n] = N(N+1)/2$ for all positive integers N.
2. Show that $\sum_{n \leqslant x}(2n-1)[x/n] = \sum_{m \leqslant x}[x/m]^2$.
3. Show that $\sum_{n \leqslant x}\sigma(n)/n = (\pi^2/6)x + O(\log x)$ for $x \geqslant 2$.
4. Show that $\sum_{n \leqslant x}\Omega(n) = x\log\log x + O(x)$ for $x \geqslant 5$.
5. Let k be a fixed integer, $k > 1$. Show that the number of kth power-free numbers $n \leqslant x$ is $x/\zeta(k) + O(x^{1/k})$.
6. Let δ_n be defined as in (8.50). Show that $\delta_n = \int_{n-1}^{n}\{u\}/u^2\,du$. Deduce that $\gamma = 1 - \int_1^{\infty}\{u\}/u^2\,du$.
7. Find the least constant $C_{1/3}$ such that $d(n) \leqslant C_{1/3}n^{1/3}$ for all positive integers n. For which n does equality hold?

8. Let q be an integer, $q > 1$. Put $((u)) = \{u\} - 1/2$. Show that the number of integers n, $1 \leqslant n \leqslant x$, for which $(n, q) = 1$ is $x\phi(q)/q + E_q(x)$, where $E_q(x) = -\sum_{d|q}\mu(d)((x/d))$. Show that $|E_q(x)| \leqslant 2^{\omega(q)-1}$.

9. Adopt the notation of the preceding problem, suppose that every prime divisor p of q is $\equiv 3 \pmod 4$, that $\omega(q)$ is even, and that q is squarefree. Show that $\mu(d)\left(\left(\frac{q}{4d}\right)\right) = -1/4$ for all divisors d of q. Deduce that $E_q(q/4) = 2^{\omega(q)-2}$, and that $E_q(3q/4) = -2^{\omega(q)-2}$.

10. Show that $1 \leqslant \Omega(n) \leqslant (\log n)/\log 2$ for every integer $n > 1$. Show also that there are infinitely many integers for which $\Omega(n) = 1$, and infinitely many integers for which $\Omega(n) = (\log n)/\log 2$.

11. Show that $(6/\pi^2)n^2 < \sigma(n)\phi(n) < n^2$ for all positive integers n. Deduce that $\sum_{n \leqslant x} n/\phi(n) = O(x)$, and hence that $\sum_{n \leqslant x} 1/\phi(n) = O(\log x)$. (H)

12. Use Euler products to show that $n/\phi(n) = \sum_{d|n}\mu(d)/\phi(d)$. Deduce that $\sum_{n \leqslant x} n/\phi(n) = cx + O(\log x)$ for $x \geqslant 2$, where $c = \zeta(2)\zeta(3)/\zeta(6)$.

13. Let $D(x) = \sum_{n \leqslant x} d(n)$. Show that $\sum_{n \leqslant x} d(n)/n = D(x)/x + \int_1^x D(u)/u^2\,du$. Deduce that $\sum_{n \leqslant x} d(n)/n = (1/2)(\log x)^2 + O(\log x)$.

14. Let $d_k(n)$ be defined as in Problem 15 of the preceding section. Show that $d_3(n) = \sum_{d|n} d(d)$. Deduce that $\sum_{n \leqslant x} d_3(n) = (1/2)x(\log x)^2 + O(x \log x)$.

15. Let $\mathscr{R}$ be the set of those positive integers $r \geqslant 3$ such that $\phi(r)/r < \phi(n)/n$ for every integer $n < r$. Show that $r \in \mathscr{R}$ if and only if r can be written in the form $r = \prod_{p \leqslant y} p$ for some real number $y \geqslant 3$. Show that if $r \in \mathscr{R}$ then $\phi(r)/r = c(\log\log r)^{-1}(1 + O(1/\log\log r))$ where c is the constant in Theorem 8.8(e). Deduce that $\phi(n) \geqslant cn(\log\log n)^{-1}(1 + O(1/\log\log n))$ for all integers $n \geqslant 3$.

16. Let $\psi(x, y)$ denote the number of integers n, $1 \leqslant n \leqslant x$, such that all prime factors of n are $\leqslant y$. Let p be a prime number. Show that the number of integers n, $1 \leqslant n \leqslant x$, whose largest prime factor is p, is $\psi(x/p, p)$. Deduce that $\psi(x, y) = \sum_{p \leqslant y}\psi(x/p, p)$.

17. Adopt the notation of the preceding problem. Show that if $y \geqslant x$ then $\psi(x, y) = [x]$. Show that if $\sqrt{x} \leqslant y \leqslant x$ then $\psi(x, y) = [x] - \sum_{y < p \leqslant x}[x/p]$. Deduce that $\psi(x, x^{1/u}) = (1 - \log u)x + O(x/\log x)$ uniformly for $1 \leqslant u \leqslant 2$.

***18.** Suppose that $F(n) = n\sum_{d|n} f(d)$ for all n. Show that $\sum_{n \leqslant x} F(n) = \sum_{d \leqslant x} df(d)[x/d]([x/d] + 1)/2$. Show that this latter sum is $(1/2)x^2\sum_{d \leqslant x} f(d)/d + O(x\sum_{d \leqslant x}|f(d)|)$.

***19.** Show that $\sum_{n \leqslant x}\phi(n) = (3/\pi^2)x^2 + O(x \log x)$ for $x \geqslant 2$.

20. Let $\Phi(x)$ denote the sum considered in the preceding problem. Show that the number of pairs m, n of positive integers for which $m \leqslant x$, $n \leqslant x$, g.c.d. $(m, n) = 1$ is $2\Phi(x) + 1$. Deduce that if two integers are chosen at random from the interval $[1, x]$ then the probability that they are relatively prime is approximately $6/\pi^2$ if x is large.

***21.** Show that $\sum_{n \leqslant x} \sigma(n) = (\pi^2/12)x^2 + O(x \log x)$ for $x \geqslant 2$.

***22.** Let $f(z)$ be a polynomial with integral coefficients, and let $N_f(m)$ denote the number of solutions of the congruence $f(x) \equiv 0 \pmod{m}$. Show that the number of integers n, $1 \leqslant n \leqslant x$, such that $f(n) \equiv 0 \pmod{d}$ is $xN_f(d)/d + O(N_f(d))$. Deduce that if q is a given positive integer then the number of integers n, $1 \leqslant n \leqslant x$, such that $(f(n), q) = 1$ is $x\prod_{p|q}(1 - N_f(p)/p) + O(\prod_{p|q}(1 + N_f(p)))$.

***23.** Let $f(z) = kz + a$. Show that $N_f(p) = p$ if $p|k$ and $p|a$, that $N_f(p) = 0$ if $p|k$, $p\nmid a$, and that $N_f(p) = 1$ if $p \nmid k$. Deduce that the number of integers $n \equiv a \pmod{k}$, $1 \leqslant n \leqslant x$, for which $(n, q) = 1$ is $(x/k)\prod_{p|q,\, p\nmid k}(1 - 1/p) + O(2^{\omega(q)})$ if g.c.d. $(a, k, q) = 1$.

***24.** Show that the number of ways of writing a positive integer N in the form $N = a + b$ where $a > 0$, $b > 0$, $(a, q) = (b, q) = 1$, is $N\prod_{p|q,\, p|N}(1 - 1/p)\prod_{p|q,\, p\nmid N}(1 - 2/p) + O(3^{\omega(q)})$.

***25.** Let p and q denote prime numbers. Explain why $(\sum_{p \leqslant \sqrt{x}} 1/p)^2 \leqslant \sum_{pq \leqslant x} 1/(pq) \leqslant (\sum_{p \leqslant x} 1/p)^2$. Deduce that $\sum_{pq \leqslant x} 1/(pq) = (\log\log x)^2 + O(\log\log x)$.

***26.** Show that $\sum_{n \leqslant x}(\Omega(n) - \omega(n))^2 = \sum_{p^k \leqslant x,\, k>1}(2k - 3)[x/p^k] + \sum_{p^k \leqslant x,\, q^j \leqslant x,\, k>1,\, j>1,\, p \neq q}[x/(p^k q^j)]$ where p and q denote primes. Deduce that $\sum_{n \leqslant x}(\Omega(n) - \omega(n))^2 = O(x)$, and hence that $\sum_{n \leqslant x}(\Omega(n) - \log\log n)^2 = O(x \log\log x)$.

***27.** Let c be the constant in Theorem 8.8(e). Show that $c = e^{-\gamma}$ where γ is Euler's constant, as follows. First show that

$$\sum_{\substack{p,\, k \\ p \leqslant x,\, p^k > x}} 1/(kp^k) = O(1/\log x)$$

for $x \geqslant 2$. By taking logarithms in Theorem 8.8(e), deduce that

$$\sum_{n \leqslant x} \Lambda(n)/(n \log n) = \log\log x - \log c + O(1/\log x) \quad (8.59)$$

for $x \geqslant 2$, and hence that

$$\sum_{n \leqslant x} \Lambda(n)/(n \log n) = \sum_{n \leqslant \log x} 1/n - (\gamma + \log c) + O(1/\log 2x) \quad (8.60)$$

for $x \geqslant 1$. Write this as $T_1 = T_2 + T_3 + T_4$, and for $0 < \delta \leqslant 1/2$ put

$I_i(\delta) = \delta\int_1^\infty T_i(x)x^{-1-\delta}\,dx$. Show that $I_1(\delta) = \log\zeta(1+\delta) = \log 1/\delta + O(\delta)$. Show that $I_2(\delta) = \log(1-e^{-\delta})^{-1} = \log 1/\delta + O(\delta)$. Show that $I_3(\delta) = -\gamma - \log c$. Show that $I_4(\delta) = O(\delta\log 1/\delta)$. By comparing these estimates as $\delta \to 0^+$, show that $I_3(\delta) = 0$, and thus derive the proposed identity.

***28.** Write the relation $\Sigma_{n \leqslant \log x} 1/n = \log\log x + \gamma + O(1/\log x)$ in the form $U_1 = U_2 + U_3 + U_4$, and for $0 < \delta \leqslant 1/2$ put $J_i(\delta) = \delta\int_1^\infty U_i(x)x^{-1-\delta}\,dx$. Show that $J_2(\delta) = \log 1/\delta + \int_0^\infty(\log u)e^{-u}\,du$. By comparing estimates as $\delta \to 0^+$, show that $\int_0^\infty(\log u)e^{-u}\,du = -\gamma$.

***29.** Let $((u))$ be defined as in Problem 8. Show that $\Sigma_{n \leqslant x} 1/n = \log x + \gamma - ((x))/x - \int_x^\infty((u))/u^2\,du$. Show that this integral is $O(1/x^2)$.

***30.** Write $\Sigma_{n \leqslant x} d(n) = x\log x + (2\gamma - 1)x + \Delta(x)$. Show that $\Delta(x) = -2\Sigma_{n \leqslant \sqrt{x}}((x/n)) + O(1)$.

***31.** Show that if $(a, q) = 1$ and β is real, then $\Sigma_{n=1}^q((an/q + \beta)) = ((q\beta))$.

***32.** Show that if $A \geqslant 1$, $|f'(x) - a/q| \leqslant A/q^2$ for $1 \leqslant x \leqslant q$ and $(a, q) = 1$, then $\Sigma_{n=1}^q(f(n)) = O(A)$.

***33.** Suppose that Q is an integer, $Q \geqslant 1$, that $B \geqslant 1$, and that $1/Q^3 \leqslant \pm f''(x) \leqslant B/Q^3$ for $0 \leqslant x \leqslant N$, where the choice of sign is independent of x. Show that numbers a_r, q_r, N_r can be determined, $0 \leqslant r \leqslant R$ for some R, so that (*i*) $(a_r, q_r) = 1$; (*ii*) $q_r \leqslant Q$; (*iii*) $|f'(x) - a_r/q_r| \leqslant 1/(q_rQ)$ for $N_r \leqslant x \leqslant N_{r+1}$; (*iv*) $N_0 = 0$, $N_r = N_{r-1} + q_{r-1}$ for $1 \leqslant r \leqslant R$, $N - Q \leqslant N_R \leqslant N$.

***34.** Show that under the hypotheses of Problem 33, $\Sigma_{n=0}^N((f(n))) = O(B(R+1) + Q)$.

***35.** Show that in the situation of Problem 31 that the number of s for which $a_s/q_s = a_r/q_r$ is $O(Q^2/q^2)$. Suppose that $1 \leqslant q \leqslant Q$. Show that the number of r for which $q_r = q$ is $O((Q^2/q^2)(BNq/Q^3 + 1))$. Deduce that $R = O(BN(\log 2Q)/Q + Q^2)$.

***36.** Suppose that Q is an integer, $Q \geqslant 1$, that $B \geqslant 1$, and that $1/Q^3 \leqslant \pm f''(x) \leqslant B/Q^3$ for $0 \leqslant x \leqslant N$ where the choice of sign is independent of x. Show that $\Sigma_{n=0}^N((f(n))) = O(B^2N(\log 2Q)/Q + BQ^2)$.

***37.** Show tht if $U \leqslant \sqrt{x}$ then $\Sigma_{U < n \leqslant 2U}((x/n)) = O(x^{1/3}\log x)$. Let $\Delta(x)$ be as in Problem 28. Show that $\Delta(x) = O(x^{1/3}(\log x)^2)$.

8.4 PRIMES IN ARITHMETIC PROGRESSIONS

In 1839, Dirichlet established that if $(a, q) = 1$ then there are infinitely many primes $p \equiv a \pmod q$. We have already indicated special arguments that give this result for certain special pairs of q and q (notably Problem

36 at the end of Section 2.8), but we now describe the original method of Dirichlet applied to arbitrary pairs of relatively prime integers. To provide a model for the method, we first show that there are infinitely many primes by using properties of the zeta function. Then we extend this to primes in arithmetic progressions modulo 4, and finally we outline the further ideas that are required to extend the argument to general q.

By combining the formula (8.41) of Corollary 8.22 with the estimate (8.43) of Theorem 8.23, we find that

$$\sum_{n=1}^{\infty} \frac{\Lambda(n)}{\log n} n^{-s} = \log \frac{1}{s-1} + O(s-1)$$

uniformly for $1 < s \leqslant 2$. We recall that $\Lambda(n)$ is nonzero only when n is a prime power, say $n = p^k$. The contribution made by the higher powers of the primes is

$$\sum_p \sum_{k=2}^{\infty} \frac{1}{k} p^{-ks} \leqslant \sum_p \sum_{k=2}^{\infty} p^{-ks} = \sum_p p^{-2s}/(1-p^{-s})$$

$$\leqslant \sum_p \frac{1}{p(p-1)} < \infty \tag{8.61}$$

uniformly for $s \geqslant 1$. Hence

$$\sum_p p^{-s} = \log \frac{1}{s-1} + O(1) \tag{8.62}$$

for $s > 1$. If there were only finitely many primes then the sum on the left would tend to a finite limit as s tends to 1 from above. Since the right side tends to infinity as s tends to 1 from above, we conclude that there are infinitely many primes.

In order to show that there are infinitely many primes of the forms $4k + 1$ and $4k + 3$ we introduce two arithmetic functions $\chi_0(n)$ and $\chi_1(n)$ that allow us to distinguish between these two arithmetic progressions. For even n we set $\chi_0(n) = \chi_1(n) = 0$, while if n is odd then we put $\chi_0(n) = 1$ and $\chi_1(n) = (-1)^{(n-1)/2}$. Thus $\chi_1(n) = 1$ if $n \equiv 1 \pmod 4$ and $\chi_1(n) = -1$ if $n \equiv 3 \pmod 4$. Consequently,

$$(\chi_0(n) + \chi_1(n))/2 = \begin{cases} 1 & \text{if } n \equiv 1 \pmod 4, \\ 0 & \text{otherwise;} \end{cases}$$

$$(\chi_0(n) - \chi_1(n))/2 = \begin{cases} 1 & \text{if } n \equiv 3 \pmod 4, \\ 0 & \text{otherwise.} \end{cases} \tag{8.63}$$

The advantage that these functions offer in picking out arithmetic progressions is that the functions $\chi_i(n)$ are totally multiplicative. Let $\chi(n)$ denote either one of these functions. Since $|\chi(n)| \leqslant 1$ for all n, the Dirichlet series

$$L(s,\chi) = \sum_{n=1}^{\infty} \chi(n)n^{-s}$$

is absolutely convergent for $s > 1$. Since $\chi(n)$ is totally multiplicative, by Corollary 8.19 it follows that

$$L(s,\chi) = \prod_{p} (1 - \chi(p)/p^s)^{-1}$$

for $s > 1$. Taking logarithms, and arguing as in the proof of Corollary 8.22, we deduce that

$$\log L(s,\chi) = \sum_{n=1}^{\infty} \frac{\Lambda(n)}{\log n} \chi(n)n^{-s}$$

for $s > 1$. From the estimate (8.61) it follows that

$$\sum_{p} \chi(p)/p^s = \log L(s,\chi) + O(1) \tag{8.64}$$

for $s > 1$. By the identities (8.63) we conclude that

$$\sum_{\substack{p \\ p \equiv 1 \pmod 4}} 1/p^s = \frac{1}{2} \log L(s,\chi_0) + \frac{1}{2} \log L(s,\chi_1) + O(1)$$

and

$$\sum_{\substack{p \\ p \equiv 3 \pmod 4}} 1/p^s = \frac{1}{2} \log L(s,\chi_0) - \frac{1}{2} \log L(s,\chi_1) + O(1)$$

for $s > 1$.

It remains to determine the behavior of $\log L(s,\chi_0)$ and of $\log L(s,\chi_1)$ as s tends to 1 from above. If we take $\chi = \chi_0$ in (8.64), we find that the sum on the left differs from that in (8.62) only in that the prime 2 is missing. Thus from (8.62) we deduce that

$$\log L(s, \chi_0) = \log \frac{1}{s-1} + O(1)$$

for $s > 1$.

As for $L(s, \chi_1)$, we note first the Dirichlet series $\Sigma \chi_1(n) n^{-s}$ is absolutely convergent only for $s > 1$. We now show that this series is conditionally convergent for $0 < s \leqslant 1$. To this end, observe that the coefficient sum $\Sigma_{n \leqslant x} \chi_1(n)$ takes only the values 0 and 1, and hence is uniformly bounded. If s is fixed, $s > 0$, then the sequence n^{-s} tends to 0 monotonically. Hence by Dirichlet's test the series $\Sigma \chi_1(n) n^{-s}$ converges. Indeed, this series is uniformly convergent for $s \geqslant \delta > 0$. Since each term is a continuous function of s, it follows that the sum $L(s, \chi_1) = \Sigma \chi_1(n) n^{-s}$ is a continuous function of s for $s > 0$. In particular, $L(s, \chi_1)$ tends to the finite limit $L(1, \chi_1)$ as s tends to 1. Moreover, by the alternating series test we see that $1 - 1/3 < L(1, \chi_1) < 1$. (With more work one may show that $L(1, \chi_1) = \pi/4$.) Hence $L(1, \chi_1) > 0$, so that $\log L(s, \chi_1)$ tends to the finite limit $\log L(1, \chi_1)$ as s tends to 1 from above. As $\log L(s, \chi_1) = O(1)$ uniformly for $s \geqslant 1$, on combining our estimates we find that

$$\sum_{\substack{p \\ p \equiv 1 \,(\mathrm{mod}\, 4)}} 1/p^s = \frac{1}{2} \log \frac{1}{s-1} + O(1)$$

and that

$$\sum_{\substack{p \\ p \equiv 3 \,(\mathrm{mod}\, 4)}} 1/p^s = \frac{1}{2} \log \frac{1}{s-1} + O(1)$$

for $s > 1$. Since the right side tends to infinity as s tends to 1 from above, it follows that the sums on the left contain infinitely many terms.

In general, a *Dirichlet character modulo q* is a function $\chi(n)$ from $\mathbb{Z}$ to $\mathbb{C}$ with the following properties:

(*i*) If $m \equiv n \,(\mathrm{mod}\, q)$ then $\chi(m) = \chi(n)$;

(*ii*) $\chi(mn) = \chi(m)\chi(n)$ for all integers m and n;

(*iii*) $\chi(n) = 0$ if and only if $(n, q) = 1$.

If χ is a Dirichlet character $(\mathrm{mod}\, q)$ then from (*ii*) it follows $\chi(1) = \chi(1 \cdot 1) = \chi(1)\chi(1)$, which implies that $\chi(1) = 0$ or 1. In view of (*iii*), we deduce that $\chi(1) = 1$. If $(n, q) = 1$, then $n^{\phi(q)} \equiv 1 \,(\mathrm{mod}\, q)$ by Euler's congruence, and hence by (*i*) we see $\chi(n^{\phi(q)}) = \chi(1) = 1$. Then by (*ii*) it follows that $\chi(n)^{\phi(q)} = 1$. That is, if $(n, q) = 1$ then $\chi(n)$ is one of the

$\phi(q)$th roots of unity. With more work one may show that there are precisely $\phi(q)$ Dirichlet characters (mod q), and that a linear combination of them may be formed to pick out any given reduced residue class, as was done in (8.63) for the modulus 4. Let $\chi_0(n) = 1$ when $(n, q) = 1$, $\chi_0(n) = 0$ otherwise. This is the *principal character* (mod q). The corresponding Dirichlet series $L(s, \chi_0)$ is closely related to the Riemann zeta function, and it is not hard to show that

$$\log L(s, \chi_0) = \log \frac{1}{s-1} + O(\log \log q)$$

for $s > 1$. Let $\chi(n)$ be a character (mod q), $\chi \neq \chi_0$. It may be shown that $\sum_{n=1}^{q} \chi(n) = 0$, from which it follows that coefficient sum $\sum_{n \leqslant x} \chi(n)$ is uniformly bounded. Thus by Dirichlet's test the series $L(s, \chi) = \sum \chi(n) n^{-s}$ defines a continuous function for $s > 0$, $\chi \neq \chi_0$. The final step of the proof, and the most challenging, is to show that $L(1, \chi) \neq 0$ when $\chi \neq \chi_0$.

PROBLEMS

1. Let $\chi_0(n)$ denote the principal Dirichlet character (mod 3), and put $\chi_1(n) = 1$ for $n \equiv 1 \pmod 3$, $\chi_1(n) = -1$ for $n \equiv 2 \pmod 3$, $\chi_1(n) = 0$ for $n \equiv 0 \pmod 3$. Construct an argument similar to that in the text, to show that there exist infinitely many primes of the form $3k + 1$, and of the form $3k + 2$.
2. Let $Q_0(x)$ denote the number of odd square-free numbers not exceeding x, and let $Q(x)$ denote the total number of square-free integers not exceeding x, as in Theorem 8.25. Show that $Q_0(x) + Q_0(x/2) = Q(x)$. Deduce that $Q_0(x) = Q(x) - Q(x/2) + Q(x/4) - Q(x/8) + \cdots$. Conclude that $Q_0(x) = (4/\pi)x + O(\sqrt{x})$.
3. Let $\chi_1(n)$ denote the nonprincipal character (mod 4) defined in the text. Put $S(x) = \sum_{n \leqslant x} \chi_1(n)$, and suppose that $F(n) = \sum_{d \mid n} f(d)$. Show that $\sum_{n \leqslant x} \chi_1(n) F(n) = \sum_{d \leqslant x} \chi_1(d) f(d) S(x/d)$. Show that this latter sum is $O(\sum_{d \leqslant x} |f(d)|)$.
4. Let $\chi_1(n)$ denote the nonprincipal character (mod 4) defined in the text. Show that $\sum_{n \leqslant x} \chi_1(n) |\mu(n)| = O(\sqrt{x})$. Deduce that the number of square-free integers of the form $4k + 1$ not exceeding x is $(2/\pi)x + O(\sqrt{x})$, and that the same is true with $4k + 1$ replaced by $4k + 3$.
5. Let $\chi(n)$ denote either one of the Dirichlet characters (mod 4) defined in the text. Show that $\sum_{n=1}^{\infty} \chi(n) \phi(n) n^{-s} = L(s-1, \chi)/L(s, \chi)$ for $s > 1$, and that $\sum_{n=1}^{\infty} \chi(n) d(n) n^{-s} = L(s, \chi)^2$ for $s > 1$.
6. Let $\chi_0(n)$ denote the principal Dirichlet character (mod q). Show that $L(s, \chi_0) = \zeta(s) \prod_{p \mid q} (1 - p^{-s})$ for $s > 1$.

NOTES ON CHAPTER 8

§8.1 The first proof of (8.2) was given by Chebyshev in 1852. Chebyshev used his estimates to prove Bertrand's postulate, which had been stated by J. L. F. Bertrand in 1845. It is known that for any $\varepsilon > 0$ there is a choice of the $\nu(d)$ in Chebyshev's method that gives (8.2) with $1 - \varepsilon < a$ and $b < 1 + \varepsilon$. However, the only known proof of this makes use of the Prime Number Theorem, so this does not provide a method of proving the Prime Number Theorem (as far as we know). For an account of this, see H. G. Diamond and P. Erdös, "On sharp elementary prime number estimates," *L'Enseignement math.*, 26 (1980), 313–321. A more general survey of elementary techniques in prime number theory has been given by H. G. Diamond, "Elementary methods in the study of the distribution of prime numbers," *Bull. Amer. Math. Soc.* 7 (1982), 553–589. Theorem 8.8 was proved in 1874 by F. Mertens.

The method of Problem 13 can be improved to obtain a constant larger than $(\log 5)/2$, but E. Aparicio, "Metodos para el calculo aproximado de la desviacion diofantica uniforme minima a cero en un segmento," *Rev. Mat. Hisp.-Amer.* 38 (1978), 259–270, has shown that one cannot obtain a constant arbitrarily close to 1 using non-negative polynomials $P(x)$.

§8.2 At a more advanced level, it is useful to consider Dirichlet series for complex value of s, not just real s. The deeper analytic properties of the zeta function are closely related to the asymptotic distribution of the prime numbers. Indeed, in 1859 G. F. B. Riemann showed that the error term in the prime number theorem may be expressed as a sum involving the complex numbers ρ for which $\zeta(\rho) = 0$. Since the Euler product for $\zeta(s)$ is absolutely convergent when $\mathcal{R}e\, s > 1$, it follows that $\zeta(s) \neq 0$ in this half-plane. That is, if $\rho = \beta + i\gamma$ and $\zeta(\rho) = 0$ then $\beta \leqslant 1$. From Riemann's analysis it becomes evident that to prove the Prime Number Theorem one must show further that there are no zeros for which $\beta = 1$. It was in this way that Hadamard and de la Vallée Poussin proved the Prime Number Theorem in 1896. Riemann conjectured that much more is true, namely that if $\zeta(\rho) = 0$ and $\beta > 0$ then $\beta = 1/2$. This is known as the *Riemann Hypothesis*. Riemann located the first several complex zeros of the zeta function, and confirmed that they do lie exactly on the line $\mathcal{R}e\, s = 1/2$ in the complex plane. Such calculations have been performed over successively longer ranges, so that it is now known that the first 1,500,000,000 zeros of the zeta function have real part 1/2. It is known that the Riemann Hypothesis is equivalent to a sharp quantitative version of the Prime Number Theorem, namely to the estimate $\psi(x) = x + O(x^{1/2}(\log x)^2)$.

§8.3 Theorem 8.24 was proved by F. Mertens in 1874. It is known that the error term is $O((\log x)^{\alpha})$ with $\alpha < 1$, and in the opposite direction that it is infinitely often as large as $c\sqrt{\log\log x}$. By using a quantitative form of the prime number theorem it may be shown that the error term in Theorem 8.25 is $O(x^{1/2}\exp(-\sqrt{\log x}))$. In the opposite direction it is known that the error term is as large as $cx^{1/4}$ infinitely often. Assuming the Riemann Hypothesis, the error term is $O(x^{\alpha})$ with $\alpha < 1/3$. Theorem 8.28 was first established by Dirichlet in 1849. The error term has been improved many times. The current record is held by H. Iwaniec and C. J. Mozzochi, "On the divisor and circle problems," *J. Number Theory*, 29 (1988), 60–93, who proved that it is $O(x^{7/22+\varepsilon})$. In the opposite direction it is known that the error term is infinitely often as large as $x^{1/4}$. Theorem 8.26 and Corollary 8.33 are weakened forms of estimates given in 1917 by G. H. Hardy and S. Ramanujan. Theorem 8.32, which provides a simpler path to Corollary 8.33, was proved in a more precise form by P. Turán in 1934 and generalized later by J. Kubilius.

In Problem 27 one may proceed directly from (8.59) provided that one knows that $\int_0^\infty (\log u)e^{-u}\,du = -\gamma$. We owe to D. R. Heath-Brown the observation that this integral may be avoided by considering instead the relation (8.60).

Although the simple estimate (8.48) may be improved upon in many particular cases, it is not easy to strengthen this result in general. In particular, the mere convergence of the series $\Sigma f(n)/n$, say to c, does not imply that $\lim_{x\to\infty}\dfrac{1}{x}\Sigma_{n\leqslant x}F(n) = c$. The precise relation between these two assertions involves delicate issues of summability that are discussed in an appendix of G. H. Hardy, *Divergent Series*. More recently, H. Delange, "Sur les fonctions arithmétiques multiplicatives," *Ann. Scient. Éc. Sup.*, 78 (1961), 273–304, showed that the proposed implication is valid under the additional assumptions that $F(n)$ is a multiplicative function for which $|F(n)| \leqslant 1$ for all n. Multiplicative functions for which the asymptotic mean c is 0 are more difficult to treat. Although it is not surprising that $\lim_{x\to\infty}\dfrac{1}{x}\Sigma_{n\leqslant x}\mu(n) = 0$, this estimate is essentially equivalent to the Prime Number Theorem. G. Halász, "Über die Mittelwerte multiplikativer zahlentheoretischer Funktionen," *Acta Math. Acad. Sci. Hung.*, 19 (1968), 365–403, has given a useful characterization of those multiplicative functions $F(n)$ with $|F(n)| \leqslant 1$ for all n and asymptotic mean value 0.

We say that a set $\mathscr{S}$ of positive integers has natural density $\delta(\mathscr{S})$ if

$$\lim_{x\to\infty}\frac{1}{x}\sum_{n\in\mathscr{S},\, n\leqslant x} 1 = \delta(\mathscr{S}).$$

It is not difficult to show that if $\mathscr{S}_1, \mathscr{S}_2, \cdots, \mathscr{S}_K$ are pairwise disjoint sets of positive integers that have densities, then $\delta(\cup\mathscr{S}_k) = \Sigma\,\delta(\mathscr{S}_k)$. Hence it is tempting to think of natural density as defining a probability measure on the positive integers. However, Kolmogorov's axioms specify that $\mathbf{P}(\cup\mathscr{S}_k)$

$= \Sigma \mathbf{P}(\mathscr{S}_k)$ should hold for countably infinite families of pairwise disjoint sets, not merely finite collections. To see that this fails, take $\mathscr{S}_k = \{k\}$. Then $\delta(\mathscr{S}_k) = 0$ for each k and the $\mathscr{S}_k$ are pairwise disjoint, but $\delta(\cup \mathscr{S}_k) = \delta(\mathbb{N}) = 1 \neq \Sigma\, \delta(\mathscr{S}_k)$. Nevertheless, useful insights may be gained by exploring the extent to which probabilistic predictions reflect reality. For example, let $\mathscr{S}(d) = \{n\colon d|n\}$. Then $\delta(\mathscr{S}(d)) = 1/d$. Moreover, $\mathscr{S}(d_1) \cap \mathscr{S}(d_2) = \mathscr{S}([d_1, d_2])$, so that if $(d_1, d_2) = 1$ then $\delta(\mathscr{S}(d_1) \cap \mathscr{S}(d_2)) = \delta(\mathscr{S}(d_1))\,\delta(\mathscr{S}(d_2))$. We observe that an integer n is square free if and only if there is no prime p such that $p^2|n$. That is, the set of square-free numbers is precisely $\cap_p \mathscr{S}(p^2)^c$. One might therefore anticipate that the density of square-free integers is $\prod_p (1 - 1/p^2)$, which is precisely what is established in Theorem 8.25. Since the "probability" that $p|n$ is $1/p$, we might also anticipate that the "expected" number of prime divisors of n is approximately $\Sigma_{p \leqslant n} 1/p$. This is borne out in Theorem 8.26, and Theorem 8.32 is suggested by considering the variance of a random variable. On the other hand, predictions based on probabilistic models are not so reliable when applied to sieving questions. For example, by the sieve of Eratosthenes we know that if $q = \prod_{p \leqslant \sqrt{x}} p$, then the number of integers $n \leqslant x$ such that $(n, q) = 1$ is $\pi(x) - \pi(\sqrt{x}) + 1$. This suggests that perhaps $\pi(x)$ is asymptotic to $x \prod_{p \leqslant \sqrt{x}} (1 - 1/p)$. However, by Theorem 8.8(e) in conjuction with Problem 27 at the end of Section 8.3 we see that the prediction here is that $\pi(x) \sim ax/\log x$ where $a = 2e^{-\gamma} = 1.1229 \cdots$, in conflict with the Prime Number Theorem. In more advanced work, tools of probability theory may be used to provide information concerning the statistical distribution of arithmetic functions. For example, it is known that for any number c, $0 \leqslant c \leqslant 1$, the set of integers n for which $\phi(n) \leqslant cn$ has an asymptotic distribution, say $F(c)$. Moreover, the function $F(c)$ is continuous, $F(0) = 0$, $F(1) = 1$, $F(c)$ is strictly increasing, and $F(c)$ is singular (i.e., $F'(c) = 0$ for all c outside a set of Lebesgue measure 0). The body of knowledge that has developed in this area over the past 50 years is recounted in P. D. T. A. Elliott, *Probabilistic Number Theory*, Springer-Verlag (New York), 1979.

§8.4 The first proof that if $(a, q) = 1$ then there exist infinitely many prime numbers $p \equiv a \pmod{q}$ was given by P. G. Lejeune Dirichlet in 1839. The exposition in Davenport (1980) follows the historical development quite closely. Other expositions are found in the books listed in the General References by Apostol, Borevich and Shafarevich, Hua, Landau, LeVeque (1956), and Serre.

CHAPTER 9

Algebraic Numbers

To illustrate one purpose of this chapter, we take a different approach to the equation $x^2 + y^2 = z^2$ than in Section 5.3. Factoring $x^2 + y^2$ into $(x + yi)(x - yi)$, we can write

$$x^2 + y^2 = (x + yi)(x - yi) = z^2.$$

If from this we could conclude that $x + yi$ and $x - yi$ are both squares of complex numbers of the same type, we would have

$$x + yi = (r + si)^2, \qquad x - yi = (r - si)^2.$$

Equating the real and the nonreal parts here gives

$$x = r^2 - s^2, \qquad y = 2rs$$

and so $z = r^2 + s^2$. These are precisely the equations in Theorem 5.5.

The steps in this argument are valid but not quite complete, and they need justification. We shall make the justification and complete the argument in Section 9.9. A similar factoring of $x^3 + y^3$ into three linear factors in complex numbers is used in the last section of the chapter to prove that $x^3 + y^3 = z^3$ has no solutions in positive integers. This is another case of Fermat's last theorem, $x^4 + y^4 = z^4$ having been proved impossible in positive integers in Section 5.4.

However, the analysis of Diophantine equations is just one purpose of this chapter. Algebraic integers are a natural extension of the ordinary integers and are interesting in their own right. The title of this chapter is a little pretentious, because the algebraic numbers studied here are primarily only quadratic in nature, satisfying simple algebraic equations of degree 2. The plan is to develop some general theory in the first four sections and then take up the special case of the quadratic case, where much more can be said than in the general case.

9.1 POLYNOMIALS

Algebraic numbers are the roots of certain types of polynomials, so it is natural to begin our discussion with this topic. Our plan in this chapter is to proceed from the most general results about algebraic numbers to stronger specific results about special classes of algebraic numbers. In this process of proving more and more about less and less, we have selected material of a number theoretic aspect as contrasted with the more "algebraic" parts of the theory. In other words, we are concerned with such questions as divisibility, uniqueness of factorization, and prime numbers, rather than questions concerning the algebraic structure of the groups, rings, and fields arising in the theory.

The polynomials that we shall consider will have rational numbers for coefficients. Such polynomials are called *polynomials over* $\mathbb{Q}$, where $\mathbb{Q}$ denotes the field of rational numbers. This collection of polynomials in one variable x is often denoted by $\mathbb{Q}[x]$, just as all polynomials in x with integral coefficients are denoted by $\mathbb{Z}[x]$, and the set of all polynomials in x with coefficients in any set of numbers F is denoted by $F[x]$. That the set of rational numbers forms a field can be verified from the postulates in Section 2.11. In a polynomial such as

$$f(x) = a_0x^n + a_1x^{n-1} + \cdots + a_n, \qquad a_0 \neq 0$$

the nonnegative integer n is called the *degree* of the polynomial, and a_0 is called the *leading* coefficient. If $a_0 = 1$, the polynomial is called *monic*. Since we assign no degree to the zero polynomial, we can assert without exception that the degree of the product of two polynomials is the sum of the degrees of the polynomials.

A polynomial $f(x)$ is said to be *divisible* by a polynomial $g(x)$, not identically zero, if there exists a polynomial $q(x)$ such that $f(x) = g(x)q(x)$ and we write

$$g(x) \mid f(x).$$

Also, $g(x)$ is said to be a *divisor* or *factor* of $f(x)$. The degree of $g(x)$ here does not exceed that of $f(x)$, unless $f(x)$ is identically zero, written $f(x) \equiv 0$. This concept of divisibility is not the same as the divisibility that we have considered earlier. In fact $3 \mid 7$ holds if 3 and 7 are thought of as polynomials of degree zero, whereas it is not true that the integer 3 divides the integer 7.

Theorem 9.1 *To any polynomials $f(x)$ and $g(x)$ over $\mathbb{Q}$ with $g(x) \not\equiv 0$, there correspond unique polynomials $q(x)$ and $r(x)$ such that $f(x) = g(x)q(x) + r(x)$, where either $r(x) \equiv 0$ or $r(x)$ is of lower degree than $g(x)$.*

This result is the *division algorithm for polynomials* with rational coefficients, analogous to the division algorithm for integers in Theorem 1.2. Most of the theorems in this section have analogues in Chapter 1, and the methods used earlier can often be adapted to give proofs here. Although it is stated explicitly in Theorem 9.1 that $f(x)$ and $g(x)$ belong to $\mathbb{Q}[x]$, as do $q(x)$ and $r(x)$, this assumption will be taken for granted implicitly in subsequent theorems.

Proof In case $f(x) \equiv 0$ or $f(x)$ has lower degree than $g(x)$, define $q(x) \equiv 0$ and $r(x) = f(x)$. Otherwise divide $g(x)$ into $f(x)$ to get a quotient $q(x)$ and a remainder $r(x)$. Clearly $q(x)$ and $r(x)$ are polynomials over $\mathbb{Q}$, and either $r(x) \equiv 0$ or the degree of $r(x)$ is less than the degree of $g(x)$ if the division has been carried to completion. If there were another pair, $q_1(x)$ and $r_1(x)$, then we would have

$$f(x) = g(x)q_1(x) + r_1(x), \qquad r(x) - r_1(x) = g(x)\{q_1(x) - q(x)\}.$$

Thus $g(x)$ would be a divisor of the polynomial $r(x) - r_1(x)$, which, unless identically zero, has lower degree than $g(x)$. Hence $r(x) - r_1(x) \equiv 0$, and it follows that $q(x) = q_1(x)$.

Theorem 9.2 *Any polynomials $f(x)$ and $g(x)$, not both identically zero, have a common divisor $h(x)$ that is a linear combination of $f(x)$ and $g(x)$. Thus $h(x) | f(x)$, $h(x) | g(x)$, and*

$$h(x) = f(x)F(x) + g(x)G(x) \tag{9.1}$$

for some polynomials $F(x)$ and $G(x)$.

Proof From all the polynomials of the form (9.1) that are not identically zero, choose any one of least degree and designate it by $h(x)$. If $h(x)$ were not a divisor of $f(x)$, Theorem 9.1 would give us $f(x) = h(x)q(x) + r(x)$ with $r(x) \not\equiv 0$ and $r(x)$ of degree lower than $h(x)$. But then $r(x) = f(x) - h(x)q(x) = f(x)\{1 - f(x)q(x)\} - g(x)\{G(x)q(x)\}$, which is of the form (9.1) in contradiction with the choice of $h(x)$. Thus $h(x) | f(x)$ and similarly $h(x) | g(x)$.

Theorem 9.3 *To any polynomials $f(x)$ and $g(x)$, not both identically zero, there corresponds a unique monic polynomial $d(x)$ having the properties*

(1) $d(x) | f(x)$, $d(x) | g(x)$;

(2) $d(x)$ is a linear combination of $f(x)$ and $g(x)$, as in (9.1);

(3) any common divisor of $f(x)$ and $g(x)$ is a divisor of $d(x)$, and thus there is no common divisor having higher degree than that of $d(x)$.

Proof Define $d(x) = c^{-1}h(x)$, where c is the leading coefficient of $h(x)$, so that $d(x)$ is monic. Properties (1) and (2) are inherited from $h(x)$ by $d(x)$. Equation (9.1) implies $d(x) = c^{-1}f(x)F(x) + c^{-1}g(x)G(x)$, and this equation shows that if $m(x)$ is a common divisor of $f(x)$ and $g(x)$, then $m(x)|d(x)$. Finally, to prove that $d(x)$ is unique, suppose that $d(x)$ and $d_1(x)$ both satisfy properties (1), (2), (3). We then have $d(x)|d_1(x)$ and $d_1(x)|d(x)$, hence $d_1(x) = q(x)d(x)$ and $d(x) = q_1(x)d_1(x)$ for some polynomials $q(x)$ and $q_1(x)$. This implies $q(x)q_1(x) = 1$, from which we see that $q(x)$ and $q_1(x)$ are of degree zero. Since both $d(x)$ and $d_1(x)$ are monic, we have $q(x) = 1$, $d_1(x) = d(x)$.

Definition 9.1 *The polynomial $d(x)$ is called the* greatest common divisor *of $f(x)$ and $g(x)$. We write $(f(x), g(x)) = d(x)$.*

Definition 9.2 *A polynomial $f(x)$, not identically zero, is* irreducible, *or* prime, *over $\mathbb{Q}$ if there is no factoring, $f(x) = g(x)h(x)$, of $f(x)$ into two polynomials $g(x)$ and $h(x)$ of positive degrees over $\mathbb{Q}$.*

For example $x^2 - 2$ is irreducible over $\mathbb{Q}$. It has the factoring $(x - \sqrt{2})(x + \sqrt{2})$ over the field of real numbers, but it has no factoring over $\mathbb{Q}$.

Theorem 9.4 *If an irreducible polynomial $p(x)$ divides a product $f(x)g(x)$, then $p(x)$ divides at least one of the polynomials $f(x)$ and $g(x)$.*

Proof If $f(x) \equiv 0$ or $g(x) \equiv 0$ the result is obvious. If neither is identically zero, let us assume that $p(x) \nmid f(x)$ and prove that $p(x)|g(x)$. The assumption that $p(x) \nmid f(x)$ implies that $(p(x), f(x)) = 1$, and hence by Theorem 9.3 there exist polynomials $F(x)$ and $G(x)$ such that $1 = p(x)F(x) + f(x)G(x)$. Multiplying by $g(x)$ we get

$$g(x) = p(x)g(x)F(x) + f(x)g(x)G(x).$$

Now $p(x)$ is a divisor of the right member of this equation because $p(x)|f(x)g(x)$, and hence $p(x)|g(x)$.

Theorem 9.5 *Any polynomial $f(x)$ over $\mathbb{Q}$ of positive degree can be factored into a product $f(x) = cp_1(x)p_2(x) \cdots p_k(x)$ where the $p_j(x)$ are irreducible monic polynomials over $\mathbb{Q}$. This factoring is unique apart from order.*

Proof Clearly $f(x)$ can be factored repeatedly until it becomes a product of irreducible polynomials, and the constant c can be adjusted to make all the factors monic. We must prove uniqueness. Let us consider another factoring, $f(x) = cq_1(x)q_2(x) \cdots q_j(x)$, into irreducible monic polynomials. According to Theorem 9.4, $p_1(x)$ divides some $q_i(x)$, and we can reorder the $q_m(x)$ to make $p_1(x)|q_1(x)$. Since $p_1(x)$ and $q_1(x)$ are irreducible and monic, we have $p_1(x) = q_1(x)$. A repetition of this argument yields

$$p_2(x) = q_2(x), \qquad p_3(x) = q_3(x), \cdots, \qquad \text{and } k = j.$$

Definition 9.3 *A polynomial $f(x) = a_0x^n + \cdots + a_n$ with integral coefficients a_j is said to be* primitive *if the greatest common divisor of its coefficients is* 1. *Obviously, here we mean the greatest common divisor of integers as defined in Definition* 1.2.

Theorem 9.6 *The product of two primitive polynomials is primitive.*

Proof Let $a_0x^n + \cdots + a_n$ and $b_0x^m + \cdots + b_m$ be primitive polynomials and denote their product by $c_0x^{n+m} + \cdots + c_{n+m}$. Suppose that this product polynomial is not primitive, so that there is a prime p that divides every coefficient c_k. Since $a_0x^n + \cdots + a_n$ is primitive, at least one of its coefficients is not divisible by p. Let a_i denote the first such coefficient and let b_j denote the first coefficient of $b_0x^m + \cdots + b_m$, not divisible by p. Then the coefficient of $x^{n+m-i-j}$ in the product polynomial is

$$c_{i+j} = \sum a_k b_{i+j-k} \tag{9.2}$$

summed over all k such that $0 \leqslant k \leqslant n$, $0 \leqslant i + j - k \leqslant m$. In this sum, any term with $k < i$ is a multiple of p. Any term with $k > i$ that appears in the sum will have the factor b_{i+j-k} with $i + j - k < j$ and will also be a multiple of p. The term a_ib_j, for $k = i$, appears in the sum, and we have $c_{i+j} \equiv a_jb_j \pmod{p}$. But this is in contradiction with $p|c_{i+j}$, $p \nmid a_i$, $p \nmid b_j$.

Theorem 9.7 *Gauss's lemma. If a monic polynomial $f(x)$ with integral coefficients factors into two monic polynomials with rational coefficients, say $f(x) = g(x)h(x)$, then $g(x)$ and $h(x)$ have integral coefficients.*

Proof Let c be the least positive integer such that $cg(x)$ has integral coefficients; if $g(x)$ has integral coefficients take $c = 1$. Then $cg(x)$ is a primitive polynomial, because if p is a divisor of its coefficients, then $p|c$ because c is the leading coefficient, and $(c/p)g(x)$ would have integral coefficients contrary to the minimal property of c. Similarly let c_1 be least positive integer such that $c_1h(x)$ has integral coefficients, and hence

$c_1h(x)$ is also primitive. Then by Theorem 9.6 the product $\{cg(x)\}\{c_1h(x)\} = cc_1f(x)$ is primitive. But since $f(x)$ has integral coefficients, it follows that $cc_1 = 1$ and $c = c_1 = 1$.

PROBLEMS

1. If $f(x)|g(x)$ and $g(x)|f(x)$, prove that there is a rational number c such that $g(x) = cf(x)$.
2. If $f(x)|g(x)$ and $g(x)|h(x)$, prove that $f(x)|h(x)$.
3. If $p(x)$ is irreducible and $g(x)|p(x)$, prove that either $g(x)$ is a constant or $g(x) = cp(x)$ for some rational number c.
4. If $p(x)$ is irreducible, prove that $cp(x)$ is irreducible for any rational $c \neq 0$.

*5. If a polynomial $f(x)$ with integral coefficients factors into a product $g(x)h(x)$ of two polynomials with coefficients in $\mathbb{Q}$, prove that there is a factoring $g_1(x)h_1(x)$ with integral coefficients.

6. If $f(x)$ and $g(x)$ are primitive polynomials, and if $f(x)|g(x)$ and $g(x)|f(x)$, prove that $f(x) = \pm g(x)$.
7. Let $f(x)$ and $g(x)$ be polynomials in $\mathbb{Z}[x]$, that is, polynomials with integral coefficients. Suppose that $g(m)|f(m)$ for infinitely many positive integers m. Prove that $g(x)|f(x)$ in $\mathbb{Q}[x]$, that is, there exists a quotient polynomial $q(x)$ with rational coefficients such that $f(x) = g(x)q(x)$. (*Remark*: The example $g(x) = 2x + 2$, $f(x) = x^2 - 1$ with m odd shows that $q(x)$ need not have integral coefficients.) (H)
8. Let $f(x)$ and $g(x)$ be primitive nonconstant polynomials in $\mathbb{Z}[x]$ such that the greatest common divisor $(f(m), g(m)) > 1$ for infinitely many positive integers m. Construct an example to show that such polynomials exist with g.c.d.$(f(x), g(x)) = 1$ in the polynomial sense.
9. Given any nonconstant polynomial $f(x)$ with integral coefficients, prove that there are infinitely many primes p such that $f(x) \equiv 0 \pmod{p}$ is solvable. (H)

9.2 ALGEBRAIC NUMBERS

Definition 9.4 *A complex number ξ is called an* algebraic number *if it satisfies some polynomial equation $f(x) = 0$ where $f(x)$ is a polynomial over $\mathbb{Q}$.*

Every rational number r is an algebraic number because $f(x)$ can be taken as $x - r$ in this case.

Any complex number that is not algebraic is said to be *transcendental*. Perhaps the best known examples of transcendental numbers are the familiar constants π and e. At the end of this section, we prove the existence of transcendental numbers by exhibiting one, using a very simple classical example.

Theorem 9.8 *An algebraic number ξ satisfies a unique irreducible monic polynomial equation $g(x) = 0$ over $\mathbb{Q}$. Furthermore, every polynomial equation over $\mathbb{Q}$ satisfied by ξ is divisible by $g(x)$.*

Proof From all polynomial equations over $\mathbb{Q}$ satisfied by ξ, choose one of lowest degree, say $G(x) = 0$. If the leading coefficient of $G(x)$ is c, define $g(x) = c^{-1}G(x)$, so that $g(\xi) = 0$ and $g(x)$ is monic. The polynomial $g(x)$ is irreducible, for if $g(x) = h_1(x)h_2(x)$, then one at least of $h_1(\xi) = 0$ and $h_2(\xi) = 0$ would hold, contrary to the fact that $G(x) = 0$ and $g(x) = 0$ are polynomial equations over $\mathbb{Q}$ of least degree satisfied by ξ.

Next let $f(x) = 0$ be any polynomial equation over $\mathbb{Q}$ have ξ as a root. Applying Theorem 9.1, we get $f(x) = g(x)q(x) + r(x)$. The remainder $r(x)$ must be identically zero, for otherwise the degree of $r(x)$ would be less than that of $g(x)$, and ξ would be a root of $r(x)$ since $f(\xi) = g(\xi) = 0$. Hence $g(x)$ is a divisor of $f(x)$.

Finally, to prove that $g(x)$ is unique, suppose that $g_1(x)$ is an irreducible monic polynomial such that $g_1(\xi) = 0$. Then $g(x) | g_1(x)$ by the argument above, say $g_1(x) = g(x)q(x)$. But the irreducibility of $g_1(x)$ then implies that $q(x)$ is a constant, in fact $q(x) = 1$ since $g_1(x)$ and $g(x)$ are monic. Thus we have $g_1(x) = g(x)$.

Definition 9.5 *The* minimal equation *of an algebraic number ξ is the equation $g(x) = 0$ described in Theorem 9.8. The minimal polynomial of ξ is $g(x)$. The degree of an algebraic number is the degree of its minimal polynomial.*

Definition 9.6 *An algebraic number ξ is an* algebraic integer *if it satisfies some monic polynomial equation*

$$f(x) = x^n + b_1 x^{n-1} + \cdots + b_n = 0 \tag{9.3}$$

with integral coefficients.

Theorem 9.9 *Among the rational numbers, the only ones that are algebraic integers are the integers* $0, \pm 1, \pm 2, \cdots$.

Proof Any integer m is an algebraic integer because $f(x)$ can be taken as $x - m$. On the other hand, if any rational number m/q is an algebraic integer, then we may suppose $(m, q) = 1$, and we have

$$\left(\frac{m}{q}\right)^n + b_1\left(\frac{m}{q}\right)^{n-1} + \cdots + b_n = 0,$$

$$m^n + b_1 q m^{n-1} + \cdots + b_n q^n = 0.$$

Thus $q|m^n$, so that $q = \pm 1$, and m/q is an integer.

The work "integer" in Definition 9.6 is thus simply a generalization of our previous usage. In algebraic number theory, $0, \pm 1, \pm 2, \cdots$ are often referred to as "rational integers" to distinguish them from the other algebraic integers that are not rational. For example, $\sqrt{2}$ is an algebraic integer but not a rational integer.

Theorem 9.10 *The minimal equation of an algebraic integer is monic with integral coefficients.*

Proof The equation is monic by definition, so we need prove only that the coefficients are integers. Let the algebraic integer ξ satisfy $f(x) = 0$ as in (9.3), and let its minimal equation be $g(x) = 0$, monic and irreducible over $\mathbb{Q}$. By Theorem 9.8, $g(x)$ is a divisor of $f(x)$, say $f(x) = g(x)h(x)$, and the quotient $h(x)$, like $f(x)$ and $g(x)$, is monic and has coefficients in $\mathbb{Q}$. Applying Theorem 9.7, we see that $g(x)$ has integral coefficients.

Theorem 9.11 *Let n be a positive rational integer and ξ a complex number. Suppose that the complex numbers $\theta_1, \theta_2, \cdots, \theta_n$, not all zero, satisfy the equations*

$$\xi\theta_j = a_{j,1}\theta_1 + a_{j,2}\theta_2 + \cdots + a_{j,n}\theta_n, \qquad j = 1, 2, \cdots, n \tag{9.4}$$

where the n^2 coefficients $a_{j,i}$ are rational. Then ξ is an algebraic number. Moreover, if the $a_{j,i}$ are rational integers, ξ is an algebraic integer.

Proof Equations (9.4) can be thought of as a system of homogeneous linear equations in $\theta_1, \theta_2, \cdots, \theta_n$. Since the θ_i are not all zero, the determinant of coefficients must vanish:

$$\begin{vmatrix} \xi - a_{1,1} & -a_{1,2} & \cdots & -a_{1,n} \\ -a_{2,1} & \xi - a_{2,2} & \cdots & -a_{2,n} \\ \cdots & \cdots & \cdots & \cdots \\ -a_{n,1} & -a_{n,2} & \cdots & \xi - a_{n,n} \end{vmatrix} = 0.$$

Expansion of this determinant gives an equation $\xi^n + b_1\xi^{n-1} + \cdots + b_n$

$= 0$, where the b_i are polynomials in the $a_{j,k}$. Thus the b_i are rational, and they are rational integers if the $a_{j,k}$ are.

Theorem 9.12 *If α and β are algebraic numbers, so are $\alpha + \beta$ and $\alpha\beta$. If α and β are algebraic integers, so are $\alpha + \beta$ and $\alpha\beta$.*

Proof Suppose that α and β satisfy

$$\alpha^m + a_1\alpha^{m-1} + \cdots + a_m = 0$$

$$\beta^r + b_1\beta^{r-1} + \cdots + b_r = 0$$

with rational coefficients a_i and b_j. Let $n = mr$, and define the complex numbers $\theta_1, \cdots, \theta_n$ as the numbers

$$\begin{array}{ccccc} 1, & \alpha, & \alpha^2, & \cdots, & \alpha^{m-1}, \\ \beta, & \alpha\beta, & \alpha^2\beta, & \cdots, & \alpha^{m-1}\beta, \\ \cdots & \cdots & \cdots & \cdots & \cdots \\ \beta^{r-1}, & \alpha\beta^{r-1}, & \alpha^2\beta^{r-1}, & \cdots, & \alpha^{m-1}\beta^{r-1} \end{array}$$

in any order. Thus $\theta_1, \cdots, \theta_n$ are the numbers $\alpha^s\beta^t$ with $s = 0, 1, \cdots, m-1$ and $t = 0, 1, \cdots, r-1$. Hence for any θ_j,

$$\alpha\theta_j = \alpha^{s+1}\beta^t = \begin{cases} \text{some } \theta_k & \text{if } s+1 \leqslant m-1 \\ (-a_1\alpha^{m-1} - a_2\alpha^{m-2} - \cdots - a_m)\beta^t & \text{if } s+1 = m \end{cases}$$

In either case we see that there are rational constants $h_{j,1}, \cdots, h_{j,n}$ such that $\alpha\theta_j = h_{j,1}\theta_1 + \cdots + h_{j,n}\theta_n$. Similarly there are rational constants $k_{j,1}, \cdots, k_{j,n}$ such that $\beta\theta_j = k_{j,1}\theta_1 + \cdots + k_{j,n}\theta_n$, and hence $(\alpha + \beta)\theta_j = (h_{j,1} + k_{j,1})\theta_1 + \cdots + (h_{j,n} + k_{j,n})\theta_n$. These equations are of the form (9.4), so we conclude that $\alpha + \beta$ is algebraic. Furthermore, if α and β are algebraic integers, then the a_j, b_j, $h_{j,i}$, $k_{j,i}$ are all rational integers, and $\alpha + \beta$ is an algebraic integer.

We also have $\alpha\beta\theta_j = \alpha(k_{j,1}\theta_1 + \cdots + k_{j,n}\theta_n) = k_{j,1}\alpha\theta_1 + \cdots + k_{j,n}\alpha\theta_n$ from which we find $\alpha\beta\theta_j = c_{j,1}\theta_1 + \cdots + c_{j,1}\theta_n$ where $c_{j,i} = k_{j,1}h_{1,i} + k_{j,2}h_{2,i} + \cdots + k_{j,n}h_{n,i}$. Again we apply Theorem 9.11 to conclude that $\alpha\beta$ is algebraic, and that it is an algebraic integer if α and β are.

This theorem states that the set of algebraic numbers is closed under addition and multiplication, and likewise for the set of algebraic integers. The following result states a little more.

Theorem 9.13 *The set of all algebraic numbers forms a field. The set of all algebraic integers forms a ring.*

Proof Rings and fields are defined in Definition 2.12. The rational numbers 0 and 1 serve as the zero and unit for the system. Most of the postulates are easily seen to be satisfied if we remember that algebraic numbers are complex numbers, whose properties we are familiar with. The only place where any difficulty arises is in proving the existence of additive and multiplicative inverses. If $\alpha \neq 0$ is a solution of

$$a_0x^n + a_1x^{n-1} + \cdots + a_n = 0$$

then $-\alpha$ and α^{-1} are solutions of

$$a_0x^n - a_1x^{n-1} + a_2x^{n-2} - \cdots + (-1)^n a_n = 0$$

and

$$a_0 + a_1x + a_2x^2 + \cdots + a_nx^n = 0$$

respectively. Therefore, if α is an algebraic number, then so are $-\alpha$ and α^{-1}. If α is an algebraic integer, then so is $-\alpha$, but not necessarily α^{-1}. Therefore the algebraic numbers form a field, the algebraic integers a ring.

Example of a Transcendental Number To demonstrate that not all real numbers are algebraic, we prove that the number

$$\beta = \sum_{j=1}^{\infty} 10^{-j!} = 0.110001000\cdots$$

is trancendental. (This was one of the numbers used by Liouville in 1851 in the first proof of the existence of transcendental numbers.) Suppose β is algebraic, so that it satisfies some equation

$$f(x) = \sum_{j=0}^{n} c_jx^j = 0$$

with integral coefficients. For any x satisfying $0 < x < 1$, we have by the triangle inequality

$$|f'(x)| = \left| \sum_{j=1}^{n} jc_jx^{j-1} \right| < \sum |jc_j| = C,$$

where the constant C, defined by the last equation, depends only on the coefficients of $f(x)$. Define $\beta_k = \sum_{j=1}^{k} 10^{-j!}$ so that

$$\beta - \beta_k = \sum_{j=k+1}^{\infty} 10^{-j!} < 2 \cdot 10^{-(k+1)!}.$$

By the mean value theorem,

$$|f(\beta) - f(\beta_k)| = |\beta - \beta_k| \cdot |f'(\theta)|$$

for some θ between β and β_k. We get a contradiction by proving that the right side is smaller than the left, if k is chosen sufficiently large. The right side is smaller than $2C/10^{(k+1)!}$. Since $f(x)$ has only n zeros, we can choose k sufficiently large so that $f(\beta_k) \neq 0$. Using $f(\beta) = 0$ we see that

$$|f(\beta) - f(\beta_k)| = |f(\beta_k)| = \left| \sum_{j=0}^{n} c_j \beta_k^j \right| \geqslant 1/10^{n \cdot k!},$$

because $c_j \beta_k^j$ is a rational number with denominator $10^{j \cdot k!}$. Finally we observe that $1/10^{n \cdot k!} > 2C/10^{(k+1)!}$ if k is sufficiently large.

PROBLEMS

1. Find the minimal polynomial of each of the following algebraic numbers: 7, $\sqrt[3]{7}$, $(1 + \sqrt[3]{7})/2$, $1 + \sqrt{2} + \sqrt{3}$. Which of these are algebraic integers?
2. Prove that if α is algebraic of degree n, then $-\alpha$, α^{-1}, and $\alpha - 1$ are also of degree n, assuming $\alpha \neq 0$ in the case of α^{-1}.
3. Prove that if α is algebraic of degree n, and β is algebraic of degree m, then $\alpha + \beta$ is of degree $\leqslant mn$. Prove a similar result for $\alpha\beta$.
4. Prove that the set of real algebraic numbers (i.e., algebraic numbers that are real) forms a field, and the set of all real algebraic integers forms a ring.

9.3 ALGEBRAIC NUMBER FIELDS

The field discussed in Theorem 9.13 contains the totality of algebraic numbers. In general, an *algebraic number field* is any subset of this total collection that is a field itself. For example, if ξ is an algebraic number, then it can be readily verified that the collection of all numbers of the form $f(\xi)/h(\xi)$, $h(\xi) \neq 0$, f and h polynomials over $\mathbb{Q}$, constitutes a field. This field is denoted by $\mathbb{Q}(\xi)$, and it is called the *extension* of $\mathbb{Q}$ by ξ.

(Some authors prefer a more restrictive definition of algebraic number field than the one just given. Without going into technical details here, suffice it to say that, in effect, the restriction imposed puts an upper bound on the degrees of the algebraic numbers in the field.)

Theorem 9.14 *If ξ is an algebraic number of degree n, then every number in $\mathbb{Q}(\xi)$ can be written uniquely in the form*

$$a_0 + a_1\xi + \cdots + a_{n-1}\xi^{n-1} \tag{9.5}$$

where the a_i are rational numbers.

Proof Consider any number $f(\xi)/h(\xi)$ of $\mathbb{Q}(\xi)$. If the minimal polynomial of ξ is $g(x)$, then $g(x) \nmid h(x)$ since $h(\xi) \neq 0$. But $g(x)$ is irreducible, so the greatest common polynomial divisor of $g(x)$ and $h(x)$ is 1, so by Theorem 9.3 there exist polynomials $G(x)$ and $H(x)$ such that $1 = g(x)G(x) + h(x)H(x)$. Replacing x by ξ and using the fact that $g(\xi) = 0$, we get $1/h(\xi) = H(\xi)$ and $f(\xi)/h(\xi) = f(\xi)H(\xi)$. Let $k(x) = f(x)H(x)$ so that $f(\xi)/h(\xi) = k(\xi)$. Dividing $k(x)$ by $g(x)$, we get $k(x) = g(x)q(x) + r(x)$, and hence $f(\xi)/h(\xi) = k(\xi) = r(\xi)$ where $r(\xi)$ is of the form (9.5).

To prove that the form (9.5) is unique, suppose $f(\xi)$ and $r_1(\xi)$ are expressions of the form (9.15). If $r(x) - r_1(x)$ is not identically zero, then it is a polynomial of degree less than n. Since the minimal polynomial of ξ has degree n, we have $r(\xi) - r_1(\xi) \neq 0$, $r(\xi) \neq r_1(\xi)$, unless $r(x)$ and $r_1(x)$ are the same polynomial.

The field $\mathbb{Q}(\xi)$ can be looked at in a different way, by consideration of congruences modulo the polynomial $g(x)$. That is, in analogy with Definition 2.1, for any polynomial $G(x)$ of degree at least one we write

$$f_1(x) \equiv f_2(x) \pmod{G(x)}$$

if $G(x) \mid (f_1(x) - f_2(x))$. Ultimately, in order to get back to $\mathbb{Q}(\xi)$ we take the minimal polynomial $g(x)$ of ξ for $G(x)$. However, the theory of congruences is more general, and we start with the polynomial $G(x)$ over $\mathbb{Q}$ irreducible or not. The properties of congruences in Theorem 2.1 can be extended at once to the polynomial case. For example, part (iii) of the theorem has the analogue: If $f_1(x) \equiv f_2(x) \pmod{G(x)}$ and $h_1(x) \equiv h_2(x) \pmod{G(x)}$, then $f_1(x)h_1(x) \equiv f_2(x)h_2(x) \pmod{G(x)}$.

By the division algorithm Theorem 9.1, any polynomial $f(x)$ over $\mathbb{Q}$ is mapped by division by $G(x)$ onto a unique polynomial $r(x)$ modulo $G(x)$;

$$f(x) = G(x)q(x) + r(x), \qquad f(x) \equiv r(x) \pmod{G(x)}.$$

Thus the set of polynomials $r(x)$ consisting of 0 and all polynomials over $\mathbb{Q}$ of degree less than n constitute a "complete residue system modulo $G(x)$" in the sense of Definition 2.2. Of course the present residue system has infinitely many members, whereas the residue system modulo m contained precisely m elements.

Theorem 9.15 *Let $G(x)$ be a polynomial over $\mathbb{Q}$ of degree $n \geqslant 1$. The totality of polynomials*

$$r(x) = a_0 + a_1 x + \cdots + a_{n-1} x^{n-1} \tag{9.6}$$

with coefficients in $\mathbb{Q}$, and with addition and multiplication modulo $G(x)$, forms a ring.

Proof This theorem is the analogue of the first part of Theorem 2.33, and its proof is virtually the same. First we note that the polynomials (9.6) form a group under addition, with identity element 0, the additive inverse of $r(x)$ being $-r(x)$. Next, the polynomials (9.6) are closed under multiplication modulo $G(x)$, and the associative property of multiplication comes from the corresponding property for polynomials over $\mathbb{Q}$ with ordinary multiplication, that is

$$\{r_1(x)r_2(x)\}r_3(x) = r_1(x)\{r_2(x)r_3(x)\}$$

implies

$$\{r_1(x)r_2(x)\}r_3(x) \equiv r_1(x)\{r_2(x)r_3(x)\} \pmod{G(x)}.$$

Similarly, the distributive property modulo $G(x)$ is inherited from the distributive property of polynomials over $\mathbb{Q}$.

Before stating the next theorem, we extend Definition 2.10 to the concept of isomorphism between fields. Two fields F and F' are *isomorphic* if there is a one-to-one correspondence between the elements of F and the elements of F' such that if a and b in F correspond respectively to a' and b' in F', then $a + b$ and ab in F correspond respectively to $a' + b'$ and $a'b'$ in F'. A virtually identical definition is used for the concept of isomorphism between rings. The following result is a direct analogue of the second part of Theorem 2.33.

Theorem 9.16 *The ring of polynomials modulo $G(x)$ described in Theorem 9.15 is a field if and only if $G(x)$ is an irreducible polynomial. If $G(x)$ is the minimal polynomial of the algebraic number ξ, then this field is isomorphic to $\mathbb{Q}(\xi)$.*

Proof If the polynomial $G(x)$ is reducible over $\mathbb{Q}$, say $G(x) = G_1(x)G_2(x)$ where $G_1(x)$ and $G_2(x)$ have degrees between 1 and $n - 1$, then $G_1(x)$ and $G_2(x)$ are of the form (9.6). But then $G_1(x)$ has no multiplicative inverse modulo $G(x)$ since $G_1(x)f(x) \equiv 1 \pmod{G(x)}$ implies

$$G(x)|\{G_1(x)f(x) - 1\}, G_1(x)|\{G_1(x)f(x) - 1\}, G_1(x)|1.$$

Hence the ring of polynomials modulo $G(x)$ is not a field.

On the other hand, if $G(x)$ is irreducible over $\mathbb{Q}$, then every polynomial $r(x)$ of the form (9.6) has a unique multiplicative inverse $r_1(x)$ modulo $G(x)$, of the form (9.6). To show this we note that the greatest common divisor of $G(x)$ and $r(x)$ is 1, and so by Theorem 9.3 there exist polynomials $f(x)$ and $h(x)$ such that

$$1 = r(x)f(x) + G(x)h(x). \tag{9.7}$$

Applying Theorem 9.1 to $f(x)$ and $G(x)$ we get $f(x) = G(x)q(x) + r_1(x)$ where $r_1(x)$ is of the form (9.6). Thus (9.7) can be written

$$1 = r(x)r_1(x) + G(x)\{h(x) + r(x)q(x)\},$$

$$r(x)r_1(x) \equiv 1 \pmod{G(x)}$$

so $r_1(x)$ is a multiplicative inverse of $r(x)$ of the form (9.6). This inverse is unique because if $r(x)r_2(x) \equiv 1 \pmod{G(x)}$ then

$$r(x)r_1(x) \equiv r(x)r_2(x) \pmod{G(x)}, G(x)|r(x)\{r_1(x) - r_2(x)\}.$$

Since $G(x) \nmid r(x)$ we have $G(x)|\{r_1(x) - r_2(x)\}$ by Theorem 9.4. But the polynomial $r_1(x) - r_2(x)$ is either identically zero or is of degree less than n, the degree of $G(x)$. Hence $r_1(x) - r_2(x) = 0$, $r_1(x) = r_2(x)$.

Finally, if $G(x)$ is the minimal polynomial $g(x)$ of the algebraic number ξ, we must show that the field is isomorphic to $\mathbb{Q}(\xi)$. To each $r(x)$ of the form (9.6) we let correspond the number $r(\xi)$ of $\mathbb{Q}(\xi)$. Theorem 9.14 shows that this correspondence is one-to-one. If

$$r_1(x)r_2(x) \equiv r_3(x), \qquad r_1(x) + r_2(x) \equiv r_4(x) \pmod{G(x)}$$

then

$$r_1(x)r_2(x) = r_3(x) + q_1(x)G(x),$$

$$r_1(x) + r_2(x) = r_4(x) + q_2(x)G(x),$$

and hence

$$r_1(\xi)r_2(\xi) = r_3(\xi), \qquad r_1(\xi) + r_3(\xi) = r_4(\xi),$$

since $G(\xi) = 0$. Therefore the correspondence preserves multiplication and addition.

The theorem we have just proved is significant in that it makes possible the development of the theory of algebraic numbers from the consideration of polynomials without any reference to the roots of the polynomials. The *fundamental theorem of algebra* states that every polynomial of positive degree over $\mathbb{Q}$ has a root that is a complex number. Therefore the algebraic number fields obtained by means of Theorem 9.16 are essentially the same—isomorphic to—the fields $\mathbb{Q}(\xi)$ of Theorem 9.14, but one does not need a knowledge of the fundamental theorem of algebra to use the method of Theorem 9.16.

The fundamental theorem of algebra implies, and is sometimes stated in the form, that every polynomial $f(x)$ of degree n over $\mathbb{Q}$ has n complex roots. If $f(x)$ is irreducible over $\mathbb{Q}$, then the n roots, say $\xi_1, \cdots, \xi_n$, are called *conjugate algebraic numbers*, and the conjugates of any one of them are simply all the others. Now Theorem 9.16 does not make any distinction between conjugates, whereas Theorem 9.14 allows for such a distinction. For example, let $g(x)$ be the irreducible polynomial $x^3 - 2$. In Theorem 9.14 we can take ξ to be any one of the three algebraic numbers that are solutions of $x^3 - 2 = 0$, namely $\sqrt[3]{2}, \omega\sqrt[3]{2}, \omega^2\sqrt[3]{2}$ where $\omega = (-1 + i\sqrt{3})/2$. Thus there are three fields

$$\mathbb{Q}\left(\sqrt[3]{2}\right), \qquad \mathbb{Q}\left(\omega\sqrt[3]{2}\right), \qquad \mathbb{Q}\left(\omega^2\sqrt[3]{2}\right) \tag{9.8}$$

The first of these consists of real numbers, whereas the other two contain nonreal elements. Therefore, the first is certainly a different field from the others. It is not so apparent, but can be proved, that the last two differ from each other. On the other hand, if we apply Theorem 9.16 to the polynomial $x^3 - 2$, we obtain a single field consisting of all polynomials $a_0 + a_1x + a_2x^2$ over $\mathbb{Q}$ modulo $x^3 - 2$. According to Theorem 9.16, this field is isomorphic to each of the fields (9.8). Since isomorphism is a transitive property, the fields (9.8) are isomorphic to each other. They differ in that they contain different elements, but they are essentially the same except for the names of their elements.

PROBLEMS

1. Prove that the fields of (9.8), although isomorphic, are distinct. (H)

2. Prove that the field $\mathbb{Q}(i)$, where $i^2 = -1$, is isomorphic to the field of all polynomials $a + bx$ with a and b in $\mathbb{Q}$, taken modulo $x^2 + 1$.

3. Prove that any algebraic number field contains $\mathbb{Q}$ as a subfield.

4. Assuming the fundamental theorem of algebra, prove Theorem 9.10 by the following procedure. Let the algebraic integer ξ satisfy some monic polynomial equation $f(x) = 0$ with integral coefficients. Then we can factor $f(x)$ in the field of complex numbers, say

$$f(x) = (x - \xi)(x - \xi_2)(x - \xi_3) \cdots (x - \xi_n).$$

If $g(x)$ is the minimal polynomial of ξ, then $g(x)|f(x)$ by Theorem 9.8, and so

$$g(x) = (x - \xi)(x - \theta_2) \cdots (x - \theta_r)$$

where $\theta_2 \cdots \theta_r$ is a subset of $\xi_2, \cdots, \xi_n$. Thus $\xi, \theta_2, \cdots, \theta_r$ are algebraic integers and by Theorem 9.12 the coefficients of $g(x)$ are algebraic integers. Then apply Theorem 9.9.

9.4 ALGEBRAIC INTEGERS

Any algebraic number field contains the elements 0 and 1, and so, by the postulates for a field, must contain all the rational numbers. Thus any algebraic number field contains at least some algebraic integers, the rational integers $0, +1, \pm 2, \cdots$. The following result shows that, in general, an algebraic number field also contains other algebraic integers.

Theorem 9.17 *If α is any algebraic number, there is a rational integer b such that $b\alpha$ is an algebraic integer.*

Proof Let $f(x)$ be a polynomial over $\mathbb{Q}$ such that $f(\alpha) = 0$. We may presume that the coefficients of $f(x)$ are rational integers, since we can multiply by the least common multiple of the denominators of the coefficients. Thus we can take $f(x)$ in the form

$$f(x) = bx^n + a_1x^{n-1} + \cdots + a_n = bx^n + \sum_{j=1}^{n} a_jx^{n-j}$$

with rational integers b and a_j. Then $b\alpha$ is a zero of

$$b^{n-1}f\left(\frac{x}{b}\right) = x^n + \sum_{j=1}^{n} a_jb^{j-1}x^{n-j}$$

and hence $b\alpha$ is an algebraic integer.

Theorem 9.18 *The integers of any algebraic number field form a ring.*

Proof If α and β are integers in such a field F, then $\alpha + \beta$ and $\alpha\beta$ are in F since F is a field. But by Theorems 9.12 and 9.13, $\alpha + \beta$, $\alpha\beta$, and $-\alpha$ are algebraic integers. Thus the integers of F form a ring with 0 and 1 as the identity elements of addition and multiplication.

Definition 9.7 *In any algebraic number field F an integer $\alpha \neq 0$ is said to be* a divisor *of an integer β if there exists an integer γ such that $\beta = \alpha\gamma$. In this case we write $\alpha | \beta$. Any divisor of the integer* 1 *is called a* unit *of F. Nonzero integers α and β are called* associates *if α/β is a unit.*

This definition of associates does not appear to be symmetrical in α and β, but we shall establish that the property really is symmetric.

Theorem 9.19 *The reciprocal of a unit is a unit. The units of an algebraic number field form a multiplicative group.*

Proof If ε_1 is a unit, then there exists an integer ε_2 such that $\varepsilon_1\varepsilon_2 = 1$. Hence ε_2 is also a unit, and it is the reciprocal of ε_1. If, similarly, ε_3 is any unit with reciprocal ε_4, then the product $\varepsilon_1\varepsilon_3$ is a unit because $(\varepsilon_1\varepsilon_3)(\varepsilon_2\varepsilon_4) = 1$. Hence the units of an algebraic number field form a multiplicative group where the identity element is 1, and the inverse of ε is the reciprocal of ε.

If α and β are associates, then α/β is a unit by definition, and by Theorem 9.19 β/α is also a unit. Hence the definition of associates is symmetric: if α and β are associates, then so are β and α.

PROBLEMS

1. Prove that the units of the rational number field $\mathbb{Q}$ are ± 1, and that integers α and β are associates in this field if and only if $\alpha = \pm\beta$.
2. For any algebraic number α, define m as the smallest positive rational integer such that $m\alpha$ is an algebraic integer. Prove that if $b\alpha$ is an algebraic integer, where b is a rational integer, then $m | b$.
3. Let $\alpha = \alpha_1 + \alpha_2 i$ be an algebraic number, where α_1 and α_2 are real. Does it follow that α_1 and α_2 are algebraic numbers? If α is an algebraic integer, would α_1 and α_2 necessarily be algebraic integers?

9.5 QUADRATIC FIELDS

A *quadratic field* is one of the form $\mathbb{Q}(\xi)$ where ξ is a root of an irreducible quadratic polynomial over $\mathbb{Q}$. By Theorem 9.14 the elements of such a field are the totality of numbers of the form $a_0 + a_1\xi$, where a_0

and a_1 are rational numbers. Since ξ is of the form $(a + b\sqrt{m})/c$ where a, b, c, m are integers, we see that

$$\mathbb{Q}(\xi) = \mathbb{Q}\left(\frac{a + b\sqrt{m}}{c}\right) = \mathbb{Q}(a + b\sqrt{m}) = \mathbb{Q}(b\sqrt{m}) = \mathbb{Q}(\sqrt{m}).$$

Here we have presumed that $c \neq 0$ and that m is square-free, $m \neq 1$. On the other hand, if m and n are two different square-free rational integers, neither of which is 1, then $\mathbb{Q}(\sqrt{m}) \neq \mathbb{Q}(\sqrt{n})$ since $\sqrt{m}$ is not in $\mathbb{Q}(\sqrt{n})$. That is, it is impossible to find rational numbers a and b such that $\sqrt{m} = \mathrm{a} + \mathrm{b}\sqrt{n}$.

Theorem 9.20 *Every quadratic field is of the form $\mathbb{Q}(\sqrt{m})$ where m is a square-free rational integer, positive or negative but not equal to 1. Numbers of the form $a + b\sqrt{m}$ with rational integers a and b are integers of $\mathbb{Q}(\sqrt{m})$. These are the only integers of $\mathbb{Q}(\sqrt{m})$ if $m \equiv 2$ or 3 (mod 4). If $m \equiv 1$ (mod 4), the numbers $(a + b\sqrt{m})/2$, with odd rational integers a and b, are also integers of $\mathbb{Q}(\sqrt{m})$, and there are no further integers.*

Proof We have already proved the first part of the theorem. All that remains is to identify the algebraic integers. Any number in $\mathbb{Q}(\sqrt{m})$ is of the form $\alpha = (a + b\sqrt{m})/c$ where a, b, c are rational integers with $c > 0$. There is no loss in generality in assuming that $(a, b, c) = 1$ so that α is in its lowest terms. If $b = 0$, then α is rational and, by Theorem 9.9, is an algebraic integer if and only if it is a rational integer, that is $c = 1$. If $b \neq 0$, then α is not rational, and its minimal equation is quadratic,

$$\left(x - \frac{a + b\sqrt{m}}{c}\right)\left(x - \frac{a - b\sqrt{m}}{c}\right) = x^2 - \frac{2a}{c}x + \frac{a^2 - b^2 m}{c^2} = 0.$$

According to Theorem 9.10, α will then be an algebraic integer if and only if this equation is monic with integral coefficients. Thus α is an algebraic integer if and only if

$$c \mid 2a \quad \text{and} \quad c^2 \mid (a^2 - b^2 m), \tag{9.9}$$

and this includes the case $b = 0$, since $(a, b, c) = 1$. If $(a, c) > 1$ and $c \mid 2a$, then a and c have some common prime factor, say p, and $p \nmid b$ since $(a, b, c) = 1$. Then $p^2 \mid a^2$ and $p^2 \mid c^2$, and if $c^2 \mid (a^2 - b^2 m)$, we would have $p^2 \mid b^2 m$, $p^2 \mid m$, which is impossible since m is square-free. Therefore (9.9) can hold only if $(a, c) = 1$. If $c \mid 2a$ and $c > 2$ then $(a, c) > 1$, so that (9.9) can hold only if $c = 1$ or $c = 2$. It is obvious that (9.9) holds for $c = 1$. For $c = 2$ condition (9.9) becomes $a^2 \equiv b^2 m$ (mod 4) and we also have a odd since $(a, c) = 1$. Then (9.9) becomes $b^2m \equiv a^2 \equiv 1 \pmod 4$, which requires that b be odd, and then reduces to $m \equiv b^2m \equiv 1 \pmod 4$. To sum up: (9.9) is satisfied if and only if either $c = 1$ or $c = 2$, a odd, b odd, $m \equiv 1 \pmod 4$, and this completes the proof.

Definition 9.8 *The* norm $N(\alpha)$ *of a number* $\alpha = (a + b\sqrt{m})/c$ *in* $\mathbb{Q}(\sqrt{m})$ *is the product of* α *and its conjugate,* $\bar{\alpha} = (a - b\sqrt{m})/c$,

$$N(\alpha) = \alpha\bar{\alpha} = \frac{a + b\sqrt{m}}{c}\,\frac{a - b\sqrt{m}}{c} = \frac{a^2 - b^2m}{c^2}.$$

Note that by Theorem 9.20 the number α is an integer in $\mathbb{Q}(\sqrt{m})$ if and only if its conjugate $\bar{\alpha}$ is an integer, and that if α is a rational number then $\bar{\alpha} = \alpha$.

Theorem 9.21 *The norm of a product equals the product of the norms,* $N(\alpha\beta) = N(\alpha N(\beta)$. $N(\alpha) = 0$ *if and only if* $\alpha = 0$. *The norm of an integer in* $\mathbb{Q}(\sqrt{m})$ *is a rational integer. If* γ *is an integer in* $\mathbb{Q}(\sqrt{m})$, *then* $N(\gamma) = \pm 1$ *if and only if* γ *is a unit.*

Proof For α and β in $\mathbb{Q}(\sqrt{m})$ it is easy to verify that $(\overline{\alpha\beta}) = \bar{\alpha}\bar{\beta}$. Then we have $N(\alpha\beta) = \alpha\beta\bar{\alpha}\bar{\beta} = \alpha\bar{\alpha}\beta\bar{\beta} = N(\alpha)N(\beta)$. If $\alpha = 0$, then $\bar{\alpha} = 0$ and $N(\alpha) = 0$. Conversely if $N(\alpha) = 0$, then $\alpha\bar{\alpha} = 0$ so that $\alpha = 0$ or $\bar{\alpha} = 0$; but $\bar{\alpha} = 0$ implies $\alpha = 0$.

Next, if γ is an algebraic integer in $\mathbb{Q}(\sqrt{m})$, it has degree either 1 or 2. If it has degree 1, then γ is a rational integer by Theorem 9.9, and $N(\gamma) = \gamma\bar{\gamma} = \gamma^2$ so that $N(\gamma)$ is a rational integer. If γ is of degree 2, then the minimal equation of γ, $x^2 - (\gamma + \bar{\gamma})x + \gamma\bar{\gamma} = 0$, has rational integer coefficients, and again $N(\gamma) = \gamma\bar{\gamma}$ is a rational integer.

If $N(\gamma) = \pm 1$ and γ is an integer, then $\gamma\bar{\gamma} = \pm 1$, $\gamma \mid 1$, so that γ is a unit. To prove the converse, let γ be a unit. Then there is an integer ε such that $\gamma\varepsilon = 1$. This implies $N(\gamma)N(\varepsilon) = N(1) = 1$, so that $N(\gamma) = \pm 1$ since $N(\gamma)$ and $N(\varepsilon)$ are rational integers.

Remark The integers of $\mathbb{Q}(i)$ are often called *Gaussian integers*.

PROBLEMS

1. If an integer α in $\mathbb{Q}(\sqrt{m})$ is neither zero nor a unit, prove that $|N(\alpha)| > 1$.
2. If $m \equiv 1 \pmod 4$, prove that the integers of $\mathbb{Q}(\sqrt{m})$ are all numbers of the form
$$a + b\frac{1 + \sqrt{m}}{2}$$
where a and b are rational integers.

3. If α is any integer, and ε any unit, in $\mathbb{Q}(\sqrt{m})$, prove that $\varepsilon | \alpha$.
4. If α and $\beta \neq 0$ are integers in $\mathbb{Q}(\sqrt{m})$, and if $\alpha | \beta$, prove that $\bar{\alpha} | \bar{\beta}$ and $N(\alpha) | N(\beta)$.
5. If α is an algebraic number in $\mathbb{Q}(\sqrt{m})$ with $m < 0$, prove that $N(\alpha) \geqslant 0$. Show that this is false if $m > 0$.
6. Prove that the following assertion is false in $\mathbb{Q}(i)$: If $N(\alpha)$ is a rational integer, then α is an algebraic integer.
7. Prove that the assertion of the preceding problem is false in every quadratic field. (H)

9.6 UNITS IN QUADRATIC FIELDS

A quadratic field $\mathbb{Q}(\sqrt{m})$ is called *imaginary* if $m < 0$, and it is called *real* if $m > 1$. There are striking differences between these two sorts of quadratic fields. We shall see that an imaginary quadratic field has only a finite number of units; in fact for most of these fields ± 1 are the only units. On the other hand, every real quadratic field has infinitely many units.

Theorem 9.22 *Let m be a negative square-free rational integer. The field $\mathbb{Q}(\sqrt{m})$ has units ± 1, and these are the only units except in the cases $m = -1$ and $m = -3$. The units for $\mathbb{Q}(i)$ are ± 1 and $\pm i$. The units for $\mathbb{Q}(\sqrt{-3})$ are $\pm 1, (1 \pm \sqrt{-3})/2$, and $(-1 \pm \sqrt{-3})/2$.*

Proof Taking note of Theorem 9.21, we look for all integers α in $\mathbb{Q}(\sqrt{m})$ such that $N(\alpha) = \pm 1$. According to Theorem 9.20 we can write α in one of the two forms $x + y\sqrt{m}$ and $(x + y\sqrt{m})/2$ where x and y are rational integers and where, in the second form, x and y are odd and $m \equiv 1 \pmod 4$. Then $N(\alpha) = x^2 - my^2$ or $N(\alpha) = (x^2 - my^2)/4$ respectively. Since m is negative we have $x^2 - my^2 \geqslant 0$ so there are no α with $N(\alpha) = -1$. For $m < -1$ we have $x^2 - my^2 \geqslant -my^2 \geqslant 2y^2$ and the only solutions of $x^2 - my^2 = 1$ are $y = 0$, $x = \pm 1$ in this case. For $m = -1$, the equation $x^2 - my^2 = 1$ has the solutions $x = 0$, $y = \pm 1$, and $x = \pm 1$, $y = 0$ and no others. For $m \equiv 1 \pmod 4$, $m < -3$ there are no solutions of $(x^2 - my^2)/4 = 1$ with odd x and y since $x^2 - my^2 \geqslant 1 - m > 4$. Finally, for $m = -3$, we see that the solutions of the equation $(x^2 + 3y^2)/4 = 1$ with odd x and y are just $x = 1$, $y = \pm 1$, and $x = -1$, $y = \pm 1$. These solutions give exactly the units described in the theorem.

Theorem 9.23 *There are infinitely many units in any real quadratic field.*

Proof The numbers $\alpha = x + y\sqrt{m}$ with integers x, y are integers in $\mathbb{Q}(\sqrt{m})$ with norms $N(\alpha) = x^2 - my^2$. If $x^2 - my^2 = 1$, then α is a unit. But the equation $x^2 - my^2 = 1$, $m > 1$, was treated in Theorems 7.25 and 7.26 where it was proved that it has infinitely many solutions.

PROBLEM

1. Prove that the units of $\mathbb{Q}(\sqrt{2})$ are $\pm(1 + \sqrt{2})^n$ where n ranges over all integers.

9.7 PRIMES IN QUADRATIC FIELDS

Definition 9.9 *An algebraic integer α, not a unit, in a quadratic field $\mathbb{Q}(\sqrt{m})$ is called a* prime *if it is divisible only by its associates and the units of the field.*

This definition is almost the same as the definition of primes among the rational integers. There is this difference, however. In $\mathbb{Q}$ all primes are positive, whereas in $\mathbb{Q}(\sqrt{m})$ no such property is required. Thus if π is a prime and ε is a unit in $\mathbb{Q}(\sqrt{m})$, then $\varepsilon\pi$ is an associated prime in $\mathbb{Q}(\sqrt{m})$. For example, $-\pi$ is an associated prime of π.

Theorem 9.24 *If the norm of an integer α in $\mathbb{Q}(\sqrt{m})$ is $\pm p$, where p is a rational prime, then α is a prime.*

Proof Suppose that $\alpha = \beta\gamma$ where β and γ are integers in $\mathbb{Q}(\sqrt{m})$. By Theorem 9.21 we have $N(\alpha) = N(\beta)N(\gamma) = \pm p$. Then since $N(\beta)$ and $N(\gamma)$ are rational integers, one of them must be ± 1, so that either β or γ is a unit and the other an associate of α. Thus α is a prime.

Theorem 9.25 *Every integer in $\mathbb{Q}(\sqrt{m})$, not zero or a unit, can be factored into a product of primes.*

Proof If α is not a prime, it can be factored into a product $\beta\gamma$ where neither β nor γ is a unit. Repeating the procedure, we factor β and γ if they are not primes. The process of factoring must stop since otherwise we could get α in the form $\beta_1\beta_2 \cdots \beta_n$ with n arbitrarily large, and no factor β_j a unit. But this would imply that

$$N(\alpha) = \prod_{j=1}^{n} N(\beta_j), \quad |N(\alpha)| = \prod_{j=1}^{n} |N(\beta_j)| \geqslant 2^n, \qquad n \text{ arbitrary}$$

since $|N(\beta_j)|$ is an integer > 1.

Although we have established that there is factorization into primes, this factorization may not be unique. In fact, we showed in Section 1.3 that factorization in the field $\mathbb{Q}(\sqrt{-6})$ is not unique. In the next section we prove that factorization is unique in the field $\mathbb{Q}(i)$. The general question of the values of m for which $\mathbb{Q}(\sqrt{m})$ has the unique factorization property is an unsolved problem. There is, however, a close connection between unique factorization and the Euclidean algorithm, as we now show.

Just as in the case of the rational field, a unique factorization theorem will have to disregard the order in which the various prime factors appear. But now a new ambiguity arises due to the existence of associated primes. The two factorings

$$\alpha = \pi_1\pi_2 \cdots \pi_r = (\varepsilon_1\pi_1)(\varepsilon_2\pi_2) \cdots (\varepsilon_r\pi_r)$$

where the ε_j are units with product 1, should be considered as being the same.

Definition 9.10 *A quadratic field* $\mathbb{Q}(\sqrt{m})$ *is said to have the* unique factorization property *if every integer* α *in* $\mathbb{Q}(\sqrt{m})$, *not zero or a unit, can be factored into primes uniquely, apart from the order of the primes and ambiguities between associated primes.*

Definition 9.11 *A quadratic field* $\mathbb{Q}(\sqrt{m})$ *is said to be* Euclidean *if the integers of* $\mathbb{Q}(\sqrt{m})$ *satisfy a Euclidean algorithm, that is, if* α *and* β *are integers of* $\mathbb{Q}(\sqrt{m})$ *with* $\beta \neq 0$, *there exist integers* γ *and* δ *of* $\mathbb{Q}(\sqrt{m})$ *such that* $\alpha = \beta\gamma + \delta$, $|N(\delta)| < |N(\beta)|$.

Theorem 9.26 *Every Euclidean quadratic field has the unique factorization property.*

Proof The proof of this theorem is similar to the procedure used in establishing the fundamental theorem of arithmetic, Theorem 1.16. First we establish that if α and β are any two integers of $\mathbb{Q}(\sqrt{m})$ having no common factors except units, then there exist integers γ_0 and μ_0 in $\mathbb{Q}(\sqrt{m})$ such that $\alpha\lambda_0 + \beta\mu_0 = 1$. Let $\mathscr{S}$ denote the set of integers of the form $\alpha\lambda + \beta\mu$ where λ and μ range over all integers of $\mathbb{Q}(\sqrt{m})$. The norm $N(\alpha\lambda + \beta\mu)$ of any integer in $\mathscr{S}$ is a rational integer, so we can choose an integer, $\alpha\lambda_1 + \beta\mu_1 = \varepsilon$ say, such that $|N(\varepsilon)|$ is the least positive value taken on by $|N(\alpha\lambda + \beta\mu)|$. Applying the Euclidean algorithm to α and ε we get

$$\alpha = \varepsilon\gamma + \delta, \qquad |N(\delta)| < |N(\varepsilon)|.$$

Then we have

$$\delta = \alpha - \varepsilon\gamma = \alpha - \gamma(\alpha\lambda_1 + \beta\mu_1) = \alpha(1 - \gamma\lambda_1) + \beta(-\gamma\mu_1)$$

so that δ is an integer in $\mathscr{S}$. Now this requires $|N(\delta)| = 0$ by the definition of ε, and we have $\delta = 0$ by Theorem 9.21. Thus $\alpha = \varepsilon\gamma$ and hence $\varepsilon|\alpha$. Similarly we find $\varepsilon|\beta$, and therefore ε is a unit. Then ε^{-1} is also a unit by Theorem 9.19, and we have,

$$1 = \varepsilon^{-1}\varepsilon = \varepsilon^{-1}(\alpha\lambda_1 + \beta\mu_1) = \alpha(\varepsilon^{-1}\lambda_1) + \beta(\varepsilon^{-1}\mu_1) = \alpha\lambda_0 + \beta\mu_0.$$

Next we prove that if π is a prime in $\mathbb{Q}(\sqrt{m})$ and if $\pi|\alpha\beta$, then $\pi|\alpha$ or $\pi|\beta$. For if $\pi \nmid \alpha$, then π and α have no common factors except units, and hence there exist integers λ_0 and μ_0 such that $1 = \pi\lambda_0 + \alpha\mu_0$. Then $\beta = \pi\beta\lambda_0 + \alpha\beta\mu_0$ and $\pi|\beta$ because $\pi|\alpha\beta$. This can be extended by mathematical induction to prove that if $\pi|(\alpha_1\alpha_2 \cdots \alpha_n)$, then π divides at least one factor α_j of the product.

From this point on the proof is identical with the first proof of Theorem 1.16, and there is no need to repeat the details.

PROBLEMS

1. If π is a prime and ε a unit in $\mathbb{Q}(\sqrt{m})$, prove that $\varepsilon\pi$ is a prime.
2. Prove that $1 + i$ is a prime in $\mathbb{Q}(i)$.
3. Prove that $11 + 2\sqrt{6}$ is a prime in $\mathbb{Q}(\sqrt{6})$.
4. Prove that 3 is a prime in $\mathbb{Q}(i)$, but not a prime in $\mathbb{Q}(\sqrt{6})$.
5. Prove that there are infinitely many primes in any quadratic field $\mathbb{Q}(\sqrt{m})$.

9.8 UNIQUE FACTORIZATION

In this section we shall apply Theorem 9.26 to various quadratic fields, namely $\mathbb{Q}(i)$, $\mathbb{Q}(\sqrt{-2})$, $\mathbb{Q}(\sqrt{-3})$, $\mathbb{Q}(\sqrt{-7})$, $\mathbb{Q}(\sqrt{2})$, $\mathbb{Q}(\sqrt{3})$. We shall show that these fields have the unique factorization property by proving that they are Euclidean fields. There are other Euclidean quadratic fields, but we focus our attention on these few for which the Euclidean algorithm is easily established.

Theorem 9.27 *The fields $\mathbb{Q}(\sqrt{m})$ for $m = -1, -2, -3, -7, 2, 3,$ are Euclidean and so have the unique factorization property.*

Proof Consider any integers α and β of $\mathbb{Q}(\sqrt{m})$ with $\beta \neq 0$. Then $\alpha/\beta = u + v\sqrt{m}$ where u and v are rational numbers, and we choose rational integers x and y that are closest to u and v, that is, so that

$$0 \leqslant |u - x| \leqslant \tfrac{1}{2}, \qquad 0 \leqslant |v - y| \leqslant \tfrac{1}{2}. \tag{9.10}$$

If we denote $x + y\sqrt{m}$ by γ and $\alpha - \beta\gamma$ by δ, then γ and δ are integers in $\mathbb{Q}(\sqrt{m})$ and $N(\delta) = N(\alpha - \beta\gamma) = N(\beta)N(\alpha/\beta - \gamma) = N(\beta)N((u - x) + (v - y)\sqrt{m}) = N(\beta)\{(u - x)^2 - m(v - y)^2\}$,

$$|N(\delta)| = |N(\beta)|\,|(u - x)^2 - m(v - y)^2|. \tag{9.11}$$

By equations (9.10) we have

$$-\frac{m}{4} \leqslant (u - x)^2 - m(v - y)^2 \leqslant \frac{1}{4} \text{ if } m > 0$$

$$0 \leqslant (u - x)^2 - m(v - y)^2 \leqslant \frac{1}{4} + \frac{1}{4}(-m) \text{ if } m < 0$$

and hence, by (9.11), $|N(\delta)| < |N(\beta)|$ if $m = 2, 3, -1, -2$. Therefore $\mathbb{Q}(\sqrt{m})$ is Euclidean for these values of m.

For the case $m = -3$ and $m = -7$ we must choose γ in a different way. With u and v defined as above, we choose a rational integer s as close to $2v$ as possible and then choose a rational integer r, such that $r \equiv s \pmod 2$, as close to $2u$ as possible. Then we have $|2v - s| \leqslant \frac{1}{2}$ and $|2u - r| \leqslant 1$, and the number $\gamma = (r + s\sqrt{m})/2$ is an integer of $\mathbb{Q}(\sqrt{m})$ by Theorem 9.20, since $m \equiv 1 \pmod 4$ in the cases under discussion. As before, $\delta = \alpha - \beta\gamma$ is an integer in $\mathbb{Q}(\sqrt{m})$ and

$$N(\delta) = N(\beta)N\left(\frac{\alpha}{\beta} - \gamma\right) = N(\beta)\left\{\left(u - \frac{r}{2}\right)^2 - m\left(v - \frac{s}{2}\right)^2\right\},$$

$$|N(\delta)| \leqslant |N(\beta)|\left\{\frac{1}{4} + \frac{1}{16}(-m)\right\} < |N(\beta)|$$

for $m = -3$ and $m = -7$.

PROBLEMS

1. Prove that $\mathbb{Q}(\sqrt{-11})$ has the unique factorization property.
2. Prove that $\mathbb{Q}(\sqrt{5})$ has the unique factorization property.

3. Prove that in $\mathbb{Q}(i)$ the quotient γ and remainder δ obtained in the proof of Theorem 9.27 are not necessarily unique. That is, prove that in $\mathbb{Q}(i)$ there exist integers $\alpha, \beta, \gamma, \delta, \gamma_1, \delta_1$ such that

$$\alpha = \beta\gamma + \delta = \beta\gamma_1 + \delta_1, \qquad N(\delta) < N(\beta),$$

$$N(\delta_1) < N(\beta), \qquad \gamma \neq \gamma_1, \qquad \delta \neq \delta_1.$$

***4.** If α and β are integers of $\mathbb{Q}(i)$, not both zero, say that γ is a *greatest common divisor* of α and β if $N(\gamma)$ is greatest among norms of all common divisors of α and β. Prove that there are exactly four greatest common divisors of any fixed pair α, β, and that each of the four is divisible by any common divisor.

9.9 PRIMES IN QUADRATIC FIELDS HAVING THE UNIQUE FACTORIZATION PROPERTY

If a field $\mathbb{Q}(\sqrt{m})$ has the unique factorization property, we can say much more about the primes of the fields than we did in Section 9.7.

Theorem 9.28 *Let $\mathbb{Q}(\sqrt{m})$ have the unique factorization property. Then to any prime π in $\mathbb{Q}(\sqrt{m})$ there corresponds one and only one rational prime p such that $\pi | p$.*

Proof The prime π is a divisor of the rational integer $N(\pi)$, and hence there exist positive rational integers divisible by π. Let n be the least of these. Then n is a rational prime. For otherwise $n = n_1 n_2$, and we have, by the unique factorization property, $\pi | n$, $\pi | (n_1 n_2)$, $\pi | n_1$ or $\pi | n_2$, a contradiction since $0 < n_1 < n$, $0 < n_2 < n$. Hence n is a rational prime, call it p. And, if π were a divisor of another rational prime q, we could find rational integers by Theorem 1.3 such that $1 = px + qy$. Since $\pi | (px + qy)$ this implies $\pi | 1$, which is false, and hence the prime p is unique.

Theorem 9.29 *Let $\mathbb{Q}(\sqrt{m})$ have the unique factorization property. Then:*

(1) Any rational prime p is either a prime π of the field or a product $\pi_1\pi_2$ of two primes, not necessarily distinct, of $\mathbb{Q}(\sqrt{m})$.

(2) The totality of primes π, π_1, π_2 obtained by applying part 1 to all rational primes, together with their associates, constitute the set of all primes of $\mathbb{Q}(\sqrt{m})$.

(3) *An odd rational prime p satisfying* $(p, m) = 1$ *is a product* $\pi_1\pi_2$ *of two primes in* $\mathbb{Q}(\sqrt{m})$ *if and only if* $\left(\frac{m}{p}\right) = 1$. *Furthermore if* $p = \pi_1\pi_2$, *the product of two primes, then* π_1 *and* π_2 *are not associates, but* π_1 *and* $\bar{\pi}_2$ *are, and* π_2 *and* $\bar{\pi}_1$ *are.*

(4) *If* $(2, m) = 1$, *then 2 is the associate of a square of a prime if* $m \equiv 3 \pmod 4$; *2 is a prime if* $m \equiv 5 \pmod 8$; *and 2 is the product of two distinct primes if* $m \equiv 1 \pmod 8$.

(5) *Any rational prime p that divides m is the associate of the square of a prime in* $\mathbb{Q}(\sqrt{m})$.

Proof (*1*) If the rational prime p is not a prime in $\mathbb{Q}(\sqrt{m})$, then $p = \pi\beta$ for some prime π and some integer β of $\mathbb{Q}(\sqrt{m})$. Then we have $N(\pi)N(\beta) = N(p) = p^2$. Since $N(\pi) \neq \pm 1$, we must have either $N(\beta) = \pm 1$ or $N(\beta) = \pm p$. If $N(\beta) = \pm 1$, then β is a unit by Theorem 9.21, and π is an associate of p, which then must be a prime in $\mathbb{Q}(\sqrt{m})$. If $N(\beta) = \pm p$ then β is a prime by Theorem 9.24, and so p is a product $\pi\beta$ of two primes in $\mathbb{Q}(\sqrt{m})$.

(*2*) The statement (2) now follows directly from Theorem 9.28 and statement (1).

(*3*) If p is an odd rational prime such that $(p, m) = 1$ and $\left(\frac{m}{p}\right) = 1$, there exists a rational integer x satisfying

$$x^2 \equiv m \pmod p, \qquad p \mid (x^2 - m), \qquad p \mid (x - \sqrt{m})(x + \sqrt{m}).$$

If p were a prime of $\mathbb{Q}(\sqrt{m})$, it would divide one of the factors $x - \sqrt{m}$ and $x + \sqrt{m}$, so that one of

$$\frac{x}{p} - \frac{\sqrt{m}}{p}, \qquad \frac{x}{p} + \frac{\sqrt{m}}{p}$$

would be an integer in $\mathbb{Q}(\sqrt{m})$. But this is impossible by Theorem 9.20, and hence p is not a prime in $\mathbb{Q}(\sqrt{m})$. Therefore, by statement (1), $p = \pi_1\pi_2$ if $\left(\frac{m}{p}\right) = 1$.

Now suppose that p is an odd rational prime, that $(p, m) = 1$, and that p is not a prime in $\mathbb{Q}(\sqrt{m})$. Then from the proof of statement (1) we see that $p = \pi\beta$, $N(\beta) = \pm p$, and $N(\pi) = \pm p$. We can write $\pi = a + b\sqrt{m}$ where a and b are rational integers or, if $m \equiv 1 \pmod 4$, halves of odd rational integers. Then $a^2 - mb^2 = N(\pi) = \pm p$, and we have $(2a)^2 - m(2b)^2 = \pm 4p$, $(2a)^2 \equiv m(2b)^2 \pmod p$. Here $2a$ and $2b$ are rational integers and neither is a multiple of p, for if p divided either one it would divide the other and we would have $p^2|4a^2$, $p^2|4b^2$, $p^2|(4a^2 - 4mb^2)$, $p^2|4p$. Therefore $(2b, p) = 1$, and there is a rational integer w such that $2bw \equiv 1 \pmod{p}$,

$$(2aw)^2 \equiv m(2bw)^2 \equiv m \pmod{p}, \text{ and we have } \left(\frac{m}{p}\right) = 1.$$

Furthermore, with the notation of the preceding paragraph we prove that π and β are not associates, but π and $\bar{\beta}$ are, and $\bar{\pi}$ and β are. From $p = \pi\beta$ and $N(\pi) = a^2 - mb^2 = \pm p$ we have

$$\beta = \frac{p}{\pi} = \frac{p}{a + b\sqrt{m}} = \pm(a - b\sqrt{m}), \qquad \bar{\beta} = \pm(a + b\sqrt{m})$$

so π and $\bar{\beta}$ are associates. On the other hand we note that

$$\frac{\pi}{\beta} = \pm\frac{a + b\sqrt{m}}{a - b\sqrt{m}} = \frac{(2a)^2 + m(2b)^2}{4p} + \frac{8ab\sqrt{m}}{4p}$$

and this is not an integer, and so not a unit, because p does not divide $8ab$. Thus π and β are not associates.

(4) If $m \equiv 3 \pmod{4}$, then

$$m^2 - m = 2\frac{m^2 - m}{2} = (m + \sqrt{m})(m - \sqrt{m})$$

and $2 \nmid (m \pm \sqrt{m})$, so 2 cannot be a prime of $\mathbb{Q}(\sqrt{m})$. Hence 2 is divisible by a prime $x + y\sqrt{m}$ and this prime must have norm ± 2. Therefore $x^2 - my^2 = \pm 2$. But this implies that

$$\pm\frac{x - y\sqrt{m}}{x + y\sqrt{m}} = \pm\frac{x^2 + my^2 - 2xy\sqrt{m}}{x^2 - my^2} = \frac{x^2 + my^2}{2} - xy\sqrt{m}$$

and, similarly

$$\pm\frac{x + y\sqrt{m}}{x - y\sqrt{m}} = \frac{x^2 + my^2}{2} + xy\sqrt{m}$$

and therefore $(x - y\sqrt{m})(x + y\sqrt{m})^{-1}$ and its inverse are integers of $\mathbb{Q}(\sqrt{m})$. Hence $(x - y\sqrt{m})(x + y\sqrt{m})^{-1}$ is a unit, and $x - y\sqrt{m}$ and $x + y\sqrt{m}$ are associates.

If $m \equiv 1 \pmod 4$ and if 2 is not a prime in $\mathbb{Q}(\sqrt{m})$ then 2 is divisible by a prime $\frac{1}{2}(x + y\sqrt{m})$ having norm ± 2. This would mean that there are rational integers x and y, both even or both odd such that

$$x^2 - my^2 = \pm 8. \tag{9.12}$$

If x and y are even, say $x = 2x_0$, $y = 2y_0$, then (9.12) would require $x_0^2 - my_0^2 = \pm 2$. But, since $m \equiv 1 \pmod 4$, $x_0^2 - my_0^2$ is either odd or a multiple of 4. Thus (9.12) can have solutions only with odd x and y. Then $x^2 \equiv y^2 \equiv 1 \pmod 8$, and (9.12) implies

$$x^2 - my^2 \equiv 1 - m \equiv 0, \qquad m \equiv 1 \pmod 8.$$

It follows that 2 is a prime in $\mathbb{Q}(\sqrt{m})$ if $m \equiv 5 \pmod 8$.

Now if $m \equiv 1 \pmod 8$ we observe that

$$\frac{1-m}{4} = 2\frac{1-m}{8} = \frac{1-\sqrt{m}}{2} \cdot \frac{1+\sqrt{m}}{2}$$

and $2 \nmid (1 \pm \sqrt{m})/2$, so 2 cannot be a prime in $\mathbb{Q}(\sqrt{m})$. Hence (9.12) has solutions in odd integers x and y. Now the primes $\frac{1}{2}(x + y\sqrt{m})$ and $\frac{1}{2}(x - y\sqrt{m})$ are not associates in $\mathbb{Q}(\sqrt{m})$ because their quotient is not a unit. In fact their quotient is

$$\frac{x + y\sqrt{m}}{x - y\sqrt{m}} = \pm \frac{x^2 + my^2}{8} \pm \frac{xy\sqrt{m}}{4}$$

which is not even an integer in $\mathbb{Q}(\sqrt{m})$.

(5) Let p be a rational prime divisor of m. If $p = |m|$ then $p = \pm\sqrt{m} \cdot \sqrt{m}$ and hence p is the associate of the square of a prime in $\mathbb{Q}(\sqrt{m})$ by Theorem 9.24. If $p < |m|$, we note that

$$m = p\frac{m}{p} = \sqrt{m} \cdot \sqrt{m}. \tag{9.13}$$

But p is not a divisor of $\sqrt{m}$ in $\mathbb{Q}(\sqrt{m})$ by Theorem 9.20 and hence p is not a prime in $\mathbb{Q}(\sqrt{m})$. Therefore p is divisible by a prime π, with $N(\pi) = \pm p$, and hence is not a divisor of m/p. But, by (9.13), π is also a divisor of $\sqrt{m}$, π^2 is a divisor of m, and hence π^2 is a divisor of p.

The theorem we have just proved provides a method for determining the primes of a quadratic field having the unique factorization property. For such $\mathbb{Q}(\sqrt{m})$ we look at all the rational primes p. Those p for which

$(p, 2m) = 1$ and $\left(\dfrac{m}{p}\right) = -1$, together with all their associates in $\mathbb{Q}(\sqrt{m})$, are primes in $\mathbb{Q}(\sqrt{m})$. Those p for which $(p, 2m) = 1$ and $\left(\dfrac{m}{p}\right) = +1$ will factor into $p = \pi_1\pi_2$, a product of two primes of $\mathbb{Q}(\sqrt{m})$, with $N(\pi_1) = N(\pi_2) = \pm p$. Any other factoring of p will merely replace π_1 and π_2 by associates. The primes p for which $(p, 2m) > 1$ will either be primes of $\mathbb{Q}(\sqrt{m})$ or products of two primes of $\mathbb{Q}(\sqrt{m})$.

Suppose that α is an integer in $\mathbb{Q}(\sqrt{m})$ and that $N(\alpha) = \pm p$, p a rational prime. Then $\bar{\alpha}$ is also an integer in $\mathbb{Q}(\sqrt{m})$ and $\alpha\bar{\alpha} = N(\alpha) = \pm p$, and this necessitates that $\bar{\alpha}$ be a prime in $\mathbb{Q}(\sqrt{m})$. If $m \not\equiv 1 \pmod 4$, we can write $\alpha = x + y\sqrt{m}$, $N(\alpha) = x^2 - my^2$, with integers x and y. If $m \equiv 1 \pmod 4$, we can write $\alpha = (x + y\sqrt{m})/2$, $4N(\alpha) = x^2 - my^2$, with x and y integers, both odd or both even.

Combining these facts we have following. Let $\mathbb{Q}(\sqrt{m})$ have the unique factorization property, and let p be a rational prime such that $(p, 2m) = 1$, $\left(\dfrac{m}{p}\right) = +1$. Then if $m \not\equiv 1 \pmod 4$, one at least of the two equations $x^2 - my^2 = \pm p$ has a solution. Let $x = a$, $y = b$ be such a solution. Then the numbers $\alpha = a + b\sqrt{m}$, $\bar{\alpha} = a - b\sqrt{m}$, and the associates of α and $\bar{\alpha}$ are primes in $\mathbb{Q}(\sqrt{m})$, and these are the only primes in $\mathbb{Q}(\sqrt{m})$ that divide p. On the other hand, if $m \equiv 1 \pmod 4$, one at least of the two equations $x^2 - my^2 = \pm 4p$ has a solution with x and y both odd or both even. Again denoting such a solution by $x = a$, $y = b$, we can say that the numbers $\alpha = (a + b\sqrt{m})/2$, $\bar{\alpha} = (a - b\sqrt{m})/2$, and their associates are primes in $\mathbb{Q}(\sqrt{m})$, and these are the only primes in $\mathbb{Q}(\sqrt{m})$ that divide p. It is worth noting that our consideration of algebraic number fields has thus given us information concerning Diophantine equations.

It must be remembered that these results apply only to those $\mathbb{Q}(\sqrt{m})$ that have the unique factorization property.

Example 1 $m = -1$. Gaussian primes. The field is $\mathbb{Q}(i)$ and we have

$$2m = -2, \qquad 1^2 + 1^2 = 2, \qquad \overline{1 + i} = 1 - i$$

$$\left(\frac{m}{p}\right) = \begin{cases} +1 \text{ if } p = 4k + 1 \\ -1 \text{ if } p = 4k + 3. \end{cases}$$

For each rational prime p of the form $4k + 1$ the equation $x^2 + y^2 = p$ has a solution since $x^2 + y^2 = -p$ is clearly impossible. For each such p choose a solution $x = a_p$, $y = b_p$.

The primes in $\mathbb{Q}(i)$ are $1 + i$, all rational primes $p = 4k + 3$, all $a_p + ib_p$, all $a_p - ib_p$, together with all their associates. Note that $1 - i = \overline{1 + i}$ has not been included since $1 - i = -i(1 + i)$, i is a unit of $\mathbb{Q}(i)$, and hence $1 - i$ is an associate of $1 + i$.

Example 2 $m = -3$. The field is $\mathbb{Q}(\sqrt{-3})$ and we have

$$2m = -6, \qquad x^2 + 3y^2 = \pm 4 \cdot 2 \text{ has no solution}$$

$$3^2 + 3 \cdot 1^2 = 4 \cdot 3, \qquad \overline{\frac{3+\sqrt{-3}}{2}} = \frac{3-\sqrt{-3}}{2}$$

$$\left(\frac{m}{p}\right) = \begin{cases} +1 \text{ if } p = 3k+1, (p,6) = 1 \\ -1 \text{ if } p = 3k+2, (p,6) = 1. \end{cases}$$

For each odd $p = 3k + 1$, choose a_p, b_p such that $a_p^2 + 3b_p^2 = 4p$.

The primes in $\mathbb{Q}(\sqrt{-3})$ are 2, $(3 + \sqrt{-3})/2$, all odd rational primes $p = 3k + 2$, all $(a_p + b_p\sqrt{-3})/2$, all $(a_p - b_p\sqrt{-3})/2$, together with all their associates. Here, again, we omit $(3 - \sqrt{-3})/2$ because it can be shown to be an associate of $(3 + \sqrt{-3})/2$. We could have included 2 among the $p = 3k + 2$ by just omitting the word "odd."

Example 3 Prove that the field $\mathbb{Q}(\sqrt{-14})$ does not have the unique factorization property.

By Theorem 9.29, part 5, the integer 2 factors into two primes if this field has the unique factorization property. So it suffices to prove that 2 is a prime. Suppose that 2 is not a prime in the field, so that $2 = \pm(a + b\sqrt{-14})(a - b\sqrt{-14})$ for some integers a and b. This gives $2 = \pm(a^2 + 14b^2)$, which is easily shown to be impossible in integers.

Applications to Diophantine Equations The problem of finding all solutions of $x^2 + y^2 = z^2$ in rational integers was settled in Theorem 5.5. In the introduction to the present chapter, this equation is reexamined by use of the factoring $(x + yi)(x - yi) = z^2$. We now look a little more carefully at the steps used. It is presumed that $(x, y, z) = 1$, so that *primitive* solutions are sought. We now prove that there is no prime α in $\mathbb{Q}(i)$ that divides both $x + yi$ and $x - yi$. If there were such a prime divisor, it would divide the sum $2x$ and the difference $2yi$. But $(x, y, z) = 1$ implies $(x, y) = 1$ and hence $\alpha \mid 2$. This means that $\alpha = 1 + i$. It is very easy to prove that $1 + i$ is a divisor of $x + yi$ if and only if x and y are both even or both odd. But $(x, y) = 1$, so this leads to the conclusion that x and y are odd, and then

$$z^2 \equiv x^2 + y^2 \equiv 1 + 1 \equiv 2 \pmod 4$$

which is impossible because any square is of the form $4k$ or the form $4k + 1$.

Thus $x + yi$ and $x - yi$ have no common prime factor in $\mathbb{Q}(i)$, and since their product is z^2, it follows that $x + yi$ is the product of a unit and a perfect square,

$$x + yi = \pm(r + si)^2 \qquad \text{or} \qquad x + yi = \pm i(r + si)^2.$$

It is easy to finish this analysis by equating the real and nonreal parts here. The first equation, for example, implies that

$$x = \pm(r^2 - s^2), \qquad y = \pm 2rs.$$

We do not pursue the details here, because what emerges is just a variation on the solutions found in Theorem 5.5.

As a second example, consider the equation

$$x^2 + y^2 = 2z^2.$$

Again we seek primitive solutions in rational integers, so that $(x, y, z) = 1$. It follows that $(x, y) = 1$, and since $x^2 + y^2$ is even, we see that x and y are odd. From this we conclude, as noted above, that $x + yi$ is divisible by $1 + i$, say $x + yi = (1 + i)(u + vi)$. Equating the real and nonreal parts gives $x = u - v$, $y = u + v$. The equation $x^2 + y^2 = 2z^2$ is thereby reduced to $u^2 + v^2 = z^2$. Now we are on familiar ground, because this equation is analyzed completely in Theorem 5.5. Hence the solutions of $x^2 + y^2 = 2z^2$ can be obtained from those of $u^2 + v^2 = z^2$ by the use of $x = u - v$, $y = u + v$. The details are omitted.

For a third example of the application of the theory to Diophantine equations, we prove that the only solutions of

$$y^2 + 2 = x^3$$

in rational integers are $x = 3$, $y = \pm 5$. First we note that x and y must be odd, since if y is even, then x is even, and the equation is impossible modulo 4. The equation is now studied in the field $\mathbb{Q}(\sqrt{-2})$, where it can be written as

$$(y + \sqrt{-2})(y - \sqrt{-2}) = x^3.$$

Since x is odd, it is not divisible by the prime $\sqrt{-2}$, and so $\sqrt{-2}$ is not a divisor of $y + \sqrt{-2}$ or $y - \sqrt{-2}$. Note that here we are using the unique factorization property of the field $\mathbb{Q}(\sqrt{-2})$, by Theorem 9.27. What we want to establish from this equation is that $y + \sqrt{-2}$ and $y - \sqrt{-2}$ are perfect cubes. Since by Theorem 9.20 neither $y + \sqrt{-2}$

nor $y - \sqrt{-2}$ is divisible by any rational integer $k > 1$, it follows that any prime divisor of $y + \sqrt{-2}$ is of the form $r + s\sqrt{-2}$, where r and s are nonzero rational integers. Then $r - s\sqrt{-2}$ is also a prime, not an associate of $r + s\sqrt{-2}$ by part 3 of Theorem 9.29. Although $r - s\sqrt{-2}$ is a divisor of $y - \sqrt{-2}$, we prove that $r + s\sqrt{-2}$ is not such a divisor. If it were, then the product $(r + s\sqrt{-2})(r - s\sqrt{-2})$ would also be a divisor of $y - \sqrt{-2}$. But the product is $r^2 + 2s^2$, a rational integer > 1, and we have already seen that such a divisor is not possible.

Now the prime divisor $r + s\sqrt{-2}$ of $y + \sqrt{-2}$ is also a divisor of x, and so $(r + s\sqrt{-2})^3$ is a divisor of $y + \sqrt{-2}$. Grouping all the prime divisors of $y + \sqrt{-2}$, we can write

$$y + \sqrt{-2} = (a + b\sqrt{-2})^3$$

for some rational integers a and b, because the units of the field are the perfect cubes ± 1 by Theorem 9.20. Equating the coefficients of $\sqrt{-2}$ here, we get $1 = b(3a^2 - 2b^2)$, the only solutions of which are $b = 1$, $a = \pm 1$, giving $x = 3$, $y = \pm 5$.

The unique factorization property is of central importance in the argument just given. For example, if an analysis similar to that above is applied to $y^2 + 47 = x^3$, assuming unique factorization in $\mathbb{Q}(\sqrt{-47})$, the procedure does not turn up all solutions in integers. The reason for this is that $\mathbb{Q}(\sqrt{-47})$ does *not* have the unique factorization property, as can be seen by examining 2 as a possible prime in the field, exactly as in the case of $\mathbb{Q}(\sqrt{-14})$ in the preceding example.

PROBLEMS

1. In Example 2, where $m = -3$, we know from the theory that if p is any prime of the form $3k + 1$, then there are integers x and y such that $x^2 + 3y^2 = 4p$. Let $x = 2u - y$ and establish that any such prime can be expressed in the form $u^2 - uy + y^2$.
2. The rational prime 13 can be factored in two ways in $\mathbb{Q}(\sqrt{-3})$,
$$13 = \frac{7 + \sqrt{-3}}{2} \cdot \frac{7 - \sqrt{-3}}{2} = (1 + 2\sqrt{-3})(1 - 2\sqrt{-3}).$$
Prove that this is not in conflict with the fact that $\mathbb{Q}(\sqrt{-3})$ has the unique factorization property.
3. Prove that $\sqrt{3} - 1$ and $\sqrt{3} + 1$ are associates in $\mathbb{Q}(\sqrt{3})$.
4. Prove that the primes of $\mathbb{Q}(\sqrt{3})$ are $\sqrt{3} - 1$, $\sqrt{3}$, all rational primes $p \equiv \pm 5 \pmod{12}$, all factors $a + b\sqrt{3}$ of rational primes $p \equiv \pm 1 \pmod{12}$, and all associates of these primes.

5. Prove that the primes of $\mathbb{Q}(\sqrt{2})$ are $\sqrt{2}$, all rational primes of the form $8k + 3$, and all factors $a + b\sqrt{2}$ of rational primes of the form $8k \pm 1$, and all associates of these primes.

***6.** Prove that if m is square-free, $m < -1$, $|m|$ not a prime, then $\mathbb{Q}(\sqrt{m})$ does not have the unique factorization property. (H)

7. Find all solutions of $y^2 + 1 = x^3$ in rational integers.

9.10 THE EQUATION $x^3 + y^3 = z^3$

We shall prove that $x^3 + y^3 = z^3$ has no solutions in positive rational integers x, y, z. Even more, it will be established that $\alpha^3 + \beta^3 + \gamma^3 = 0$ has no solutions in nonzero integers in the quadratic field $\mathbb{Q}(\sqrt{-3})$. Note that this amounts to proving that $\alpha^3 + \beta^3 = \gamma^3$ has no solutions in nonzero integers of $\mathbb{Q}(\sqrt{-3})$, because this equation can be written as $\alpha^3 + \beta^3 + (-\gamma)^3 = 0$.

For convenience throughout this discussion we denote $(-1 + \sqrt{-3})/2$ by ω, which satisfies the equations $\omega^2 + \omega + 1 = 0$ and $\omega^3 = 1$. In this notation the units of $\mathbb{Q}(\sqrt{-3})$ are $\pm 1, \pm\omega, \pm\omega^2$, as given in Theorem 9.22. Also, in this field the integer $\sqrt{-3}$ is a prime, by Theorem 9.24. Because this prime plays a central role in the discussion we denote it by θ. Multiplying θ by the six units, we observe that the associates of θ are

$$\pm(1-\omega), \qquad \pm(1-\omega^2), \qquad \pm(\omega - \omega^2) = \pm\theta = \pm\sqrt{-3}. \quad (9.14)$$

Lemma 9.30 *Every integer in* $\mathbb{Q}(\sqrt{-3})$ *is congruent to exactly one of* 0, $+1$, -1 *modulo* θ.

Proof Consider any integer $(a + b\theta)/2$ in $\mathbb{Q}(\sqrt{-3})$, where a and b are rational integers, both even or both odd. Then $(b + a\theta)/2$ is also an integer, and so

$$\tfrac{1}{2}(a + b\theta) = \tfrac{1}{2}(b + a\theta)\theta + 2a \equiv 2a \pmod{\theta}.$$

Now the rational integer $2a$ is congruent to 0, 1, or -1 modulo 3, and $\theta | 3$, so the lemma is proved.

Lemma 9.31 *Let* ξ *and* η *be integers of* $\mathbb{Q}(\sqrt{-3})$, *not divisible by* θ. *If* $\xi \equiv 1 \pmod{\theta}$ *then* $\xi^3 \equiv 1 \pmod{\theta^4}$. *If* $\xi \equiv -1 \pmod{\theta}$ *then* $\xi^3 \equiv -1 \pmod{\theta^4}$. *If* $\xi^3 + \eta^3 \equiv 0 \pmod{\theta}$ *then* $\xi^3 + \eta^3 \equiv 0 \pmod{\theta^4}$. *Finally if* $\xi^3 - \eta^3 \equiv 0 \pmod{\theta}$ *then* $\xi^3 - \eta^3 \equiv 0 \pmod{\theta^4}$.

Proof From Lemma 9.30 it follows that $\xi \equiv \pm 1 \pmod{\theta}$. First if $\xi \equiv +1 \pmod{\theta}$ then $\xi = 1 + \beta\theta$ for some integer β. Then

$$\xi^3 = (1 + \beta\theta)^3 = 1 + 3\beta\theta - 9\beta^2 + \beta^3\theta^3 \equiv 1 + 3\beta\theta + \beta^3\theta^3 \pmod{\theta^4}$$

because $\theta^4 = 9$. Also we note that

$$3\beta\theta + \beta^3\theta^3 = \theta^3(\beta^3 - \beta) = \theta^3(\beta)(\beta - 1)(\beta + 1).$$

But θ is a divisor of $\beta(\beta - 1)(\beta + 1)$ by Lemma 9.30 and hence $\xi^3 \equiv 1 \pmod{\theta^4}$. Second if $\xi \equiv -1 \pmod{\theta}$ then $(-\xi) \equiv 1 \pmod{\theta}$, $(-\xi)^3 \equiv 1 \pmod{\theta^4}$ and $\xi^3 \equiv -1 \pmod{\theta^4}$.

Now $\xi^3 \equiv \xi \pmod{\theta}$ because θ is a divisor of $\xi(\xi - 1)(\xi + 1)$, so $\xi^3 + \eta^3 \equiv 0 \pmod{\theta}$ implies $\xi + \eta \equiv 0 \pmod{\theta}$. If $\xi \equiv 1 \pmod{\theta}$ then $\eta \equiv -1 \pmod{\theta}$ and hence $\xi^3 + \eta^3 \equiv 1 - 1 \equiv 0 \pmod{\theta^4}$. Finally if $\xi^3 - \eta^3 \equiv 0 \pmod{\theta}$ then $\xi^3 + (-\eta)^3 \equiv 0 \pmod{\theta}$ and so $\xi^3 + (-\eta)^3 \equiv 0 \pmod{\theta^4}$.

Lemma 9.32 *Suppose there are integers α, β, γ of $\mathbb{Q}(\sqrt{-3})$ such that $\alpha^3 + \beta^3 + \gamma^3 = 0$. If g.c.d. $(\alpha, \beta, \gamma) = 1$ then θ divides one and only one of α, β, γ.*

Proof Suppose that θ divides none of α, β, γ. Then by Lemma 9.31,

$$0 = \alpha^3 + \beta^3 + \gamma^3 \equiv \pm 1 \pm 1 \pm 1 \pmod{\theta^4}.$$

Considering all possible combinations of signs, we conclude that θ^4 is a divisor of 3, 1, -1, or -3. But $\theta^4 = 9$, and hence we conclude that θ divides at least one of α, β, γ.

Furthermore if θ divides any two of them, it must divide the third, contrary to hypothesis.

Lemma 9.33 *Suppose there are nonzero integers α, β, γ of $\mathbb{Q}(\sqrt{-3})$, with $\theta \nmid \alpha\beta\gamma$, and units $\varepsilon_1, \varepsilon_2$, and a positive rational integer r such that*

$$\alpha^3 + \varepsilon_1\beta^3 + \varepsilon_2(\theta^r\gamma)^3 = 0.$$

Then $\varepsilon_1 = \pm 1$ and $r \geqslant 2$.

Proof Since $r > 0$ we see that $\alpha^3 + \varepsilon_1\beta^3 \equiv 0 \pmod{\theta^3}$. Using Lemma 9.31 we see that $\alpha^3 + \varepsilon_1\beta^3 \equiv \pm 1 + \varepsilon_1(\pm 1) \equiv 0 \pmod{\theta^3}$. The unit ε_1 is one of $\pm 1, \pm\omega, \pm\omega^2$, so $\pm 1 + \varepsilon_1(\pm 1)$ is one of $2, 0, -2, \pm(1 \pm \omega)$, $\pm(1 \pm \omega^2)$ with all possible combinations of signs. But θ^3 divides none of these except 0, because $1 - \omega$ and $1 - \omega^2$ are associates of θ, $1 + \omega =$

$-\omega^2$ and $1 + \omega^2 = -\omega$ are units, and $N(\pm 2) = 4$ whereas $N(\theta^3) = 27$. It follows that $\pm 1 + \varepsilon_1(\pm 1) = 0$, so $\varepsilon_1 = \pm 1$.

By Lemma 9.31, $\alpha^3 + \varepsilon_1\beta^3 \equiv 0 \pmod{\theta^3}$ implies $\alpha^3 + \varepsilon_1\beta^3 \equiv 0 \pmod{\theta^4}$. From this it follows that θ^4 is a divisor of $\varepsilon_2(\theta^r\gamma)^3$ and $r \geqslant 2$.

Lemma 9.34 *There do not exist nonzero integers α, β, γ in $\mathbb{Q}(\sqrt{-3})$, a unit ε, and a rational integer $r \geqslant 2$ such that*

$$\alpha^3 + \beta^3 + \varepsilon(\theta^r\gamma)^3 = 0. \tag{9.15}$$

Proof We may presume that g.c.d.$(\alpha, \beta, \theta^r\gamma) = 1$, and that $\theta \nmid \gamma$. Furthermore, θ does not divide both α and β, and so, interchanging α and β if necessary, we may presume that $\theta \nmid \beta$. If there are integers satisfying (9.15) select a set such that

$$N(\alpha^3\beta^3\theta^{3r}\gamma^3) \tag{9.16}$$

is a minimum. This can be done because every norm in $\mathbb{Q}(\sqrt{-3})$ is a nonnegative integer. Note that ε in (9.15) is omitted in (9.16) because $N(\varepsilon) = +1$. We now construct a solution of (9.15) with a smaller norm in (9.16), and this will establish the lemma.

Since $r \geqslant 2$, we have $\alpha^3 + \beta^3 \equiv 0 \pmod{\theta^6}$. Also

$$\alpha^3 + \beta^3 = (\alpha + \beta)(\alpha + \omega\beta)(\alpha + \omega^2\beta). \tag{9.17}$$

We first prove that if any prime π divides any two of $\alpha + \beta$, $\alpha + \omega\beta$, and $\alpha + \omega^2\beta$, it must be an associate of θ. First if $\pi|(\alpha + \beta)$ and $\pi|(\alpha + \omega\beta)$ then $\pi|\beta(1 - \omega)$ and $\pi|\alpha(1 - \omega)$. But g.c.d.$(\alpha, \beta) = 1$ and $1 - \omega$ is an associate of θ by (9.14). Second if $\pi|(\alpha + \beta)$ and $\pi|(\alpha + \omega^2\beta)$ then $\pi|\beta(1 - \omega^2)$ and $\pi|\alpha(1 - \omega^2)$. Again we see that $\pi|(1 - \omega^2)$ and so $\pi|\theta$ by (9.14). Third if $\pi|(\alpha + \omega\beta)$ and $\pi|(\alpha + \omega^2\beta)$ then $\pi|\beta(\omega - \omega^2)$ and $\pi|\alpha(\omega - \omega^2)$, and again by (9.14) we get $\pi|\theta$.

Furthermore, because of (9.14) and the fact that $\theta \nmid \beta$, we notice that the differences between $\alpha + \beta$, $\alpha + \omega\beta$, and $\alpha + \omega^2\beta$ are divisible by θ, but not by θ^2. The product of these three is divisible by θ^6, as in (9.17). Hence if $\theta^a, \theta^b, \theta^c$ are the highest powers of θ dividing $\alpha + \beta$, $a + \omega\beta$, and $\alpha + \omega^2\beta$, respectively, then from this argument and (9.15) we conclude that a, b, c are $1, 1, 3r - 2$ in some order, and

$$\frac{\alpha + \beta}{\theta^a}, \qquad \frac{\alpha + \omega\beta}{\theta^b}, \qquad \frac{\alpha + \omega^2\beta}{\theta^c}$$

are integers with no common prime factor in $\mathbb{Q}(\sqrt{-3})$. And (9.15) can be written as

$$\frac{\alpha + \beta}{\theta^a} \cdot \frac{\alpha + \omega\beta}{\theta^b} \cdot \frac{\alpha + \omega^2\beta}{\theta^c} = -\varepsilon\gamma^3 \tag{9.18}$$

so each of the factors on the left is an associate of the cube of an integer, say

$$\alpha + \beta = \varepsilon_1\theta^a\lambda_1^3, \qquad \alpha + \omega\beta = \varepsilon_2\theta^b\lambda_2^3, \qquad \alpha + \omega^2\beta = \varepsilon_3\theta^c\lambda_3^3 \tag{9.19}$$

where $\varepsilon_1, \varepsilon_2, \varepsilon_3$ are units. Also we note that

$$(\alpha + \beta) + \omega(\alpha + \omega\beta) + \omega^2(\alpha + \omega^2\beta) = (\alpha + \beta)(1 + \omega + \omega^2) = 0,$$

and so

$$\varepsilon_1\theta^a\lambda_1^3 + \varepsilon_4\theta^b\lambda_2^3 + \varepsilon_5\theta^c\lambda_3^3 = 0 \tag{9.20}$$

where $\varepsilon_4 = \omega\varepsilon_2$ and $\varepsilon_5 = \omega^2\varepsilon_3$.

Thus ε_4 and ε_5 are units, and (9.20) is symmetric in the three terms on the left side of the equation. Thus we can assign the values 1, 1, $3r - 2$ to a, b, c in any order, say $a = 1$, $b = 1$, $c = 3r - 2$. Substituting these values in (9.20) and dividing by $\varepsilon_1\theta$ we get

$$\lambda_1^3 + \varepsilon_6\lambda_2^3 + \varepsilon_7(\theta^{r-1}\lambda_3)^3 = 0 \tag{9.21}$$

where ε_6 and ε_7 are the units $\varepsilon_4/\varepsilon_1$ and $\varepsilon_5/\varepsilon_1$. Since $\gamma \neq 0$ we see that $\lambda_1\lambda_2\lambda_3 \neq 0$ from (9.18) and (9.19). Also $\theta \nmid (\lambda_1\lambda_2\lambda_3)$ so by Lemma 9.33 we conclude that $\varepsilon_6 = \pm 1$ and $r - 1 \geqslant 2$. But (9.21) is of the form (9.15) because $\varepsilon_6\lambda_2^3$ is either λ_2^3 or $(-\lambda_2)^3$. Taking the norm analogous to (9.16) we have by (9.19), (9.18), and $a + b + c = 3r$,

$$\begin{aligned} N(\lambda_1^3\lambda_2^3\theta^{3r-3}\lambda_3^3) &= N(\theta^{-3}(\alpha + \beta)(\alpha + \omega\beta)(\alpha + \omega^2\beta)) \\ &= N(\theta^{3r-3}\gamma^3) < N(\alpha^3\beta^3\theta^{3r}\gamma^3) \end{aligned}$$

because $N(\theta) = 3$ and $N(\alpha) \geqslant 1$, $N(\beta) \geqslant 1$.

This complete the proof of Lemma 9.34.

Theorem 9.35 *There are no nonzero integers α, β, γ in $\mathbb{Q}(\sqrt{-3})$ such that $\alpha^3 + \beta^3 + \gamma^3 = 0$. There are no positive rational integers x, y, z such that $x^3 + y^3 = z^3$.*

Proof The second assertion follows from the first. To prove the first, suppose there are nonzero integers α, β, γ such that $\alpha^3 + \beta^3 + \gamma^3 = 0$. We may presume that g.c.d.$(\alpha, \beta, \gamma) = 1$. Then by Lemma 9.32, θ divides exactly one of α, β, γ, say $\theta | \gamma$. Let θ^r be the highest power of θ dividing γ, say $\gamma = \theta^r \gamma_1$ where $\theta \nmid \gamma_1$. Then by Lemma 9.33 we conclude that $r \geqslant 2$, and

$$\alpha^3 + \beta^3 + \left(\theta^r \gamma_1\right)^3 = 0.$$

But this contradicts Lemma 9.34.

PROBLEMS

1. Suppose there are nonzero integers α, β, γ in $\mathbb{Q}(\sqrt{-3})$ and units $\varepsilon_1, \varepsilon_2, \varepsilon_3$ such that $\varepsilon_1\alpha^3 + \varepsilon_2\beta^3 + \varepsilon_3\gamma^3 = 0$. Since $\varepsilon_1\alpha^3$ can be written $-\varepsilon_1(-\alpha)^3$ we may presume that $\varepsilon_1 = 1, \omega$, or ω^2. Likewise for ε_2 and ε_3. Prove that $\varepsilon_1, \varepsilon_2, \varepsilon_3$ are $1, \omega, \omega^2$ in some order.
2. Prove that there *are* nonzero integers and units as in Problem 1 such that $\varepsilon_1\alpha^3 + \varepsilon_2\beta^3 + \varepsilon_3\gamma^3 = 0$.

NOTES ON CHAPTER 9

It can be noted that after Sections 9.1 to 9.4 on algebraic numbers in general, we turned our attention to quadratic fields. Many of our theorems can be extended to fields of algebraic numbers of higher degree, but of course it is not possible to obtain results as detailed as those for quadratic fields. Our brief survey of algebraic numbers has omitted not only these generalizations but also many other aspects of algebraic number theory that have been investigated.

§9.2 A complex number is said to be *nonalgebraic* or *transcendental* if it is not algebraic. The basic mathematical constants π and e are transcendental numbers; proofs are given in the books by Hardy and Wright, LeVeque, and Niven listed in the General References.

§9.8 The only fields $\mathbb{Q}(\sqrt{m})$ with $m < 0$ having unique factorization are the cases $m = -1, -2, -3, -7, -11, -19, -43, -67, -163$. The history of the problem of finding all such fields is recounted by D. Goldfeld, "Gauss' class number problem for imaginary quadratic fields," *Bull. Amer. Math. Soc.*, 13 (1985), 23–37.

For further readings on the subject of this chapter, see the books listed in the General References by Borevich and Shafarevich, Hua, Ireland and Rosen, Pollard and Diamond, Ribenboim, and Robinson.

CHAPTER 10

The Partition Function

10.1 PARTITIONS

Definition 10.1 *The partition function $p(n)$ is defined as the number of ways that the positive integer n can be written as a sum of positive integers, as in $n = a_1 + a_2 + \cdots + a_r$. The summands a_j are called the parts of the partition. Although the parts need not be distinct, two partitions are not considered as different if they differ only in the order of their parts. It is convenient to define $p(0) = 1$.*

For example $5 = 5 = 4 + 1 = 3 + 2 = 3 + 1 + 1 = 2 + 2 + 1 = 2 + 1 + 1 + 1 = 1 + 1 + 1 + 1 + 1$, and $p(5) = 7$. Similarly, $p(1) = 1$, $p(2) = 2$, $p(3) = 3$, $p(4) = 5$.

We shall also discuss some other partition functions in which the parts must satisfy special restrictions, as follows.

Definition 10.2

$p_m(n)$: *the number of partitions of n into parts no larger than m.*
$p^o(n)$: *the number of partitions of n into odd parts.*
$p^d(n)$: *the number of partitions of n into distinct parts.*
$q^e(n)$: *the number of partitions of n into an even number of distinct parts.*
$q^o(n)$: *the number of partitions of n into an odd number of distinct parts.*

We make the convention $p_m(0) = p^o(0) = p^d(0) = q^e(0) = 1$, $q^o(0) = 0$.

Since $5 = 2 + 2 + 1 = 2 + 1 + 1 + 1 = 1 + 1 + 1 + 1 + 1$ we have $p_2(5) = 3$. Also $5 = 5 = 3 + 1 + 1 = 1 + 1 + 1 + 1 + 1$, and $5 = 5 = 4 + 1 = 3 + 2$, and $5 = 4 + 1 = 3 + 2$, and $5 = 5$, so that $p^o(5) = 3$, $p^d(5) = 3$, $q^e(5) = 2$, $q^o(5) = 1$.

Theorem 10.1 *We have*

(1) $p_m(n) = p(n)$ *if* $n \leqslant m$,

(2) $p_m(n) \leqslant p(n)$ *for all* $n \geqslant 0$,

(3) $p_m(n) = p_{m-1}(n) + p_m(n-m)$ *if* $n \geqslant m > 1$,

(4) $p^d(n) = q^e(n) + q^o(n)$.

Proof With the possible exception of (*3*), these are all obvious from the definitions. To prove (*3*) we note that each partition of n counted by $p_m(n)$ either has or does not have a summand equal to m. The partitions of the second sort are counted by $p_{m-1}(n)$. The partitions of the first sort are obtained by adding a summand m to each partition of $n - m$ into summands less than or equal to m, and hence are $p_m(n-m)$ in number. If $n = m$, the term $p_m(n-m) = 1$ counts the single partition $n = m$.

Theorem 10.2 *For* $n \geqslant 1$ *we have* $p^d(n) = p^o(n)$.

Proof We establish a one-to-one correspondence between the partitions counted by $p^d(n)$ and those counted by $p^o(n)$. Let $n = a_1 + a_2 + \cdots + a_r$ be a partition of n into distinct parts. We convert this into a partition of n with odd parts. For any natural number m define $f(m)$ as the last integer in the sequence $m, m/2, m/4, m/8, \cdots$, so that $m = 2^j f(m)$, where 2^j is the highest power of 2 dividing m. Suppose there are s distinct odd integers among $f(a_1), f(a_2), \cdots, f(a_r)$. Rearrange the subscripts if necessary so that $f(a_1), f(a_2), \cdots, f(a_s)$ are distinct, and $f(a_{s+1}), f(a_{s+2}), \cdots, f(a_r)$ are duplicates of these. Collecting terms, we can write $n = \sum_{i=1}^{s} c_i f(a_i)$, with positive integer coefficients c_i. The final step is to write each $c_i f(a_i)$ in the form $f(a_i) + f(a_i) + \cdots + f(a_i)$ with c_i terms in the sum. Thus n is expressed as a sum of odd integers, a partition with $\sum c_i$ parts.

Conversely, these steps can be reversed as follows. Start with any partition of n with odd parts, say $n = b_1 + b_2 + \cdots + b_t$. Among these t odd integers, suppose there are s distinct ones, say $b_1, b_2, \cdots, b_s$ by rearranging notation if necessary. Collecting like terms in the partition of n, we get $n = e_1 b_1 + e_2 b_2 + \cdots + e_s b_s$. Write each coefficient e_i as a unique sum of distinct powers of 2, and so write each $e_i b_i$ as a sum of terms of the type $2^k b_i$. This gives n as a partition with distinct parts. Thus we have the one-to-one correspondence and the theorem is proved.

PROBLEMS

1. With $1 \leqslant j \leqslant n$, prove that the number of partitions of n containing the part 1 at least j times is $p(n-j)$.

2. With $1 \leqslant j \leqslant n$, prove that the number of partitions of n containing j as a part is $p(n-j)$.
*3. For every partition π of a fixed integer n, define $F(\pi)$ as the number of occurrences (if any) of 1 as a summand, and define $G(\pi)$ as the number of distinct summands in the partition. Prove that $\Sigma F(\pi) = \Sigma G(\pi)$, where each sum is taken over all partitions of n. (H)
4. If $p(n, 2)$ denotes the number of partitions of n with parts $\geqslant 2$, prove that $p(n, 2) > p(n-1, 2)$ for all $n \geqslant 8$, that $p(n) = p(n-1) + p(n, 2)$ for all $n \geqslant 1$, and that $p(n+1) + p(n-1) > 2p(n)$ for all $n \geqslant 7$.

10.2 FERRERS GRAPHS

A partition of n can be represented graphically. If $n = a_1 + a_2 + \cdots + a_r$, we may presume that $a_1 \geqslant a_2 \geqslant \cdots \geqslant a_r$. Then the graph of this partition is the array of points having a_1 points in the top row, a_2 in the next row, and so on down to a_r in the bottom row.

```
.  .  .  .  .  .
.  .  .  .  .
.  .  .  .  .
.  .
.
```

$$19 = 6 + 5 + 5 + 2 + 1.$$

If we read the graph vertically instead of horizontally, we obtain a possibly different partition. For example, from $19 = 6 + 5 + 5 + 2 + 1$ we get the *conjugate partition* $19 = 5 + 4 + 3 + 3 + 3 + 1$. The conjugate of the conjugate partition is the original partition. Given a partition $n = a_1 + a_2 + \cdots + a_r$ consisting of r parts with the largest part a_1, the conjugate partition of n has a_1 parts with largest part r. Since this correspondence is reversible, we have the following theorem.

Theorem 10.3 *The number of partitions of n into m parts is the same as the number of partitions of n having largest part m. Similarly, the number of partitions of n into at most m parts is equal to $p_m(n)$, the number of partitions of n into parts less than or equal to m.*

The next theorem has a more subtle proof by the graphic method.

Theorem 10.4 *If $n \geqslant 0$, then*

$$q^e(n) - q^o(n) = \begin{cases} (-1)^j & \text{if } n = (3j^2 \pm j)/2 \text{ for some } j = 0, 1, 2, \cdots \\ 0 & \text{otherwise.} \end{cases}$$

Proof For $n = 0$ we have $j = 0$ and $q^e(0) - q^o(0) = 1$. We now suppose $n \geqslant 1$ and consider a partition $n = a_1 + a_2 + \cdots + a_r$ into distinct parts. In the graph of this partition we let A_1 denote the point farthest to the right in the first row. Since the parts are distinct, there will be no point directly below A_1. If $a_2 = a_1 - 1$, there will be a point A_2 directly below the point that is immediately to the left of A_1. If $a_2 < a_1 - 1$, there will be no such point A_2. If $a_3 = a_1 - 2$, then $a_2 = a_1 - 1$ and there will be a point A_3 directly below the point that is immediately to the left of A_2. If $a_2 = a_1 - 1$ and $a_3 < a_2 - 1$, there will be no point A_3. We continue this process as far as possible, thus obtaining a set of points $A_1, A_2, \cdots, A_s$, $s \geqslant 1$, lying on a line through A_1 with slope 1. We also label the points of the bottom row $B_1, B_2, \cdots, B_t$, $t = a_r$. Notice that B_t and A_s may be the same point.

$$
\begin{array}{cccccccc}
\bullet & \bullet & \bullet & \bullet & \bullet & \bullet & \bullet & A_1 \\
\bullet & \bullet & \bullet & \bullet & \bullet & \bullet & A_2 & \\
\bullet & \bullet & \bullet & \bullet & \bullet & A_3 & & \\
\bullet & \bullet & \bullet & & & & & \\
\bullet & \bullet & & & & & & \\
B_1 & B_2 & & & & & &
\end{array}
$$

Now we wish to change the graph into the graph of another partition of n into distinct parts. First, we try taking the points $B_1, B_2, \cdots, B_t$ and placing them to the right of $A_1, A_2, \cdots, A_t$; B_1 to the right of A_1, B_2 to the right of A_2, and so on. It is obvious that we cannot do this if $t > s$ or if $t = s$ and $B_t = A_s$. However we can do it if $t < s$ or if $t = s$ and $B_t \neq A_s$, and we obtain a graph of a partition into distinct parts. Second, we try the reverse process, putting $A_1, A_2, \cdots, A_s$ underneath $B_1, B_2, \cdots, B_s$. This will give a proper graph if and only if $s < t - 1$ or $s = t - 1$ and $B_t \neq A_s$.

To refine this description, the transformation just described acts in one of three different ways on a partition π of the fixed integer n, say $n = \varepsilon_1 + a_2 + \cdots + a_r$. If the partition π has $t < s$, or $t = s$ with distinct points A_s and B_t, the transformation removes the entire bottom row of the graph, $B_1, B_2, \cdots, B_t$, and extends the first t rows of the graph by one point each. If the partition π has $s < t - 1$, or $s = t - 1$, again with $A_s \neq B_t$, the transformation moves the points $A_1, A_2, \cdots, A_s$ to form an additional bottom row in the graph. The third type of partition π has $A_s = B_t$ with $s = t$ or $s = t - 1$; for partitions π of this type, the transformation leaves π unchanged. The three types account for all possible partitions of n with distinct parts.

Examples of the three types for $n = 22$ are $22 = 8 + 7 + 6 + 1$, $22 = 9 + 8 + 5$, and $22 = 7 + 6 + 5 + 4$. The first two partitions π are changed into π', namely $22 = 9 + 7 + 6$ and $22 = 8 + 7 + 5 + 2$, whereas the transformation leaves the partition $7 + 6 + 5 + 4$ unchanged. The transformation changes a partition of the first type into one of the second, and vice versa. Moreover, a second application of the transformation brings any partition back to its original form.

With the first and second types of partition P, the partition P' also has distinct parts, but has one fewer or one more part than P. Thus, apart from partitions of the third type, we have paired off partitions with an odd number of parts and those with an even number.

Now consider the exceptional partitions of the third type, with $A_s = B_t$ and $s = t$ or $s = t - 1$. Since A_s and B_t are identical points, it follows that $s = r$, and $a_1, a_2, \cdots, a_r$ are consecutive integers, with a_1 largest. Since $t = a_r$ in all cases, the partition has the form

$$n = (t + s - 1) + (t + s - 2) + \cdots + (t + 1) + t.$$

If $s = t$ we have $n = (3s^2 - s)/2$, whereas if $s = t - 1$, then $n = (3s^2 + s)/2$. It is not difficult to verify that positive integers of the form $(3s^2 - s)/2$ do not overlap those of the form $(3s^2 + s)/2$. Hence if $n = (3s^2 \pm s)/2$ for some natural number s, that is, if n is one of the numbers $1, 2, 5, 7, 12, 15, 22, 26, \cdots$, then $q^e(n)$ exceeds $q^o(n)$ by 1 if s is even, but $q^o(n)$ is larger by 1 if s is odd. For all other values of n, there are no partitions of the third type, and $q^e(n) = q^o(n)$.

PROBLEMS

1. Let π be the partition $n = a_1 + a_2 + \cdots + a_r$, $a_1 \geqslant a_2 \geqslant \cdots \geqslant a_r > 0$ and let π' be the partition $n = b_1 + b_2 + \cdots + b_s$, $b_1 \geqslant b_2 \geqslant \cdots \geqslant b_s > 0$. Prove that π' is the conjugate of π if and only if $r = b_1$ and $s = a_1$, and b_j is the number of parts in the partition π that are $\geqslant j$ for $j = 1, 2, \cdots, s$. (These conditions are of course equivalent to: $r = b_1$ and $s = a_1$, and $b_j - b_{j+1}$ is the number of parts in the partition π that are equal to j, for $j = 1, 2, \cdots, s - 1$, and b_s is the number of parts equal to s, that is, equal to a_1. These results give alternative definitions of the conjugate partition that are independent of the idea of a graph. If the partition π is given and we want to construct the conjugate π', it is quicker to use the first result above than to draw a graph, especially if π has 20 or more parts.)
2. Let $F(n)$ denote the number of partitions of n with every part appearing at least twice. Let $G(n)$ denote the number of partitions of n into parts larger than 1 such that no two parts are consecutive integers. Use conjugate partitions to prove that $F(n) = G(n)$.
3. (Notation as in Problem 1.) Prove that the number of partitions π of n with $a_r = 1$ and $a_j - a_{j+1} = 0$ or 1 for $1 \leqslant j \leqslant r - 1$ equals $p^d(n)$, the number of partitions of n into distinct parts.

4. A partition is said to be self-conjugate if it is identical with its conjugate, as in the examples $18 = 5 + 4 + 4 + 4 + 1$ and $15 = 6 + 3 + 3 + 1 + 1 + 1$. Prove that the number of self-conjugate partitions of n equals the number of partitions of n into distinct odd parts, by using the idea suggested by the accompanying graph of the self-conjugate partition $25 = 6 + 6 + 5 + 3 + 3 + 2$ and the natural transformation into $25 = 11 + 9 + 5$ taken from the right-angle batches. Prove that this is the same as the number of partitions of n whose parts (except for a special case) are all the consecutive integers from 1 to some j, with all parts appearing an even number of times except j, which appears an odd number of times; the special case is $n = 1 + 1 + 1 + \cdots + 1 + 1$, with n odd. (By the use of Problem 1, it is easy to decide whether a given partition is self-conjugate, or to create self-conjugate partitions.)

The next two problems outline a proof that

$$T(n) = \left[\frac{n^2 + 6}{12}\right] - \left[\frac{n}{4}\right] \cdot \left[\frac{n + 2}{4}\right],$$

where $T(n)$ denotes the number of triangles with integral sides and perimeter n, with no two triangles congruent. (Curiously enough, $T(n)$ is not a monotonic increasing function: $T(7) = 2$, but $T(8) = 1$, for example.)

5. Let $P_2(n)$ denote the number of partitions of n into 2 parts, and $P_3(n)$ the number into 3 parts. Verify that $P_2(n) = [n/2]$. Next, by Theorem 10.3 we see that $P_3(n)$ equals the number of partitions of n with largest part 3. That is, $P_3(n)$ equals the number of solutions of $3x + 2y + z = n$ in integers $x > 0$, $y \geqslant 0$, and $z \geqslant 0$, since every such solution corresponds to a partition of n with x parts equal to 3, y parts equal to 2, and z parts equal to 1. Now the number of solutions of $2y + z = k \geqslant 0$ in non-negative integers is $1 + [k/2]$ or $[(k + 2)/2]$. Hence we can add the numbers of solutions of $2y + z = n - 3$, $2y + z = n - 6$, $2y + z = n - 9, \cdots$ to get

$$P_3(n) = \left[\frac{n - 1}{2}\right] + \left[\frac{n - 4}{2}\right] + \left[\frac{n - 7}{2}\right] + \cdots \quad \text{(positive terms only)}.$$

Prove that $P_3(n) = \left[\dfrac{n^2 + 6}{12}\right]$ by induction from n to $n + 6$, that is, prove that $[(n + 5)/2] + [(n + 2)/2] = [((n + 6)^2 + 6)/12] - [(n^2 + 6)/12]$.

6. To count the number $T(n)$ of triangles with perimeter n and integral sides, we can start with $P_3(n)$, the number of partitions of n into 3 parts, $n = a + b + c$, $a \geqslant b \geqslant c > 0$. But this gives no triangle if $b + c \leqslant a$, that is, if $b + c = 2$, or 3, or $4, \cdots$, or $[n/2]$. These equations have, respectively, $P_2(2), P_2(3), \cdots, P_2([n/2])$ solutions in positive integers b and c with $b \geqslant c$. By induction or any other method, prove that $P_2(2) + P_2(3) + \cdots + P_2([n/2]) = [n/4] \cdot [(n + 2)/4]$ and hence that $T(n) = [(n^2 + 6)/12] - [n/4] \cdot [(n + 2)/4]$.

7. Consider n dots in a row, with a separator between adjacent dots, so $n - 1$ separators in all. By choosing $j - 1$ separators to be left in place while the others are removed, and then counting the number of dots between adjacent separators prove that the equation

$$x_1 + x_2 + \cdots + x_j = n$$

has $\binom{n-1}{j-1}$ solutions in positive integers, where two solutions $x_1, x_2, \cdots, x_j$ and $x_1', x_2', \cdots, x_j'$ are counted as distinct if $x_k \neq x_k'$ for at least one subscript k. (Note that the order of summands is taken into account here, so these are *not* partitions of n.)

8. By taking $j = 1, 2, 3, \cdots$ in the preceding problem, prove that the number of ways of writing n as a sum of positive integers is 2^{n-1}, where again the order of summands is taken into account. For example, if $n = 4$ the sums being counted are

$$1 + 1 + 1 + 1, 1 + 1 + 2, 1 + 2 + 1,$$

$$2 + 1 + 1, 1 + 3, 3 + 1, 2 + 2, 4.$$

10.3 FORMAL POWER SERIES, GENERATING FUNCTIONS, AND EULER'S IDENTITY

In the first two sections, combinatorial methods have been used, including arguments with graphs. In this section formal power series and generating functions are introduced, and in the next and subsequent sections we use analytic methods.

The power series we use are of the form $a_0 + a_1x + a_2x^2 + a_3x^3 + \cdots$ where $a_0 \neq 0$. Such a power series is treated formally if no numerical values are ever substituted for x. Thus x is a dummy variable, and the power series is just a way of writing an infinite sequence of constants $a_0, a_1, a_2, a_3, \cdots$. However, it is very convenient to retain the x for easy identification of the general term. Two power series $\Sigma a_j x^j$ and $\Sigma b_j x^j$ are said to be equal if $a_j = b_j$ for all subscripts j. The product of these two power series is defined to be

$$a_0b_0 + (a_0b_1 + a_1b_0)x + (a_0b_2 + a_1b_1 + a_2b_0)x^2 + \cdots$$
$$+\left(\sum_{j=0}^{n} a_jb_{n-j}\right)x^n + \cdots.$$

With these definitions of equality and multiplication of formal power series, the set of all power series with real coefficients with $a_0 \neq 0$ (and $b_0 \neq 0$) forms an abelian group. The associative property is easy to prove; in fact it follows from the associative property for polynomials in x because the coefficient of x^n in any product is determined by the terms up to x^n so that all terms in higher powers of x can be discarded in all power series in proving that the coefficients of x^n are identical.

The identity element of the group is 1 or $1 + 0x + 0x^2 + 0x^3 + \cdots$. The inverse of any power series Σa_jx^j with $a_0 \neq 0$ is the power series Σb_jx^j such that

$$\left(\sum_{j=0}^{\infty} a_jx^j\right)\left(\sum_{j=0}^{\infty} b_jx^j\right) = 1$$

holds. The definition of multiplication of power series gives at once an infinite sequence of equations

$$a_0b_0 = 1, \qquad a_0b_1 + a_1b_0 = 0$$
$$a_0b_2 + a_1b_1 + a_2b_0 = 0, \cdots, \qquad \sum_{j=0}^{n} a_jb_{n-j} = 0, \cdots$$

that can be solved serially for $b_0, b_1, b_2, \cdots$. Thus the inverse power series exists, is unique, and can be calculated directly. Finally, the group is abelian because of the symmetry of the definition of multiplication.

The inverse of $1 - x$ is readily calculated to be $1 + x + x^2 + x^3 + \cdots$. As in analysis, this is called the power series expansion of $(1 - x)^{-1}$.

Under suitable circumstances an infinite number of power series can be multiplied. An illustration of this is

$$(1 + x)(1 + x^2)(1 + x^3)(1 + x^4) \cdots = \prod_{n=1}^{\infty} (1 + x^n),$$

a product that will be used in what follows. The reason that this infinite product is well defined is that the coefficient of x^m for any positive integer

m depends on only a finite number of factors, in fact it depends on

$$(1+x)(1+x^2)(1+x^3)\cdots(1+x^m) = \prod_{n=1}^{m}(1+x^n)$$

In general let $P_1, P_2, P_3, \cdots$ be an infinite sequence of power series each with leading term 1. Then the infinite product $P_1P_2P_3 \cdots$ is well defined if for every positive integer k the power x^k occurs in only a finite number of the power series. For if this condition is satisfied it is clear that the x^m term in the product is determined by a finite product $P_1P_2P_3 \cdots P_r$ where r is chosen so that none of the power series $P_{r+1}, P_{r+2}, P_{r+3}, \cdots$ has any term of degree m or lower, except of course the constant term 1 in each series.

The function $(1-x^n)^{-1}$ has the expansion $\Sigma_{j=0}^{\infty}x^{jn}$. Taking $n = 1, 2, \cdots, m$ and multiplying we find

$$\begin{aligned}\prod_{n=1}^{m}(1-x^n)^{-1} &= (1+x^{1\cdot 1}+x^{2\cdot 1}+\cdots)(1+x^{1\cdot 2}+x^{2\cdot 2}+\cdots)\\ &\quad\times(1+x^{1\cdot 3}+x^{2\cdot 3}+\cdots)\cdots(1+x^{1\cdot m}+x^{2\cdot m}+\cdots)\\ &= \sum_{j_1=0}^{\infty}\sum_{j_2=0}^{\infty}\cdots\sum_{j_m=0}^{\infty}x^{j_1\cdot 1+j_2\cdot 2+\cdots+j\cdot m}\\ &= \sum_{j=0}^{\infty}c_jx^j\end{aligned}$$

where c_j is the number of solutions of $j_1\cdot 1 + j_2\cdot 2 + \cdots + j_m \cdot m = j$ in nonnegative integers $j_1, j_2, \cdots, j_m$. That is $c_j = p_m(j)$, and we have

$$\sum_{n=0}^{\infty}p_m(n)x^n = \prod_{n=1}^{m}(1-x^n)^{-1}.$$

This can be written as

$$\sum_{n=0}^{m}p(n)x^n + \sum_{n=m+1}^{\infty}p_m(n)x^n = \prod_{n=1}^{m}(1-x^n)^{-1}$$

since $p_m(k) = p(k)$ if $k \leqslant m$. Since these equations have only the restricted meaning in formal power series that coefficients of the same power of x are equal, we can let m increase beyond bound to get

$$\sum_{n=0}^{\infty}p(n)x^n = \prod_{n=1}^{\infty}(1-x^n)^{-1}.$$

The function $\prod_{n=1}^{\infty}(1-x^n)^{-1}$ is called the *generating function* for $p(n)$, and it will be used to derive information about the partition function. The generating function for $p_m(n)$ is $\prod_{n=1}^{m}(1-x^n)^{-1}$. Similarly the generating function of $p^o(n)$ is found to be

$$\sum_{n=0}^{\infty} p^o(n)x^n = \prod_{n=1}^{\infty}(1-x^{2n-1})^{-1}$$

and the generating function for $p^d(n)$ is

$$\sum_{n=0}^{\infty} p^d(n)x^n = \prod_{n=1}^{\infty}(1+x^n).$$

Theorem 10.2 is equivalent to $\prod_{n=1}^{\infty}(1+x^n) = \prod_{n=1}^{\infty}(1-x^{2n-1})^{-1}$. This formula is now proved directly, giving us a proof of Theorem 10.2 by the use of generating functions. We multiply two factors at a time in the following infinite product to get

$$\begin{aligned}
&(1-x)(1+x)(1+x^2)(1+x^4)(1+x^8)(1+x^{16})\cdots \\
&\quad = (1-x^2)(1+x^2)(1+x^4)(1+x^8)(1+x^{16})\cdots \\
&\quad = (1-x^4)(1+x^4)(1+x^8)(1+x^{16})\cdots \\
&\quad = (1-x^8)(1+x^8)(1+x^{16})\cdots = \cdots = 1.
\end{aligned}$$

Similarly we see that

$$(1-x^3)(1+x^3)(1+x^6)(1+x^{12})(1+x^{24})\cdots = 1,$$

$$(1-x^5)(1+x^5)(1+x^{10})(1+x^{20})(1+x^{40})\cdots = 1,$$

and so forth, where the first factor runs through all odd powers $1-x^7$, $1-x^9, 1-x^{11}, \cdots$. Multiplying all these we get

$$\prod_{n=1}^{\infty}(1-x^{2n-1})\prod_{j=1}^{\infty}(1+x^j) = 1$$

and

$$\prod_{j=1}^{\infty}(1+x^j) = \prod_{n=1}^{\infty}(1-x^{2n-1})^{-1}.$$

In a similar way we can multiply out $\prod_{n=1}^{\infty}(1 - x^n)$ formally to get

$$\prod_{n=1}^{\infty}(1 - x^n) = \sum_{n=0}^{\infty}(q^e(n) - q^o(n))x^n.$$

Then Theorem 10.4 implies

$$\prod_{n=1}^{\infty}(1 - x^n) = 1 + \sum_{j=1}^{\infty}(-1)^j(x^{(3j^2+j)/2} + x^{(3j^2-j)/2}).$$

This is known as *Euler's formula*. Whereas here we have proved it only in the formal sense that the coefficients of the power series are identical, an analytic proof is given in Theorem 10.9 with convergence indicated for suitable values of x. Since a variable is never assigned a numerical value in formal power series, questions of convergence never arise.

Theorem 10.5 *Euler's identity. For any positive integer n,*

$$\begin{aligned} p(n) &= p(n-1) + p(n-2) - p(n-5) - p(n-7) \\ &\quad + p(n-12) + p(n-15) - \cdots \\ &= \sum_j (-1)^{j+1} p\left(n - \tfrac{1}{2}(3j^2 + j)\right) + \sum_j (-1)^{j+1} p\left(n - \tfrac{1}{2}(3j^2 - j)\right) \end{aligned}$$

where each sum extends over all positive integers j for which the arguments of the partition functions are non-negative.

Proof From Euler's formula and the fact that $\prod(1 - x^n)^{-1}$ is the generating function for $p(n)$ we can write

$$\left\{1 + \sum_{j=1}^{\infty}(-1)^j\{x^{(3j^2+j)/2} + x^{(3j^2-j)/2}\}\right\}\sum_{k=0}^{\infty}p(k)x^k = 1$$

or

$$\{1 - x - x^2 + x^5 + x^7 - x^{12} - x^{15} + \cdots\}\sum_{k=0}^{\infty}p(k)x^k = 1.$$

Equating coefficients of x^n on the two sides we get

$$\begin{aligned} p(n) - p(n-1) + p(n-2) + p(n-5) + p(n-7) \\ - p(n-12) - p(n-15) + \cdots = 0, \end{aligned}$$

and thus the theorem is established.

PROBLEMS

1. Show that the infinite product

$$(1+x_1)(1+x_1x_2)(1+x_1x_2x_3)\cdots = 1+\sum x_1^{a_1}x_2^{a_2}\cdots x_k^{a_k}$$

where $a_i - a_{i+1}$ is 0 or 1, and $a_k = 1$. Count the number of terms in the expansion that are of degree n. Set $x_1 = x_2 = x_3 = \cdots = x$ to show that $(1+x)(1+x^2)(1+x^3)\cdots$ is the generating function for $p'(n)$ of Problem 1, Section 10.2.

2. Compute a short table of the values of $p(n)$, from $n = 1$ to $n = 20$, by use of Theorem 10.5. (Recall that $p(0) = 1$.)

3. By writing the inverse of $1 - x$ as an infinite product $(1+x)(1+x^2)(1+x^4)(1+x^8)\cdots$ and also as an infinite series, use these generating functions to prove that every positive integer can be expressed uniquely as a sum of distinct powers of 2 (cf. Problem 44, Section 1.2).

10.4 EULER'S FORMULA; BOUNDS ON $p(n)$

We open the section by proving Euler's formula as an equality between two functions, not just in the formal sense. Formal power series arguments have serious limitations, so it is convenient now to use a few rudimentary facts concerning infinite series and limits. A reader familiar with the theory of analytic functions will recognize that our functions are analytic in $|x| < 1$, and will be able to shorten our proofs.

Theorem 10.6 *Suppose* $0 \leqslant x < 1$ *and let* $\phi_m(x) = \prod_{n=1}^{m}(1-x^n)$. *Then* $\sum_{n=0}^{\infty} p_m(n)x^n$ *converges and*

$$\sum_{n=0}^{\infty} p_m(n)x^n = \frac{1}{\phi_m(x)}.$$

Proof By Theorem 10.3, $p_m(n)$ is equal to the number of partitions of n into at most m summands. This is the same as the number of partitions into exactly m summands if we allow zero summands. Then each summand is 0 or 1 or 2 or $\cdots$ or n, and we have $p_m(n) \leqslant (n+1)^m$. The series $\sum_{n=0}^{\infty}(n+1)^m x^n$ converges, by the ratio test, and hence so does $\sum_{n=0}^{\infty} p_m(n)x^n$, by the comparison test.

Now

$$(1 - x^{m!k})^m \phi_m(x)^{-1} = \prod_{n=1}^{m} \frac{1 - x^{m!k}}{1 - x^n} = \prod_{n=1}^{m} \frac{1 - (x^m)^{(m!/n)k}}{1 - x^n}$$

$$= \prod_{n=1}^{m} \sum_{j=0}^{(m!/n)k-1} x^{jn} = \sum_h c_h x^h$$

where the last sum is a finite sum and $0 \leqslant c_h \leqslant p_m(h)$ for all $h = 0, 1, 2, \cdots$, and $c_h = p_m(h)$ if $h < m!\,k$. Therefore we have

$$\sum_{h=0}^{m!k-1} p_m(h)x^h \leqslant (1 - x^{m!k})^m \phi_m(x)^{-1} \leqslant \sum_{h=0}^{\infty} p_m(h)x^h.$$

As $k \to \infty$ we have

$$\sum_{h=0}^{m!k-1} p_m(h)x^h \to \sum_{h=0}^{\infty} p_m(h)x^h, \qquad (1 - x^{m!k})^m \to 1$$

and hence

$$\phi_m(x)^{-1} = \sum_{h=0}^{\infty} p_m(h)x^h.$$

Theorem 10.7 *For* $0 \leqslant x < 1$, $\lim_{m\to\infty} \phi_m(x)$ *exists and is different from zero. We let* $\phi(x) = \lim_{m\to\infty} \phi_m(x)$ *and define* $\prod_{n=1}^{\infty}(1 - x^n)$ *to be* $\phi(x)$.

Proof Since $\phi_m(0) = 1$ the result is obvious for $x = 0$. For $x > 0$ we apply the mean value theorem to the function $\log z$ to obtain a y such that $1 - x^n < y < 1$ and

$$\frac{\log 1 - \log(1 - x^n)}{1 - (1 - x^n)} = \frac{1}{y}.$$

Therefore

$$-\log(1 - x^n) = \frac{x^n}{y}, \qquad -\log(1 - x^n) \leqslant \frac{x^n}{1 - x^n} \leqslant \frac{x^n}{1 - x}$$

and hence

$$-\log \phi_m(x) = \sum_{n=1}^{m} -\log(1-x^n)$$

$$\leqslant \sum_{n=1}^{m} \frac{x^n}{1-x} \leqslant \frac{1-x^{m+1}}{(1-x)^2} < \frac{1}{(1-x)^2}.$$

This shows that $-\log \phi_m(x)$, and hence $\phi_m(x)^{-1}$, is bounded for x fixed as $m \to \infty$.

But

$$\phi_m(x)^{-1} = \prod_{n=1}^{m} \frac{1}{1-x^n}$$

increases monotonically for x fixed as $m \to \infty$. Since $\phi_1(x)^{-1} = 1/(1-x) > 0$ this shows that $\lim_{m\to\infty} \phi_m(x)^{-1}$ exists and is different from zero. Therefore $\lim_{m\to\infty} \phi_m(x)$ exists and is also different from zero.

Theorem 10.8 *For $0 \leqslant x < 1$ the series $\sum_{n=0}^{\infty} p(n)x^n$ converges, and*

$$\sum_{n=0}^{\infty} p(n)x^n = \phi(x)^{-1}.$$

Proof We have, using Theorem 10.6,

$$\sum_{n=0}^{m} p(n)x^n = \sum_{n=0}^{m} p_m(n)x^n \leqslant \sum_{n=0}^{\infty} p_m(n)x^n = \phi_m(x)^{-1} \leqslant \phi(x)^{-1}.$$

For x fixed, $\sum_{n=0}^{m} p(n)x^n$ increases as $m \to \infty$. Therefore $\sum_{n=0}^{\infty} p(n)x^n = \lim_{m\to\infty} \sum_{n=0}^{m} p(n)x^n$ exists and is $\leqslant \phi(x)^{-1}$.

But now

$$\sum_{n=0}^{\infty} p(n)x^n \geqslant \sum_{n=0}^{\infty} p_m(n)x^n = \phi_m(x)^{-1}.$$

Letting $m \to \infty$ we have $\sum_{n=0}^{\infty} p(n)x^n \geqslant \phi(x)^{-1}$, and hence $\sum_{n=0}^{\infty} p(n)x^n = \phi(x)^{-1}$.

Theorem 10.9 *Euler's formula. For $0 \leqslant x < 1$, we have*

$$\phi(x) = 1 + \sum_{j=1}^{\infty} (-1)^j \left(x^{(3j^2+j)/2} + x^{(3j^2-j)/2}\right).$$

Proof The ratio test shows that $\Sigma_{j=1}^{\infty} x^{(3j^2 \pm j)/2}$ converges; therefore so does the above series. Let $q_m^e(n)$ be the number of partitions of n into an even number of distinct summands no greater than m, and let $q_m^o(n)$ be the number of partitions of n into an odd number of distinct summands no greater than m. As in Definition 10.2 we will take $q_m^e(0) = 1$, $q_m^o(0) = 0$. Then

$$\begin{aligned}\phi_m(x) &= (1-x)(1-x^2)(1-x^3)\cdots(1-x^m)\\ &= \sum_n (q_m^e(n) - q_m^o(n))x^n, \qquad (10.1)\end{aligned}$$

a finite sum. But for $n \leqslant m$ we have $q_m^e(n) = q^e(n)$, $q_m^o(n) = q^o(n)$, and we also have $q_m^e(n) + q_m^o(n) \leqslant p(n)$ for all n. Therefore

$$\begin{aligned}&\left|\phi_m(x) - \sum_{n=0}^{m} (q^e(n) - q^o(n))x^n\right|\\ &\qquad \leqslant \sum_{n>m} |q_m^e(n) - q_m^o(n)|x^n \leqslant \sum_{n=m+1}^{\infty} p(n)x^n.\end{aligned}$$

Since $\Sigma_{n=m+1}^{\infty} p(n)x^n \to 0$ as $m \to \infty$, we get $\Sigma_{n=0}^{\infty}(q^e(n) - q^o(n))x^n = \phi(x)$ by letting $m \to \infty$. Using Theorem 10.4, we have the present theorem.

For convenient reference we now state two needed results on power series. Proofs of these propositions are given in standard books on advanced calculus or elementary function theory.

Lemma 10.10 *Let $\Sigma_{j=0}^{\infty} a_j x^j$ and $\Sigma_{k=0}^{\infty} b_k x^k$ be absolutely convergent for $0 \leqslant x < 1$. Then $\Sigma_{h=0}^{\infty}(\Sigma_{j=0}^{h} a_j b_{h-j})x^h$ converges and has the value $\Sigma_{j=0}^{\infty} a_j x^j \Sigma_{k=0}^{\infty} b_k x^k$ for $0 \leqslant x < 1$. Moreover, if $\Sigma_{j=0}^{\infty} a_j x^j = \Sigma_{j=0}^{\infty} b_j x^j$ for $0 \leqslant x < 1$, then $a_j = b_j$ for all $j = 0, 1, 2, \cdots$.*

The next theorem gives for the sum of divisors functions, $\sigma(n)$, an identity similar to that for $p(n)$ in Theorem 10.5.

Theorem 10.11 *For $n \geqslant 1$ we have*

$$\begin{aligned}&\sigma(n) - \sigma(n-1) - \sigma(n-2) + \sigma(n-5) + \sigma(n-7)\\ &\qquad - \sigma(n-12) - \sigma(n-15) + \cdots = \begin{cases} (-1)^{j+1} n & \text{if } n = \dfrac{3j^2 \pm j}{2} \\ 0 & \text{otherwise,} \end{cases}\end{aligned}$$

where the sum extends as far as the arguments are positive.

Proof Taking the derivative of $\log \phi_m(x) = \log \prod_{n=1}^{m}(1 - x^n)$ we get

$$\frac{\phi_m'(x)}{\phi_m(x)} = \sum_{n=1}^{m} \frac{-nx^{n-1}}{1 - x^n} = \sum_{n=1}^{m} \sum_{j=1}^{\infty} -nx^{jn-1} = \sum_{n=1}^{m} \sum_{k=1}^{\infty} c_{n,k} x^{k-1}$$

for $0 \leqslant x < 1$, where

$$c_{n,k} = \begin{cases} -n & \text{if } n \mid k \\ 0 & \text{otherwise.} \end{cases}$$

There are m series $\sum_{k=1}^{\infty} c_{n,k} x^{k-1}$ each of which converges absolutely. They can be added term by term to give

$$\frac{\phi_m'(x)}{\phi_m(x)} = \sum_{k=1}^{\infty} \left(\sum_{n=1}^{m} c_{n,k} \right) x^{k-1}. \tag{10.2}$$

Using (10.1) we have $\phi_m'(x) = \sum_n n(q_m^e(n) - q_m^o(n))x^{n-1}$ since $\phi_m(x)$ is a finite sum, a polynomial in x. But we can also write (10.1) in the form of an infinite series,

$$\phi_m(x) = \sum_{n=0}^{\infty} (q_m^e(n) - q_m^o(n))x^n$$

in which all the terms from a certain n on are zero. Then equation (10.2) can be put in the form

$$\begin{aligned} \sum_n n(q_m^e(n) - q_m^o(n))x^{n-1} &= \sum_{n=0}^{\infty} (q_m^e(n) - q_m^o(n))x^n \sum_{j=0}^{\infty} \left(\sum_{i=1}^{m} c_{i,j+1} \right) x^j \\ &= \sum_{h=0}^{\infty} \left(\sum_{n=0}^{h} (q_m^e(n) - q_m^o(n)) \sum_{i=1}^{m} c_{i,h-n+1} \right) x^h \end{aligned}$$

by the first part of Lemma 10.10. The second part gives us

$$k(q_m^e(k) - q_m^o(k)) = \sum_{n=0}^{k-1} (q_m^e(n) - q_m^o(n)) \sum_{i=1}^{m} c_{i,k-n}.$$

For any given k we can choose $m > k$. Then $q_m^e(k) = q^e(k)$, $q_m^o(k) = q^o(k)$, $q_m^e(n) = q^e(n)$, $q_m^o(n) = q^o(n)$, and $\sum_{i=1}^{m} c_{i,k-n} = -\sum_{d \mid k-n} d = -\sigma(k-n)$ for $n \leqslant k-1$. This with Theorem 10.4 gives us

$$\begin{aligned} &-\sigma(k) + \sigma(k-1) + \sigma(k-2) - \sigma(k-5) - \sigma(k-7) + \cdots \\ &= \begin{cases} (-1)^j k & \text{if } k = \dfrac{3j^2 \pm j}{2} \\ 0 & \text{otherwise,} \end{cases} \end{aligned}$$

and the theorem is proved.

Theorem 10.12 *Bounds on $p(n)$. The inequalities*

$$2^{\sqrt{n}} < p(n) < e^{c\sqrt{n}}$$

with $c = \pi\sqrt{2/3}$ hold for $n > 3$ (first inequality) and for $n \geqslant 1$ (second inequality).

Proof To prove the first inequality, we define k as the unique positive integer satisfying $k^2 > n \geqslant (k-1)^2$. Consider the 2^k partitions of n,

$$n = \varepsilon_1 + 2\varepsilon_2 + 3\varepsilon_3 + \cdots + k\varepsilon_k + x,$$

where each ε_i may be 0 or 1, and the integer x (which we show to be positive) is chosen to balance the equation. These are distinct partitions of n because $x > k$ by the following argument:

$$x \geqslant n - 1 - 2 - 3 - \cdots - k$$

$$\geqslant (k-1)^2 - k(k+1)/2 > k \text{ if } k^2 - 7k + 2 > 0.$$

This holds for $k > 6$ or $n \geqslant 36$. Thus we have 2^k distinct partitions, and hence for $n \geqslant 36$ we have $p(n) \geqslant 2^k > 2^{\sqrt{n}}$.

For smaller values of n we argue as follows. We know that $p(12) = 77$ (from Problem 2 of the preceding section, for example) and so for $12 \leqslant n < 36$ we have $p(n) \geqslant p(12) = 77 > 2^6 > 2^{\sqrt{n}}$. Finally, the inequality can be verified readily for the cases $4 \leqslant n \leqslant 11$.

To prove the second inequality in the theorem, we need the result $\sum_{n=1}^{\infty} 1/n^2 = \pi^2/6$, which is proved in Appendix A.3. From Theorem 10.8 we have, for $0 \leqslant x < 1$,

$$P(x) = \sum_{n=0}^{\infty} p(n)x^n = \prod_{k=1}^{\infty} (1 - x^k)^{-1},$$

where the first equation defines $P(x)$. Using the power series expansion

$$\log(1-w)^{-1} = \sum_{m=1}^{\infty} \frac{1}{m} w^m$$

we see that $P(x)$ can be written in the form

$$\exp\left(\sum_{m=1}^{\infty} \frac{1}{m} \sum_{k=1}^{\infty} x^{mk}\right) = \exp\left(\sum_{m=1}^{\infty} \frac{1}{m} \cdot \frac{x^m}{1-x^m}\right).$$

We are now in a position to show that for any positive δ

$$P(e^{-\delta}) < \exp(\pi^2/6\delta)$$

To see this, note that if $x = e^{-\delta}$ then

$$\frac{x^m}{1 - x^m} = \frac{1}{x^{-m} - 1} = \frac{1}{e^{m\delta} - 1} < \frac{1}{m\delta}$$

where the last step follows from the inequality $e^u > 1 + u$ for any positive real number u. Hence we have

$$P(e^{-\delta}) < \exp\left(\sum_{m=1}^{\infty} \frac{1}{m} \cdot \frac{1}{m\delta} \right) = \exp\left(\frac{1}{\delta} \sum_{m=1}^{\infty} \frac{1}{m^2} \right) = \exp \frac{\pi^2}{6\delta}.$$

From the definition of $P(x)$, we get

$$p(n)e^{-n\delta} < P(e^{-\delta}) < \exp(\pi^2/6\delta)$$

so that $p(n) < \exp(n\delta + \pi^2/6\delta)$. We choose $\delta = \pi/\sqrt{6n}$ to minimize the right side, and thus the theorem is proved.

PROBLEMS

1. Compute a short table of the values of $\sigma(n)$, from $n = 1$ to $n = 20$, by means of Theorem 10.11. Verify the entries by computing $\sigma(n) = \sum_{d|n} d$ directly.
2. Verify that the first part of the proof of Theorem 10.12 establishes a little more than the theorem claims, namely that $2^{\sqrt{n}} < p^d(n)$ for $n \geqslant 36$.

10.5 JACOBI'S FORMULA

Theorem 10.13 *Jacobi's formula. For* $0 \leqslant x < 1$,

$$\phi(x)^3 = \sum_{j=0}^{\infty} (-1)^j (2j + 1) x^{(j^2+j)/2}.$$

Proof The formula is obvious for $x = 0$, so we can suppose $0 < x < 1$. For $0 < q < 1$, $0 < z < 1$, we define

$$f_n(z) = \prod_{k=1}^{n} \{(1 - q^{2k-1}z^2)(1 - q^{2k-1}z^{-2})\} = \sum_{j=-n}^{n} a_j z^{2j} \quad (10.3)$$

where the a_j are polynomials in q. Since $f_n(1/z) = f_n(z)$ we have $a_{-j} = a_j$, and it is easy to see that

$$a_n = (-1)^n q^{1+3+5+\cdots+(2n-1)} = (-1)^n q^{n^2}. \tag{10.4}$$

In order to obtain the other a_j we replace z by qz in (10.3) and find

$$f_n(qz) = \prod_{k=1}^{n} \{(1 - q^{2k+1}z^2)(1 - q^{2k-3}z^{-2})\}$$

$$= \prod_{k=2}^{n+1} (1 - q^{2k-1}z^2) \prod_{j=0}^{n-1} (1 - q^{2j-1}z^{-2})$$

and hence

$$qz^2(1 - q^{2n-1}z^{-2})f_n(qz)$$

$$= (1 - q^{2n+1}z^2)\left\{ \prod_{k=2}^{n} (1 - q^{2k-1}z^2)\right\}$$

$$\times qz^2(1 - q^{-1}z^{-2}) \prod_{j=1}^{n} (1 - q^{2j-1}z^{-2})$$

$$= -(1 - q^{2n+1}z^2) \prod_{k=1}^{n} (1 - q^{2k-1}z^2) \prod_{j=1}^{n} (1 - q^{2j-1}z^{-2})$$

$$= (q^{2n+1}z^2 - 1)f_n(z).$$

If we write the functions f_n in terms of the a_j, using (10.3), and equate the coefficients of z^{2k}, we find

$$qa_{k-1}q^{2k-2} - q^{2n}a_k q^{2k} = q^{2n+1}a_{k-1} - a_k$$

and then

$$a_{k-1} = \frac{-(1 - q^{2n+2k})}{q^{2k-1}(1 - q^{2n-2k+2})} a_k.$$

This, along with (10.4) allows us to find $a_{n-1}, a_{n-2}, \cdots$, in turn. In fact, for $0 < j \leqslant n$ we find

$$a_{n-j} = \frac{(-1)^j(1 - q^{4n})(1 - q^{4n-2}) \cdots (1 - q^{4n-2j+2})}{(1 - q^2)(1 - q^4) \cdots (1 - q^{2j})} (-1)^n q^{(n-j)^2}$$

and hence

$$a_k = \frac{\prod_{h=n+k+1}^{2n} (1 - q^{2h})}{\prod_{h=1}^{n-k} (1 - q^{2h})} (-1)^k q^{k^2} = \frac{\phi_{2n}(q^2)}{\phi_{n+k}(q^2)\phi_{n-k}(q^2)} (-1)^k q^{k^2}. \tag{10.5}$$

This formula is valid for $0 \leqslant k \leqslant n$ if we agree to take $\phi_0(q^2) = 1$.

Returning to (10.3), we see that $f_n(z)$ is a product of $2n$ factors, one of which is $(1 - qz^{-2})$, which has the value 0 at $z = q^{1/2}$. Therefore taking the derivative and then setting $z = q^{1/2}$ we have

$$f_n'(q^{1/2}) = \prod_{k=1}^{n} (1 - q^{2k-1}q)\left\{\prod_{j=2}^{n} (1 - q^{2j-1}q^{-1})\right\} 2qq^{-3/2}$$

$$= \frac{2q^{-1/2}}{1 - q^{2n}} \phi_n(q^2)^2.$$

On the other hand, we also have, from (10.3),

$$f_n'(q^{1/2}) = \sum_{j=-n}^{n} 2ja_j q^{j - 1/2} = \sum_{j-1}^{n} 2ja_j q^{-1/2}(q^j - q^{-j}).$$

Thus we find

$$\phi_n(q^2)^2 = (1 - q^{2n}) \sum_{j=1}^{n} ja_j(q^j - q^{-j})$$

and hence, by (10.5),

$$\phi_n(q^2)^3 = (1 - q^{2n}) \sum_{j=1}^{n} (-1)^j jq^{j^2}(q^j - q^{-j}) \frac{\phi_{2n}(q^2)\phi_n(q^2)}{\phi_{n+j}(q^2)\phi_{n-j}(q^2)}.$$

Now

$$0 \leqslant \frac{\phi_{2n}(q^2)\phi_n(q^2)}{\phi_{n+j}(q^2)\phi_{n-j}(q^2)} = \prod_{h=n+j+1}^{2n} (1 - q^{2h}) \prod_{k=n-j+1}^{n} (1 - q^{2k}) \leqslant 1,$$

and $\sum_{j=1}^{\infty} jq^{j^2}|q^j - q^{-j}|$ converges, so we have for $n > m$

$$\left|\phi_n(q^2)^3 - (1 - q^{2n}) \sum_{j=1}^{m} (-1)^j jq^{j^2}(q^j - q^{-j}) \frac{\phi_{2n}(q^2)\phi_n(q^2)}{\phi_{n+j}(q^2)\phi_{n-j}(q^2)}\right|$$

$$\leqslant \sum_{j=m+1}^{n} jq^{j^2}|q^j - q^{-j}| \leqslant \sum_{j=m+1}^{\infty} jq^{j^2}|q^j - q^{-j}|.$$

We keep m fixed but arbitrary and let $n \to \infty$. By Theorem 10.7 we have

$$\lim_{n \to \infty} \frac{\phi_{2n}(q^2)\phi_n(q^2)}{\phi_{n+j}(q^2)\phi_{n-j}(q^2)} = \frac{\phi(q^2)^2}{\phi(q^2)^2} = 1$$

and $\lim_{n \to \infty} \phi_n(q^2)^3 = \phi(q^2)^3$ so that we get

$$\left|\phi(q^2)^3 - \sum_{j=1}^{m} (-1)^j jq^{j^2}(q^j - q^{-j})\right| \leqslant \sum_{j=m+1}^{\infty} jq^{j^2}|q^j - q^{-j}|.$$

Now letting $m \to \infty$ we find

$$\phi(q^2)^3 = \sum_{j=1}^{\infty} (-1)^j jq^{j^2}(q^j - q^{-j})$$

$$= \sum_{j=1}^{\infty} (-1)^j jq^{j^2+j} + \sum_{j=1}^{\infty} (-1)^{j-1} jq^{j^2-j}$$

where we can make the last step because both series converge. Changing j to $j + 1$, we write the last series $\sum_{j=0}^{\infty}(-1)^j(j + 1)q^{j^2+j}$ and can then add it to the first series to obtain

$$\phi(q^2)^3 \sum_{j=0}^{\infty} (-1)^j(2j + 1)q^{j^2+j}.$$

This is our theorem with x replaced by q^2.

PROBLEM

1. Replace z by $q^{1/6}$ in (10.3), multiply by $\phi_n(q^2)$, and use (10.5) to obtain a proof of Euler's formula.

10.6 A DIVISIBILITY PROPERTY

Theorem 10.14 *If p is a prime and $0 \leqslant x < 1$, then*

$$\frac{\phi(x^p)}{\phi(x)^p} = 1 + p\sum_{j=1}^{\infty} a_j x^j$$

where the a_j are integers.

Proof For $0 \leqslant u < 1$ we have the expansion

$$(1-u)^{-p} = 1 \sum_{j=1}^{\infty} (-1)^j \frac{(-p)(-p-1)\cdots(-p-j+1)}{j!} u^j$$

$$= 1 + \sum_{j=1}^{\infty} \frac{(p+j-1)!}{j!\,(p-1)!} u^j = \sum_{j=0}^{\infty} b_j u^j$$

say, and therefore

$$\frac{1-u^p}{(1-u)^p} = (1-u)^{-p} - u^p(1-u)^{-p} = \sum_{j=0}^{\infty} b_j u^j - \sum_{j=0}^{\infty} b_j u^{j+p}$$

$$= \sum_{j=0}^{p-1} b_j u^j + \sum_{j=p}^{\infty} (b_j - b_{j-p}) u^j = \sum_{j=0}^{\infty} c_j u^j,$$

say. But

$$b_j = \frac{(j+1)(j+2)\cdots(j+p-1)}{(p-1)!} \equiv \begin{cases} 1 & (\text{mod } p) \text{ if } j \equiv 0 \ (\text{mod } p) \\ 0 & (\text{mod } p) \text{ if } j \not\equiv 0 \ (\text{mod } p) \end{cases}$$

and $b_0 < b_1 < b_2 < \cdots$, so that we have $c_0 = b_0 = 1$, $c_j > 0$, $c_j \equiv 0 \pmod p$ for $j > 0$.

Now, for $0 \leqslant x < 1$,

$$\frac{\phi_m(x^p)}{\phi_m(x)^p} = \prod_{n=1}^{m} \frac{1 - x^{pn}}{(1-x^n)^p} = \sum_{j=0}^{\infty} a_j^{(m)} x^j$$

where $a_j^{(1)} = c_j$ and, by Lemma 10.10,

$$\sum_{h=0}^{\infty} a_h^{(m)} x^h = \sum_{j=0}^{\infty} c_j x^{mj} \sum_{k=0}^{\infty} a_k^{(m-1)} x^k = \sum_{h=0}^{\infty} \sum_{j=0}^{[h/m]} c_j a_{h-mj}^{(m-1)} x^h.$$

By Lemma 10.10 we then have

$$a_h^{(m)} = \sum_{j=0}^{[h/m]} c_j a_{h-mj}^{(m-1)}$$

and hence

$$a_h^{(m)} \equiv a_h^{(m-1)} \equiv a_h^{(1)} \equiv c_h \pmod{p}$$

$$a_h^{(m)} \geqslant a_h^{(m-1)} \geqslant a_h^{(1)} = c_h > 0$$

$$a_h^{(m)} = a_h^{(m-1)} \text{ if } h \leqslant m - 1.$$

Therefore

$$\sum_{h=0}^{m} a_h^{(h)} x^h = \sum_{h=0}^{m} a_h^{(m)} x^h \leqslant \sum_{h=0}^{\infty} a_h^{(m)} x^h = \frac{\phi_m(x^p)}{\phi_m(x)^p}.$$

Since the sum on the left increases as $m \to \infty$, we see that $\sum_{h=0}^{\infty} a_h^{(h)} x^h$ converges and

$$\sum_{h=0}^{\infty} a_h^{(h)} x^h \leqslant \frac{\phi(x^p)}{\phi(x)^p}.$$

But we also have

$$\sum_{h=0}^{\infty} a_h^{(h)} x^h = \sum_{h=0}^{m} a_h^{(m)} x^h + \sum_{h=m+1}^{\infty} a_h^{(h)} x^h$$

$$\geqslant \sum_{h=0}^{m} a_h^{(m)} x^h + \sum_{h=m+1}^{\infty} a_h^{(m)} x^h = \frac{\phi_m(x^p)}{\phi_m(x)^p},$$

$$\sum_{h=0}^{\infty} a_h^{(h)} x^h \geqslant \frac{\phi(x^p)}{\phi(x)^p}$$

and finally

$$\sum_{h=0}^{\infty} a_h^{(h)} x^h = \frac{\phi(x^p)}{\phi(x)^p}.$$

Since $a_0^{(o)} = c_0 = 1$ and $a_h^{(h)} \equiv c_h \equiv 0 \pmod{p}$ for $h \geqslant 1$, the theorem is proved.

Theorem 10.15 *For $0 \leqslant x < 1$ we have $x\phi(x)^4 = \sum_{m=1}^{\infty} b_m x^m$ where the b_m are integers and $b_m \equiv 0 \pmod 5$ if $m \equiv 0 \pmod 5$.*

Proof We can write Theorem 10.9 in the form

$$\phi(x) = \sum_{k=0}^{\infty} c_k x^k, \qquad c_k = \begin{cases} (-1)^j & \text{if } k = (3j^2 \pm j)/2 \\ 0 & \text{otherwise,} \end{cases}$$

and Theorem 10.13 as

$$\phi(x)^3 = \sum_{n=0}^{\infty} d_n x^n, \qquad d_n = \begin{cases} (-1)^j(2j+1) & \text{if } n = (j^2+j)/2 \\ 0 & \text{otherwise,} \end{cases}$$

and then apply Lemma 10.10 to obtain

$$\begin{aligned} x\phi(x)^4 &= x\phi(x)\phi(x)^3 \\ &= x \sum_{h=0}^{\infty} \left(\sum_{k=0}^{h} c_k d_{h-k} \right) x^h = \sum_{m=1}^{\infty} b_m x^m. \end{aligned}$$

Then $b_m = \sum_{k=0}^{m-1} c_k d_{m-1-k}$ can be written as $\sum c_k d_n$ summed over all $k \geqslant 0$, $n \geqslant 0$, such that $k + n = m - 1$. But d_n is 0 unless $n = (j^2+j)/2$, $j = 0, 1, 2, \cdots$, in which case it is $(-1)^j(2j+1)$. Furthermore we can describe c_k by saying that it is 0 unless $k = (3i^2+i)/2$, $i = 0, \pm 1, \pm 2, \cdots$, in which case it is $(-1)^i$. Then we can write

$$b_m = \sum (-1)^i (-1)^j (2j+1) = \sum (-1)^{i+j} (2j+1) \tag{10.6}$$

summed over all i and j such that $j \geqslant 0$ and $(3i^2+i)/2 + (j^2+j)/2 = m - 1$. But

$$2(i+1)^2 + (2j+1)^2 = 8\left(1 + \frac{3i^2+i}{2} + \frac{j^2+j}{2}\right) + 10i^2 - 5$$

so that if $m \equiv 0 \pmod 5$, the terms in (10.6) will have to be such that $2(i+1)^2 + (2j+1)^2 \equiv 0 \pmod 5$. That is $(2j+1)^2 \equiv -2(i+1)^2 \pmod 5$. However, -2 is a quadratic nonresidue modulo 5, so this condition implies $2j+1 \equiv 0 \pmod 5$, and hence $b_m \equiv 0 \pmod 5$ if $m \equiv 0 \pmod 5$.

Theorem 10.16 *We have* $p(5m+4) \equiv 0 \pmod 5$.

Proof By Theorems 10.15, 10.14, and 10.8 we have

$$\sum_{n=0}^{\infty} p(n)x^{n+1} = \frac{x}{\phi(x)} = x\phi(x)^4 \frac{\phi(x^5)}{\phi(x)^5} \frac{1}{\phi(x^5)}$$

$$= \sum_{m=1}^{\infty} b_m x^m \left(1 + 5\sum_{j=1}^{\infty} a_j x^j\right) \sum_{k=0}^{\infty} p(k) x^{5k}$$

where the a_j and b_m are integers and $b_m \equiv 0 \pmod 5$ for $m \equiv 0 \pmod 5$. Using Lemma 10.10 we find that

$$p(n-1) \equiv \sum_{k=0}^{[n/5]} p(k) b_{n-5k} \pmod 5$$

and hence $p(5m+4) \equiv 0 \pmod 5$ since $b_{5m+5-5k} \equiv 0 \pmod 5$.

This theorem is only one of several divisibility properties of the partition function. The methods of this section can be used to prove that $p(7n+5) \equiv 0 \pmod 7$. With the aid of more extensive analysis, it can be shown that $p(5^k n + r) \equiv 0 \pmod{5^k}$ if $24r \equiv 1 \pmod{5^k}$, $k = 2, 3, 4, \cdots$, and there are still other congruences related to powers of 5. There are somewhat similar congruences related to powers of 7, but it is an interesting fact that $p(7^k n + r) \equiv 0 \pmod{7^k}$ if $24r \equiv 1 \pmod{7^k}$ is valid for $k = 1, 2$ but is false for $k = 3$. There are also divisibility properties related to the number 11. An identity typical of several connected with the divisibility properties is

$$\sum_{n=0}^{\infty} p(5n+4)x^n = 5\frac{\phi(x^5)^5}{\phi(x)^6}, \qquad |x| < 1.$$

PROBLEMS

1. Write Euler's formula as

$$\phi(x) = \sum_{j=-\infty}^{\infty} (-1)^j x^{(3j^2+j)/2}$$

Use Jacobi's formula as in Theorem 10.13, multiply $x\phi(x)\phi(x)^3$ out formally and verify (10.6).

2. Obtain a congruence similar to that in Theorem 10.16 but for the modulus 35, using Theorem 10.16 and $p(7n+5) \equiv 0 \pmod 7$.

NOTES ON CHAPTER 10

For a comprehensive survey of the entire subject of partitions, see the book by Andrews listed in the General References.

The proof given of Theorem 10.4, by F. Franklin, has been called by George E. Andrews "one of the truly remarkable achievements of nineteenth-century American mathematics."

Problem 6 of Section 10.3, giving a formula for the number of incongruent triangles of perimeter n, has been adapted from a short paper by George E. Andrews, "A note on partitions and triangles with integer sides," *Amer. Math. Monthly* 86 (1979), 477–478.

For a fuller discussion of the methods of Section 10.3, including, for example, a proof avoiding all questions of convergence of the basic recurrence formula (Theorem 10.11) for the sum of divisors function $\sigma(n)$, see Ivan Niven, "Formal power series," *Amer. Math. Monthly* 76 (1969), 871–889.

Theorem 10.16 is due to S. Ramanujan. For further congruence properties of partitions, see M. I. Knopp, *Modular Functions in Analytic Number Theory*, Markham, Chicago (1970), Chapters 7 & 8.

Consider the question of the number of abelian groups of order q^n, where q is a prime and n is positive. The answer is $p(n)$, the number of partitions of n. For a proof, see Herstein, *Topics in Algebra*, p. 114.

A clear expository account of identities has been given by Henry L. Alder, "Partition identities ... from Euler to the present," *Amer. Math. Monthly*, 76 (1969), 733–746.

Some interesting historical aspects of partition theory are discussed by G. E. Andrews in an article "J. J. Sylvester, Johns Hopkins and Partitions" in *A Century of Mathematics in America*, Part I, P. Duren, editor, Amer. Math. Society, Providence, R.I. (1988), pp. 21–40. The bibliography in this paper cites other basic articles on partitions, including several quite accessible expository papers.

CHAPTER 11

The Density of Sequences of Integers

In order even to define what is meant by the density of a sequence of integers, it is necessary to use certain concepts from analysis. In this chapter, it is assumed that the reader is familiar with the ideas of the limit inferior of a sequence of real numbers and the greatest lower bound, or infimum, of a set of real numbers.

Two common types of density are considered in this chapter, asymptotic density and Schnirelmann density. The first is discussed in Section 11.1 and the second in Section 11.2. Density will be defined for a set $\mathscr{A}$ of distinct positive integers. We will think of the elements of $\mathscr{A}$ as being arranged in a sequence according to size,

$$a_1 < a_2 < a_3 < \cdots \tag{11.1}$$

and we will also denote $\mathscr{A}$ by $\{a_i\}$. Furthermore we will use both the terms set and sequence to describe $\mathscr{A}$. The set $\mathscr{A}$ may be infinite or finite. That is, it may contain infinitely many elements or only a finite number of elements. It may even be empty, in which case it will be denoted by $\varnothing$. If an integer m is an element of $\mathscr{A}$ we write $m \in \mathscr{A}$; if not, we write $m \notin \mathscr{A}$. The set $\mathscr{A}$ is contained in $\mathscr{B}$, $\mathscr{A} \subset \mathscr{B}$ or $\mathscr{B} \supset \mathscr{A}$, if every element of $\mathscr{A}$ is an element of $\mathscr{B}$. We write $\mathscr{A} = \mathscr{B}$ if $\mathscr{A} \subset \mathscr{B}$ and $\mathscr{B} \subset \mathscr{A}$, that is if $\mathscr{A}$ and $\mathscr{B}$ have precisely the same elements. The *union* $\mathscr{A} \cup \mathscr{B}$ of two sets $\mathscr{A}$ and $\mathscr{B}$ is the set of all elements m such that $m \in \mathscr{A}$ or $m \in \mathscr{B}$. The *intersection* $\mathscr{A} \cap \mathscr{B}$ of $\mathscr{A}$ and $\mathscr{B}$ is the set of all m such that $m \in \mathscr{A}$ and $m \in \mathscr{B}$. Thus, for example $\mathscr{A} \cup \mathscr{A} = \mathscr{A} \cap \mathscr{A} = \mathscr{A}$, $\mathscr{A} \cup \varnothing = \mathscr{A}$, $\mathscr{A} \cap \varnothing = \varnothing$. If $\mathscr{A}$ and $\mathscr{B}$ have no element in common, $\mathscr{A} \cap \mathscr{B} = \varnothing$, $\mathscr{A}$ and $\mathscr{B}$ are said to be *disjoint*. By the *complement* $\overline{\mathscr{A}}$ *of* $\mathscr{A}$ *we mean the set of all positive integers that are not elements of* $\mathscr{A}$. *Thus* $\mathscr{A} \cap \overline{\mathscr{A}} = \varnothing$ *and* $\overline{\varnothing}$ *is the set of all positive integers.*

11.1 ASYMPTOTIC DENSITY

The number of positive integers in a set $\mathscr{A}$ that are less than or equal to x is denoted by $A(x)$. For example, if $\mathscr{A}$ consists of the even integers $2, 4, 6, \cdots$, then $A(1) = 0$, $A(2) = 1$, $A(6) = 3$, $A(7) = 3$, $A(15/2) = 3$; in fact $A(x) = [x/2]$ if $x \geqslant 0$. On the other hand, for any set $\mathscr{A} = \{a_i\}$ we have $A(a_j) = j$.

Definition 11.1 *The* asymptotic density *of a set* $\mathscr{A}$ *is*

$$\delta_1(\mathscr{A}) = \liminf_{n \to \infty} \frac{A(n)}{n}. \tag{11.2}$$

In case the sequence $A(n)/n$ *has a limit, we say that* $\mathscr{A}$ *has a* natural density, $\delta(\mathscr{A})$. *Thus*

$$\delta(\mathscr{A}) = \delta_1(\mathscr{A}) = \lim_{n \to \infty} \frac{A(n)}{n} \tag{11.3}$$

if A has a natural density.

If $\mathscr{A}$ is a finite sequence, it is clear that $\delta(\mathscr{A}) = 0$.

Theorem 11.1 *If* $\mathscr{A}$ *is an infinite sequence, then*

$$\delta_1(\mathscr{A}) = \liminf_{n \to \infty} \frac{n}{a_n}.$$

If $\delta(\mathscr{A})$ *exists, then* $\delta(\mathscr{A}) = \lim_{n \to \infty} n/a_n$.

Proof The sequence k/a_k is a subsequence of $A(n)/n$ and hence

$$\liminf_{n \to \infty} \frac{A(n)}{n} \leqslant \liminf_{k \to \infty} \frac{k}{a_k}.$$

If n is any integer $\geqslant a_1$ and a_k is the smallest integer in $\mathscr{A}$ that exceeds n, then $a_{k-1} \leqslant n < a_k$ and

$$\frac{k}{a_k} - \frac{A(n)}{n} = \frac{k}{a_k} - \frac{k-1}{n} < \frac{k}{n} - \frac{k-1}{n} = \frac{1}{n}.$$

It follows that

$$\frac{k}{a_k} < \frac{A(n)}{n} + \frac{1}{n}, \qquad \liminf_{k \to \infty} \frac{k}{a_k} \leqslant \liminf_{n \to \infty} \frac{A(n)}{n}$$

and so the theorem is proved.

Although we have proved in Theorem 8.25 that the set of square-free integers has density $6/\pi^2$, this information alone does not imply the following additive property.

Theorem 11.2 *Every integer greater than* 1 *can be written as a sum of two square-free integers.*

The proof of this is based on the following preliminary result.

Lemma 11.3 *For every positive integer* n, *if* $Q(n)$ *denotes the number of square-free integers among* $1, 2, \cdots, n$, *then* $Q(n) > n/2$.

Proof Let $\mathscr{A}_{p,n}$ denote the set of integers k with the properties $p^2 | k$ and $1 \leqslant k \leqslant n$, where p is any prime. Let $\mathscr{B}_n$ denote the union over all primes p of the sets $\mathscr{A}_{p,n}$. The elements of $\mathscr{B}_n$ are precisely the positive integers $\leqslant n$ that are not square-free. It follows that $Q(n) + |\mathscr{B}_n| = n$, where $|\mathscr{B}_n|$ is the number of elements in $\mathscr{B}_n$. The number of elements in $\mathscr{A}_{p,n}$ is $[n/p^2]$, and hence

$$|\mathscr{B}_n| \leqslant \sum_p [n/p^2] \leqslant n \sum_p 1/p^2 \qquad \text{and} \qquad Q(n) \geqslant n - n\sum 1/p^2.$$

We prove that $\Sigma 1/p^2 < 1/2$, and this gives $Q(n) > n - n/2 = n/2$. Since all primes $p > 2$ are odd, the sum in question is $< 1/4 + \Sigma_{k=1}^{\infty}(2k+1)^{-2}$. But $(2k+1)^2 = 4k^2 + 4k + 1 > 4k^2 + 4k = 4k(k+1)$, so that

$$\sum_{k=1}^{\infty} \frac{1}{(2k+1)^2} < \frac{1}{4} \sum_{k=1}^{\infty} \frac{1}{k(k+1)} = \frac{1}{4} \sum_{k=1}^{\infty} \left(\frac{1}{k} - \frac{1}{k+1} \right) = \frac{1}{4}.$$

This gives the stated estimate.

Proof of Theorem 11.2 This is an easy consequence of the lemma, by the following argument. Let $\mathscr{A}$ denote the set of those integers a, $1 \leqslant a \leqslant n-1$, such that a is square-free, and let $\mathscr{A}'$ denote the set of those integers a', $1 \leqslant a' \leqslant n-1$, such that $n - a'$ is square-free. Then $|\mathscr{A}| = |\mathscr{A}'| = Q(n-1)$. Since $|\mathscr{A}| + |\mathscr{A}'| + 2Q(n-1) > n - 1$, it follows that $\mathscr{A}$ and $\mathscr{A}'$ cannot be disjoint. That is, there is an integer a such that a

and $n - a$ are both square-free. Since $n = a + (n - a)$, this is the desired result.

PROBLEMS

1. Prove that each of the following sets has a natural density, and find its value:

(a) the set of even positive integers;
(b) the set of odd positive integers;
(c) the positive multiples of 3;
(d) the positive integers of the form $4k + 2$;
(e) all positive integers a satisfying $a \equiv b \pmod{m}$, where b and $m > 1$ are fixed;
(f) the set of primes;
(g) the set $\{ar^n\}$ with $n = 1, 2, 3, \cdots$ and fixed $a \geqslant 1$, fixed $r > 1$;
(h) the set of all perfect squares;
(i) the set of all positive cubes;
(j) the set of all positive powers, that is, all numbers of the form a^n with $a \geqslant 1$, $n \geqslant 2$.

2. If the natural density $\delta(\mathscr{A})$ exists, prove that $\delta(\overline{\mathscr{A}})$ also exists and that $\delta(\mathscr{A}) + \delta(\overline{\mathscr{A}}) = 1$.

3. Prove that $\delta(\mathscr{A})$ exists if and only if $\delta_1(\mathscr{A}) + \delta_1(\overline{\mathscr{A}}) = 1$.

4. For any set $\mathscr{A}$, prove that $\delta_1(\mathscr{A}) + \delta_1(\overline{\mathscr{A}}) \leqslant 1$.

5. Define $\mathscr{A}_n$ as the set of all a such that $(2n)! \leqslant a < (2n + 1)!$ and let $\mathscr{A}$ be the union of all sets $\mathscr{A}_n$, $n = 1, 2, 3, \cdots$. Prove that $\delta_1(\mathscr{A}) + \delta_1(\overline{\mathscr{A}}) = 0$.

6. Let $\mathscr{A}^*$ be the set remaining after a finite number of integers are deleted from a set $\mathscr{A}$. Prove that $\delta_1(\mathscr{A}) = \delta_1(\mathscr{A}^*)$, and that $\delta(\mathscr{A})$ exists if and only if $\delta(\mathscr{A}^*)$ exists.

7. If two sets $\mathscr{A}$ and $\mathscr{B}$ are identical beyond a fixed integer n, prove that $\delta_1(\mathscr{A}) = \delta_1(\mathscr{B})$.

8. Given any set $\mathscr{A} = \{a_j\}$ and any integer $b \geqslant 0$, define $\mathscr{B} = \{b + a_j\}$. Prove that $\delta_1(\mathscr{A}) = \delta_1(\mathscr{B})$.

9. Let $\mathscr{A}$ be the set of all even positive integers, $\mathscr{B}_1$ the set of all even positive integers with an even number of digits to base ten, and $\mathscr{B}_2$ the set of all odd positive integers with an odd number of digits. Define $\mathscr{B} = \mathscr{B}_1 \cup \mathscr{B}_2$, and prove that $\delta(\mathscr{A})$ and $\delta(\mathscr{B})$ exist, but that $\delta(\mathscr{A} \cup \mathscr{B})$ and $\delta(\mathscr{A} \cap \mathscr{B})$ do not exist.

10. If $\mathscr{A} \cap \mathscr{B} = 0$, prove that $\delta_1(\mathscr{A} \cup \mathscr{B}) \geqslant \delta_1(\mathscr{A}) + \delta_1(\mathscr{B})$.

11. Let $\mathscr{S}$ denote any finite set of positive integers $a_1, a_2, \cdots, a_m$. Prove that the set $\mathscr{A}$ of all positive integers not divisible by any member of $\mathscr{S}$ has natural density

$$1 - \sum_{i=1}^{m} \frac{1}{a_i} + \sum_{i<j} \frac{1}{[a_i, a_j]}$$

$$- \sum_{i<j<k} \frac{1}{[a_i, a_j, a_k]} + \cdots + \frac{(-1)^m}{[a_1, a_2, \cdots, a_m]}.$$

Suggestion: Use the inclusion-exclusion principle of Section 4.5.

12. Let $\mathscr{A}$ be a set of positive integers such that for every integer m, the equation $x + y = m$ has at most one solution not counting order, with x and y in $\mathscr{A}$. Prove that $\mathscr{A}$ has density zero. Even more, prove that $A(n) \leqslant 2\sqrt{n}$.

13. Define $\mathscr{A} = \{a_j\}$ as follows. With $a_1 = 1$, define a_{k+1} as the least positive integer that is different from all the numbers $a_h + a_i - a_j$, with $1 \leqslant h \leqslant k$, $1 \leqslant i \leqslant k$, $1 \leqslant j \leqslant k$. prove that $\mathscr{A}$ satisfies the inequality of the preceding problem, and that $A(n) \geqslant \sqrt[3]{n} - 1$.

***14.** Let $\mathscr{P}$ be the set of integers $\{m^k\}$ with $m = 1, 2, 3, \cdots$ and $k = 2, 3, 4, \cdots$. Let $\mathscr{P}_1$ be the subset with $k = 3, 4, \cdots$. Prove that

$$\lim_{n \to \infty} \frac{P_1(n)}{P(n)} = 0.$$

15. Find the asymptotic density of the set of positive integers having an odd number of digits in base 10 representation.

16. If $\mathscr{A} = \{a_1, a_2, a_3, \cdots\}$ is an increasing sequence of positive integers with positive natural density, prove that $\lim (a_n - a_{n-1})/a_n = 0$ as n tends to infinity.

11.2 SCHNIRELMANN DENSITY AND THE αβ THEOREM

Definition 11.2 *The* Schnirelmann density $d(\mathscr{A})$ *of a set* $\mathscr{A}$ *of non-negative integers is*

$$d(\mathscr{A}) = \inf_{n \geqslant 1} \frac{A(n)}{n}$$

where $A(n)$ *is the number of positive integers* $\leqslant n$ *in the set* $\mathscr{A}$.

Comparing this with Definition 11.1 we immediately see that $0 \leqslant d(\mathscr{A}) \leqslant \delta_1(A) \leqslant 1$. Schnirelmann density differs from asymptotic density in that it is sensitive to the first terms in the sequence. Indeed if $1 \notin \mathscr{A}$

then $d(\mathscr{A}) = 0$, if $2 \notin \mathscr{A}$ then $d(\mathscr{A}) \leqslant \frac{1}{2}$, whereas it is easy to see that $\delta_1(\mathscr{A})$ is unchanged if the numbers 1 or 2 are removed from or adjoined to $\mathscr{A}$. Also, $d(\mathscr{A}) = 1$ if and only if $\mathscr{A}$ contains all the positive integers.

Until now we have been considering sets $\mathscr{A}$ consisting only of positive integers. However, Definition 11.2 is worded in such a way that $\mathscr{A}$ can contain 0, but it should be noted that the number 0 is not counted by $A(n)$.

Definition 11.3 *Assume that $0 \in \mathscr{A}$ and $0 \in \mathscr{B}$. The* sum $\mathscr{A} + \mathscr{B}$ *of the sets A and B is the collection of all integers of the form $a + b$ where $a \in A$ and $b \in B$.*

Note that $\mathscr{A} \subset \mathscr{A} + \mathscr{B}$, $\mathscr{B} \subset \mathscr{A} + \mathscr{B}$. As an example let us take $\mathscr{S}$ to be the set of squares $0, 1, 4, 9, \cdots$ and $\mathscr{I}$ the set of all non-negative integers. Then by Theorem 6.26 we see that $\mathscr{S} + \mathscr{S} + \mathscr{S} + \mathscr{S} = \mathscr{I}$.

The sum $\mathscr{A} + \mathscr{B}$ has not been defined unless $0 \in \mathscr{A}$ and $0 \in \mathscr{B}$. We shall assume that 0 is in both $\mathscr{A}$ and $\mathscr{B}$ in the rest of this chapter. However, the sum could be defined for all $\mathscr{A}$ and $\mathscr{B}$ as the sum of the sets obtained from $\mathscr{A}$ and $\mathscr{B}$ by adjoining the number 0 to each. This is equivalent to defining the sum as the collection $\{a, b, a + b\}$ with $a \in \mathscr{A}$, $b \in \mathscr{B}$.

The result that is proved in the remainder of this section is the $\alpha\beta$ theorem of H. B. Mann, which was conjectured about 1931, with proofs attempted subsequently by many mathematicians. The theorem states that if $\mathscr{A}$ and $\mathscr{B}$ are sets of non-negative integers, each containing 0, and if α, β, γ are the Schnirelmann densities of $\mathscr{A}, \mathscr{B}, \mathscr{A} + \mathscr{B}$, then $\gamma \geqslant \min(1, \alpha + \beta)$. In other words $\gamma \geqslant \alpha + \beta$ unless $\alpha + \beta > 1$, in which case $\gamma = 1$.

Actually we shall prove a somewhat stronger result, Theorem 11.9, from which we shall deduce the $\alpha\beta$ theorem. We start by considering any positive integer g and two sets $\mathscr{A}_1$ and $\mathscr{B}_1$ of non-negative integers not exceeding g. We assume throughout that $\mathscr{A}_1$ and $\mathscr{B}_1$ are such sets and that 0 belongs to both $\mathscr{A}_1$ and $\mathscr{B}_1$. Denoting $\mathscr{A}_1 + \mathscr{B}_1$ by $\mathscr{C}_1$, we observe that $\mathscr{C}_1$ may have elements $> g$ even though $\mathscr{A}_1$ and $\mathscr{B}_1$ do not. We also assume that for some θ, $0 < \theta \leqslant 1$,

$$A_1(m) + B_1(m) \geqslant \theta m, \qquad m = 1, 2, \cdots, g. \tag{11.4}$$

Our idea is to first replace $\mathscr{A}_1$ and $\mathscr{B}_1$ by two new sets, $\mathscr{A}_2$ and $\mathscr{B}_2$, in such a way that (11.4) holds for $\mathscr{A}_2$ and $\mathscr{B}_2$, that $\mathscr{C}_2 = \mathscr{A}_2 + \mathscr{B}_2 \subset \mathscr{C}_1$, and that $B_2(g) < B_1(g)$.

Lemma 11.4 *Let $\mathscr{A}_1$ and $\mathscr{B}_1$ satisfy (11.4). If $\mathscr{B}_1 \not\subset \mathscr{A}_1$, then there exist sets $\mathscr{A}_2$ and $\mathscr{B}_2$ with $\mathscr{C}_2 = \mathscr{A}_2 + \mathscr{B}_2$ such that $\mathscr{C}_2 \subset \mathscr{C}_1$, $B_2(g) < B_1(g)$ and $A_2(m) + B_2(m) \geqslant \theta m$ for $m = 1, 2, \cdots, g$.*

Proof We merely shift to $\mathscr{A}_1$ all elements of $\mathscr{B}_1$ that are not already in $\mathscr{A}_1$. Define $\mathscr{B}' = \mathscr{B}_1 \cap \overline{\mathscr{A}}_1$, $\mathscr{A}_2 = \mathscr{A}_1 \cup \mathscr{B}'$, $\mathscr{B}_2 = \mathscr{B}_1 \cap \overline{\mathscr{B}}'$, where by $\overline{\mathscr{A}}_1$ we mean the complement of $\mathscr{A}_1$, now the set of all non-negative integers not in $\mathscr{A}_1$. Thus 0 belongs to both $\mathscr{A}_2$ and $\mathscr{B}_2$. Then $\mathscr{A}_2(m) = \mathscr{A}_1(m) + \mathscr{B}'(m)$ and $B_2(m) = B_1(m) - B'(m)$, so we have $A_2(m) + B_2(m) = A_1(m) + B_1(m) \geqslant \theta m$ for $m = 1, 2, \cdots, g$. Now consider any $h \in \mathscr{C}_2$. Then $h = a + b$ with $a \in \mathscr{A}_2$ and $b \in \mathscr{B}_2$. Noting that $\mathscr{B}_2$ is contained in both $\mathscr{A}_1$ and $\mathscr{B}_1$ and that $\mathscr{A}_2 = \mathscr{A}_1 \cup \mathscr{B}'$, we have either $a \in \mathscr{A}_1$ or $a \in \mathscr{B}' \subset \mathscr{B}_1$. In the first case we can write $h = a + b$, $a \in \mathscr{A}_1$, $b \in \mathscr{B}_1$; in the second case $h = b + a$, $b \in \mathscr{A}_1$, $a \in \mathscr{B}_1$; hence in both cases we have $h \in \mathscr{C}_1$. Thus we have $\mathscr{C}_2 \subset \mathscr{C}_1$. Since it is obvious that $B_2(g) < B_1(g)$, the lemma is proved.

We shall get a similar result for the case $\mathscr{B}_1 \subset \mathscr{A}_1$, but it is a little more complicated. We assume $B_1(g) > 0$, which implies that there is some integer $b > 0$ in $\mathscr{B}_1$. Then if a is the largest integer in $\mathscr{A}_1$, the sum $a + b$ is certainly not in $\mathscr{A}_1$. There may be other pairs $a \in \mathscr{A}_1$, $b \in \mathscr{B}_1$ such that $a + b \notin \mathscr{A}_1$. We let a_0 denote the smallest $a \in \mathscr{A}_1$ such that there is a $b \in \mathscr{B}_1$ for which $a + b \notin \mathscr{A}_1$. Since $\mathscr{B}_1 \subset \mathscr{A}_1$ we see that $a_0 \neq 0$. Before defining $\mathscr{A}_2$ and $\mathscr{B}_2$ we shall obtain two preliminary results.

Lemma 11.5 *Let $\mathscr{A}_1$ and $\mathscr{B}_1$ satisfy $\mathscr{B}_1 \subset \mathscr{A}_1$ and $B_1(g) > 0$. Let a_0 be defined as above. Suppose that there are integers b and z such that $b \in \mathscr{B}_1$ and $z - a_0 < b \leqslant z \leqslant g$. Then for each $a \in \mathscr{A}_1$ such that $1 \leqslant a \leqslant z - b$, we have $a + b \in \mathscr{A}_1$, and*

$$A_1(z) \geqslant A_1(b) + A_1(z - b) \tag{11.5}$$

Proof We have $a \leqslant z - b < a_0$ and $a + b \leqslant z \leqslant g$, hence $a + b \in \mathscr{A}_1$ because a_0 is minimal. Now there are $A_1(z - b)$ positive integers a belonging to $\mathscr{A}_1$ with $a \leqslant z - b$, and to each such a the corresponding $a + b$ also belongs to $\mathscr{A}_1$. Furthermore, each such $a + b$ satisfies $b < a + b \leqslant z$, and hence $A_1(z) - A_1(b) \geqslant A_1(z - b)$, and we have (11.5).

Lemma 11.6 *Let $\mathscr{A}_1$ and $\mathscr{B}_1$ satisfy (11.4), $\mathscr{B}_1 \subset \mathscr{A}_1$, and $B_1(g) > 0$. Define a_0 as before. If there is an integer $y \leqslant g$ such that $A_1(y) < \theta y$, then $y > a_0$.*

Proof Let z be the least integer such that $A_1(z) < \theta z$. Then $y \geqslant z \geqslant 1$. Since $A_1(z) + B_1(z) \geqslant \theta z$ we have $B_1(z) > 0$, and hence there is a $b \in \mathscr{B}_1$ such that $0 < b \leqslant z \leqslant g$. If $z \leqslant a_0$, we would have $z - a_0 < b \leqslant z \leqslant g$, and we could apply Lemma 11.5 to get $A_1(z) \geqslant A_1(b) + A_1(z-b)$. Now $b \in \mathscr{B}_1 \subset \mathscr{A}_1$, so we have $A_1(b) = A_1(b-1) + 1 \geqslant \theta(b-1) + 1$ since $b - 1 < z$. Also, $A_1(z-b) \geqslant \theta(z-b)$, and we are led to the contradiction $A_1(z) \geqslant \theta(b-1) + 1 + \theta(z-b) = \theta(z-1) + 1 \geqslant \theta z$. Therefore, we have $z > a_0$, and hence $y > a_0$.

Lemma 11.7 *Let $\mathscr{A}_1$ and $\mathscr{B}_1$ satisfy $\mathscr{B}_1 \subset \mathscr{A}_1$ and $B_1(g) > 0$. Let $\mathscr{B}'$ denote the set of all $b \in \mathscr{B}_1$ such that $a_0 + b \notin \mathscr{A}_1$, and let $\mathscr{A}'$ denote the set of all integers $a_0 + b$ such that $b \in \mathscr{B}'$ and $a_0 + b \leqslant g$. Finally let $\mathscr{A}_2 = \mathscr{A}_1 \cup \mathscr{A}'$ and $\mathscr{B}_2 = \mathscr{B}_1 \cap \overline{\mathscr{B}'}$. Then $\mathscr{C}_2 \subset \mathscr{C}_1$ and $\mathscr{B}_2(g) < B_1(g)$.*

Proof Note that $0 \in \mathscr{A}_2$ and $0 \in \mathscr{B}_2$, so that the sum $\mathscr{C}_2$ is well defined. If $h \in \mathscr{C}_2$, then $h = a + b$, $a \in \mathscr{A}_1 \cup \mathscr{A}'$, $b \in \mathscr{B}_1 \cap \overline{\mathscr{B}'}$. If $a \in \mathscr{A}_1$, then $h = a + b \in \mathscr{C}_1$ since $a \in \mathscr{A}_1$, $b \in \mathscr{B}_1$. If $a \in \mathscr{A}'$, then $a = a_0 + b_1$ for some $b_1 \in \mathscr{B}'$, and we have $h = a_0 + b + b_1$. Here $a_0 + b \in \mathscr{A}_1$ since otherwise we would have $b \in \mathscr{B}'$. Since $b_1 \in \mathscr{B}_1$, we again have $h \in \mathscr{C}_1$. Finally $B_2\,(g) < B_1(g)$, since the definition of a_0 ensures that $B'(g) > 0$.

Lemma 11.8 *For $\mathscr{A}_1$, $\mathscr{B}_1$, $\mathscr{A}_2$, $\mathscr{B}_2$ as in Lemma 11.7, if $\mathscr{A}_1$, $\mathscr{B}_1$ satisfy (11.4) then*

$$A_2(m) + B_2(m) \geqslant \theta m \qquad \text{for} \qquad m = 1, 2, \cdots, g. \tag{11.6}$$

Proof From the way A', B', A_2, B_2 were chosen we have

$$A_2(m) = A_1(m) + A'(m)$$

$$B_2(m) = B_1(m) - B'(m)$$

$$A'(m) = B'(m - a_0)$$

$$A_2(m) + B_2(m) = A_1(m) + B_1(m) - (B'(m) - B'(m - a_0))$$

for $m = 1, 2, \cdots, g$. Therefore (11.6) holds for all m for which $B'(m) = B'(m - a_0)$. Consider any $m \leqslant g$ for which $B'(m) > B'(m - a_0)$. Then $B_1(m) - B_1(m - a_0) \geqslant B'(m) - B'(m - a_0) > 0$, and we let b_0 denote the smallest element of B_1 such that $m - a_0 < b_0 \leqslant m$. Therefore

$$\begin{aligned} A_2(m) + B_2(m) &\geqslant A_1(m) + B_1(m) - (B_1(m) - B_1(m - a_0)) \\ &= A_1(m) + B_1(m - a_0) \\ &= A_1(m) + B_1(b_0 - 1). \end{aligned} \tag{11.7}$$

Now $m - a_0 < b_0 \leqslant m \leqslant g$, so we can apply Lemma 11.5 with $b = b_0$ and $z = m$ to get

$$A_1(m) \geqslant A_1(b_0) + A_1(m - b_0).$$

We also have $m - b_0 < a_0$ so Lemma 11.6 shows that

$$A_1(m - b_0) \geqslant \theta(m - b_0).$$

Thus we can reduce (11.7) to

$$A_2(m) + B_2(m) \geqslant A_1(b_0) + \theta(m - b_0) + B_1(b_0 - 1).$$

But $b_0 \in \mathscr{B}_1 \subset \mathscr{A}_1$, so we have $A_1(b_0) = A_1(b_0 - 1) + 1$. Using this and (11.4) we have,

$$\begin{aligned} A_2(m) + B_2(m) &\geqslant A_1(b_0 - 1) + B_1(b_0 - 1) + 1 + \theta(m - b_0) \\ &\geqslant \theta(b_0 - 1) + 1 + \theta(m - b_0) \\ &\geqslant \theta m. \end{aligned}$$

Theorem 11.9 *For any positive integer g let $\mathscr{A}_1$ and $\mathscr{B}_1$ denote fixed sets of non-negative integers $\leqslant g$. Let 0 belong to both sets $\mathscr{A}_1$ and $\mathscr{B}_1$, and write $\mathscr{C}_1$ for $\mathscr{A}_1 + \mathscr{B}_1$. If for some θ such that $0 < \theta \leqslant 1$,*

$$A_1(m) + B_1(m) \geqslant \theta m, \qquad m = 1, 2, \cdots, g$$

then $C_1(g) \geqslant \theta g$.

Proof If $B_1(g) = 0$, then $\mathscr{B}_1$ consists of the single integer 0, $\mathscr{C}_1 = \mathscr{A}_1$, and $C_1(g) = A_1(g) = A_1(g) + B_1(g) \geqslant \theta g$. We prove the theorem for general sets by mathematical induction. Suppose $k \geqslant 1$ and that the theorem is true for all $\mathscr{A}_1, \mathscr{B}_1$ with $B_1(g) < k$. If $A_1(m) + B_1(m) \geqslant \theta$ for $m = 1, 2, \cdots, g$, and if $B_1(g) = k$, then Lemma 11.4 or Lemmas 11.7 and 11.8 supply us with sets $\mathscr{A}_2, \mathscr{B}_2$ such that $B_2(g) < k$, $\mathscr{C}_2 \subset \mathscr{C}_1$, and $A_2(m) + B_2(m) \geqslant \theta m$ for $m = 1, 2, \cdots, g$. Therefore, by our induction hypothesis, we have $\mathscr{C}_2(g) \geqslant \theta g$, which implies $C_1(g) \geqslant \theta g$.

Theorem 11.10 *The $\alpha\beta$ theorem. Let $\mathscr{A}$ and $\mathscr{B}$ be any sets of non-negative integers, each containing 0, and let α, β, γ denote the Schnirelmann densities of $\mathscr{A}$, $\mathscr{B}$, $\mathscr{A} + \mathscr{B}$ respectively. Then $\gamma \geqslant \min(1, \alpha + \beta)$.*

Proof Let $\mathscr{A}_1$ and $\mathscr{B}_1$ consist of the elements of $\mathscr{A}$ and $\mathscr{B}$, respectively, that do not exceed g, an arbitrary positive integer. Then $A_1(m) \geqslant \alpha m$ and $B_1(m) \geqslant \beta m$ for $m = 1, 2, \cdots, g$. If we take $\theta = \min(1, \alpha + \beta)$, the conditions of Theorem 11.9 are satisfied and we conclude that $C_1(g) \geqslant \theta g$.

Since $C_1(g) \geqslant \theta g$ for every positive integer g, we have $\gamma \geqslant \theta = \min(1, \alpha + \beta)$.

PROBLEMS

1. What is the Schnirelmann density of the set of positive odd integers? The set of positive even integers? The set of positive integers $\equiv 1 \pmod 3$? The set of positive integers $\equiv 1 \pmod m$?
2. Prove that the analogue of Theorem 11.1 for Schnirelmann density, namely, $d(\mathscr{A}) = \inf n/a_n$, is false.
3. Prove that the analogue of Theorem 11.10 for asymptotic density is false. *Suggestion*: Take $\mathscr{A}$ as the set of all positive even integers, and consider $\mathscr{A} + \mathscr{A}$.
4. Prove that if $d(\mathscr{A}) = \alpha$, then $A(n) \geqslant \alpha n$ for every positive integer n. Prove that the analogue of this for asymptotic density is false.
5. Establish that Theorem 11.10 does not imply Theorem 11.9 by considering the sets $\mathscr{A} = \{0, 1, 2, 4, 6, 8, 10, \cdots\}$, $\mathscr{B} = \{0, 2, 4, 6, 8, 10, \cdots\}$. Theorem 11.10 asserts that the density of $\mathscr{A} + \mathscr{B}$ is $\geqslant \frac{1}{2}$, whereas Theorem 11.9 says much more.
6. Exhibit two sets $\mathscr{A}$ and $\mathscr{B}$ such that $d(\mathscr{A}) = d(\mathscr{B}) = 0$, $d(\mathscr{A} + \mathscr{B}) = 1$.
7. For any two sets $\mathscr{A}$ and $\mathscr{B}$ of non-negative integers, write $\alpha = d(\mathscr{A})$, $\beta = d(\mathscr{B})$, $\gamma = d(\mathscr{A} + \mathscr{B})$. Prove that $\gamma \geqslant \alpha + \beta - \alpha\beta$.
8. Consider a set $\mathscr{A}$ with positive Schnirelmann density. Prove that for some positive integer n
$$n\mathscr{A} = (n+1)\mathscr{A} = (n+2)\mathscr{A} = \cdots = \mathscr{I}$$
where $\mathscr{I}$ is the set of all non-negative integers, and $n\mathscr{A} = \mathscr{A} + \mathscr{A} + \cdots + \mathscr{A}$ with n summands.

NOTES ON CHAPTER 11

For further reading on the subject of this chapter, see the book by Halberstam and Roth listed in the General References and the following:

E. Artin and P. Scherk, "On the sums of two sets of integers," *Ann. Math.* (2) 44 (1943), 138–142.

F. J. Dyson, "A theorem on the densities of sets of integers," *J. London Math. Soc.* 20 (1945), 8–14.

H. B. Mann, "A proof of the fundamental theorem on the density of sums of sets of positive integers," *Ann. Math.* (2) 43 (1942), 523–527.

H. B. Mann, *Addition Theorems*, Wiley-Interscience (New York), 1965.

Appendices

We present a number of disconnected topics of algebra and analysis which are useful at various points in the book, and with which the reader might not be familiar.

A.1 THE FUNDAMENTAL THEOREM OF ALGEBRA

A simple proof of this theorem can be given using the argument principle in the theory of analytic functions of a complex variable, but we give here an elementary proof that depends on compactness and on the simplest algebraic properties and inequalities concerning complex numbers. We begin with a basic lemma.

Lemma A.1 *Let $P(z)$ be a polynomial of degree at least one, whose coefficients are complex numbers. If $P(z_0) \neq 0$, then the point z_0 is not a local minimum of $|P(z)|$.*

Since the real numbers form a subset of the complex numbers, the coefficients of $P(z)$ may in fact all be real, or even integers.

Proof Let n be the degree of $P(z)$, and put $Q(z) = P(z_0 + z)/P(z_0)$. On expanding the binomials $(z_0 + z)^k$, we find that $Q(z)$ is a polynomial of degree n, say

$$Q(z) = c_n z^n + \cdots + c_0.$$

We note that $c_0 = Q(0) = 1$. We have to show that $|Q(z)|$ does not have a local minimum at $z = 0$. Let k be the least positive number for which

$c_k \neq 0$, and suppose that the real number r is so small that

$$|c_n|r^{n-k} + \cdots + |c_{k+1}|r < |c_k|/2. \tag{A.1}$$

This inequality holds for all small r, since the left side tends to 0 with r, while the right side is a positive constant. If $|z| = r$, then by the triangle inequality

$$|Q(z)| \leqslant |c_n|r^n + \cdots + |c_{k+1}|r^{k+1} + |c_k z^k + 1|,$$

and by (A.1) this is

$$\leqslant \frac{1}{2}|c_k|r^k + |c_k z^k + 1|. \tag{A.2}$$

Now write $c_k = Ce^{2\pi i\theta}$ where $C > 0$ and $0 \leqslant \theta < 1$. If $z = re^{2\pi i(-\theta+1/2)/k}$ then $c_k z^k + 1 = -Cr^k + 1$. We assume that r is so small that this quantity is positive. Then the expression (A.2) is

$$1 - Cr^k/2.$$

Since this is < 1, we conclude that the point $z = 0$ is not a local minimum of $|Q(z)|$, and the proof is complete.

Theorem A.2 *Let $P(z)$ be a polynomial of degree at least one, whose coefficients are complex numbers. Then there is at least one complex number r for which $P(r) = 0$.*

A complex number r with this property is referred to as a *root* or *zero* of $P(z)$.

By dividing the polynomial $z - r$ into $P(z)$, we find that we may write $P(z) = (z - r)Q(z) + s$, where s is some complex constant. If $P(r) = 0$, then on substituting $z = r$ in the above we deduce that $s = 0$. That is, we may write $P(z) = (z - r)Q(z)$. This process may be repeated, so that we may write

$$P(z) = a_n(z - r_1)(z - r_2)\cdots(z - r_n)$$

where $a_n \neq 0$ is the leading coefficient of $P(z)$, and n is its degree. This representation of $P(z)$ is unique, apart from permutations of the roots r_i. Thus we see that a polynomial of degree $n > 0$ has precisely n roots, provided that repeated roots are counted according to their multiplicity.

Proof Suppose that $P(z)$ is of degree n, and write $P(z)$ explicitly as

$$P(z) = a_n z^n + a_{n-1} z^{n-1} + \cdots + a_0.$$

If $a_0 = 0$, then $P(0) = 0$, and we are finished. Henceforth we assume that $a_0 \neq 0$. Let m denote the greatest lower bound of the real numbers $|P(z)|$, where z is allowed to take on any complex value. We show that there is a complex number r for which $|P(r)| = m$. It then follows from the Lemma that $m = 0$, and hence $P(r) = 0$.

When $|z|$ is large, the leading term of $P(z)$ dominates all the other terms, so that $|P(z)|$ is large. More precisely, let R be chosen so large that

$$|a_n| \geqslant |a_{n-1}|/R + \cdots + |a_1|/R^{n-1} + 3|a_0|/R^n. \tag{A.3}$$

We may write

$$P(z) = z^n\left(a_n + a_{n-1}/z + a_{n-2}/z^2 + \cdots + a_0/z^n\right).$$

Hence by the triangle inequality,

$$|P(z)| \geqslant |z|^n\left(|a_n| - |a_{n-1}|/|z| - |a_{n-2}|/|z|^2 - \cdots - |a_0|/|z|^n\right).$$

If $|z| > R$, then this is

$$> |z|^n\left(|a_n| - |a_{n-1}|/R - \cdots - |a_0|/R^n\right),$$

which by (A.3) is

$$\geqslant |z|^n\left(2|a_0|/R^n\right) \geqslant 2|a_0|.$$

Since $m \leqslant |P(0)| = |a_0|$, we deduce that if $|z| > R$ then $|P(z)| > m + |a_0|$. That is, if $|P(z)| \leqslant m + |a_0|$, then $|z| \leqslant R$. Consequently, the greatest lower bound m of all values of $|P(z)|$ is the same as the greatest lower bound of those values of $|P(z)|$ for which $|z| \leqslant R$. But $|P(z)|$ is a continuous function, and the disc $|z| \leqslant R$ is closed and bounded, so that by the compactness principle $|P(z)|$ must assume its greatest lower bound m at some point, say $|P(r)| = m$. By the Lemma it follows that $m = 0$, so that $P(r) = 0$.

A.2 SYMMETRIC FUNCTIONS

A polynomial $P(r_1, \cdots, r_n)$ in the variables $r_1, \cdots, r_n$ is called *symmetric* if all permutations of the variables produce the same polynomial. Among the symmetric polynomials are the *elementary symmetric polynomials*

$\sigma_1, \sigma_2, \cdots, \sigma_n$, defined as follows: σ_1 is the sum of all the r_i, σ_2 is the sum of all products $r_{i_1} r_{i_2}$ with $1 \leqslant i_1 < i_2 \leqslant n$, and in general σ_k is the sum of all products $r_{i_1} r_{i_2} \cdots r_{i_k}$ with $1 \leqslant i_1 < \cdots < i_k \leqslant n$. Thus σ_k is a sum of $\binom{n}{k}$ products, and in particular, $\sigma_n = r_1 r_2 \cdots r_n$. On forming a monic polynomial whose roots are the r_i, we find that

$$(z - r_1)(z - r_2) \cdots (z - r_n) = z^n - \sigma_1 z^{n-1} + \sigma_2 z^{n-2} - \cdots + (-1)^n \sigma_n.$$

Indeed, this identity may be used to define the σ_k. By the fundamental theorem of algebra (Theorem A.2), the general polynomial

$$P(z) = a_n z^n + a_{n-1} z^{n-1} + \cdots + a_0$$

can be written in factored form, and on comparing the two expressions we see that

$$\sigma_k = (-1)^k a_{n-k}/a_n. \tag{A.4}$$

We now show that all polynomials symmetric in the r_i can be expressed in terms of the σ_k.

Theorem A.3 *The fundamental theorem of symmetric polynomials. Let $F(r_1, \cdots, r_n)$ be a symmetric polynomial in the indeterminates $r_1, \cdots, r_n$. Then there is a polynomial $P(z_1, \cdots, z_n)$ such that $F(r_1, \cdots, r_n) - P(\sigma_1, \cdots, \sigma_n)$. The coefficients of P can be expressed as linear combinations, with integral coefficients, of the coefficients of F. The degree of P is equal to the highest power of r_1 occurring in F.*

The assertion concerning the coefficients of P implies that if the coefficients of F lie in a certain ring, then those of P will lie in the same ring. In particular, if the coefficients of F are integers, then the coefficients of P are also integers. Concerning the degrees of polynomials in several variables, we note that the degree of a monomial $c z_1^{m_1} z_2^{m_2} \cdots z_n^{m_n}$ is defined to be $m_1 + m_2 + \cdots + m_n$, and the degree of a polynomial is the maximum of the degrees of the monomial terms with nonzero coefficients. A *homogeneous* polynomial (also called a *form*) is a polynomial all of whose monomial terms are of the same degree. In symbols, the last phrase of Theorem A.3 would be written $\deg P = \deg_{r_1} F$.

Proof We introduce a lexicographic ordering of monomials as follows: Assuming that $a \neq 0$ and $b \neq 0$, we say that

$$a z_1^{j_1} z_2^{j_2} \cdots z_n^{j_n} \succ b z_1^{k_1} z_2^{k_2} \cdots z_n^{k_n}$$

if the first nonzero term in the sequence $j_1 - k_1, j_2 - k_2, j_3 - k_3, \cdots, j_n - k_n$ is positive. Note that this ordering is independent of the nonzero coefficients a and b, so that $-3z_1^2z_2^4z_3^4 \succ 100\, z_1^2z_2^3z_3^5$. The *leading term* of a polynomial F is the monomial term of F that is largest with respect to this ordering, and we say that $F \succ G$ if the leading term of F is greater than the leading term of G. This does not totally order polynomials, since two distinct polynomials might have the same leading term, or the same leading term but with different coefficients. The relation $\succ$ has the property that if $F \succ G$ and $G \succ H$, then $F \succ H$. That is, the relation $\succ$ is a partial ordering.

Let $a_1, a_2, \cdots, a_n$ be non-negative integers, and consider

$$\sigma_1^{a_1}\sigma_2^{a_2} \cdots \sigma_n^{a_n}$$

as a polynomial in $r_1, \cdots, r_n$. The leading term of this polynomial is

$$r_1^{a_1+a_2+\cdots+a_n} r_2^{a_2+\cdots+a_n} \cdots r_n^{a_n}.$$

Suppose that $c_1 r_1^{m_1} r_2^{m_2} \cdots r_n^{m_n}$ is the leading term of F. Since F is symmetric, it is clear that $m_1 \geqslant m_2 \geqslant \cdots \geqslant m_n$. On taking $a_1 = m_1 - m_2, a_2 = m_2 - m_3, \cdots, a_n = m_n$, we see that the leading term of

$$G_1 = \sigma_1^{m_1 - m_2}\sigma_2^{m_2-m_3} \cdots \sigma_{n-1}^{m_{n-1}-m_n}\sigma_n^{m_n}$$

is $r_1^{m_1}r_2^{m_2} \cdots r_n^{m_n}$. Put $F_1 = F - c_1G_1$. Since the leading terms cancel, we see that $F \succ F_1$. We note also that the coefficients of F_1 are linear combinations of the coefficients of F. As F_1 is also symmetric, we may repeat this process, obtaining a further symmetric polynomial $F_2 = F - c_1G_1 - c_2G_2$ where c_2 is the coefficient of the leading term of F_1. The coefficients of F_2 are linear combinations of the coefficients of F_1, and hence are linear combinations of the coefficients of F. Continuing in this way, we construct a sequence $F \succ F_1 \succ F_2 \succ \cdots$. It is necessary to show that this method terminates, that is, that F_k is identically 0 for some k. Suppose that $c_k r_1^{q_1} r_2^{q_2} \cdots r_n^{q_n}$ is the leading term of F_k. Since F_k is symmetric, we know that $q_1 \geqslant q_2 \geqslant \cdots \geqslant q_n$. As $F \succ F_k$, we also have $m_1 \geqslant q_1$. Hence $0 \leqslant q_i \leqslant m_1$ for all i. But there are only $(m_1 + 1)^n$ such n-tuples $(q_1, q_2, \cdots, q_n)$, so the reduction must terminate in at most this many steps.

From this construction we find that each coefficient of P is a linear combination of the coefficients of F. In passing from F to F_1, we introduced a monomial of degree m_1 in the variables $\sigma_1, \cdots, \sigma_n$. Since subsequent monomials will have at most this degree, and will not cancel this first monomial term, we observe that $\deg P = \deg_{r_1} F$.

Example 1 Express $F = \sum_{i \neq j} r_i^2 r_j$ in terms of elementary symmetric polynomials.

Solution The leading term of F is $r_1^2 r_2$. On taking $F_1 = F - \sigma_1\sigma_2$, we find that

$$F_1 = -3 \sum_{i<j<k} r_i r_j r_k = -3\sigma_3.$$

That is, $F = \sigma_1\sigma_2 - 3\sigma_3$. Here we are assuming that $n \geqslant 3$. If $n = 2$, then $F = \sigma_1\sigma_2$.

The fundamental theorem of symmetric polynomials (Theorem A.3) has many important applications. One of them is to provide information concerning the discriminant of a polynomial.

Definition A.1 *Let $f(z)$ be a polynomial of degree n with leading coefficient a_n and roots $r_1, \cdots, r_n$. The* discriminant *of f is*

$$D(f) = a_n^{2n-2} \prod_{1 \leqslant i < j \leqslant n} (r_i - r_j)^2.$$

Clearly $D(f) = 0$ if and only if f has a repeated root. In the case that f is a quadratic polynomial, $f(z) = az^2 + bz + c$, we know how to write the roots explicitly in terms of a, b, and c, and we find that the expression above reduces to the familiar quantity $b^2 - 4ac$. For polynomials of higher degree it is in general not possible to express the roots in such explicit form in terms of the coefficients. Thus it is useful that the discriminant can still be calculated.

Theorem A.4 *Let*

$$f(z) = a_n z^n + a_{n-1} z^{n-1} + \cdots + a_0$$

be a polynomial of degree n. There is a homogeneous polynomial $F(w_0, w_1, \cdots, w_n)$ of degree $2n - 2$ with integral coefficients such that

$$D(f) = F(a_0, a_1, \cdots, a_n).$$

Moreover, if $f(z) = (z - r_1)g(z)$, then $D(f) = D(g)g(r_1)^2$.

Thus we see that if the coefficients of f are integers then $D(f)$ is also an integer. By determining this integer we are able to determine whether f has a repeated root.

Proof By the fundamental theorem of symmetric polynomials (Theorem A.3), there is a polynomial P with integral coefficients such that the product over i and j in Definition A.1 is $P(\sigma_1, \sigma_2, \cdots, \sigma_n)$. When the product over the roots is expanded, the highest power of r_1 that occurs is r_1^{2n-2}. Thus $\deg P = 2n - 2$. By (A.4) we see that

$$D(f) = a_n^{2n-2} P\big(-a_{n-1}/a_n, a_{n-2}/a_n, -a_{n-3}/a_n, \cdots, (-1)^n a_0/a_n\big).$$

Here the right side is a form of degree $2n - 2$ in the coefficients a_i. The last clause of the theorem is a direct consequence of the definition of the discriminant.

Remark on Calculation For polynomials of higher degree it is not an easy matter to derive the form F explicitly. Even for polynomials of degree $n = 3$ it is a challenging exercise to show that

$$D(f) = -27a_3^2a_0^2 + 18a_3a_2a_1a_0 - 4a_3a_1^3 - 4a_2^3a_0 + a_2^2a_1^2.$$

For practical purposes it is often easier to appeal to the determinant formula

$$D(f) = (-1)^{(n-2)(n-1)/2} \det(\Delta)/n^{n-2}$$

where $\Delta = [\delta(i, j)]$ is a $(2n - 2) \times (2n - 2)$ matrix whose entries are as follows: if $1 \leqslant i \leqslant n - 1$ and $i \leqslant j \leqslant i + n - 1$, then $\delta(i, j) = (j + 1 - i)a_{n+i-j-1}$ and $\delta(n - 1 + i, j) = (n + i - j)a_{n+i-j}$. All other entries are 0. From this formula (whose proof we omit) it is immediate that $D(f)$ is a form of degree $2n - 2$, but the other properties of the discriminant are not so evident.

We now apply the properties of the discriminant developed above to answer a question that arose in Section 2.6 concerning the problem of lifting a singular solution of a congruence to higher powers of p.

Theorem A.5 *Let $f(x)$ be a polynomial with integral coefficients and suppose that $p^\delta \| D(f)$. If $f(a) \equiv 0 \pmod{p^j}$, $p^\tau \| f'(a)$, and $j > \delta$, then $j \geqslant 2\tau + 1$.*

From this we see in particular that if $p \nmid D(f)$ and $f(a) \equiv \pmod p$ then $f'(a) \not\equiv 0 \pmod p$. In any case, it follows that if $j > \delta$ then Theorem 2.24 applies.

Proof Write $f(x) = (x - a)g(x) + p^j r$, where $g(x)$ is a polynomial with integral coefficients and r is an integer. Let $c_0, c_1, \cdots, c_n$ denote the coefficients of $f(x)$. Since $D(f)$ is a polynomial in the c_i with integral coefficients, we see that $D(f) \equiv D((x-a)g(x)) \pmod{p^j}$. By Theorem A.4 we know that $D((x-a)g(x)) = D(g)g(a)^2$. As $f'(x) = g(x) + (x-a)g'(x)$, we find that $f'(a) = g(a)$. Hence $D(f) \equiv D(g)f'(a)^2 \pmod{p^j}$. The inequality $j > \delta$ is equivalent to the assertion that $D(f) \not\equiv 0 \pmod{p^j}$. This implies that $f'(a)^2 \not\equiv 0 \pmod{p^j}$, which is to say that $j > 2\tau$.

If $f(x)$ has a repeated factor then $D(f) = 0$ and Theorem A.5 is of no use. To avoid this difficulty one may first factor $f(x)$ and search for roots $\pmod{p^j}$ of the irreducible factors.

PROBLEMS

1. Suppose that $f(z) = \Sigma a_i z^i = a_n \Pi(z - r_i)$ is a polynomial of degree n with integral coefficients, and that $g(z)$ is a polynomial of degree m with integral coefficients. Show that $a_n^m \Pi g(r_i)$ is an integer.
2. Suppose that $f(z) = a_n(z - r_1) \cdots (z - r_n)$. Show that
$$f'(r_1) = a_n(r_1 - r_2)(r_1 - r_3) \cdots (r_1 - r_n).$$
Deduce that $D(f) = (-1)^n a_n^{n-2} f'(r_1) f'(r_2) \cdots f'(r_n)$.
3. Suppose that f is a polynomial of degree n with real coefficients and distinct roots, and that $n = r + 2s$, where r is the number of real roots of f, and s is the number of pairs of complex conjugate roots. Show that $\operatorname{sgn} D(f) = (-1)^s$. Deduce that if $n = 3$ then $D(f) > 0$ for those polynomials f with three distinct real roots, and $D(f) < 0$ for those f with one real root and a pair of complex conjugate roots.
4. Let $f(x) = a_3x^3 + a_2x^2 + a_1x + a_0$. Show that the roots of f lie in geometric progression if and only if $a_2^3 a_0 = a_3 a_1^3$.
5. Suppose that $f(x)$ is of degree n, $f(0) \neq 0$, and that $g(x) = x^n f(1/x)$ is the polynomial obtained by reversing the order of the coefficients of f. Show that $D(f) = D(g)$.
6. Suppose that f is as in Problem 1, and that $g(x) = f(x + a)$. Show that $D(g) = D(f)$. Express D as a polynomial in $a_1, a_2, \cdots, a_n$, and show that
$$na_n \frac{\partial D}{\partial a_{n-1}} + (n-1)a_{n-1} \frac{\partial D}{\partial a_{n-2}} + \cdots + a_1 \frac{\partial D}{\partial a_0} = 0.$$
7. Let polynomials in the variables $r_1, r_2, \cdots, r_n$ be ordered lexicographically, as in the proof of the fundamental theorem on symmetric polynomials. Note that $r_1 \succ r_2^k$ for all k. Show that, despite this, any nonempty set of polynomials contains a minimal element. For each polynomial F in these variables, let $P(F)$ be a proposition. Suppose that $P(F)$ is true

whenever $P(F_1)$ is true for all F_1 such that $F \succ F_1$. Show that $P(F)$ is true for all F. Let F be given. Show that a decreasing sequence $F \succ F_1 \succ F_2 \succ \cdots$ may be arbitrarily long, but not infinitely long.

A.3 A SPECIAL VALUE OF THE RIEMANN ZETA FUNCTION

Theorem A.6 $\sum_{n=1}^{\infty} \frac{1}{n^2} = \frac{\pi^2}{6}$.

Proof This formula is an easy consequence of the identity

$$\sum_{n=1}^{N} \cot^2 \frac{n\pi}{2N+1} = \frac{N(2N-1)}{3} \tag{A.5}$$

which holds for all positive integers N. We first derive the theorem from this formula, and then prove (A.5). It is well-known that $\sin \theta < \theta < \tan \theta$ for $0 < \theta < \pi/2$. Taking reciprocals and squaring, we find that $\cot^2 \theta < 1/\theta^2 < \operatorname{cosec}^2 \theta = 1 + \cot^2 \theta$. We take $\theta = n\pi/(2N+1)$ and observe that this number lies in the interval $(0, \pi/2)$ for $n = 1, 2, \cdots, N$. Hence

$$\cot^2 \frac{n\pi}{2N+1} < \frac{(2N+1)^2}{n^2\pi^2} < 1 + \cot^2 \frac{n\pi}{2N+1}.$$

Summing these inequalities of n, and using (A.5), we get

$$\frac{N(2N-1)}{3} < \sum_{n=1}^{N} \frac{(2N+1)^2}{n^2\pi^2} < N + \frac{N(2N-1)}{3}.$$

We multiply each of these expressions by $\pi^2/(2N+1)^2$, giving

$$\frac{\pi^2}{3} \cdot \frac{2N^2 - 2N}{4N^2 + 4N + 1} < \sum_{n=1}^{N} \frac{1}{n^2} < \frac{\pi^2}{3} \cdot \frac{2N^2 + 2N}{4N^2 + 4N + 1}.$$

As $N \to \infty$, the limit of the first and last of these expressions is $\frac{\pi^2}{3} \cdot \frac{1}{2}$ and hence we have

$$\lim_{n \to \infty} \sum_{n=1}^{N} \frac{1}{n^2} = \frac{\pi^2}{6}. \tag{A.6}$$

We complete the proof by establishing the identity (A.5). De Moivre's Theorem states that $(\cos\theta + i\sin\theta)^m = \cos m\theta + i\sin m\theta$. Since $\cos\theta + i\sin\theta = \sin\theta(\cot\theta + i)$, we can write

$$\cos m\theta + i\sin m\theta = \sin^m\theta(\cot\theta + i)^m,$$

and by the binomial theorem we see that this is

$$= \sin^m\theta\left\{\cot^m\theta + i\binom{m}{1}\cot^{m-1}\theta - \binom{m}{2}\cot^{m-2}\theta - \cdots\right\}.$$

Equating imaginary parts here, and using $i^3 = -i$, $i^5 = i$, and so on, we get

$$\sin m\theta = \sin^m\theta\left[\binom{m}{1}\cot^{m-1}\theta - \binom{m}{3}\cot^{m-3}\theta + \binom{m}{5}\cot^{m-5}\theta - \cdots\right].$$

We take $m = 2N + 1$, with N as in (A.5), and observe that the expression in square brackets is a polynomial in $\cot^2\theta$, say $F(\cot^2\theta)$, so that

$$\sin(2N+1)\theta = \sin^{2N+1}\theta\cdot F(\cot^2\theta) \tag{A.7}$$

where

$$F(x) = \binom{2N+1}{1}x^N - \binom{2N+1}{3}x^{N-1} + \binom{2N+1}{5}x^{N-2} - \cdots + (-1)^N. \tag{A.8}$$

If θ is one of the N numbers $\theta = n\pi/(2N+1)$, $n = 1, 2, \cdots, N$, then $\sin(2N+1)\theta = 0$, but $\sin\theta \neq 0$. Thus from (A.7) we see that $F(\cot^2\theta) = 0$ for each of these N values of θ. That is, the N roots of the equation $F(x) = 0$ are precisely the N terms in the sum (A.5). By taking $k = 1$ in (A.4) we deduce that

$$\sum_{n=1}^{N}\cot^2\frac{n\pi}{2N+1} = \binom{2N+1}{3}\Big/\binom{2N+1}{1} = \frac{N(2N-1)}{3},$$

and the proof is complete.

PROBLEMS

1. Show that for any positive integer n there is a polynomial $G(x)$ of degree $N - 1$ such that $\sin 2N\theta = \sin^{2N}\theta\cdot\cot\theta\cdot G(\cot^2\theta)$. Show

that the roots of $G(x) = 0$ are the $N - 1$ numbers $\cot^2 \theta$ where $\theta = n\pi/(2N)$ and $n = 1, 2, \cdots, N - 1$. Prove that

$$\sum_{n=1}^{N-1} \cot^2 \frac{n\pi}{2N} = \frac{(N-1)(2N-1)}{3}.$$

2. Prove that for any positive integer N,

$$\sum_{n=1}^{2N-1} \cot^2 \frac{n\pi}{2N} = \frac{2(N-1)(2N-1)}{3}.$$

(H)

3. Show that for any positive integer M,

$$\sum_{m=1}^{M-1} \cot^2 \frac{m\pi}{M} = \frac{(M-1)(M-2)}{3}, \tag{A.9}$$

and that

$$\sum_{m=1}^{M-1} \operatorname{cosec}^2 \frac{m\pi}{M} = \frac{M^2 - 1}{3}.$$

4. Prove that if N is a positive integer then

$$\prod_{n=1}^{N} \tan \frac{n\pi}{2N+1} = \sqrt{2N+1}.$$

5. Prove that if M is a positive integer then

$$z^M - 1 = \prod_{m=0}^{M} (z - e^{2\pi i m/M})$$

for all real or complex numbers z.

6. Show that if M is a positive integer then for any real number θ

$$\prod_{m=0}^{M-1} \sin(\theta + \pi m/M) = 2^{1-M} \sin M\theta.$$

(H)

7. Show that if M is a positive integer, and θ is a real number for which $M\theta/\pi$ is not an integer, then

$$\sum_{m=0}^{M-1} \cot(\theta + \pi m/M) = M \cot M\theta.$$

(H)

8. Show that if M is a positive integer, and θ is a real number for which $M\theta/\pi$ is not an integer, then

$$\sum_{m=0}^{M-1} \operatorname{cosec}^2(\theta + \pi m/M) = M^2 \operatorname{cosec}^2 M\theta.$$

(H)

9. Show that $\lim_{x \to 0}(\operatorname{cosec}^2 x - x^{-2}) = 1/3$. Use this in the preceding problem to provide a second solution of Problem 3, and hence a second proof of the identity (A.5) used to prove Theorem A.6.

10. Show that if M is a positive integer then

$$\prod_{m=1}^{M-1} \sin \pi m/M = M2^{1-M}.$$

A.4 LINEAR RECURRENCES

Definition A.2 *Let k be a positive integer. We say that the sequence $u_0, u_1, u_2, \cdots$ of real or complex numbers satisfies a* linear recurrence of order k *if there exist real or complex numbers $b_1, b_2, \cdots, b_k$ such that*

$$u_n = b_1 u_{n-1} + b_2 u_{n-2} + \cdots + b_k u_{n-k} \tag{A.10}$$

for all integers $n \geqslant k$.

Linear recurrences of order 2 were discussed in Section 4.4. A sequence may satisfy a linear recurrence of order k and also other recurrences of other orders. For example, the sequence $u_n = (-1)^n$ satisfies the linear recurrence $u_n = -u_{n-1}$ of order 1, but it also satisfies the linear recurrence $u_n = u_{n-2}$ of order 2.

Suppose that the sequence $u_0, u_1, u_2, \cdots$ satisfies the linear recurrence (A.10) for all $n \geqslant k$. Let $B = |b_1| + |b_2| + \cdots + |b_k|$, and for each non-negative integer n let $M_n = \max(|u_0|, |u_1|, \cdots, |u_n|)$. We see by the triangle inequality that if $n \geqslant k$ then $M_n \leqslant BM_{n-1}$. By induction it follows that if $\{u_n\}$ satisfies (A.10) then there is a constant A such that $|u_n| \leqslant AB^n$ for $n = 0, 1, 2, \cdots$. A sequence satisfying a bound of this kind is said to have "at most exponential growth." For such a sequence, the associated power series generating function

$$f(z) = \sum_{n=0}^{\infty} u_n z^n \tag{A.11}$$

has positive radius of convergence. More precisely, if $|z| \leqslant r < 1/B$, then the above series is absolutely convergent by comparison with the convergent geometric series $\sum_{n=0}^{\infty} AB^n r^n$.

Theorem A.7 *Let $\{u_n\}$ be a sequence of real or complex numbers. The following two assertions are equivalent:*

(i) u_n satisfies a linear recurrence of order k;

(ii) the power series (A.11) *has positive radius of convergence and $f(z)$ is a rational function, say $f(z) = P(z)/Q(z)$ where $P(z)$ and $Q(z)$ are polynomials with real or complex coefficients*, $\deg(P) < k$, *and* $\deg(Q) \leqslant k$.

Proof Suppose that (i) holds. More specifically, we suppose that (A.10) holds for all $n \geqslant k$. We have already shown that the power series (A.11) has positive radius of convergence. Let

$$Q(z) = 1 - b_1 z - b_2 z^2 - \cdots - b_k z^k. \tag{A.12}$$

By grouping terms appropriately, we may write $Q(z)f(z)$ as a power series,

$$Q(z)f(z) = \sum_{n=0}^{\infty} c_n z^n. \tag{A.13}$$

This new power series has positive radius of convergence because $f(z)$ does. By direct calculation we find that $c_0 = u_0$, $c_1 = u_1 - b_1 u_0$, and $c_2 = u_2 - b_1 u_1 - b_2 u_2$. The number of terms required to write c_n continues to increase with n until $n = k$. For $n \geqslant k$ we find that the number of terms is constant, and that

$$c_n = u_n - b_1 u_{n-1} - b_2 u_{n-2} - \cdots - b_k u_{n-k}. \tag{A.14}$$

From (A.10) we deduce that $c_n = 0$ for all $n \geqslant k$. That is, the power series in (A.13) turns out to be only a polynomial, say $P(z)$, whose degree is strictly less than k. Then $f(z) = P(z)/Q(z)$, and we have (ii), since $\deg(Q) \leqslant k$.

We now suppose that (ii) holds, and derive (i). We write $f(z) = P(z)/Q(z)$. If $P(0) = 0$ and $Q(0) = 0$, then we may divide both P and Q by an appropriate power of z so that at least one of $P(0)$ and $Q(0)$ is nonzero. If it were the case that $P(0) \neq 0$ and $Q(0) = 0$ then $|f(z)|$ would tend to ∞ as $z \to 0$, contrary to our hypothesis that the power series (A.11) has positive radius of convergence. Thus we see that $f(z)$ may be expressed as the quotient of two polynomials, $f(z) = P(z)/Q(z)$, with $Q(0) \neq 0$. By dividing $P(z)$ and $Q(z)$ by the nonzero constant $Q(0)$, we deduce that $f(z)$ may be written as such a quotient with $Q(0) = 1$. These two polynomials may not be the ones we started with, but their degrees are no larger than they were originally, so that $\deg(P) < k$ and $\deg(Q) \leqslant k$. Hence $Q(z)$ may be expressed in the form (A.12). Then (A.13) and (A.14) follow as before. Since $Q(z)f(z) = P(z)$ is a polynomial of degree less than k, it follows that $c_n = 0$ for all $n \geqslant k$. Then (A.14) gives (i) and the proof is complete.

In the examples considered in Section 4.4, the solution u_n of a linear recurrence was written as a linear combination of exponential functions. To do this in general, we express the rational function $f(z)$ in terms of partial fractions.

Lemma A.8 *Let k be a positive integer, and suppose that $f(z) = P(z)/Q(z)$ is a rational function with $\deg(P) < k$ and $\deg(Q) = k$, and that when $Q(z)$ is factored it takes the form*

$$Q(z) = c\prod_{j=1}^{J}(z - r_j)^{m_j} \tag{A.15}$$

where $c \neq 0$, the r_j are distinct real or complex numbers, and $\sum_{j=1}^{J} m_j = k$. Then there exist real or complex numbers, α_{ij} such that

$$f(z) = \sum_{j=1}^{J}\sum_{i=1}^{m_j}\frac{\alpha_{ij}}{(z - r_j)^i}. \tag{A.16}$$

Proof We proceed by induction on k. Suppose first that $k = 1$. If $P(z)$ is identically 0 then the representation is obtained by taking all the α_{ij} to be 0. Otherwise $\deg(P) = 0$, which is to say that $P(z)$ is a nonzero constant, say p. Since $Q(z) = c(z - r_1)$, we observe that if $\alpha_{11} = p/c$ then

$$\frac{P(z)}{Q(z)} = \frac{\alpha_{11}}{z - r_1},$$

which is (A.16) in this case.

Now suppose that $k > 1$, and that the representation (A.16) can be found for polynomials of degree $k - 1$. Let r be a root of $Q(z)$ and let m denote its multiplicity, so that $Q(z) = (z - r)^m T(z)$ with $T(r) \neq 0$. Put $\alpha = P(r)/T(r)$. Then

$$\frac{P(z)}{Q(z)} = \frac{\alpha}{(z - r)^m} + \frac{P(z) - \alpha T(z)}{Q(z)}. \tag{A.17}$$

In the second term on the right, the numerator vanishes when $z = r$, and thus it has the factor $z - r$. That is, the numerator may be written as

$(z-r)P_1(z)$, say. Put $Q_1(z) = T(z)(z-r)^{m-1}$, so that $Q(z) = Q_1(z)$ $(z-r)$. Then the second term on the right is $P_1(z)/Q_1(z)$, where $\deg(P_1)$ $< k-1$ and $\deg(Q_1) = k-1$. By the inductive hypothesis, the expansion (A.16) is already known for $P_1(z)/Q_1(z)$. This with (A.17) gives (A.16) for $P(z)/Q(z)$, and the proof is complete.

All the roots of the polynomial $Q(z)$ in (A.12) are nonzero, since $Q(0) = 1$. Suppose that $b_k \neq 0$, so that $\deg(Q) = k$. We may write $Q(z)$ in the form

$$Q(z) = \prod_{j=1}^{k} (1 - \lambda_j z).$$

In this notation, the roots of $Q(z)$ are the numbers $1/\lambda_1, 1/\lambda_2, \cdots, 1/\lambda_k$. These roots are not necessarily distinct, but in case they are, the partial fraction expansion of $f(z) = P(z)/Q(z)$ may be written more simply as

$$f(z) = \frac{P(z)}{Q(z)} = \sum_{j=1}^{k} \frac{\beta_j}{1 - \lambda_j z}. \tag{A.18}$$

Theorem A.9 *Suppose that the sequence* $u_0, u_1, \cdots$ *satisfies the linear recurrence* (A.10), *and that the polynomial* $Q(z)$ *in* (A.12) *has k distinct roots, so that there exist real or complex numbers* β_j *and* λ_j *for which* (A.18) *holds. Then*

$$u_n = \sum_{j=1}^{k} \beta_j \lambda_j^n \tag{A.19}$$

for all non-negative integers n.

Proof If $|z| < 1/|\lambda|$ then

$$\frac{1}{1 - \lambda z} = \sum_{n=0}^{\infty} \lambda^n z^n. \tag{A.20}$$

Thus if $|z| < 1/|\lambda_j|$ for $j = 1, 2, \cdots, k$ then

$$\sum_{j=1}^{k} \frac{\beta_j}{1 - \lambda_j z} = \sum_{n=0}^{\infty} \left(\sum_{j=1}^{k} \beta_j \lambda_j^n \right) z^n.$$

Since the power series expansion (A.11) is unique, the stated formula (A.19) follows for all sufficiently large n.

We now consider the general case, in which the polynomial $Q(z)$ may have repeated roots.

Theorem A.10 *Suppose that the sequence* $u_0, u_1, \cdots$ *satisfies the linear recurrence* (A.10), *and that the polynomial* $Q(z)$ *in* (A.12) *has the factorization*

$$Q(z) = \prod_{j=1}^{J} (1 - \lambda_j z)^{m_j} \tag{A.21}$$

where the numbers $\lambda_1, \lambda_2, \cdots, \lambda_J$ *are distinct and nonzero. Then there exist polynomials* $B_j(x)$ *with* $\deg(B_j) < m_j$ *such that*

$$u_n = \sum_{j=1}^{J} B_j(n) \lambda_j^n \tag{A.22}$$

for all non-negative integers n. *Conversely, any sequence of the form* (A.22) *with* $\deg(B_j) < m_j$ *satisfies the linear recurrence* (A.10).

The possibility that one or more of the polynomials B_j may vanish identically is not excluded.

Proof From Theorem A.7 and Lemma A.8 we see that if $|z|$ is sufficiently near 0 then

$$f(z) = \sum_{n=0}^{\infty} u_n z^n = \frac{P(z)}{Q(z)} = \sum_{j=1}^{J} \sum_{i=1}^{m_i} \frac{\beta_{ij}}{(1 - \lambda_j z)^i}.$$

By taking $\lambda = 1$ in (A.20), and then repeatedly differentiating both sides, we find that

$$\frac{1}{(1 - z)^i} = \sum_{n=0}^{\infty} \binom{n + i - 1}{i - 1} z^n$$

for $|z| < 1$. Alternatively, this follows from the binomial theorem in the form of identity (1.13). Thus we see that

$$f(z) = \sum_{n=0}^{\infty} \left(\sum_{j=1}^{J} B_j(n) \lambda_j^n \right) z^n$$

when $|z|$ is sufficiently small, where

$$B_j(x) = \sum_{i=0}^{m_i} \beta_{ij} \binom{x + i - 1}{i - 1} \tag{A.23}$$

is a polynomial in x of degree $< m_i$. Then (A.22) follows by the uniqueness of the power series expansion.

Suppose, conversely, that u_n is given by the formula (A.22), where $\deg(B_j) < m_j$ for $j = 1, 2, \cdots, J$. Then there exist numbers β_{ij} for which (A.23) holds, and hence

$$\sum_{n=0}^{\infty} u_n z^n = \sum_{j=1}^{J} \sum_{i=1}^{m_j} \frac{\beta_{ij}}{(1 - \lambda_j z)^i}.$$

We suppose that $Q(z)$ is defined by (A.21) and deduce that the right side above may be written in the form $P(z)/Q(z)$, with $\deg(P) < k$. Then the stated result follows by Theorem A.7.

One may note that our use of power series to analyze linear recurrences is analogous to the use of Laplace transforms in the study of solutions of linear differential equations with constant coefficients.

PROBLEMS

1. Show that there is an integer n_0 such that the sequence u_n satisfies the linear recurrence (A.10) for all integers $n \geqslant n_0$, if and only if the power series (A.11) is a rational function.

2. Let $\mathscr{S}$ be a finite set of real or complex numbers. Suppose that $u_n \in \mathscr{S}$ for each n, and that u_n satisfies the linear recurrence (A.10) for all $n \geqslant n_0$. Show that u_n is eventually periodic. That is, there is a positive integer q such that $u_{n+q} = u_n$ for all $n \geqslant n_1$.

3. Suppose that for each n, $u_n = 0$ or $u_n = 1$. Let $f(z)$ be the power series (A.11). Show that $f(z)$ is a rational function if and only if $f(1/2)$ is a rational number.

4. Prove that if $|z| < (\sqrt{5} - 1)/2$ then $\dfrac{1}{1 - z - z^2} = \sum_{n=0}^{\infty} F_n z^n$, where F_n is the nth Fibonacci number, as defined in Section 4.4.

5. Let $B(z)$ and $C(z)$ be polynomials with real or complex coefficients that have no common root. Show that there exist polynomials $X(z)$ and $Y(z)$ such that $B(z)X(z) + C(z)Y(z) = 1$, $\deg X < \deg C$, and $\deg Y < \deg B$.

6. Use the result of the preceding problem to give a proof of Lemma A.8.

***7.** Show that if $u_0, u_1, \cdots$ is a sequence of real or complex numbers for which $u_n = \dfrac{1}{q} \sum_{k=1}^{q} u_{n-k}$ for all $n \geqslant q$, then $\lim_{n \to \infty} u_n$ exists, and is a certain weighted average of $u_0, u_1, \cdots, u_{q-1}$.

***8.** Let $\lambda_1, \lambda_2, \cdots, \lambda_q$ be non-negative real numbers whose sum is 1. Suppose that $u_0, u_1, \cdots$ is a sequence of real numbers for which $u_n = \sum_{k=1}^{q} \lambda_k u_{n-k}$ when $n \geqslant q$. Show that $\lim_{n \to \infty} u_n$ exists, and has the value

$$\frac{\lambda_1 u_0 + (\lambda_1 + \lambda_2) u_1 + (\lambda_1 + \lambda_2 + \lambda_3) u_2 + \cdots + u_{q-1}}{q\lambda_1 + (q-1)\lambda_2 + (q-2)\lambda_3 + \cdots + \lambda_q}.$$

9. Let $r_1, r_2, \cdots, r_n$ be given real or complex numbers, and for non-negative integers k let s_k be the symmetric function $s_k = r_1^k + r_2^k + \cdots + r_n^k$ of the r_i. Show that if $|z|$ is sufficiently small, then

$$\frac{1}{1 - r_1 z} + \frac{1}{1 - r_2 z} + \cdots + \frac{1}{1 - r_n z} = \sum_{k=0}^{\infty} s_k z^k. \quad \text{(A.24)}$$

***10.** Let $r_1, r_2, \cdots, r_n$ be given real or complex numbers, as in the preceding problem. Put $P(z) = (1 - r_1 z)(1 - r_2 z) \cdots (1 - r_n z)$. Show that $P(z) = 1 - \sigma_1 z + \sigma_2 z^2 - \cdots + (-1)^n \sigma_n z^n$, where the σ_k are the elementary symmetric polynomials of the r_i. Show also that

$$-P'(z) = \frac{r_1 P(z)}{1 - r_1 z} + \frac{r_2 P(z)}{1 - r_2 z} + \cdots + \frac{r_n P(z)}{1 - r_n z}.$$

Conclude that the left side of (A.24) is $n - zP'(z)/P(z)$.

***11.** Using the two preceding problems, or otherwise, establish the *Newton-Girard identities:* If $1 \leqslant k \leqslant n$, then

$$s_k = (-1)^k \sigma_k - \sum_{j=1}^{k-1} (-1)^j \sigma_j s_{k-j},$$

while if $k > n$, then $s_k = -\sum_{j=1}^{n} (-1)^j \sigma_j s_{k-j}$.

***12.** Let k be a fixed positive integer. Show that the elementary symmetric function σ_k of the integers $1, 2, \cdots, n$ is a polynomial in n of degree $2k$, and with leading coefficient $1/(k!2^k)$.

General References

G. E. Andrews, *The Theory of Partitions*, Addison-Wesley (Reading), 1976.

T. M. Apostol, *Introduction to Analytic Number Theory*, Springer-Verlag (New York), 1976.

A. Baker, *Transcendental Number Theory*, Cambridge University Press (London), 1975.

Z. I. Borevich and I. R. Shafarevich, *Number Theory*, Academic Press (New York), 1966.

D. M. Bressoud, *Factorization and primality testing,* Springer-Verlag, (New York), 1989.

J. W. S. Cassels, *An Introduction to Diophantine Approximation*, Cambridge Tract 45, Cambridge University Press (Cambridge), 1957.

J. W. S. Cassels, *An Introduction to the Geometry of Numbers*, Springer-Verlag (Berlin), 1971.

J. W. S. Cassels, *Rational Quadratic Forms*, Academic Press (London), 1978.

J. S. Chahal, *Topics in Number Theory*, Plenum (New York), 1988.

H. Davenport, *The Higher Arithmetic*, 5th ed., Cambridge University Press (Cambridge), 1982.

H. Davenport, *Multiplicative Number Theory*, 2nd ed., Springer-Verlag (New York), 1980.

L. E. Dickson, *History of the Theory of Numbers*, Carnegie Institution of Washington (Washington), 1919; reprinted, New York, Chelsea (New York), 1950.

W. Fulton, *Algebraic Curves: An Introduction to Algebraic Geometry*, Addison-Wesley (Reading), 1989.

C. F. Gauss, *Disquisitiones Arithmeticae* (*Discourses on Arithmetic*), English ed., Yale University Press (New Haven), 1966.

E. Grosswald, *Representations of Integers as Sums of Squares*, Springer-Verlag (New York), 1985.

P. M. Gruber and C. G. Lekkerkerker, *Geometry of Numbers*, North-Holland (Amsterdam), 1987.

H. Halberstam and K. F. Roth, *Sequences*, Clarendon Press (Oxford), 1966.

G. H. Hardy and E. M. Wright, *An Introduction to the Theory of Numbers*, 5th ed., Clarendon Press (Oxford), 1979.

I. N. Herstein, *Topics in Algebra*, 2nd ed., Xerox (Lexington), 1975.

L. K. Hua, *Introduction to Number Theory*, Springer-Verlag (Berlin), 1982.

D. Husemöller, *Elliptic Curves*, Springer-Verlag (New York), 1987.

K. F. Ireland and M. Rosen, *A Classical Introduction to Modern Number Theory*, Springer-Verlag (New York), 1982.

B. W. Jones, *The Arithmetic Theory of Quadratic Forms*, Carus Monograph 10, Wiley (New York), 1950.

A. Y. Khinchin, *Continued Fractions*, University of Chicago Press (Chicago), 1964.

D. E. Knuth, *The Art of Computer Programming*, Vol. 2: *Seminumerical Algorithms*, Addison-Wesley (Reading), 1969.

N. Koblitz, *A Course in Number Theory and Cryptography*, Springer-Verlag (New York), 1987.

N. Koblitz, *Introduction to Elliptic Curves and Modular Forms*, Springer-Verlag (New York), 1984.

L. Kuipers and H. Niederreiter, *Uniform Distribution of Sequences*, Wiley (New York), 1974.

E. Landau, *Elementary Number Theory*, Chelsea (New York), 1958.

D. H. Lehmer, *Guide to Tables in the Theory of Numbers*, Bulletin, National Research Council (Washington), No. 105, 1941.

H. W. Lenstra, Jr. and R. Tijdeman (ed.), *Computational Methods in Number Theory*, Parts I, II, Math. Centre Tracts 154, 155, Amsterdam, 1982.

W. J. LeVeque, *Fundamentals of Number Theory*, Addison-Wesley (Reading), 1977.

W. J. LeVeque, ed., *Reviews in Number Theory*, Vols. 1–6, Amer. Math. Society (Providence), 1974.

W. J. LeVeque, ed., *Studies in Number Theory*, Math. Assoc. of America (Washington), 1969.

W. J. LeVeque, *Topics in Number Theory*, Vol. II, Addison-Wesley (Reading), 1956.

L. J. Mordell, *Diophantine Equations*, Academic Press (New York), 1969.

T. Nagell, *Introduction to Number Theory*, Chelsea (New York), 1981.

M. Newman, *Integral Matrices*, Academic Press (New York), 1972.

I. Niven, *Irrational Numbers*, Carus Monograph 11, Wiley (New York), 1956.

O. T. O'Meara, *Introduction to Quadratic Forms*, Springer-Verlag (Berlin), 1973.

O. Ore, *Number Theory and Its History*, McGraw-Hill (New York), 1949.

H. Pollard and H. Diamond, *The Theory of Algebraic Numbers*, Carus Monograph 9, Math. Assoc. of America (Washington), 1975.

G. Pólya and G. Szegö, *Problems and Theorems in Analysis*, Vol. I, Springer-Verlag (New York), 1972.

G. Pólya and G. Szegö, *Problems and Theorems in Analysis*, Vol. II, Springer-Verlag (New York), 1976.

C. Pomerance, *Lecture Notes on Primality Testing and Factoring*, MAA Notes 4, Math. Assoc. of America (Washington), 1984.

H. Rademacher, *Lectures in Elementary Number Theory*, Blaisdell (New York), 1964.

M. Reid, *Undergraduate Algebraic Geometry*, Cambridge University Press (Cambridge), 1988.

P. Ribenboim, *Algebraic Numbers*, Wiley (New York), 1972.

P. Ribenboim, *The Book of Prime Number Records*, 2nd ed., Springer-Verlag (New York), 1989.

P. Ribenboim, *13 Lectures on Fermat's Last Theorem*, Springer-Verlag (New York), 1979.

H. Riesel, *Prime Numbers and Computer Methods for Factorization*, Second Edition, Birkhäuser (Boston), 1994.

A. Robinson, *Numbers and Ideals*, Holden-Day (San Francisco), 1965.

K. H. Rosen, *Elementary Number Theory and Its Applications*, Addison-Wesley (Reading), 1984.

J. P. Serre, *A Course In Arithmetic*, Springer-Verlag (New York), 1973.

D. Shanks, *Solved and Unsolved Problems in Number Theory*, Fourth Edition, Chelsea (New York), 1993.

J. H. Silverman, *The Arithmetic of Elliptic Curves*, Springer-Verlag (New York), 1986.

T. Skolem, *Diophantische Gleichungen*, Chelsea (New York), 1950.

N. J. A. Sloane, *Handbook of Integer Sequences*, Academic Press (New York), 1974.

R. J. Walker, *Algebraic Curves*, Dover (New York), 1962.

A. Weil, *Number Theory, An Approach Through History*, Birkhäuser (Boston), 1984.

Hints

SECTION 1.2

32. $n^k = ((n - 1) + 1)^k$.

40. Use induction.

41. Use Theorem 1.10. Deal separately with the case in which one of b and c is 0.

SECTION 1.3

8. If the units digit of n is j, then n has the form $10k + j$, and we see that $r = k - 2j$. So the problem is to prove that if either $10k + j$ or $k - 2j$ is divisible by 7, so is the other one.

26. Use a variation of the proof of Theorem 1.17 and recall Problem 10.

31. If $f(j) = p$, then $f(j + kp) - f(j)$ is a multiple of p for every k, so $f(j + kp)$ has the same property.

42. If k is odd, $x^k + 1$ has a factor $x + 1$.

48. Use Problem 46 of Section 1.2.

51. Consider the highest power of 2, of 3, of 5, of 7 less than the square root.

SECTION 1.4

5. Consider the number of ways of partitioning a set of ab elements into b disjoint subsets each containing a elements.

SECTION 2.1

28. $3^4 \equiv 1 \pmod 5$ by Fermat's theorem, and this with $3^4 \equiv 1 \pmod 2$ implies that $3^4 \equiv 1 \pmod{10}$. Hence $3^{4n} \equiv 1 \pmod{10}$ for any $n \geqslant 1$.

30. Use Theorem 2.8 to establish that $3^{20} \equiv 1 \pmod{25}$. In addition, $3^2 \equiv 1 \pmod 4$, whence $3^{20} \equiv 1 \pmod{100}$.

36. If $p > 5$, $(p-1)!$ has factors 2, $p-1$ and $(p-1)/2$, and so $(p-1)!$ is divisible by $(p-1)^2$. Then recall Problem 32 in Section 1.2.

38. If there are only finitely many such primes, let P be their product, and consider any prime factor of $4P^2 + 1$ in the light of Lemma 2.14.

41. If a, b, c is such a set, so also is ka, kb, kc for any positive integer k. Hence it suffices to determine all "primitive" sets with the property $(a, b, c) = 1$. Also there is no loss in generality in assuming that $a \leqslant b \leqslant c$.

52. Consider congruences $(\text{mod } p - 1)$.

55. Find a small modulus m for which the given determinant is $\not\equiv 0 \pmod m$.

SECTION 2.2

10. In case $a \not\equiv 0 \pmod p$, show that if $c \not\equiv 0 \pmod p$ is given, then there is a unique solution (x, y) for which $x - y \equiv c \pmod p$.

14. Use the identity $\binom{p^\alpha}{k} = \binom{p^\alpha - 1}{k - 1}\dfrac{p^\alpha}{k}$.

15. Use the identity $\binom{p^\alpha - 1}{k} = \binom{p^\alpha}{k} - \binom{p^\alpha - 1}{k - 1}$.

SECTION 2.3

36. Use Theorem 2.8 with $m = 4$ and $m = 25$.

SECTION 2.4

1. Verify that $bx + cy \equiv 1 \pmod{p_i}$, $i = 1, 2, \cdots, 5$, where the p_i are 5 distinct five digit primes. Then use Part 3 of Theorem 2.3.

22. Show that $1 - v \leqslant e^{-v}$ for all real numbers v. Derive a corresponding lower bound from inequality (1.9) in Section 1.3.

SECTION 2.5

4. Let $m_1 = (a, m)$, $m_2 = m/m_1$, then apply the preceding problem to show that $a^{k\bar{k}} \equiv a \pmod{m_i}$, $i = 1, 2$.

SECTION 2.7

4. Begin by showing that there is a $q_1(x)$ such that $f(x) \equiv (x - a_1)q_1(x) \pmod p$ and that $q_1(x) \equiv 0 \pmod p$ has solutions $x \equiv a_2$, $x \equiv a_3, \cdots, x \equiv a_j \pmod p$. Then use induction.

SECTION 2.8

17. Show that if $p > 2$ then a belongs to the exponent $2^{n+1} \pmod p$.

32. Recall Lemma 2.22.

33. Show that k is the order of a modulo m where $m = a^k - 1$.

35. Let $q_1, q_2, \cdots, q_r$ be a collection of such primes. Take $a = pq_1q_2 \cdots q_r$, $k = p$ in Problem 33, and then apply Problem 34.

37. Note that if p is the least prime divisor of n then $(p - 1, n) = 1$.

SECTION 2.11

22. Interpret the sum in Z_p, and use the result $1^3 + 2^3 + \cdots + n^3 = n^2(n + 1)^2/4$.

23. After the first k columns of A have been determined, choose the $(k + 1)$st column so that it lies outside the column space of the first k columns.

SECTION 3.1

13. Use the fact that there is some integer a such that $r \equiv a^2 \pmod m$.

14. The sum of the squares of the first n natural numbers is $n(n + 1) \cdot (2n + 1)/6$.

17. Denoting the first given product by P, and $(2k + 1)!$ by Q, prove that $P \equiv (-1)^k Q \pmod p$ by using $2j \equiv -(p - 2j) \pmod p$. Similarly, relate Q to the product of the quadratic residues modulo p by replacing any nonresidue n in Q by the quadratic residue $-n$, and use the preceding problem.

19. Use Theorem 2.37.

20. Use Theorem 2.37 for the case $p \mid (x^4 + 1)$.

SECTION 3.2

13. First determine the primes p such that $\left(\dfrac{-3}{p}\right) = 1$.

18. Note that $1001 = 7 \cdot 11 \cdot 13$.

SECTION 3.3

11. Use Corollary 2.44.

13. Use a primitive root modulo p.

15. Consider cases according to the values of $\left(\dfrac{2}{p}\right)$ and $\left(\dfrac{5}{p}\right)$.

17. Show that if $(a, p) = 1$ then $s(a, p)$ is unchanged if n is replaced by an.

18. Show first that $N_{++}(p) = \dfrac{1}{4}\sum_{n=1}^{p-2}\left(1 + \left(\dfrac{n}{p}\right)\right)\left(1 + \left(\dfrac{n+1}{p}\right)\right)$. Then use the results of Problems 5 and 17.

23. Show that if $a \in G$ then $a^{m-1} \equiv 1 \pmod{m}$, and recall Problem 26 in Section 2.8.

SECTION 3.4

9. Suppose that $a \neq 0$. Show that there are rational numbers r_1 and r_2 such that $f(x, y) = a(x - r_1 y)(x - r_2 y)$. Argue that $ar_1 r_2 \in \mathbb{Z}$, and hence that there exist integers h_1 and h_2 such that $h_1 h_2 = a$, $h_1 r_1 \in \mathbb{Z}$, $h_2 r_2 \in \mathbb{Z}$. Treat the case $a = 0$ separately.

SECTION 3.5

2. First find all $M \in \Gamma$ that commute with $\begin{bmatrix} 1 & 0 \\ 1 & 1 \end{bmatrix}$.

7. Recall Problem 3 in this section and Problem 9 of Section 3.4.

9. Use (3.3).

SECTION 3.7

2. Consider the form $x^2 + xy + 4y^2$.
6. Find h such that $h^2 \equiv -23 \pmod{4 \cdot 139}$, and then reduce the form $139x^2 + hxy + ky^2$.

SECTION 4.1

21. Use the identity $\{(u+v)/2\}^2 - \{(u-v)/2\}^2 = uv$ to get bounds on the integer $(u+v)/2$.
24. $f(\alpha, \beta, \gamma)$ is related to the number of solutions of $\alpha x + \beta y \leqslant \gamma$ in positive integer pairs x, y.
25. Denote $a - b$ by c. For any prime p dividing both c and n, if p^k is the highest power of p dividing n, prove that p^k divides every term in the expansion of $\{(b+c)^n - b^n\}/c$.
27. It suffices to take $0 \leqslant \alpha < 1$, $0 \leqslant \beta < 1$, and g.c.d. $(j, k) = 1$.

SECTION 4.2

19. Assume that $2^{n-1}q$ is a perfect number, where $n > 1$ and q is odd. Write $\sigma(q) = q + k$ and so deduce from $\sigma(2^{n-1}q) = 2^n q$ that $q = k(2^n - 1)$. Thus $k|q$ and $k < q$.
22. For part (a) prove that the largest prime divisor of m is a divisor of n to the same power.

SECTION 4.3

11. Separate the integers $\leqslant n$ into classes, so that all integers k such that $(k, n) = d$ are in the same class.
14. Use (4.1).
27. $F(n)$ is the sum of the roots of the polynomial $x^n - 1$. Thus one may appeal to the case $k = 1$ of the identity (A.4) in Appendix A.3.

SECTION 4.4

4. Recall (1.15).
7. Let $n = mq$, and induct on q.

SECTION 5.1

14. For the first part, use Theorem 4.1. For the second part, note that one may take $x_1 = c/a$, $y_1 = 0$.

SECTION 5.3

11. Write the equation in the form $(x + y)^2 + (x - y)^2 = (2z)^2$.

13. Any solution has y even, because y odd implies $z^2 - x^2 \equiv 2 \pmod 8$, which is impossible. Hence x and z are odd, and the proof of Theorem 5.5 may be used as a model.

14. After replacing x by $-x$, if necessary, argue as in the proof of Lemma 5.4 that there exist odd integers s and t such that $z - x = 5s^2$, $z + x = t^2$. Then choose r so that $t = s + 2r$.

SECTION 5.4

3. Remove powers of 2 common to x and y, then argue (mod 16).

4. Consider powers of 2.

12. Write the equation as $x^2 + 4 = (y + 3)(y^2 - 3y + 9)$.

SECTION 5.5

8. Recall the Remark on p. 132 and Problem 17 at the end of Section 3.3.

SECTION 5.7

10. Recall Problem 14 at the end of Section 5.4, and argue as in the proof of Theorem 5.24.

11. See the preceding hint.

SECTION 5.8

1. Treat $p = 2$, $p = 3$ by separate arguments.

SECTION 6.3

6. If $(a/b)^{1/n}$ is rational, so is $b(a/b)^{1/n}$, which is a root of the equation $x^n = ab^{n-1}$.

9. Use the infinite series of cos x and adapt the ideas of the two preceding problems.

11. If $n = 3$, the area of such a triangle can be shown to be rational by the use of one standard elementary formula, but irrational by another. For values of n other then 3, 4, or 6, a similar contradiction can be obtained by applying the law of cosines to a triangle formed by two adjacent vertices and the center of the polygon.

SECTION 6.4

14. If $|x| + |y| \leqslant c$, then $|xy| \leqslant c^2/4$.

17. Recall the method developed in Section 5.2.

SECTION 7.3

1. By Lemma 7.8, we see that $\theta = 1 + 1/\theta$ in this case. This gives a quadratic equation, only one of whose roots is positive.

2. Usc thc result of the preceding problem along with Lemma 7.8.

SECTION 7.5

2. Use $\xi = \pi^{-1}$ and $n = 1$.

SECTION 7.8

2. Use the identity $(x_1^2 - dy_1^2)(x_2^2 - dy_2^2) = (x_1x_2 - dy_1y_2)^2 - d(x_1y_2 - x_2y_1)^2$.

6. Use Theorem 5.1 and Corollary 7.23.

SECTION 8.1

3. Recall Theorem 4.1(3), (5).

4. The product is $e^{\vartheta(x)}$.

6. Expand $(1 + 1)^{2n}$ using the binomial theorem.

8. Use Bertrand's postulate.

SECTION 8.2

4. For the conditional convergence use the alternating series test.
5. Consider the limit as $s \to +\infty$.
23. The integral is a sum of terms of the form $n\int_n^{n+1} u^{-s-1}\,du$.

SECTION 8.3

11. For the last assertion, note that

$$\sum_{U<n\leqslant 2U} 1/\phi(n) = O\Big((1/U)\sum_{n\leqslant 2U} n/\phi(n)\Big) = O(1).$$

Then put $U = x/2^k$ and sum over k.

SECTION 9.1

7. After applying Theorem 9.1 to get polynomials $q(x)$ and $r(x)$ in $\mathbb{Q}[x]$, multiply by a suitable positive integer k so that $kq(x)$ and $kr(x)$ have integral coefficients, and use the fact that $g(m) > kr(m)$ for sufficiently large integers m.
9. If there were only finitely many such primes p, let P be their product, define $x_0 = P^n f(0)$, and examine $f(x_0)$ with n large.

SECTION 9.3

1. To prove that $\mathbb{Q}(\omega^2\sqrt[3]{2})$ is different from $\mathbb{Q}(\omega\sqrt[3]{2})$, assume that $\omega^2\sqrt[3]{2}$ is an element of the latter field, that is, assume that there are rational numbers a, b, c such that $\omega^2\sqrt[3]{2} = a + b\omega\sqrt[3]{2} + c(\omega\sqrt[3]{2})^2$. Prove that no such numbers exist.

SECTION 9.5

7. Define $\alpha = (x - 2\sqrt{m})/y$, so that $N(\alpha)$ is certainly an integer if x and y satisfy $x^2 - y^2 = 4m$. Choose $x = m + 1$, $y = m - 1$ so that α is not an integer if $|m - 1| > 4$. The cases $|m - 1| \leqslant 4$ can be treated specially.

SECTION 9.9

6. Use part (5) of Theorem 9.29.

SECTION 10.1

3. Use Problems 1 and 2 to show that the common value is $\Sigma_{j=1}^{n} p(n-j)$.

APPENDIX A.3

2. Recall that $\cot\theta = -\cot(\pi - \theta)$, and that $\cot \pi/2 = 0$.

4. Consider the product of the roots of the polynomial $F(x)$ in (A.8).

6. Recall that $\sin\phi = (e^{i\phi} - e^{-i\phi})/(2i)$, and use the identity of the preceding problem.

7. Take the derivative of the logarithm of the absolute value of both sides of the identity in the preceding problem.

8. Recall that $\dfrac{d}{du}\cot u = -\operatorname{cosec}^2 u$.

Answers

Section 1.2, p. 17

1. (a) 77, (b) 1, (c) 7, (d) 1.
2. $g = 17$; $x = 71$, $y = -36$.
3. (a) $x = 9$, $y = -11$, (b) $x = 31$, $y = 44$, (c) $x = 3$, $y = -2$, (d) $x = 7$, $y = 8$, (e) $x = 1$, $y = 1$, $z = -1$.
4. (a) 3374, (b) 3660.
5. 128.
7. 6, 10, 15.
17. 1, $n(n + 1)$.
18. a, b.
25. $x = 100n + 5$, $y = 95 - 100n$, $n = 1, 2, 3, \cdots$, will do.
27. $a = 10$, $b = 100$ is a solution in positive integers. All solutions are given by $a = \pm 10$, $b = \pm 100$; $a = \pm 20$, $b = \pm 50$; $a = \pm 50$, $b = \pm 20$; $a = \pm 100$, $b = \pm 10$, with all arrangements of signs. There are 16 solutions in all.
28. $a = 10$, $b = 100$, $c = 10, 20, 50$, or 100; $a = 20$, $b = 50$, $c = 10, 20, 50$, or 100; and all permutations of these, 36 answers in all.

Section 1.3, p. 28

1. For every prime p, at least one of $\alpha(p), \beta(p)$ is 0.
2. 3; 7.
13. $p, p^2; p, p^2, p^3; p^2, p^3$.
14. p^3, p.
15. $3|\alpha(p)$ for all p; $\alpha(p) \leqslant \beta(p)$ for all p.
16. $2^{15}3^{10}5^6$.
22. Counterexamples for false statements are
    (1) $a = 1$, $b = 2$, $c = 3$.
    (8) $a = 8$, $c = 4$.
    (10) $p = 5$, $a = 2$, $b = 1$, $c = 3$.
    (13) $a = 2$, $b = 5$.

Section 2.1, p. 56

1. 7, 24, 41, 58, 75, 92.
2. 0, 3, 6, 9, 12, 15, 18, 21, 24, 27, 30, 33, 36, 39, 42, 45, 48.
3. 1, 5, 7, 11 (mod 12); 1, 7, 11, 13, 17, 19, 23, 29 (mod 30).
4. $y \equiv 1 \pmod 2$; $z \equiv 1 \pmod 6$.

5. $x \equiv 5 \pmod{12}$.

10. $m = 1, 2, 3, 4, 5, 6, 7, 8, 9, 10, 11, 12.$
$\phi(m) = 1, 1, 2, 2, 4, 2, 6, 4, 6, 4, 10, 4.$

11. $x = 5$.

13. 1, 3, 9, 27, 81, 243.

15. 0, 0, 1, 1, 11.

28. 1.

29. 6.

30. 01.

39. One example is $a = 17$, $b = 8$, $m = 15$.

41. Primitive solutions with $a \leqslant b \leqslant c$ are $a = b = 1$, c any positive integer.

42. Solutions such that $(a, b, c) = 1$, $c \geqslant |b| \geqslant |a|$ are
$a = -b = \pm 1$, $c = 1$ or 2;
$a = -1$, $b = 2$, $c = 3$;
$a = b = \pm 1$ with any $c > 0$;
$a = 1$, $b = 1 - c$ with any $c > 2$;
$a = 2$, $b = -2n + 1$, $c = 2n + 1$ with any $n > 1$.

Section 2.2, p. 62

4. $x(x + 1)(x + 2) \cdots (x + m - 1) \equiv 0 \pmod{m}$.

5. (*a*) No solution; (*b*) No solution; (*c*) $x \equiv 318 \pmod{400}$; (*d*) $x \equiv 31, 66, 101 \pmod{105}$; (*e*) $x \equiv 62 \pmod{105}$; (*f*) $x \equiv 17 + 43t \pmod{817}$ with $0 \leqslant t \leqslant 18$; (*g*) $x \equiv 836 \pmod{999}$.

6. (*a*) 5; (*b*) 0; (*c*) 5.

7. 73/105; 4/7.

13. $x \equiv 42 \pmod{125}$.

Section 2.3, p. 71

1. $x = 106$.

2. $23 + 30j$.

3. $x \equiv 33 \pmod{84}$.

4. $-2 + 60j$.

7. No solution.

8. 1732.

9. 1, 2.

10. 960.

11. 2640.

12. 1920.

13. 6720.

14. $x \equiv 1, 2, 6 \pmod 9$; $x \equiv 1, 3 \pmod 5$; $x \equiv 1, 6, 11, 28, 33, 38 \pmod{45}$.

15. No solution.

16. $x \equiv 1, 3, 5 \pmod{503}$.

17. $x \equiv 1, 3, 5, 14, 16, 27, 122, 133, 135 \pmod{143}$.

29. n odd.

30. n even.

31. $n = 5^k$, $k = 1, 2, \cdots$ will do.

32. 35, 39, 45, 52, 56, 70, 72, 78, 84, 90.

33. 3, 1, 2, 4.

36. 76; 01.

42. $n = 1$, 2^j or 2^j3^k with j and k positive.

45. 2^c where c is the number of distinct primes dividing m.

Section 2.4, p. 82

4. $2^{140} \equiv 67 \pmod{561}$ but $2^{280} \equiv 1 \pmod{561}$.

5. $2^{1023} \equiv 1 \pmod{2047}$.

6. $3^{1023} \equiv 1565 \pmod{2047}$.

10. $(33, 341) = 11$; $(31, 341) = 31$.

13. 14.

14. (*a*) 173; (*b*) 41; (*c*) 37; (*d*) 83; (*e*) Method fails. Taking $u_0 = 3$ gives the divisor 43; (*f*) 16193.

16. 4; 5.

17. 461333.

Section 2.5, p. 86

1. $\bar{k} = 43$; $a = 53$.

2. $m = 3989 \cdot 9839$.

Section 2.6, p. 91

1. $x \equiv 4, 13, 22 \pmod{27}$; no solution $\pmod{81}$.

2. No solution.

3. $x \equiv 4 \pmod{5^3}$.

4. 7, 15, 16, 24 (mod 36).

5. $15 \pmod{3^3}$.

6. No solution.

7. $23 \pmod{7^3}$.

8. $308060 \pmod{101^3}$.

Section 2.7, p. 96

1. (*a*) $x^5 + x^2 + 5 \equiv 0 \pmod 7$; (*b*) $x^2 + 3x - 2 \equiv 0 \pmod 7$; (*c*) $x^4 - x^3 - 4x + 3 \equiv 0 \pmod 7$.

9. 10, 35, 50, 24.

Section 2.8, p. 106

1. 2, 2, 3, 2, 2.

2. 5.

3. 4.

4. Modulo 7: 1, 3, 6, 3, 6, 2. Modulo 11: 1, 10, 5, 5, 5, 10.

7. $p - 1$. 0.

8. (*a*) 4; (*b*) 0; (*c*) 4; (*d*) 1.

10. (*a*) 9, 15, 8, 2; (*c*) 3, 5, 14, 12; (*d*) 15.

11. $x^2 \equiv 1$, $x^2 \equiv 2$, $x^2 \equiv 4$, $x^2 \equiv 8$, $x^2 \equiv 9$, $x^2 \equiv 13$, $x^2 \equiv 15$, $x^2 \equiv 16 \pmod{17}$.

20. $2 + 101t$ is a primitive root $\pmod{101^2}$ if and only if $t \not\equiv 83 \pmod{101}$.

Section 2.9, p. 114

1. (*a*) $(x-1)^2 \equiv 2 \pmod 5$; (*b*) $(x+1)^2 \equiv 4 \pmod 7$; (*c*) $(x-1)^2 \equiv 6 \pmod{11}$; (*d*) $(x-6)^2 \equiv 11 \pmod{13}$.

4. (*a*) $x \equiv \pm 6 \pmod{13}$; (*b*) $x \equiv \pm 9 \pmod{19}$; (*c*) $x \equiv \pm 5 \pmod{11}$; (*d*) $x \equiv \pm 6 \pmod{29}$.

Section 2.10, p. 119

1. (*a*), (*e*), (*f*), (*h*), (*i*).

3.

	7	−2	17	30	8	3
7	8	17	30	7	3	−2
−2	17	8	3	−2	30	7
17	30	3	−2	17	7	8
30	7	−2	17	30	8	3
8	3	30	7	8	−2	17
3	−2	7	8	3	17	30

	1	4	5	0	2	3
1	2	5	0	1	3	4
4	5	2	3	4	0	1
5	0	3	4	5	1	2
0	1	4	5	0	2	3
2	3	0	1	2	4	5
3	4	1	2	3	5	0

Section 2.11, p. 126

6. 8.

13.

⊕	0	1	2	3	4	5	6
0	0	1	2	3	4	5	6
1	1	2	3	4	5	6	0
2	2	3	4	5	6	0	1
3	3	4	5	6	0	1	2
4	4	5	6	0	1	2	3
5	5	6	0	1	2	3	4
6	6	0	1	2	3	4	5

⊙	0	1	2	3	4	5	6
0	0	0	0	0	0	0	0
1	0	1	2	3	4	5	6
2	0	2	4	6	1	3	5
3	0	3	6	2	5	1	4
4	0	4	1	5	2	6	3
5	0	5	3	1	6	4	2
6	0	6	5	4	3	2	1

20. (*a*) is an integral domain; (*b*) is an integral domain if and only if m is a prime.

Section 3.1, p. 135

1. 1, -2, 3, -7, 0.

4. $\left(\frac{-1}{11}\right) = -1$, $\left(\frac{-1}{13}\right) = +1$, $\left(\frac{-1}{17}\right) = +1$,
$\left(\frac{2}{11}\right) = -1$, $\left(\frac{2}{13}\right) = -1$, $\left(\frac{2}{17}\right) = +1$,
$\left(\frac{-2}{11}\right) = +1$, $\left(\frac{-2}{13}\right) = -1$, $\left(\frac{-2}{17}\right) = +1$,
$\left(\frac{3}{11}\right) = +1$, $\left(\frac{3}{13}\right) = +1$, $\left(\frac{3}{17}\right) = -1$.

5. $x \equiv \pm 1$, $x \equiv \pm 5$, $x \equiv \pm 2$, $x \equiv \pm 4$, $x \equiv \pm 3 \pmod{11}$.
$x \equiv \pm 1$, $x \equiv \pm 27$, $x \equiv \pm 2$, $x \equiv \pm 48$, $x \equiv \pm 3 \pmod{11^2}$.

6. (*a*) $1, 2, 4 \pmod 7$, $\pm 1, \pm 3, \pm 4 \pmod{13}$, $\pm 1, \pm 2, \pm 4, \pm 8 \pmod{17}$, $\pm 1, \pm 4, \pm 5, \pm 6, \pm 7, \pm 9, \pm 13 \pmod{29}$, $\pm 1, \pm 3, \pm 4, \pm 7, \pm 9, \pm 10, \pm 11, \pm 12, \pm 16 \pmod{37}$.

7. (*d*) 2, (*h*) 2.

8. (*a*) 2, (*b*) 0, (*c*) 4, (*d*) 0, (*e*) 2, (*f*) 0.

Section 3.2, p. 140

4. (*b*), (*c*), (*d*), (*e*), (*f*).

5. $\left(\frac{7}{227}\right) = +1$, $\left(\frac{7}{229}\right) = -1$, $\left(\frac{7}{1009}\right) = +1$,
$\left(\frac{11}{227}\right) = +1$, $\left(\frac{11}{229}\right) = +1$, $\left(\frac{11}{1009}\right) = -1$,
$\left(\frac{13}{227}\right) = -1$, $\left(\frac{13}{229}\right) = -1$, $\left(\frac{13}{1009}\right) = -1$.

6. Yes.

7. $p = 2$, $p = 13$, and $p \equiv 1, 3, 4, 9, 10, 12 \pmod{13}$.

8. $p \equiv \pm 1, \pm 3, \pm 9, \pm 13 \pmod{40}$.

9. Odd primes $q \equiv \pm 2 \pmod 5$.

10. $p \equiv 1, 3 \pmod 8$.

11. No.

Section 3.3, p. 147

1. -1, -1, $+1$, $+1$.

2. (*b*).

3. (*c*).

4. No solution.

7. $p = 2$ and $p \equiv 1 \pmod 4$.

8. 2 and p^a for $p \equiv 1 \pmod 4$ and $a = 1, 2, 3, \cdots$.

9. $n = 2^\alpha \prod p^\beta$ where $\alpha = 0$ or 1, the primes p in the product are all $\equiv 1 \pmod 4$, and $\beta = \beta(p) = 0, 1, 2, \cdots$.

22. $a \equiv 1 \pmod{21}$.

Section 3.4, p. 154

1. (*a*) Positive definite, (*b*) Negative definite, (*c*) Indefinite, (*d*) Positive definite, (*e*) Indefinite, (*f*) Positive definite.

2. The perfect squares, including 0.

Section 3.5, p. 162

1. $x^2 + xy + 5y^2$.

Section 3.6, p. 169

1. 6, 7, 8, 9.

2. 24.

3. 32.

4. $292^2 + 67^2 = 89753$.

Section 3.7, p. 176

6. Two representations by each of f_1 and f_2.

Section 4.1, p. 184

1. 529, 263, 263, 263, 87.

2. 24.

3. (*a*) All x such that $\{x\} < 1/2$. (*b*) All x. (*c*) All integers. (*d*) All x such that $\{x\} \geqslant 1/2$. (*e*) All x such that $1 \leqslant x < 10/9$.

5. (*a*) $e = \sum_{j \geqslant 1}[n/p^j]$ if p is odd, $e = n + \sum_{j \geqslant 1}[n/2^j]$ for $p = 2$. (*b*) $e = \sum_{j \geqslant 1}([2n/p^j] - [n/p^j])$ if p is odd, $e = 0$ for $p = 2$.

12. $a \quad m[(a \quad 1)/m]$.

29. 1. $(n - 1)/2$.

Section 4.2, p. 191

1. 7.

2. 12.

3. 2, 1, 12, 24.

4. 6.

8. $\sigma_k(n) = \prod_p \dfrac{p^{k(\alpha+1)} - 1}{p^k - 1}$ where $n = \prod_p p^\alpha$.

10. If $f(n) = 1$ for all n, then $f(n)$ is totally multiplicative, but then $F(n) = d(n)$ is not.

13. Take $x = p^{n-1}$ where p is any prime.

16. 6, 28, 496.

22. $m = 12$, $n = 14$.

Section 4.3, p. 195

1. $n = 33$ will do.

3. 1.

7. $\sum_{d|n} \mu(d)\sigma(d) = (-1)^{\omega(n)} \prod_{p|n} p$.

Section 4.4, p. 204

1. $u_n = n$, $u_n = 1$.

11. $u_n = 1 + 2^{n-2} - (-2)^{n-2}$.

23. $u_n = \alpha 2^n + \beta 3^n$, (b) $u_n = \alpha 2^n + \beta 3^n + 1/2$, (c) $u_n = \alpha 2^n + \beta 3^n + \frac{1}{2}n + 7/4$.

Section 4.5, p. 210

15. $n - n/(pq) - n/(qr) - n/(rp) + 2n/(pqr)$.

Section 5.1, p. 218

2. $(1 + 7t, -1 + 10t)$ for integral t.

3. (a) $(8 + 17t, -7 - 21t)$, (b) no solution; (c) $(-29 + 99t, 34 - 101t)$.

4. (a) (2, 14), (5, 9), (8, 4); (b) (6, 3); (c) (2, 7); (d) (2, 5); (e) no solution; (f) no solution; (g) no solution.

5. (14, 65).

Section 5.2, p. 229

1. $(-2 - 2t, 3, 1, t)$ will do.

2. All a, b, c such that both $a \equiv b \equiv c \pmod 2$ and $a \equiv c \pmod 3$. $(1 - 4t, 6t, -4t, t)$ will do.

Section 5.3, p. 233

1. (3, 4, 5), (4, 3, 5), (5, 12, 13), (12, 5, 13), (15, 8, 17), (8, 15, 17), (7, 24, 25), (24, 7, 25), (21, 20, 29), (20, 21, 29).

3. (a) $(3k, 4k, 5k)$, $(4k, 3k, 5k)$. (b) None.

6. $u = dr^2$, $v = es^2$ where d and e are positive integers such that $de = 6$.

7. $n \not\equiv 2 \pmod 4$.

Section 5.5, p. 248

2. 13.

3. It has a nontrivial solution.

4. No solution.

5. It has a nontrivial solution.

Section 5.6, p. 260

1. $\left(\dfrac{2m^2 + 1}{2m^2 - 1}, \dfrac{2m}{2m^2 - 1}\right)$.

2. $\left(\dfrac{6m^2 - 8m + 3}{2m^2 - 1}, \dfrac{-4m^2 + 6m - 2}{2m^2 - 1}\right)$.

3. $m_0 = \pm\infty$, $m_1 = 1$, $m_2 = -1$.

5. $(m^2 - 2, m^3 - 2m)$.

6. $(m^2 + 2, m^3 + 3m)$.

10. Tangent through $(1, 2)$ intersects the curve at $(0, -1)$. Tangent through $(0, 1)$ intersects the curve a third time at $(0, 1)$, that is, $(0, 1)$ is an inflection point. The chord through $(0, 1)$ and $(1, 2)$ intersects the curve at $(-1, 0)$. The chord joining $(0, 1)$ and $(0, -1)$ intersects the curve at infinity. Likewise for the chord joining $(1, 2)$ and $(1, -2)$.

14. $X_{n+1} = X_n(X_n^3 + 2Y_n^3)$, $Y_{n+1} = -Y_n(2X_n^3 + Y_n^3)$, $Z_{n+1} = Z_n(X_n^3 - Y_n^3)$.

Section 5.7, p. 278

5. $c = 0$, $c = \pm 1$.

8. $(1, 1) \in \mathscr{C}_f(\mathbb{Q})$.

9. $(0, 0)$.

10. $(0, 0)$, $(\pm 1, 0)$.

11. $(0, 0)$, $(2, \pm 4)$.

18. $a = 0$, $b = a + 1$.

Section 6.1, p. 300

6. $a = b = d = 1$, $c = 0$ will do.

Section 7.1, p. 327

1. $17/3 = \langle 5, 1, 2 \rangle$, $3/17 = \langle 0, 5, 1, 2 \rangle$, $8/1 = \langle 8 \rangle$.

3. $\langle 2, 1, 4 \rangle = 14/5$, $\langle -3, 2, 12 \rangle = -63/25$, $\langle 0, 1, 1, 100 \rangle = 101/201$.

Section 7.2, p. 329

1. The following conditions are necessary and sufficient. In case $a_j = b_j$ for $0 \leqslant j \leqslant n$, then n must be even. Otherwise define r as the least value of j such that $a_j \neq b_j$. In case $r \leqslant n - 1$, then for r even we require $a_r < b_r$, but for r odd, $a_r > b_r$. In case $r = n$, then for n even we require $a_n < b_n$, but for n odd we require $a_n > 1 + b_n$, or $a_n = 1 + b_n$ with $b_{n+1} > 1$.

Section 7.3, p. 333

1. $(1 + \sqrt{5})/2$.

2. $(3 + \sqrt{5})/2$, $(25 - \sqrt{5})/10$.

3. (*a*) $1 + \sqrt{2}$, (*b*) $(1 + \sqrt{3})/2$, (*c*) $1 + \sqrt{3}$, (*d*) $3 - \sqrt{3}$.

4. $h_n/h_{n-1} = \begin{cases} \langle a_n, a_{n-1}, \cdots, a_0 \rangle & \text{if } a_0 \neq 0, \\ \langle a_n, a_{n-1}, \cdots, a_2 \rangle & \text{if } a_0 = 0. \end{cases}$

Section 7.4, p. 336

1. $\sqrt{2} = \langle 1, 2, 2, 2, \cdots \rangle$, $\sqrt{2} - 1 = \langle 0, 2, 2, 2, \cdots \rangle$, $\sqrt{2}/2 = \langle 0, 1, 2, 2, 2, \cdots \rangle$, $\sqrt{3} = \langle 1, 1, 2, 1, 2, 1, 2, \cdots \rangle$, $1/\sqrt{3} = \langle 0, 1, 1, 2, 1, 2, 1, 2, \cdots \rangle$.

Section 7.6, p. 344

1. $1/1$, $3/2$ will do.

2. $3/1$, $22/7$ will do.

Section 7.7, p. 351

1. $c = 1, 2, \cdots, 2[\sqrt{d}]$.
2. $\sqrt{83}$.
3. $\langle 3, \overline{1, 6} \rangle$.

Section 7.8, p. 356

8. No solution for $x^2 - 18y^2 = -1$; $x = 17$, $y = 4$.
9. $x = 70$, $y = 13$; $x = 9801$, $y = 1820$.
10. $x = 29718$, $y = 3805$; $x = 1766319049$, $y = 226153980$.

Section 8.2, p. 387

1. No. If $f(n) = g(n) = 1$ for all n, then $f * g(n) = d(n)$, which is not totally multiplicative.
17. $(\log 9)/(\log 10)$.

Section 9.2, p. 419

1. $x - 7$, $x^3 - 3x^2/2 + 3x/4 - 1$, $x^4 - 4x^3 - 4x^2 + 16x - 8$.
   7, $\sqrt[3]{7}$, $1 + \sqrt{2} + \sqrt{3}$ are algebraic integers.

Section 9.4, p. 425

3. Yes; no, for example $\alpha = (1 + i\sqrt{3})/2$.

Section 9.5, p. 427

6. $\alpha = (1 + 7i)/5$ will do.
7. The Hint also works in case $m = -2$. The other special cases can be handled by such numbers as $(1 + 4\sqrt{-3})/7$, $(9 + 4\sqrt{2})/7$, $(27 + \sqrt{3})/11$, $(4 + 10\sqrt{5})/11$.

Section 9.9, p. 440

7. $y = 0$, $x = 1$.

Section 10.3, p. 457

2.
$n = 1, 2, 3, 4, 5, 6, 7, 8, 9, 10, 11, 12.$
$p(n) = 1, 2, 3, 5, 7, 11, 15, 22, 30, 42, 56, 77.$
$n = 13, 14, 15, 16, 17, 18, 19, 20.$
$p(n) = 101, 135, 176, 231, 297, 385, 490, 627.$

Section 10.4, p. 463

1.
$n = 1, 2, 3, 4, 5, 6, 7, 8, 9, 10, 11, 12.$
$\sigma(n) = 1, 3, 4, 7, 6, 12, 8, 15, 13, 18, 12, 28.$
$n = 13, 14, 15, 16, 17, 18, 19, 20.$
$\sigma(n) = 14, 24, 24, 31, 18, 39, 20, 42.$

Section 10.6, p. 471

2. $p(35m + 19) \equiv 0 \pmod{35}$.

Section 11.1, p. 475

1. (a) $1/2$, (b) $1/2$, (c) $1/3$, (d) $1/4$, (e) $1/m$, (f) 0, (g) 0, (h) 0, (i) 0, (j) 0.

15. $1/11$.

Section 11.2, p. 481

1. $1/2$, 0, $1/3$, $1/m$.

Index